T0186002

The Economic Superorganism

Carey W. King

The Economic Superorganism

Beyond the Competing Narratives on Energy, Growth, and Policy

 Springer

Carey W. King
Energy Institute
University of Texas at Austin
Austin, TX, USA

ISBN 978-3-030-50294-2   ISBN 978-3-030-50295-9   (eBook)
https://doi.org/10.1007/978-3-030-50295-9

Cover illustration: Cover credit to David Bedwell

This Springer imprint is published by the registered company Springer Nature Switzerland AG
The registered company address is: Gewerbestrasse 11, 6330 Cham, Switzerland

To those that seek to understand the system.

Preface

Isaac Newton is considered one of the fathers of modern science for establishing the methods of calculus that helped explain the motion of objects on Earth as well as the planets in the solar system. His Third Law of Motion states "To every action there is an equal and opposite reaction." Energy and economic discourse seems to follow the same rule: for every narrative, there is an equal and opposite counter-narrative.

Without energy consumption, nothing moves, gets extracted or manufactured, or stays alive. In other words, without energy consumption, there is no economic activity. Because affordable energy has been so abundant for the last 100 years, most citizens might never contemplate these points. But these points are constantly on the minds of those in the energy industry, environmental organizations, and high-level political and military positions.

Nevertheless, people within these groups often disagree and promote dichotomous visions for our future energy supply using one of the two narratives: fossil fuels versus renewable energy. Both narratives claim that they best serve our societal needs with lower cost and environmental impact.

What is the difference between the energy narratives? Is one right and the other wrong? Does one have the moral high ground or run counter to the laws of physics? What is the difference between what is technically possible and what is economically and socially viable? Amid all of the political posturing on the subject, what do we know about the role of energy in the economy and our future?

Are there fundamental truths or only *energy narratives*?

These are the questions that in 2006 drove me to leave my first job, with a high-tech start-up company, and return to academic research centered on energy. I wanted to understand both the changes that were occurring in the energy system around me as well as how to sift through the rhetoric of the two basic narratives that we hear regarding the future of energy in society.

Before I began my next job at the University of Texas at Austin, where I continue to work, I performed a job talk just as most prospective employees do. In this talk, I gave a summary of research on the dynamics of the global growing population and increasing rate of energy consumption, and how these rates of growth would eventually come to an end. After all, I am educated as an engineer in the physical sciences, and everything I'd learned to that point supported my presentation. For something to grow, it needs energy and mass input, and if we're talking about energy consumption on Earth, then this can't increase forever. What is so complicated about that?

After the talk, one of the senior researchers in the audience suggested I read a book that he agreed with. That book downplays many arguments of my presentation while making the case that we'll always access more energy. (There is no need for me to name this book now, as I refer to it within the chapters that follow.) I was a little annoyed at the almost complete dismissal of my points, but I'd not worked in energy research anywhere close to as long as this person, so I thought I'd better see what I was missing.

There was a lot I was missing. But after reading that book, and ever since, I've never become convinced that the basic premise of limits to growth was misguided. I was missing both deeply insightful principles based upon scientific analysis of data and half-baked political rhetoric. Since that job talk, I've learned just how varied are the opinions regarding the types of energy resources and technologies we should use for a prosperous economy and healthy life. I also learned that it is difficult for even seasoned energy and economic professionals, much less the general public, to agree on what is rhetoric versus a scientifically and theoretically sound concept.

I realized that these disagreements weren't completely about the characteristics of energy technologies and resources. The disagreements were about why we consume energy in the first place, and how much energy consumption does or does not relate to economic prosperity and social livelihoods.

Most of us don't think much about energy until it gets expensive. Drivers in 2008 certainly thought about the cost of energy when gasoline was more than 4 dollars per gallon in much of the USA. When gasoline gets near that price, the media becomes flooded with stories of "pain at the pump" to describe the economic "pain" when filling up our car and truck fuel tanks. But these are simple news stories, largely telling us what we already know—that we're spending more of our income on fuel. They don't get to the heart of why there is some level at which energy costs simply become too high for consumers and the economy to function in the same way as before.

Getting to this heart of the issue has governed my research journey for over a decade, and this book explains what I've learned traveling down the road. I realized that simply thinking about energy prices and technologies is not enough. One needs to contemplate the concept of the whole economy as well as how energy explains its growth and the way its pieces can and do fit together. Just as there are dichotomous energy narratives, so are there *economic narratives* that vie to explain how we should perceive the economy.

The *techno-optimistic* economic narrative tells the story of *unbounded substitutability* for anything before we run out of it in the faith we can invent our way to a solution for any and all economic, social, and environmental problems. The *techno-realistic* economic narrative is the story of *biological and physical constraints*, and while it agrees we are inventive, it states we can neither break the laws of nature nor access an infinite supply of natural resources.

The Economic Superorganism sifts the energy and economic narratives into separate piles of coarse assumptions and reasoned arguments to help you separate the wheat from the chaff. You will understand the difference between historical data and future projections, between assuming past trends will continue and having something intelligent to say about likely future outcomes. In some cases, the answers are quite surprising.

Is the economy ultimately constrained by the physical resources of a finite Earth, and if so, how would we know when those constraints are imposing themselves? What data should we seek, and what trends should we measure? How would the various trends relate to each other? In short, how would we know if the finite planet is affecting our lives?

The Economic Superorganism takes a systematic approach to answer these questions to provide a valuable and viable basis for understanding how our changing energy systems have influenced our past policies and economic organization such that we can better consider future options. It is this understanding that must become more prevalent knowledge in our education. It is this understanding that must become more prevalent in our society. It is this understanding that is not sufficiently integrated into our economic modeling. It is this understanding that this book seeks to spread.

The reader should consider that this book represents my interpretations of data and research efforts to understand the historical and possible future influence of energy systems on our economy. Ultimately, this book explains my journey to understand why and how people disagree on how energy affects possible economic futures. As my journey is certainly not the first along this path, I refer to many other authors and attribute original thoughts to them as much as possible.

Because my educational training is in engineering systems modeling, it is natural for me to want to understand how the world works by integrating many pieces of information together. By practical necessity, this systems approach considers many of the most important factors rather than a deep understanding of a single concept. In doing so, I attempt to strike a balance by describing some of the more important data in detail, using charts but not equations, while only briefly introducing some concepts with references to other more authoritative works on specific subjects.

I discuss many matters as objectively as I can, but I don't claim that either I or anyone can be perfectly objective. Simply by presenting a limited set of data, examples, and stories, an author engages in inherent bias. For some quotes in the book, I find their narrative too misleading to leave them without immediately discussing why. By the end of the book, you can make your own judgment on where my position resides between the endpoints of both the energy and economic narratives, and how comfortable you are sharing this space with me. Thank you for reading.

Austin, TX, USA
April 28, 2020

Carey W. King

Acknowledgments

I thank my family members who reviewed various drafts. First, my loving wife Margaret, who not only gave feedback on drafts but also tolerated my need to seclude myself writing for many evenings and weekends. My mother, father, brother Jeff, and niece Katherine who provided editing and feedback that increased the readability of the book.

I also give thanks to the following friends and colleagues who commented on early drafts and organization of the manuscript: John Asbury, Raj Patel, Asher Price, Gary Rasp, David Spence, and Jeff Stephens.

I thank my oldest friend I have in this world, David Bedwell, for putting his heart and soul into the design of my book cover, and because of that, I hope people will judge my book by its cover (as well as its content).

I thank my colleagues and others who inspired me, educated me, and shared insights either directly in correspondence, discussions, or indirectly via their own scholarly research and writing: David Adelman, Jack Alpert, George Backus, Ross Baldick, Ugo Bardi, Garvin Boyle, Adam Brandt, Paul Brockway, James H. Brown, Michael Carbajales-Dale, Victor Court, F. Todd Davidson, Jingxuan Feng, Lianyong Feng, Felix FitzRoy, Roger Fouquet, James K. Galbraith, Gaël Giraud, Gürcan Gülen, Nathan Hagens, Robert Hebner, Carsten Herrmann-Pillath, Matthew Heun, Susan Hovorka, Jean-Marc Jancovici, Andrew Jarvis, Steve Keen, Eric Kemp-Benedict, Kent Klitgaard, Susan Krumdiek, Michael Kumhof, Michael Marder, Dennis Meadows, Safa Motessharei, Larry Lake, Michel Lepetit, Sheila Olmstead, Tadeusz "Tad" Patzek, Varun Rai, William Rees, Joshua Rhodes, Justin Ritchie, Don Ross, Bridget Scanlon, Megan Seibert, Joseph Tainter, Scott Tinker, David Tuttle, Robert Ulanowicz, Mary Wildfire, and Victor Yakovenko. To those who have passed in living body but whose living spirit continues via their writings and research on systems thinking and dynamics: Jay Forrester, Donnella Meadows, and Howard T. Odum.

I also thank my former students who performed and aided research that forms the basis for much of this book: Alyssa Donovan, Prataj Haputhanthri, Harshit Jayaswal, Qiuying "Cho" Lai, and John Maxwell.

Because of their fundamentally thoughtful analysis on the role of energy in economic activity, over the course of decades, and their openness and willingness to discuss their thoughts, a special scholarly thank you goes out to Robert Ayres, Charles A. S. Hall, Reiner Kümmel, and Vaclav Smil. Each has been an enduring champion for knowledge to foster a community of researchers and scholars working to link biophysical and energetic frameworks to describe the inner workings of our economy.

I thank my editor at Springer Nature, David Packer, who helped guide the structure of this book and who champions the integration of science and economics.

I owe particular professional gratitude to several individuals at The University of Texas at Austin. Michael E. Webber and Tom Edgar supported me professionally in a manner that afforded me stability and time to pursue the ideas and research expressed in this book over the last decade. In addition, David Allen, Charles Groat, and Ian J. Duncan provided me with early mentoring and opportunities in energy-related research. I also thank Henry Groppe for his gracious mentoring, wisdom, and feedback that improved both my personal life and professional research.

I thank the various agencies, companies, and personal donors and sponsors of research projects that I have been involved with over the last decade, all of which inform the content of the book. In particular, the Cynthia and George Mitchell Foundation provided assistance in communicating ideas among colleagues and with the public.

Contents

Praise for the Economic Superorganism

Is economic output constrained by Earth's boundaries, or limited only by our imaginations? One would hope that by now such competing economic narratives would have given way to definitive conclusions, but unfortunately most economists have not even begun to consider the links between economic activity and energy usage on a finite planet. Carey King's The Economic Superorganism convincingly explains how and why economics must be forced to confront the essential role of energy, fossil or renewable, in industrial civilisation and the dilemmas that poses for our growth-obsessed social system.

—**Professor Steve Keen**, Distinguished Research Fellow, Institute for Strategy, Resilience and Security, University College London.

Growth-addicted politicians and bureaucrats are running planet Earth on premises and principles drawn from the ecological vacuity of neoliberal economics. As Carey King deftly reveals, this is analogous flying a 787 Dreamliner using the intellectual equivalent of a 1955 Volkswagen Beetle driver's manual. Dr. King's message? There are real limits to growth; humanity has exceeded them and is dissipating the ecosphere.

—**William Rees**, Professor Emeritus and former Director of the School of Community and Regional Planning at University of British Columbia; creator and co-developer of ecological footprint analysis.

There is no aspect of sustainability more important than energy, and no topic in energy more important than depletion. The prospect of oil that is unavailable or unprofitable (in money or energy) raises questions of limits to growth. Technological optimism asserts that market forces and ingenuity will overcome all such limits. Technological realism asserts that limits are real and constraining. In exploring this debate, Carey King's The Economic Superorganism is a book that will endure in relevance.

—**Joseph Tainter**, Professor of Environment and Society, Author of The Collapse of Complex Societies.

Economics tells society what matters in the production of wealth and how to organize its distribution. In the natural sciences, theoretical errors are usually quickly revealed by experiment or smarter theoreticians and normally only damage some careers. In economic theory, however, fundamental errors that affect political practice will be revealed by experience only after they have resulted in misery for many people. In his book The Economic Superorganism Carey W. King addresses the basic facts and data on energy that are necessary for a proper understanding of energy's pivotal role in modern economies. He indicates the weak points in the narratives of mainstream economics, which hardly cares about fundamental natural laws, and he presents options for the future we may choose. The book is well written, and the reader feels the intellectual fire that moves its author. I highly recommend it, especially for those who wonder where we should go.

—**Dr. Reiner Kümmel**, Professor of Theoretical Physics, University of Würzburg.

Carey King has produced a very valuable overview of energy issues, together with their economic, social, general business and financial implications. He points to two positions which can be taken on where future energy scenarios may lead us. Position 1 is that all open discussion of science, engineering, and economic analyses of how to transition to more use of renewable energy is a good thing. That is to say, more discussion leads to more accurate analyses in the long-run, and we want the most accurate analyses as possible. If this discussion involves a lot of both good and bad information exchange during the process, then that is OK because that is just the scientific process going through its motions.

This is his position. The alternative, Position 2, is that open discussion of disagreement on the technical and/or feasibility of increasing the use of renewable energy is a bad thing because it gives ammunition to fossil fuel advocates. He goes on: quite simply, fossil fuel advocates can claim support for the view 100% renewable energy is not possible. In support of his view, King goes into basic energy-related concepts such as the laws of thermodynamics, power density and Energy Return on Investment, as well as into the history of energy resource use and its human well-being and financial implications past, present and possible future.

—**Professor Michael Jefferson**, ESCP Europe Business School, Former Chief Economist, The Royal Dutch/Shell Group.

The fate of civilization may be decided by energy policy, climate policy, economics, and physics-but each of these fields of study and domains of action is complex, and they interact in ways that are even more complicated and poorly understood. Carey King has done a magnificent job of untangling, sorting, analyzing, and explaining. Some of his conclusions may be surprising or controversial, but they are well grounded in data and logic. This book deserves to be widely read by energy and climate scientists, policy makers, reporters, and economists.

—**Richard Heinberg**, Author and Senior Fellow of the Post Carbon Institute

The Economic Superorganism offers a fresh perspective on the sometimes heated, sometimes myopic policy debate over the feasibility of a green energy transition. By backing out and taking a broader view of the interactions among physical, economic and other social systems, Carey King illuminates the important trade-offs at the heart of such a transition.

—**David Spence**, Baker Botts Chair in Law, the University of Texas at Austin School of Law

The Economic Superorganism is a deep meditation on the facts and fictions around energy, food, economic and climate systems past and future. King has a deductive approach that assumes nothing but intelligence, and calls on his prodigious skills in engineering, energy and systems science. Students in these and many other disciplines will profit from engaging with this novel and urgent analysis.

—**Raj Patel**, Research Professor, University of Texas at Austin

In The Economic Superorganism, Dr. Carey King treats the global economy as an organism, albeit a very large one, but still subject to the laws of nature as are all other organisms. He does this by discussing alternative narratives about the future of energy, the economy, and human society. The text uses non-technical language so that it should be understandable to a wide range of readers. Rather than come to strong conclusions, he lays out the basis for various narratives and invites readers to think about what these different viewpoints imply for our future society. This is a must read for those thinking seriously about our future, but be forewarned, you will likely have to rethink your own views.

—**John Day**, Distinguished Professor Emeritus, Dept. of Oceanography and Coastal Sciences, School of the Coast & Environment, Louisiana State University

We live in a time of increasing fracture and divisiveness. Younger people today probably never lived at a time when the main means of resolving political, social, economic or other issues was discussion, the marshalling and considering of facts used in support of various positions, consensus, with a certain deference to science as a means of sifting through different positions, interpretations or prejudices. Increasingly this is no longer the case. The positions, often with no connection to principles, are well understood from the political position of one party, one element of society or even one person. It seems anachronistic to read of past students in college engaging in debate clubs, and even (long ago) being forced to take courses in elocution (meaning, basically, making your argument). But today's arguments, or at least one's position, often seems to be decided before the start and is often decided by whomever has the most political power. Often science is used by one side or another to justify whatever position is maintained. That is not the proper role of science, which is, or should be according to most of the gatekeepers of scientific process, to test hypotheses, not to defend a predetermined position.

This is a book about disagreements, mostly relating to energy and its role in economies, where disagreements are rife. As such it is a refreshing perspective within a branch of science I am particularly familiar with. It is a wealth of good basic information and insights about energy and its role in society. Each chapter tends to explore a major issue where different points of view, which King calls different narratives, exist. The range is quite large, both in terms of importance (climate change), scale (global to local coal burning) and relation to different philosophical groups (economic growth vs environmental protection). It contains many perspectives that readers will find rather shocking (renewable energy sometimes uses more non-renewable resources than fossil fuels; conventional economics often leaves out critical economic information such as debt).

Perhaps most importantly, the book contrasts the difference between the physical laws of nature (which appear immutable) and the laws of humans, which may or may not be consistent with the laws of nature, setting up the most important difference, or incompatibility, of the book. This sets up one of the few conclusions in the book "not only can we use both social rules and physical laws to assess the constraints and possibilities for future energy and economic scenarios, we absolutely must".

King does not decide between the possible positions for the reader, but lets the reader decide. By giving all sides a chance to air their arguments it is a useful exercise in what young people training to enter science find later in their lives. It also demonstrates how some arguments are won or lost by changing the subject, rather than via better science or better quantitative analysis, I thought (probably naively) would be the case.

—**Charles A. S. Hall**, Professor Emeritus, SUNY College of Environmental Science and Forestry.

This book is panoramic in its vision of recasting economic discourse in biocentric terms reminiscent of the Gaia hypothesis. However, Dr. King uses naturalistic and thermodynamic metaphors not only as primers for energy policy reform but also as a critique of the predatory nature of current economic policies. By using an eclectic narrative style, speckled with theoretical excursions, as well as pithy public policy examples, the book is readable by the scholar and the informed citizen, willing to question the orthodoxy of natural resources management within contemporary economic doctrines.

—**Saleem H. Ali**, Blue & Gold Distinguished Professor of Energy and the Environment, University of Delaware

In The Economic Superorganism Carey King outlines a novel system to organize economic decision making and to evaluate outcomes. Using years of his and others' research, he points out the connections between income, consumption and the effect energy has on the economic system we operate within. Looking at the abstract data points he takes the next step, which many economist do not do, of connecting it to real life outcomes we observe in our daily lives ... our collective ability to earn less, poverty and homelessness increasing, students coming out of college unable to support themselves or build a household of their own, debt piling up and drowning the middle class ...

By taking a different analytical approach of the "economic superorganism" as King offers, policymakers can stress-test assumptions that have previously been left out of the discussion. This can lead to better qualified, more effective decisions. Knowing where to allocate capital, and what effect those allocations could have will be helpful in effecting real change and progress.

Full review and disclosure: https://jamesacox.com/2020/05/15/energy-economic-trends-and-effecting-change-a-review-of-the-economic-superorganism/

—**James Cox**, First Financial Group

Disclosure Statement Links to other sites are provided for your convenience in locating related information and services. Guardian, its subsidiaries, agents, and employees expressly disclaim any responsibility for and do not maintain, control, recommend, or endorse third-party sites, organizations, products, or services, and make no representation as to the completeness, suitability, or quality thereof.

This material contains the current opinions of the author but not necessarily those of Guardian or its subsidiaries and such opinions are subject to change without notice.

Registered Representative and Financial Advisor of Park Avenue Securities LLC (PAS). OSJ: 7101 Wisconsin Ave Suite 1200, Bethesda, MD 20814 301-907-9030 Securities products and advisory services offered through PAS, member FINRA, SIPC. CA insurance license #0I64535. First Financial Group is not an affiliate or subsidiary of PAS or Guardian. Guardian, its subsidiaries, agents, and employees do not provide tax, legal, or accounting advice. Consult your tax, legal, or accounting professional regarding your individual situation.

Part I
The Narratives and the Data

Chapter 1
Energy and Economic Narratives

Narrative: a story that connects and explains a carefully selected set of supposedly true
events, experiences, or the like, intended to support a particular viewpoint or thesis

Facts, Science, and Misinformation

In November of 2014, the citizens of Denton, Texas, 30 miles north of Fort Worth
and Dallas, became the focal point of the oil and gas industry in a state largely
defined by that industry. A combination of timeless geology, timely technology, and
a dogged petition drive led the population of about 125,000 to vote for a ban on
natural gas extraction and hydraulic fracturing within the city limits.

More specifically the people of Denton voted to ban extraction of natural gas
from a rock formation below a 5000 square mile area just to the west of Dallas,
including underneath the cities of Denton and Fort Worth. It was in this rock, known
as the Barnett Shale, that Nick Steinsberger, working for a company owned by
George Mitchell, pioneered the modern combination of hydraulic fracturing with
horizontal drilling in the 1990s.[1]

Hydraulic fracturing, or "fracking," combined with horizontal drilling, is a
process used to extract oil and natural gas from certain types of "unconventional"
reservoirs sometimes called shales, tight sands, or tight formations. Unconventional
reservoirs are loosely defined as those that require some additional technique above
that required for "conventional" reservoirs.

Conventional reservoirs have two general properties. First, they have high
permeability (think Swiss cheese or a sponge as an extreme example) such that fluids

---

[1]Gold, Russell. The Texas Well that Started a Revolution, *Wall Street Journal*, June 29, 2018,
accessed June 29, 2018 at: https://www.wsj.com/articles/the-texas-well-that-started-a-revolution-
1530270010.

© Springer Nature Switzerland AG 2021
C. W. King, *The Economic Superorganism*,
https://doi.org/10.1007/978-3-030-50295-9_1

flow easily within them. Second, they have low permeability layers above them that trap a pool of oil and gas that has risen and accumulated over time. Think of a cup of oil, but upside down, because oil and gas, being more buoyant than water, naturally rise toward the surface. Thus, once you find a conventional reservoir, it is relatively easy to extract oil and gas by drilling down through the "bottom" of the upside-down cup and inserting your drinking straw. A large amount of the hydrocarbon drink can flow through just one straw. This idea that you can drill one well and extract oil and gas from a large volume underground was popularized by Daniel Day Lewis' character in the film *There Will be Blood*, based upon Upton Sinclair's *Oil!*, when he stated to his oil competitor: "I drink your milkshake!" In this case, he and his neighbor owned land above the same very large upside-down cup, or oil reservoir, so a straw on one piece of land sucked the oil from underneath the other.

While you can drink someone's conventional milkshake, you can't drink someone else's unconventional milkshake. Whether we are talking about water, fossil fuels, or other minerals, we typically exploit the highest quality and easiest to access resources first. Higher quality means shallower and higher concentration. In that sense, conventional oil and gas reservoirs are easier than unconventional. Hydrocarbons do not readily flow through unconventional reservoirs once you drill into them. This lack of flow can be due to low permeability of the rock (think of a marble counter top rather than Swiss cheese) or high viscosity of the hydrocarbon (think of bubble gum rather than water) that does not even flow through Swiss cheese. Thus, one or more additional techniques must be employed to extract oil and gas from unconventional reservoirs. While these additional techniques require more time, materials, and money, they also bring the industry to cities, towns, and regions, like Denton, that were not conventionally associated with oil and gas production.

The Denton "Fracking Ban" first became a voter proposition after the Denton Drilling Awareness Group obtained enough signatures on a petition to force the vote. The petition claimed that fracking and related operations would "impact the City's environment, infrastructure and related public health, welfare and safety matters."[2] Documentaries such as *Gasland* promoted the idea that fracking wasn't environmentally sound or safe by showing homeowners near fracking sites light their faucet water on fire. While *Gasland* was classified as a documentary and was nominated for an Academy Award in that category, many in the energy and science communities thought the movie did not adequately provide enough background and context for its content to be considered a documentary.[3] How much does a movie focused on the downsides of oil and gas activity, for example, have to discuss other known or plausible explanations for the phenomena it shows? There is no one answer. Welcome to the narratives.

Aside from portrayal in films, whether or not a documentary, fracking-related activities created noticeable impacts that even proponents of oil and gas knew had

---

[2] State Impact Texas, https://stateimpact.npr.org/texas/tag/denton/.

[3] For example, it is possible to have methane in groundwater, such that what comes out of your sink can be lit on fire, with no connection to oil and gas extraction. In some cases, groundwater flowing from sinks can be lit on fire without nearby fracking activity.

to be addressed. The injection of wastewater from the fracking process into disposal wells led to earthquakes (seismic activity for the technically minded!) in both Texas and Oklahoma, and some seismic activity is directly related to the fracturing process itself.[4] These drove investments to increase resolution for monitoring of seismic events.[5] Also, increased oil and gas trucking activity accelerated wear and tear on country roads, and elected officials definitely hear about potholes.

Before Denton citizens voted, the Denton City Council first rejected a fracking ban. Before the council vote, Barry Smitherman, one of the elected commissioners of the Texas oil and gas regulating agency, the Texas Railroad Commission, sent a letter to the Denton city council arguing against a ban.[6] In the letter, Commissioner Smitherman told the council that economic development, tax revenues for the state and local jurisdictions, low electricity prices (because more natural gas extraction lowers its price which then reduces the cost of electricity from natural gas power plants), and enhanced national security (by reducing U.S. oil and gas imports) hinged on oil and gas production. Mentioning accusations that Russia was trying to influence anti-fracking environmental groups in Europe, he even suggested that the Denton fracking proposition might not be entirely driven by local residents.

Christi Craddick, a fellow Railroad Commissioner, lamented Denton's vote to ban hydraulic fracturing:

> We missed as far as an education process in explaining what fracking is, explaining what was going on. And I think this is the result of that, in a lot of respects, and a lot of misinformation about fracking, ...[7]—Christi Craddick (2014)

Continuing on Craddick's theme was fellow Commissioner David Porter:

> As the senior energy regulator in Texas, I am disappointed that Denton voters fell prey to scare tactics and mischaracterizations of the truth in passing the hydraulic fracturing ban.

---

[4]A 2018 study shows that "...that the shift in PW [produced water] disposal to nonproducing geologic zones related to low permeability unconventional reservoirs is a fundamental driver of induced seismicity."[13]. A 2019 study concludes that "Our results suggest some earthquakes in west Texas are more likely due to hydraulic-fracturing than saltwater disposal."[9]

[5]TexNet Seismic Monitoring Program, http://www.beg.utexas.edu/texnet. "In its 84th and 85th legislative sessions, the Texas Legislature tasked the Bureau [Bureau of Economic Geology of the University of Texas, which functions as the State Geological Survey of Texas] with helping to locate and determine the origins of earthquakes in our state and, where possibly caused by human activity, with helping to prevent earthquakes from occurring in the future. The TexNet Seismic Monitoring Program was established to accomplish these goals."

[6]Terrence Henry, State Impact Texas, "Who's Behind Denton's Fracking Ban? Head Texas Regulator Thinks It Could Be Russia," July 16, 2014, accessible April 10, 2018 at: https://stateimpact.npr.org/texas/2014/07/16/whos-behind-dentons-fracking-ban-head-texas-regulator-thinks-it-could-be-russia/ Smitherman's letter at https://assets.documentcloud.org/documents/1219669/denton-ltr-7-10-14.pdf.

[7]Dallas Morning News: November 6, 2014, Craddick: Railroad Commission will continue permitting in Denton, not ruling out action against ban. Accessed February 4, 2017 at: http://www.dallasnews.com/news/politics/2014/11/06/craddick-railroad-commission-will-continue-permitting-in-denton-not-ruling-out-action-against-ban.

Bans based on misinformation – instead of science and fact – potentially threaten this energy renaissance and as a result, the well-being of all Texans.[8]—David Porter (2014)

There are many keywords in these two quotes: "misinformation," "science and fact," "education process," and "mischaracterizations of the truth." The Railroad Commissioners must consider many tradeoffs, as stated by their mission ". . . to serve Texas by our stewardship of natural resources and the environment, our concern for personal and community safety, and our support of enhanced development and economic vitality for the benefit of Texans."[9] In stating her job, Commissioner Craddick noted "It's my job to give [drilling] permits, not Denton's . . . We're going to continue permitting up there because that's my job . . . ." I take this to mean she leaned toward the "enhanced development and economic vitality" part of the mission, whereas Denton voters might have emphasized "personal and community safety" or "stewardship of . . . the environment." Because a significant portion of Denton voters were "short-term" citizens, such as students attending the University of North Texas, they possibly played a critical role in the outcome of the vote. The vast majority of the students won't live in Denton after graduation, and they don't own local mineral rights from which to earn royalties from natural gas extraction.

Whose Choice Is It Anyway?

In 2015, 2 months after the Denton "Fracking Ban" passed, the Texas Legislature convened and wasted no time in creating legislation to override the ability of Denton to ban hydraulic fracturing and drilling activity within its border.[10] Aside from the engineering and geophysics related to horizontal drilling and hydraulic fracturing, the *fact* is that a vote from a low level political entity (a city in Texas) was overruled by a higher level political entity (the State of Texas) on a decision about energy extraction because of some qualitative and/or quantitative tradeoff of costs versus benefits. In the minds of state oil and gas regulators and state law makers, it was too costly to allow the precedent of a city restricting natural gas extraction within its borders. They were not going to give one set of constituents in one city the

---

[8]From Jim Malewitz, Texas Tribune online, November 5, 2014 article "Denton Bans Fracking, But Challenges Almost Certain," available at https://www.texastribune.org/2014/11/05/denton-bans-fracking-spurring-bigger-clashes/.

[9]Website of the Texas Railroad Commission, accessed February 4, 2017 at: http://www.rrc.state.tx.us/about-us/organization-activities/about-rrc/.

[10]Barnett, Marissa (April 17, 2015) "Texas House approves so-called 'Denton Fracking Ban' bill." Accessed April 19, 2015 at: https://www.texastribune.org/2015/04/17/texas-house-drill-denton-fracking-bill/. Malewitz, Jim (May 18, 2015) "Curbing Local Control, Abbott signs 'Denton Fracking Bill'. " Accessed May 24, 2017 at: https://www.texastribune.org/2015/05/18/abbott-signs-denton-fracking-bill/.

power to restrict natural gas extraction and thus put at risk the benefits of royalties to mineral owners, severance taxes to the State, revenues to the developers, and increased natural gas supply to U.S. consumers.

Not to be one-upped by Texas state officials, in 2015 Denton enacted a plan to source 70% of its electricity from renewables, such as wind and solar power, by 2019.[11] In 2018, the city decided to increase this goal to 100% by 2020.[12] Perhaps ironically, during the period between setting the 70% and 100% renewable energy goals, Denton signed a contract to build and own a natural gas fired power plant within the city borders.[13] It was as if Denton said we want some natural gas for our power plant just as long as it comes from somewhere else. Strictly speaking, if Denton even turns on their natural gas power plant, they are contracting less than 100% renewable electricity. Practically speaking, the city's fast-acting power plant can help it match the short-term ups and downs from wind and solar power generation to its real-time electricity consumption. Even more practically, a city like Denton does not have to match its consumption to the electricity generation from its owned or contracted power plants. This isn't how the electric grid works, but further discussion of "reliably balancing" the electric grid must wait until Chap. 3.

With regard to electricity provision, Denton and Texas are in a good position. Texas is big. It's sunny. It's windy. It's oily, and it's gassy. It has coal. Thus, Texas has an abundance of most types of natural resources used to generate electricity—notably lacking rivers with sufficient flow and elevation change to generate more than a token amount of hydropower. But who decides what sources of electricity we should use?

Like most people across the United States, those living in Denton don't each individually have a choice from whom to purchase electricity. They have to buy from their city-owned utility, Denton Municipal Electric. However, politically and economically Denton residents are indirectly in charge of the utility. They own it because they are citizens of Denton. And they elect city officials who direct the strategy of the utility. People living in cities with municipal electric utilities thus have *collective power* to influence the evolution of the energy system. This citizen-utility relationship holds for many other cities throughout the United States, but not all. In fact, not most.

Unlike in Denton, in many parts of Texas, for example, the major cities of Dallas and Houston, each individual electricity customer can choose from an array of electricity providers and plans. Many Texans have several companies competing to sell them electricity, and there are many low-cost options to supply electricity. The Texas Public Utility Commission website, *Power to Choose*, allows consumers

---

[11]Elanor Dearman, Texas Tribune, October 6, 2015: https://www.texastribune.org/2015/10/06/denton-announces-renewable-energy-plan/.

[12]Molly Evans, KERA News, February 6, 2018: http://keranews.org/post/denton-city-council-approves-energy-plan-be-all-renewable-2020.

[13]Russel Ray, Power Engineering, September 21, 2016: https://www.power-eng.com/articles/2016/09/w-rtsil-engines-used-for-225-mw-power-plant-in-texas.html. The power plant is 225 MW of reciprocating natural gas engines.

to type in their zip codes and see a list of options.[14] My brother in Dallas can buy electricity from one company this year, and switch to another company next year. He can choose a plan with 100% renewable energy this year, and 0% renewable next year. He can let the retail electric provider companies fight to have him as a customer. If his electricity provider goes bankrupt, no problem, he just chooses another company. As a customer he has both the obligation and *individual power* to choose which company he pays for *physical power* (as electricity) to flow into his home.

However, the most common situation for electricity consumers is that they neither have a choice from whom to buy electricity nor do they own their electric utility. This is the situation for customers of *investor-owned utilities*, or IOUs. These IOUs are publicly traded companies that own the power plants as well as the poles and wires that direct electricity to homes. Economically speaking, a customer of an IOU can purchase the stock of the IOU that provides its electricity and thus become partial owner of his utility. Practically speaking, most people do not have much money in the stock market in general, and thus cannot own their electric utility to have sufficient stockholder voting privileges to indirectly influence an IOU's decisions. In 2016 "...despite the fact that almost half of all households owned stock shares either directly or indirectly through mutual funds, trusts, or various pension accounts, the richest 10% of households controlled 84% of the total value of these stocks ..."[15]. Since over 80% of stocks are owned by 10% of families, how can the other 90% of families have any say over what happens in publicly traded companies whether they be energy companies or not?

One relatively recent development is that the pension funds, such as those that manage university and government endowments and collectively manage much of the stock wealth of the lower 90%, are beginning to use their size to influence investment in energy and other companies. *Divestment* is a term that describes removing investment in companies or regions that produce or sell objectionable products or practice objectionable behavior. The goal is to influence political outcomes. One of the most prominent examples was divestment from South Africa in the 1980s in protest against the racist policies of apartheid. Some now target divestment of companies that extract fossil energy with the thought that the transition to a low-carbon energy supply accelerates if fossil energy companies have less financial support. As we will find in this book it is easy, at least in the short term, to shift around money and even flows of energy. Over the longer term of decades, it is the energy system that tends to constrain the flows of money.

---

[14]Power to Choose website: http://www.powertochoose.org/. "Welcome to Power to Choose, the official and unbiased electric choice website of the Public Utility Commission of Texas. This website is available to all electric providers to list their offers for free. Compare offers and choose an array of electricity providers and plans."

The Fuzzy Boundary Between Physical and Social Processes

The Denton episode raises fundamental questions: When it comes to energy, who can and should be making choices regarding energy supply? Who owns it? How much do each of us need to know about the costs and benefits of how we obtain energy resources? What are we being told about energy?

The technologies, options, and political decisions surrounding fracking in Denton, electricity provision across the United States, and the trade of energy across the world, are all examples of the battlegrounds fought via *energy and economic narratives*. As this book discusses, these narratives compete to define the increasingly fuzzy boundary between the processes governing our physical and economic worlds—between what are scientific laws and what are merely theories, intuition, or rules of thumb. How we understand our physical world affects how we design rules governing our economic relationships, and vice versa. For example, some data indicate that over time one unit of economic value (i.e., gross domestic product) in developed countries comes from fewer physical resources (e.g., materials and energy). Thus, some techno-optimists claim that we can *decouple* economic growth from physical materials and energy. There are techno-realists who wholly disagree. When we look at data for the entire world, we do not see this decoupling at all (discussed in Chap. 9).

The internet and social networking companies, such as Google and Facebook, are examples that justify the concept of the decoupled economy. However, even in our digital age, the collecting, processing, and interpretation of data into information require materials (e.g., computers) and energy conversions (e.g., electricity to operate computers). Even blockchain-based digital currencies require electricity as an input to the servers that compute the blockchains that record transactions. Thus, there are direct limits related to decoupling information services from the physical world, but how do we figure out where these limits reside?

Indirect limits related to our social and economic domains are hard to quantify and project into the future. Because of trade, no country exists on its own. Some countries can become more oriented toward information and financial services because other countries are more focused on manufacturing and resources extraction. Consider what proportion of worldwide employment can be in the digital economy versus manufacturing and basic goods (e.g., energy and food) production. Today, someone has to produce and distribute food and energy. Even if food and energy production becomes fully automated, the people that own the related capital will want to be paid for those services provided by their investment. We can't all work for Google and Facebook since they ultimately derive revenue from advertising products they don't make (of course, maybe those companies will make and/or distribute physical stuff in the future as they battle against Amazon for world dominance!). If we did, no one would be making stuff, thus there would be no advertising and subsequent revenue, and no internet company.

The difficulties in understanding the linkages between our physical and economic worlds exist because these worlds operate as a fully integrated system. All parts evolve and adapt together such that causes and effects are difficult to determine, much less agree upon. People and machines are engines that require some amount of materials and energy to operate and survive. Given the relations that govern natural processes, human activity cannot violate these underlying principles whether or not we have knowledge of them. A child doesn't understand the concept of gravity when learning to walk, but he does have to adjust to the continuous feedback of gravitational forces. If not, he'll never learn to walk.

Further, from an energy standpoint, we cannot create a perpetual motion machine no matter how much we might want to do so. A perpetual motion machine that provides mechanical work is by definition one that breaks the laws of thermodynamics by producing more work than the energy content of the fuel input into the device. Chapter 2 summarizes the history of deriving these laws that define and describe energy, can't be rewritten by legislators, and can't be broken by even the most determined criminal.

In addition to physical and societal laws, we now also have a plethora of historical data describing both economic and energetic phenomena. The global data indicate that the more power we consume, the more people we support, and the more economic transactions we have. Most people think of cities as more "efficient" because each individual person in a larger city, for example, consumes less fuel driving a car. But the data indicate that as more people move to a city, the energy consumption and economic output of the city both rise in aggregate [1]. We cannot, however, only analyze cities and countries in isolation of the input and output flows across their and others' boundaries. The world economy is global.

Consider this: Two-hundred and fifty years ago the world population was small (fewer than one billion), we consumed renewable solar-derived energy at a slow rate (it took weeks to cross the Atlantic Ocean by sail), and the rate of economic transactions was slow. Today, the world population is large (more than seven billion), we predominately consume fossil energy resources (oil, coal, natural gas) at a fast rate (it takes hours to cross the Atlantic Ocean by jet airliner), and the rate of economic transactions is rapid.

Assuming a goal of more people and a larger economy, then all we have to do is keep doing what we've been doing for the last 200 or so years since the beginning of the industrial age, right? Can't we just use the mantra: "If it ain't broke, don't fix it."?

Given laws of thermodynamics and economic principles with which we can analyze our historical data, then surely we already know the recipe for socio-economic success. Combine the thermodynamic laws, one economic "law" of supply and demand,[15] one part human ingenuity, and one solid narrative and out comes a pot of gold at the end of a rainbow along with infinite energy: "... with the

---

[15]The "law of supply and demands" is notion that supply of goods (e.g., energy) and human demand for those goods are always equal, or matched. This of course is neither a physical nor political law, but one that economists use as rationale for interpreting the world.

rise of logic we attain the impossible—infinite energy, perpetual motion . . ." we are told in the book *The Bottomless Well* [7].

If only it were that easy. Unfortunately, we humans are a bit too complicated and ignorant. As a physical systems scientist, I can't bring myself to claim that we can use logic to create a perpetual motion machine that breaks the laws of thermodynamics. As an author, I can't make myself use such hyperbole to convince you of my worldview.

Yes, our brains might be capable of creating an infinite number of ideas whether they be perpetual motion machines, unicorns, or intergalactic space travel. The last two items are prevalent in movies such as *My Little Pony* and *Star Wars*, and some scientists entertain themselves by pointing out filmmakers lack of adherence to physical laws.[16]

But what happens when the differences between fiction and reality affect our daily lives in ways other than pure entertainment and in which it is hard to tell the difference between entertainment and a real distribution of wealth? While most of us over the age of 10 know unicorns are not real, most of us also don't know the differences between the assumptions of economic theories and those of scientific laws. If an economic or scientific theory does not accurately (enough) describe the physical and economic trends of the world, then we must work to replace it with something more accurate.

Not only is our economy a *complex system*, consisting of many connected parts, but it resides within our physical environment largely defined by Earth's boundaries (even satellites in space govern activity on Earth) [5, 6]. In addition, the purpose of a system is defined by what it does, not by what someone claims or wishes its purpose to be [10]. We did not create the physical world, and thus, do not define its purpose. While we seek to understand and describe the operation of the physical world, we *use this understanding of the physical world to define rules that govern our societal and economic relations.*

If we think the world is infinite, we might define rules without regard to limits, and vice versa. While we are in charge of defining our *socio-economic* rules, including the laws and norms that govern how people relate to each other and define allowable economic actions, the physical environment constrains the outcomes from these rules. We make choices "now," and outcomes occur "later." We experience some of these outcomes rather immediately (e.g., generating electricity from power plants automatically reacting to market signals) and others much later (e.g., population booms after curing disease). The timing of these economic outcomes are feedbacks that help confirm or discredit the rules we derive to explain the purpose of our socio-economic system.

While our socio-economic system by definition achieves its purpose, it might not be in alignment with your desires. If socio-economic outcomes *are* what you intend, then there is no need to call for change because the system achieves a purpose aligned with your desires. If socio-economic outcomes *are not* what you intend,

---

[16]Increasingly Stupid Movie Physics: http://www.intuitor.com/moviephysics/.

then you can work to change the design of the underlying system. Of course, people disagree on both the interpretations of economic outcomes and the purpose of our socio-economic system.

To understand and perhaps guide the purpose of our economy, economists and scientists think about future energy-economic options by mixing laws and theory (scientific, legal, socio-political, and economic) with data by inserting them into a computational oven of some sort that spits out results for interpretation. Depending upon the knowledge and worldview of the critic interpreting the results, as well as the quality of input ingredients, the results are described anywhere from an elegant soufflé to a half-baked pile of soggy dough.[17]

If the taste and shape of our energy and economic dessert does not meet our expectations, then how do we know whether or not the ingredients were combined incorrectly, the oven is broken, the recipe is poorly designed, or we just simply can't get what we want? If we determine we can't get the result we want, and if we just change the inputs and assumptions to get our desired result, will anyone notice? After all, "it's complicated," right?

At some point in the process of mixing and baking the ingredients that will rise into our future, we go from historical energy and economic data to *energy and economic narratives*. Somewhere between physical laws and data lies the theoretical economic oven, or "black box," that uses a mathematical equation or computational algorithm to convert inputs to outputs, data to information. We use these black boxes, or models, to help us peer into the future, to learn and project what enables future prosperity or poverty, to better understand how different the world might be if powered by fossil versus renewable energy. Sometimes people don't even use a *black box*, but only the intuition they believe they have within their *gray matter*, or brain. Sometimes a reasonable argument is made, sometimes not. For better or for worse, people with gray matter design our computational models and black boxes. These people, all of us, are affected by our cultures, our languages, our environments, and our *narratives*.

The Energy and Economic Narratives

What do narratives do for us? The introductory quote of this chapter displays a standard dictionary definition of narrative, but it doesn't explain why we have them. Narratives serve three purposes.[18] First, they tell a story of belonging. If you meet a stranger and realize you are from a common area, you more easily engage in conversation than otherwise. Second, they describe norms that guide our actions. Most people in society follow certain norms such that by doing so, they are accepted as part of the group. Part of these norms includes reciprocal obligations. If you and

---

[17] A soufflé is known as one of the most difficult pastries to master.

[18] Collier [2, pp. 32–33].

I adhere to a set of norms, then when I do something for you, you will return a favor to me. This could be as simple as opening the door for someone and that someone saying "thank you."

The third purpose of narratives most concerns the scope of this book. We use narratives to learn about how the world works. This can be good or bad. "Experiments show that we rely more on stories than on direct observation or tuition. . . . At its worst, it creates a rupture between reality and what we believe – narratives as 'fake news'."[19] Scientists spend their lives finding ways to explain how the world works, and this is usually done via mathematics confirmed by observation through experiments guided by the scientific method. The scientific method is structured to allow competing concepts, weed out those without merit, and increasingly solidify ideas that are repeatedly confirmed by new evidence, experiment, and theory. History shows it can be dangerous to challenge narratives even with the scientific evidence on your side. Whether we consider Galileo being convicted of heresy for correctly stating that the Earth revolves around the Sun, the Scopes Monkey Trial that enabled public discussion of evolution in the U.S., or contemporary discussion of climate change, it can be very hard to break down long-held narratives about how the world works.

However, the topic of this book is about *energy and economics*, and I describe narratives along those two axes (see Fig. 1.1). Because people disagree as to the costs, capabilities, and benefits of different energy technologies and resources, proponents of different visions use narratives to convince stakeholders of the validity of their positions. The energy and economic narratives don't directly concern

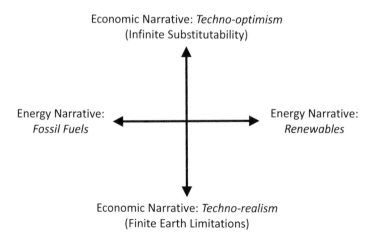

Fig. 1.1 This book discusses narratives along two dimensions: *energy*—fossil versus renewable; *economics*—technological optimism of infinitely substitutable technology versus technological realism that the finite Earth imposes limits to growth

---

[19]Collier [2, pp. 32–33].

controversies within astronomy (e.g., heliocentrism), biology (e.g., evolution), or the earth sciences (e.g., climate change), but concepts from those and other scientific fields pervade our understanding of energy and economic interactions. Many linkages and feedbacks occur between our energy and economic systems. To some, energy and economic systems are inseparable. However, any given narrative resides on a point along each axis. We obtain a comprehensive understanding of a narrative by considering both the economic worldview that leads to a position on how to provision energy, and how a position on the provision of energy leads to an economic worldview.

Let's start with the two energy narratives. These characterize the extreme views regarding the desired sources for our future energy system that best meet our future social and economic needs.

Energy Narrative: Fossil Fuels Are the Future

This narrative recognizes that fossil fuels enabled us to achieve what we have today. A proponent might say: "The physical fundamentals of fossil fuels, such as high energy-density and portability, ensure low cost and their continued dominance. Why not use them? Renewable energy technologies require subsidies to entice investment because they cannot achieve the historical or present levels of low cost and productivity of fossil fuels and related technologies. Therefore, we should promote increased fossil fuel use for the foreseeable future. Fossil fuels, and the technologies we have developed to burn them, enable us to shape and control the environment rather than the reverse situation before we invented fossil-fueled machines. Further, fossil fuels are the best hope to bring poor countries out of poverty while continuing to increase prosperity within developed countries."

Energy Narrative: Renewable Energy Is the Future

This narrative states we can use renewable energy technologies and resources to sufficiently substitute for the services currently provided by fossil fuels. A proponent might say: "Thank you fossil fuels, but we've modernized. We don't need or want you anymore. Fossil fuel production and consumption create environmental harm both locally over the short-term (e.g., air and water contamination) and globally over the long-term (e.g., climate change) to such a degree that their continued unmitigated use ensures environmental ruin that will lead to economic ruin. In addition, the concentration of fossil fuel resources means that countries and citizens have unequal ownership of them, creating geopolitical instability over extraction and distribution. Thankfully, renewable energy technologies are now cheap enough to transition from fossil fuels. Further, a renewable energy system is the best hope to bring poor countries out of poverty while continuing to increase prosperity within developed countries."

What is the difference between the energy narratives? Is one right and the other wrong? Does one have the moral high ground or better conform to the laws of physics? Is there a difference between what is technically possible and what is economically and socially viable and desirable? What is the truth about the future of energy amid all of the political posturing on the subject?

Are there fundamental truths or only *energy narratives*?

These questions drove me to leave my job with a high-tech start-up company and return to academic research. I wanted to understand both the changes that were occurring in the energy system around me as well as how to sift through the rhetoric of the two basic narratives that we hear regarding the future of energy in society.

What I've come to discover is that while both the fossil and renewable narratives have valid claims, they both often produce misleading visions.

Both energy narratives use *economic narratives* to justify their arguments, and these arguments shape energy policies that affect each one of us. Economic theory in turn informs us how to perform calculations that provide insight into the ramifications of choosing one energy pathway versus another. Unfortunately, the most common economic theory uses concepts that are incapable of explaining the energy-related changes that they are sometimes used to explore. To say the least, this is a tremendous problem when a model can't clearly distinguish between the two vastly different worlds run by fossil versus renewable energy. As you will read in this book, energy consumption is a fundamental economic driver. Thus, it is crucial that economic concepts effectively consider the role and cost of energy. If they don't, we will make policies based on flawed economic concepts, and these policies will impact people in ways we don't expect, but should expect.

The economic narratives are as follows.

Economic Narrative: Technological Optimism (There Is Infinite Substitution of Technology to Achieve Growth and Social Outcomes)

This narrative posits unbounded technological change that creates substitutes for whatever we desire. It does not necessarily deny that the Earth is finite, but it does not believe that this fact affects economic or physical outcomes that impact the overall human condition. It is the view of most mainstream economists. A proponent might say: "Technological innovation has and will always address the pressing needs for society. In order to promote seeking of solutions, we need a signal. That signal is the price of a good, or a 'bad' (e.g., air pollution), and the signal is provided by setting up a market. Therefore we must establish and promote free markets, private ownership and profits via capitalism, and business competition. This is the way toward continued growth and prosperity. With regard to energy, as long the aforementioned criteria govern the economy, its price always decreases, so there is no need to worry. Markets best address socio-economic issues because they process information better than any human regulator or government agency."[20] Got a problem? Make a market for it.

---

[20]See Postface by Philip Mirowski [12].

Economic Narrative: Technological Realism (The Finite Earth and Laws of Physics Impose Biophysical Constraints on Growth that Affect Social Outcomes)

This narrative takes to heart that the Earth is finite. It is the position of many ecologists, physical scientists, and some economists. A proponent might say: "Humans need food to survive and our economy requires energy consumption and physical resources to function. These facts very much matter for economic reasons because the feedbacks from physical growth on a finite planet will eventually force changes in structural relations within our economy and society more broadly. These changes can have positive or negative outcomes for our perception of the human condition, but to create positive outcomes, we must perceive, accept, and adjust to the physical limits of a finite Earth and relate our economy to physical laws and processes. Markets can work, but they have problems. Theoretically they can include all important pieces of information, but practically, finite time and incomplete information prevents formation of pure price signals." The narrative is summed up well by a statement attributed to economist Kenneth Boulding: "Anyone who believes that exponential growth can go on forever in a finite world is either a madman or an economist."[21]

The simplified way to think about the economic narratives is whether you believe in *infinite substitutability* or *biophysical constraints from the finite Earth*. By the term biophysical, I refer to the properties, laws, and concepts that relate to growth and maintenance of both living matter, such as animals, as well as non-living matter as physical capital, or "stuff" such as cars, buildings, and factories.

The infinite substitution narrative assumes that the finite Earth *cannot* be a descriptive factor for changes in long-term economic growth and distribution of resources, physical capital, and money. The finite Earth narrative assumes that it *can*.

Julian Simon's *The Ultimate Resource* [14] and Milton Friedman's 1980s "Free to Choose" book and 1980s PBS film series are examples of the *techno-optimism* economic narrative [4].[22]

The *techno-realism* economic narrative is emphasized at the extreme by Paul Erlich's *The Population Bomb*, and more moderately by the Club of Rome and authors of *The Limits to Growth* [11] and Tim Jackson's *Prosperity without Growth* [8].

Hydraulic fracturing and horizontal drilling for oil and gas is a great example for discussing competing narratives. One can promote the technology from the combination of the fossil and techno-optimistic narratives, or one can argue against

---

[21] Statement of John S. Steinhart attributing the quote to Kenneth Boulding in testimony to the U.S. Congress: United States. Congress. House (1973) Energy reorganization act of 1973: Hearings, Ninety-third Congress, first session, on H.R. 11510. p. 248. Viewable on January 12, 2018 at: https://babel.hathitrust.org/cgi/pt?id=mdp.39015001314395;view=1up;seq=249.

[22] Free to Choose PBS series: http://www.freetochoose.tv/broadcasts/ftc80.php.

the same technology to promote that the combined renewable and techno-realistic narratives provide the best perspective.

First consider promoting the combined "Techno-optimistic + Fossil" perspective. Here is how you could phrase it:

> In the early 2000s, U.S. oil and gas extraction had been in decline for thirty years, with the extraction from Alaska providing a significant, but temporary, increase starting in the 1970s. Yet as of 2019, after over a decade of ramped-up fracking and horizontal drilling activity, the industry had increased U.S. oil and gas extraction rates to the highest levels in history. Oh, and since burning natural gas produces less carbon dioxide than burning coal, fracking will help mitigate climate change by reducing carbon dioxide emissions from electricity generation. Further, by extracting more energy resources from the U.S., we make the world a freer place by exporting energy to our geopolitical allies such that they buy less from our enemies. Thus, motivation and "necessity" from high oil and gas prices in the 2000s were the mothers of invention that birthed yet another revolution in the oil and gas industry. *"Ask, and it shall be given to you."*[23] Human ingenuity always comes to the rescue and always will. And in the context of techno-optimism and fossil energy, it just did.

Not so fast says the "Techno-realist + renewable" perspective. Here is how you could counter:

> If fracking is so revolutionary and oil and gas resources are so infinite, then why do you have to drill under my house in the city? And why does this "revolution" need a tax break? If fossil resources were infinite, then you should always be able to go somewhere else to extract oil and gas, and you shouldn't care if I don't want you in my backyard. Half of infinity is still infinity! Plus, you want to talk about geopolitical energy security and freedom? How free are we in our own country if we aren't allowed to vote for what happens in our own city without elected officials, from some other part of the state or country, overruling us, their own constituents? Also, don't tell us that burning more natural gas reduces greenhouse gas emissions. Natural gas is a fossil fuel, and the total emissions along its supply chain aren't that much better than coal. To fight climate change, we need to transition to renewable energy systems now without investing in the gas infrastructure that becomes a long-lived bridge from coal to nowhere. We now have renewable energy technologies that can replace fossil fuels at the same or lower cost to the economy and the environment.

Notice what happened in that last sentence? I stuck some techno-optimism about renewables into an argument containing techno-realism for fossil energy. This is a common conundrum facing many arguments. My technology continues to improve (techno-optimism) but yours cannot (techno-realism). As we *knock down the narratives* in this book, we parse this contradictory statement that only certain types of energy technologies progress while others regress.

While it is rare that we are asked to decide on energy-related matters, as the Denton example shows, it does happen. Jurisdictional battles are not unique to energy resources, Texas, or the United States. Usually citizens only indirectly affect energy policy by electing officials that represent their views and enact policy accordingly. The results of energy and environmental legal and political skirmishes relate to who owns the resource, and thus benefits from the resource extraction,

---

[23] *Book of Matthew,* 7:7.

as well as who bears the burden of any local costs (environmental or otherwise). These cost and benefit situations vary greatly across the world, and the mismatches between those who benefit and bear cost create much political discontent.

A complex set of questions arise. First, how many people are affected by costs of resource extraction, and how large are these costs? Second, how many people benefit from the resource extraction, and how large are these benefits? If there are relatively many beneficiaries, should they somehow compensate the relatively few that bear the costs? If there are relatively few that bear the costs, should they have political or economic power to prevent a relatively large number from experiencing the benefits? How should we view the reverse situation where few benefit while most bear the costs?

The founding fathers of the United States were aware of this cost-benefit conundrum. The founders created a system of government to minimize the impact of small factions and allow majority or super-majority rule, at least for those with voting power. At the founding of the U.S., if you weren't a white male landowner, you didn't have a say in politics by voting or any other means. Only a subset of the population made decisions based upon their interpretation of what facts, costs, and benefits should be considered for decision making.

Thus, our existing rules, laws, values, and perspectives influence which facts are allowed for discussion and which are weighed more heavily than others. In the Denton Fracking Ban example, maybe many of the anti-fracking voters didn't know all of the facts. Maybe the Texas Railroad Commissioners also didn't know all of the facts. But a command of facts and science is required neither to vote as a citizen nor run for elected office.

In fact, some facts don't seem to drive differences in opinion. According to the University of Texas Energy Poll, when asking Americans if hydraulic fracturing should be banned on federal lands, the split was nearly equal around 38% for and 38% against, with the rest undecided.[24] On this particular issue, when all of the costs and benefits from petroleum extraction are equally shared from activities on public land, opinions are equally split.

The questions of who pays and who gains are not unique to the realm of energy. They are universal, and at first glance, they seem completely unrelated to energy. But as you will learn in this book, our economic, social, and political arrangements are fundamentally linked to the quantity and cost of consuming energy. We cannot separate them as much as we might have been taught or told.

Since the dawn of agriculture people have become increasingly effective at using natural resources to produce basic services (e.g., energy, food, clean water) and accumulate more stuff owned and consumed by more people. Thus, the Earth has become increasingly full of both man and man-made items that previously were raw

---

[24]Topline results from the March 2017 poll, asking the question: "Hydraulic fracturing on public lands ...(1) should be banned (strongly), (2) should be banned (somewhat), (3) no preference or undecided, (4) should be promoted (somewhat), or (5) should be promoted (strongly)." Formerly available at: http://www.utenergypoll.com/wp-content/uploads/2014/04/Energy-Poll-Topline-Wave-12.pdf accessed May 31, 2017.

materials. Economist Herman Daly notes: "Ways of thinking about the economy that worked well in an empty world no longer suffice in such a full world." [3] How do scientists and economists determine whether the Earth is "too full" of humans and our activities? How do they agree and disagree on the implications of our increasing human ingenuity to maintain a larger population along with increasing consumption of resources?

As any environment becomes "full" there are feedbacks that slow the rate of growth of the population. Indeed, Chap. 4 explains that the global rate of human population growth has declined since the 1970s, and it is predicted to continue to decline. These "full planet" questions are not new. We have been introduced to the issue before, in many versions. Thomas Malthus' 1798 *An Essay on the Principle of Population* suggested that human population would outgrow our food supply, eventually and inevitably leading to starvation of that outsized population. While poverty still exists in many parts of the world, we have not had mass global starvation.

We know much more today than in Malthus' time, but what can we say about his assertion? How do scientists and economists interpret data and create models to determine our energetic limits and opportunities on our one planet Earth? If you make an economic and physical model to anticipate the future, should you include the fact that the Earth is finite?

These questions might sound silly, but they are important questions for context in interpreting "answers" given by politicians, economists, and scientists. Some models include the idea that the Earth is finite, and some don't. Is either category simply wrong from the start?

Each one of us makes personal and family decisions within our local contexts that are in turn affected by constraints and opportunities at city, state, country, and global levels. We elect politicians that make decisions affecting (to varying degrees) our roles in these various levels of governance: the taxes we pay, the benefits we receive, the opportunities we have. But the world has changed dramatically in the last three generations. Many of the developed country politicians are grandparents.[25] Because so much has changed in the last 40–60 years, these politicians might not understand the fundamentally more acute constraining feedbacks facing their grandchildren. It is not enough for our politicians to tell us they are concerned about their and our grandchildren.

My parents are part of the Baby Boom generation that grew up after World War II in a United States that was the dominant economic and industrial power of the world. In 1950 world population was 2.5 billion. In the U.S. it was about 150 million. I was born in 1974 at a time when both environmental and energy constraints were causing fundamental change in the U.S. for the first time. U.S. cities could no longer dump unprocessed industrial waste into rivers, and Western oil companies were forced to take a smaller share of profits from oil extraction occurring within other countries

---

[25] Here "developed world" refers largely to the richer countries such as the U.S., Western Europe, Japan, and others within the Organization for Economic Cooperation and Development (OECD).

(e.g., the Middle East). In 1974 world and U.S. populations were 4 billion and 210 million, respectively. My nieces and nephew, born after 2000, are again born at a time of change. Almost 30 years of a "Great Moderation" of economic stability ended in 2008 with the "Great Recession," a bust in credit and commodity markets that affected people and countries across the entire globe. World and U.S. population were 6.7 billion and 300 million in the year 2000, and the world reached the cheapest energy in the history of human civilization (see Chap. 2).

Will today's children have careers during a time of "Great Contraction," "Great Revival," or "Great Volatility"? While only time will tell, this book explores the data and the theories that attempt to explain the interdependencies of the global megatrends that have shaped our present and will shape our future.

Today, there are two broad questions that address human welfare. How much is there? How should it be distributed?

That is to say if we know something about both the *size* and *structure* of the economy, then we can address important socio-economic questions. For an example of size, think of the amount of stuff around us, and the total net income of the economy, or gross domestic product. For an example of structure, consider that not every person earns the same income or has the same access to resources. As social beings who depend on physical resources (e.g., food, water, energy), we have more influence on the distribution of those resources than we do on their total production.

While the factors driving the answers to these questions of size and structure are interdependent, any given person might choose to prioritize one factor over the other. A person's prioritization on growth (i.e., how much in total) versus distribution (i.e., how much to each person) can say much about his outlook and perceptions of the role of energy in society. The techno-optimistic narrative claims we don't have to choose between growth and distribution, but that we can have both. The techno-realism narrative says that physical constraints can force us to choose which we prioritize. Use the data, history, and theories presented in this book to help make up your own mind.

Purpose and Structure of Book

The purpose of this book is to explain how physical constraints affect our social outcomes, in particular economic outcomes, more than we think. These constraints affect growth, distribution, and our existence on a finite planet. A lack of understanding of these constraints pervades our politics, policy, business decisions, and economic theory. Thus, we're too often given hollow narratives that leave us grasping at straws when we need be held firm by rigid columns of understanding. We simultaneously blame our politicians for policies that don't work yet ask for unlikely combinations of outcomes. The problem is that we don't know we're asking for too much because it is hard to see the interdependent connections within the economy. To see these connections we must consider the whole economy, or the *macro-scale economy*.

While we each have our own personal experiences that are different than others' experiences, this book focuses on aggregate or "macro" trends and data of the economy, not on the "micro" experiences and choices of individual people, or even computers running algorithms.

In pursuing the macro-scale purpose of the book, I lean on many figures, but not equations, to display important data and calculations. If we're going to understand narratives, we have to think about the data upon which they are based. These figures should not scare away readers who only want to take home the broad points.

When perusing the figures consider two principles. First, note when the data are increasing, decreasing, or staying about the same. Second, note when trends in one time series change at the same time as other time series as this is a clue that these phenomena might be related.

The book also contains a larger-than-usual number of quotes from individuals via their books, interviews, and news articles. In some cases the quotes are longer than might normally be shown. The reason is to display the energy and economic narratives in enough of their original prose such that the reader has some ability to interpret the original author's meaning.

With that said, there are many basic and undeniable facts about natural resources and the economy, but there are many more interpretations of the meaning of these facts—so many interpretations, or narratives, that I wrote this book about them. Here I list only three facts to keep in mind, and we'll revisit these throughout the book.

Fact 1 *The Earth is finite in size.* Disagreements arise on whether this fact matters for social and economic purposes.

How do we know the Earth is finite in mass and volume? Aside from deducing this via observations from Earth (e.g., sailing around and ending up at the same point without backtracking), we also sent people into space with a camera. In 1972 the astronauts of Apollo 17 took the famous Blue Marble picture which fit the entire Earth within the frame of the picture. If the Earth were infinite in size, then you could not fit it within the frame of a picture.[26] The finite Earth implies limitations to physical growth of anything on the planet either because it fills up (e.g., with people) or is limited by the need to consume or be composed of physical resources from the planet itself. My favorite joke for why we know the Earth is not infinite, and thus not a flat surface, is because cats haven't yet knocked everything off of it.[27]

---

[26]I recognize that the Earth does have mass transfer to it from such objects as meteors, meteorites, and comets, and that humans now send mass from Earth into space in the form of satellites and capsules such as during the Apollo missions. However, these mass transfers to Earth are also not infinite. Further, this book refrains from speculating regarding our ability to harvest resources from space. In addition, taking one picture of the Earth from space does not prove that the Earth is geometrically finite because in theory the Earth could be extending infinitely into the distance behind its cross section. Since we've circled the Earth and taken pictures of the Earth from multiple vantage points, we're sure of its near spherical shape.

[27]Credit to my waitress Sarah A. at the Arvada, Colorado School House Kitchen & Libations (May 31, 2017).

Frighteningly, in twenty-first century United States, people still believe the long-debunked concept that the Earth is flat. A 2018 poll of Americans indicates that 34% of 18–24 year-old are not sure that the Earth is round—poor progress on at least one basic concept.[28]

Fact 2 *The laws of nature are human constructs that describe the interactions within the natural world and are defined as being the same everywhere (per the present state of knowledge).* These human-derived concepts form the basis for our definition of energy via the laws of thermodynamics which we have not yet invalidated.

Fact 3 *The laws of society, or legal rules and social norms, are human constructs that seek to limit human interactions to a subset of all possibilities, and they are not the same everywhere.* These human-derived concepts influence how people, communities, and countries interact with one another.

Of course disagreements arise over which economic rules should govern social relations, how Facts 1 and 2 explicitly inform or constrain the socio-economic rules and norms of Fact 3, and how much we believe we're really in charge of our own decisions. As you read this book, consider how to merge these three facts into a coherent narrative.

This book has three parts. Part I defines the narratives (this chapter), presents data on energy (Chap. 2) and megatrends of societal and economic growth (Chap. 4), and demonstrates the divergence of opinions on energy by examining the major arguments for the energy narratives (Chap. 3).

Part II synthesizes the multiple sets of data to understand both how they fit together and how the energy and economic narratives can cloud this understanding. Chapter 5 summarizes systems thinking and concepts that help understand the patterns linking energy and economic phenomena. This chapter emphasizes the need to simultaneously consider the *size and structure* of the economy. The linkages between economic size and structure are highly under appreciated.

Many people readily think of an economy's size (or growth) or its structure (or distribution of income and wealth), but fewer think about how the two relate to each other. Size is important, but ignoring structure and distribution is like saying we only need to know the areas of a triangle and a circle to compare them (Fig. 1.2). Triangles and circles are shapes, or structures, defined by their geometric constraints. When we compare two circles, we only need to describe their size because we've already specified they have the same constrained shape. Unfortunately, when we compare economies and energy technologies, we don't get to constrain each to be the same shape. They are different shapes by their natures, and thus we must be careful when comparing them only by size. Chapter 5 looks to

---

[28]Hoang Nguyen, YouGov (April 2, 2018), "Most flat earthers consider themselves very religious," accessed April 7, 2018 at https://today.yougov.com/news/2018/04/02/most-flat-earthers-consider-themselves-religious/. Also see the 2018 film documentary *Behind the Curve*.

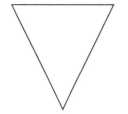 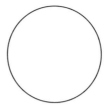

Fig. 1.2 The triangle and circle have the same area, but not the same shape. When we assess "the economy," we must consider at least two concepts: size and shape (or structure). Neglecting one or the other avoids understanding the energy and economic narratives

biology for parallels to economy size and structure. In doing so we might lose some specificity, but we gain holistic insights.

Still within Part II, Chap. 6 dives into economic theories and narratives used to understand the size and structure of the economy. Some economic concepts and models are more accurate than others, and this chapter explores the energy and economic policy implications for how people interpret concepts such as "technology" and the role of energy in economic growth. Interpretation is in full swing by Chap. 7 that summarizes the last 90 years of U.S. economic growth and structural change via three distinct phases. Part II ends with a more philosophical Chap. 8 describing a narrative of the economy that transcends the energy and economic narratives as defined in this first chapter, but that is consistent with scientific and economic understanding as well as the three facts mentioned above.

Chapter 9 begins Part III by discussing the political battles for the energy and economic narratives. It highlights the usual views of how policy is made today, such as via lobbying influence, worker union bargaining power, and economic cost-benefit analyses. While being correct and having merit on their own, these must be put into the context of the physical nature of the economy to have a more holistic view of the role of energy in the economy. The book ends with Chap. 10 outlining both ways people envision future scenarios as well as some trends and outcomes that have high probability to occur in the next several decades. Thus, the final chapter uses the content of the book to understand how to envision future change per the following sequence:

- First, physical laws and constraints describe how we can use and access energy.
- Second, energy resources physically power the economy via use in machines, buildings, and other physical capital.
- Third, our interpretations of the economy inform policy.
- Fourth, policy affects social outcomes by designing markets, regulations, and taxes that affect the distribution of money.
- Finally, the rules governing where, how, and when money is distributed affect energy resource extraction and consumption, leading back to the beginning.

One cannot discuss the future of energy without putting it into the context of economic thought. One cannot discuss economics without some modeling construct, theory, or framework that assumes how the world works. We describe how the world "works," or changes from one state to the next, via the concept of energy. Because our historical use of energy resources has invariably shaped the world we live in today, energy has invariably shaped our perceptions of the natural world and our culture.

The journey to parse the competing narratives on energy, economic growth, and the related policies begins with the history of energy, its rate of extraction, and its cost.

References

1. Bettencourt, L.M.A.: The origins of scaling in cities. Science **340**(6139), 1438–1441 (2013). https://doi.org/10.1126/science.1235823. http://science.sciencemag.org/content/340/6139/1438
2. Collier, P.: The Future of Capitalism: Facing the New Anxieties. HarperCollins, New York, NY (2018)
3. Daly, H.: Economics in a full world. Scientific American pp. 100–107 (2005)
4. Friedman, M., Friedman, R.: Fred to Choose: A Personal Statement. Harcourt Brace Jovanovich, New York and London (1980)
5. Goodland, R., Daly, H., El Serafy, S.: Environmentally sustainable economic development building on brundtland. Environment Working Paper 46, The World Bank (1991)
6. Hall, C.A.S., Klitgaard, K.A.: Energy and the Wealth of Nations: Understanding the Biophysical Economy, 1st edn. Springer (2012)
7. Huber, P.W., Mills, M.P.: The Bottomless Well: The Twilight of Fuel, the Virtue of Waste, and Why We Will Never Run Out of Energy. Basic Books, New York (2005)
8. Jackson, T.: Prosperity Without Growth: Foundations for the Economy of Tomorrow, second edition edn. Routledge, Milton, UK and New York, NY, USA (2017)
9. Lomax, A., Savvaidis, A.: Improving absolute earthquake location in west Texas using probabilistic, proxy ground-truth station corrections. Journal of Geophysical Research: Solid Earth **124**(11), 11,447–11,465 (2019). https://doi.org/10.1029/2019JB017727. https://agupubs.onlinelibrary.wiley.com/doi/abs/10.1029/2019JB017727
10. Meadows, D.H.: Thinking in Systems: A Primer. Chelsea Green Publishing, White River Junction, Vermont (2008)
11. Meadows, D.H., Meadows, D.L., Randers, J., Behrens, W.W.I.: Limits to Growth: A Report for the Club of Rome's Project on the Predicament of Mankind. Universe Books, New York (1972)
12. Mirowski, P., Plehwe, D. (eds.): The Road from Mont Pèlerin: The Making of the Neoliberal Thought Collective. Harvard University Press (2009)
13. Scanlon, B.R., Weingarten, M.B., Murray, K.E., Reedy, R.C.: Managing Basin-scale Fluid Budgets to Reduce Injection-Induced Seismicity from the Recent U.S. Shale Oil Revolution. Seismological Research Letters **90**(1), 171–182 (2018). https://doi.org/10.1785/0220180223
14. Simon, J.L.: The Ultimate Resource 2, revised edn. Princeton University Press, Princeton, N.J. (1996)
15. Wolff, E.N.: Household wealth trends in the united states, 1962 to 2016: Has middle class wealth recovered? Working Paper 24085, National Bureau of Economic Research (2017). https://doi.org/10.3386/w24085. http://www.nber.org/papers/w24085

Chapter 2
Energy and Food: The Megatrend of Megatrends

If indeed the agricultural-sector proportions of poor countries were not declining, economic growth would indeed be hampered.[1]—Julian Simon (1996)

What Is Energy?

What is energy? There are many interpretations and perspectives from that of practical engineering design to philosophical abstraction. A wonderful place to start is with the late Nobel Laureate in Physics, Richard Feynman. In his 1961 lectures he discussed the concept of energy defined as an unchanging quantity: "It states that there is a certain quantity, which we call energy, that doesn't change in the manifold changes which nature undergoes. This is a most abstract idea, because it is a mathematical principle; it says that there is a numerical quantity which does not change when something happens."[2] This idea is known as the first law of thermodynamics, or law of the conservation of energy, and *thermodynamics* is the study of the relationships among various forms of energy.

At its core, this energy concept is far from obvious. Why would there be a quantity that remains the same value before and after things change? We observe change all around us, the seasons, rain, aging. Yet the development of the concept of energy had to come by very careful observation of the world around us. We could have called this conserved "stuff" anything, but we call it *energy*. In defining energy via the first law of thermodynamics, we are not talking about "conserving" energy by running the heater less in your home during the winter, and we are not talking about burning less gasoline in your car. Energy is defined by a mathematical accounting principle stating that when you count all of the energy

---

[1] Simon [27, p. 600].
[2] http://www.feynmanlectures.caltech.edu/I_04.html.

© Springer Nature Switzerland AG 2021
C. W. King, *The Economic Superorganism*,
https://doi.org/10.1007/978-3-030-50295-9_2

residing somewhere at one time, and count it again at another time after something happens, you get the same total quantity.

To describe the concept of the conservation of energy, Feynman uses an analogy of a mother counting the number of her son's toy blocks. The mother knows her son has 28 toy blocks. Her son is a normal kid who is not very tidy, and he usually leaves the toy blocks on the floor of his room when done playing with them. He also sometimes takes his toy blocks into other rooms or places them in a toy box. When the mother comes to her son's room, often she cannot see all 28 blocks, but she knows there are 28 blocks in the house.[3] She is then forced to use clever ways of deducing where all of the blocks are located in the house. She weighs the toy box when she sees all 28 blocks (i.e., the box is empty), and she also determines that each block weighs a certain amount. She writes an equation for the conservation of blocks. On one side of the equals sign is 28 blocks. On the other side are two terms added together. The first term is the number of blocks she can see. The second term is a formula based upon the weight of the empty toy box and the weight of one block. Therefore, if she only sees 27 blocks, and if the box is too heavy by the weight of one block, she deduces that there is one block inside the box. If it is too heavy by the weight of two blocks, she knows there are two blocks in the box, and so on. One day the mother realizes that the (dirty) water level in the bathtub has increased, and she only sees 26 blocks with the toy box weight indicating it is empty. The two missing blocks are not in the box. The mother knows each block in the tub displaces the water in the tub by its volume, making the water level higher. She uses this information to develop another formula that tells her that the two missing blocks are in the bathtub.

Given these two examples, the mother has methods by which to count all of her son's toy blocks and determine where he has left them, even when she cannot directly see them. Feynman's analogy is that there are different forms of energy just like there are different places for the toy blocks to reside. In assuming that the number of toy blocks is always constant, when all toy blocks are not directly observable by sight, the mother must then come up with various ways to measure and test the location of any unaccounted blocks.

There is one critical difference between the conservation of blocks versus energy: we never directly see energy like the mother sees the toy blocks. Thus, to quantify energy in each of its different forms, we only use abstract mathematical formulas that are informed by measuring the world around us. More specifically our formulas quantify *changes* in energy. That is to say what we actually quantify is how much of each form of energy changed (e.g., increased or decreased) before and after some event. Imagine a ball resting on a table that is 1 m tall. We could say the ball has a gravitational potential energy equal to its mass times gravity times the 1 m if we consider a reference height as that of the floor. Alternatively, we could say the ball has zero potential energy at zero meters in height if we consider the reference height

---

[3]Feynman's story assumes that there is no way for the son to destroy any of the toy blocks or take them out of the house, such as by throwing them out of a window.

as that of the table. If the ball falls from the table to the floor, it will undoubtedly have fallen 1 m and the *change* in potential energy is equal to mass times gravity times the 1 m change in height regardless of whether we consider the table or the floor as our reference height.[4]

Energy is one example of a *conservation principle* that is a "...rule that some particular aspect of a phenomenon remains invariant or unaltered while the greater phenomenon undergoes certain specified transformations," Philip Mirowski writes in *More Heat than Light*, his deep history of how economics attempted to mimic the principle of energy [25].[5] As he states, because of the derivation of the concept of energy, "...did physics become the king of the sciences ..." Mirowski credits René Descartes with the first concept of a mechanical physics with a conservation principle. Descartes was trying to describe the world as an "ether" of small particles transferring motion from one to the other. Today we call this idea the conservation of momentum. Think about a game of billiards. After the moving cue ball strikes a stationary ball, the previously stationary ball now proceeds in motion while the cue ball can stay at rest and the momentum and energy from the cue ball has been transferred to the other billiard ball. For a moving mass, its momentum is its mass times its absolute velocity.[6]

Gottfried Wilhelm Leibniz, through his concept that became the basis for calculus, discovered that Descartes was incorrect in the quantity that was conserved. Instead of mass times velocity, the conserved quantity is mass times velocity times velocity, or mass times velocity squared. Today, introductory physics courses teach that the kinetic energy of a moving mass is equal to one-half of mass times velocity squared.[7] Take again the billiards example. Not only does the cue ball move linearly (e.g., from one point on the table to another) but it is also rolling. Each billiard ball has two types of kinetic energy that describe its state: kinetic energy of linear motion and kinetic energy of rotational motion. Just like the mother must have more than one method of inferring the number of toy blocks of her son, there are at least these two types of kinetic energy for a billiard ball.

But in thinking of billiard balls, we avoid an important problem. Intuitively we surmise that nearly all of the kinetic energy in the cue ball can be transferred to the other billiard balls. Physically this 100% transfer of kinetic energy does not happen. To imagine why, think not of billiard balls bouncing into each other, but instead of a baker throwing a handful of bread dough onto the counter top. When he throws a spherical ball of dough onto the counter, it flattens into a disk without bouncing

---

[4]Recall from physics that gravitational potential energy is quantified as $= mgh$, where m is the mass of an object, g is acceleration of gravity, and h is the height of the object relative to some reference height, such as the floor.

[5]Mirowski [25, p. 13].

[6]Linear momentum is mass (m) times absolute velocity (v), or $m \times |v|$.

[7]Kinetic energy of a moving mass is $\frac{1}{2}mv^2$.

back up and the counter does not move.[8] There was kinetic energy in the dough when he threw it, but after it slams onto the counter, neither the counter nor the dough are moving. Both have no kinetic energy because both have zero velocity. The same concept happens with billiard balls, but a large percentage of the kinetic energy is transferred from one ball to another. The dough transfers practically no kinetic energy into the counter.

What happened to the kinetic energy of the dough? It turned into heat and *work*, but the early philosophers of science in the late eighteenth century did not know this. The "work" done is the flattening of the dough into a disk, and the rest of the kinetic energy converted into heat. However, if kinetic energy (for example) can be converted into heat and work (reshaping things), then perhaps heat can be converted into kinetic energy and work. The practical engineering pursuit of converting heat into kinetic energy and work played perhaps the most important role in advancing the scientific concept of energy.

In 1698 Thomas Savery invented his "engine to raise water by fire."[28] In 1769 James Watt (whose name is used as a unit of power) patented an improved base design for steam engines that powered the Industrial Revolution. In these new machines, a fuel such as wood or coal was burned to generate heat. This heat then boiled water into steam, and this steam injected into the machine could cause motion and physical work to be performed.

But just what do we mean when we say "work?" The first law of thermodynamics is often expressed as the change in energy of a system is equal to the amount of useless heat dissipated minus the work performed. Thus, a change in energy (e.g., from burning wood) can be translated into a combination of *heat* and *work*. Heat quantifies the amount of energy that did not turn into anything useful. Scientists refer to discarded heat as "dissipation." From a further practical perspective, I can expand the term of work to *useful work* which is performing activity in the real world that necessitates physical exertion.

It is this useful work that we can measure in the real world. Consider pre-industrial England and United Kingdom. Before the use of steam engines, humans (many of them children) and animals performed the duties needed to extract coal from underground mines and bring it to the surface [28]. Much of this "work" was to pump water from the mines as well as lift the coal (e.g., its mass) from underground. It is easy to imagine that pre-industrial coal mining would have been "hard work" and involve much sweating and physical exhaustion.

As these steam engines started to be used for pumping water to mine coal and performing other mechanical tasks, there remained an important question. Just exactly how did these "heat" machines function?

A major leap in knowledge came from Frenchman Sadi Carnot. He is credited as the founder of the science of thermodynamics. Carnot was primarily concerned with

---

[8]Technically the counter does move and vibrate a very small amount that is generally imperceptible without scientific instruments. For the purposes of the discussion here, it is useful to imagine the counter does not move at all.

conversion of heat into mechanical motion (kinetic energy). Sadi Carnot learned from his father, Lazare, who worked at the *grand écoles* of Napoleonic France that were tasked with investigating machines for military purposes. By thinking of the impact of such things as cannonballs and the "…physics of impact …", Lazare translated his knowledge to the general idea of work.[9] He realized that a machine that was more efficient at performing work effectively minimized its internal "impact." A cannonball hitting a city wall is an extreme case of my example of the baker's dough hitting the counter. There is not enough kinetic energy in the dough to break the kitchen counter, but cannons were designed to do just that— transfer as much kinetic energy to a cannonball as possible such that it could release its kinetic energy into targeted structures and destroy them upon impact.

Carnot made the critical realization that the principle governing the function of steam engines "…was the result of the consumption or destruction of caloric [heat] from a warmer to a colder body, in direct analogy with the fall of water on a waterwheel from a higher to a lower elevation."[10] In effect, Carnot understood that for a heat engine to perform work, there had to be a transfer of heat from a high temperature source (e.g., the steam from burning wood or coal) to a low temperature sink (e.g., the ambient air or water).[11] If the temperatures are the same, then the efficiency of a heat engine to convert high temperature heat into useful work is zero because there must be a temperature *difference* to operate a heat engine. The low temperature is the reference condition for heat engines in the same way that earlier we had to think about a reference height in calculating the potential energy of the ball on a table.

This discussion of how much of a total change in energy from "before" to "after" some phenomenon becomes dissipated heat versus useful work brings us to the second law of thermodynamics. This law states that a practical device cannot take an energy input and convert all of it into useful work. Some of the energy must be ultimately converted to heat of no practical use. Using Carnot's insight into the maximum efficiency for a heat engine, we can describe how a heat engine cannot convert 100% of its input heat into useful work output. In the mid-1800s, armed with the ideas of the conservation of energy and efficiency of heat engines (not yet formalized into our current terminology or into the first and second laws of thermodynamics) scientists and engineers could perform experiments to characterize the "potential" energy from fuels (e.g., chemical energy from combustion) to perform work via machines.[12]

---

[9]Mirowski [25, p. 24].

[10]At the time the word "caloric" was used to describe a fluid that surrounded matter and was the cause of heat [25, p. 25].

[11]Carnot's famous expression quantifies the maximum theoretical efficiency for a heat engine to perform work based solely on the source (T_H) and sink (T_L) temperatures: efficiency $= 1 - \frac{T_L}{T_H}$.

[12]See Mirowski [25, pp. 35–66] for a discussion of the historical players that shaped the ideas leading to the formalization of the 1st and 2nd Laws of Thermodynamics.

Amazingly, both the engineering development of the steam engine and the concept of the second law of thermodynamics developed in ignorance of the relationship between the microscopic world of the air and water molecules in the steam and the macroscopic world in which we measure properties like temperature and pressure of the steam. It was only in the late 1800s, over half a century after Carnot's death, that Ludwig Boltzmann derived the idea that a gas, such as air, is made up of many tiny molecules banging into each other just like a continuous game of billiards. "Heat was just the combined kinetic energy of these tiny moving balls."[13] Hotter air is composed of faster moving molecules that have more kinetic energy. Since we cannot measure how fast each and every molecule is moving, Boltzmann described the statistical average of the molecules. Thus, Boltzmann's "statistical mechanical" description of gases provided a significant bridge for linking our micro to our macro descriptions of the physical world. As you will learn later in the book, physicists did not limit the application of statistical mechanics to physical phenomena. When they applied the concept to economic data, they found amazing similarity. But that will have to wait until Chap. 4.

Keep in mind that the second law of thermodynamics is not only about heat engines. First, from an engineering standpoint there are limits to converting inputs to useful work even in machines that are not based upon converting the heat energy of gases to directed mechanical motion. For example, the maximum efficiency of horizontal axis wind turbines, such as those we use to generate electricity today, have a maximum theoretical efficiency of 59% for converting kinetic energy in the wind to mechanical rotation of the turbine blades.[14] Individual solar photovoltaic (PV) cells have a maximum efficiency of 33.7% for converting sunlight into electricity.[15]

Secondly, the second law of thermodynamics also informs us how to interpret the economic process. For this statement, I quote perhaps the staunchest proponent for integrating thermodynamics and economics, Nicholas Georgescu-Roegen:

> Most important for the student of economics is the point that the Entropy Law is the taproot of economic scarcity. Were it not for this law, we could use the energy of a piece of coal over and over again, by transforming it into heat, the heat into work, and the work back into heat. [11]—Nicholas Georgescu-Roegen (1975)

The concept of energy is a great leap forward for science. With this one idea we can coherently compare the heat of moving molecules, the mass and speed of planets orbiting the sun, and a barrel of oil. If we perform our analysis right, we can design machines that take different forms of energy, as available within the environment, and convert them into useful work that replaces our physical labor, purifies materials, and transports us across the world or into space. Being able to perform thermodynamic work is one thing, but if you want to say how fast you want it done, you need to think about *power*.

---

[13] Sautoy [26, p. 89].

[14] This maximum efficiency is called the Betz Limit.

[15] The maximum efficiency of a PV cell is called the Shockley–Queisser limit, and it refers to the efficiency of a single p-n junction PV cell.

Power Is Not Energy

Now that we understand the concept of energy, we can understand *power*. It is important to understand power as related to, but distinct from, energy. Succinctly, *power describes how fast you are accumulating or consuming energy*. The simple way to remember the relationship between energy and power is to incorporate time. Power equals energy divided by the time it took to consume the energy, and energy equals power multiplied by the time during which power was consumed.[16]

Consider the following example to distinguish between energy and power. The amount of time for a NASCAR-style Toyota Camry race car to make a lap around the 2.5 mile Daytona race track is about 45 s while moving about 200 miles per hour (mph). Now imagine driving a normal Toyota Camry (one that you can buy at the dealer) around the Daytona NASCAR race track. You could reach a top speed near 100 mph, thus taking about 90 s to make a lap. Why does the race car make it around the track in shorter time? It has a more *powerful* engine. The NASCAR version of the CAMRY uses an engine rated at 700–800 horsepower, while the Camry that you and I can buy has a 300 horsepower engine. That is to say, to make it the same distance *in half the time*, the NASCAR Camry might consume about the same amount of fuel as the normal Camry, but in less time. To do that it needs a larger engine that can consume fuel at a higher rate. A higher rate of energy consumption is the same as more power.

At the scale of a country, if one country can consume energy in less time, then more work can be done each day, week, month, and year than another country that consumes its energy at a slower rate. This concept is immensely important in understanding how energy consumption, or more precisely power, relates to economic activity because economic activity is also measured as a *rate*. For example, gross domestic product (GDP) is expressed in units of money per year, not simply money. The phrase *energy consumption* implies a rate of use of energy, which is units of power. Thus, as we will discuss, GDP increases with increasing power, not energy.

For the purposes of this book, this is as complicated as we need to get to distinguish between energy and power. Now knowing this difference, some additional energy terminology will help the novice reader navigate concepts in this chapter and the rest of the book.

Energy Terminology

The highest level of accounting for human appropriation of energy is termed as *primary energy*. *Primary energy consumption* and *total primary energy supply* are

---

[16]Power (P) equals energy (E) divided by the time (t): $P = \frac{E}{t}$. Energy equals power multiplied by the time during which power was consumed: $E = P \times t$.

terms used to represent the total quantity of energy we extract from the natural environment, and ultimately dissipate as heat or convert to useful work. As you will read later in this chapter, in recent years U.S. primary energy consumption has been about 100 exajoules (EJ) of energy over the course of 1 year. Alternatively we can say that the U.S. dissipates power at the average rate of nearly 100 exajoules per year. Both statements are equivalent as when we refer to the term primary energy, the "consumed over the course of one year" is often implied.[17]

In addition to primary energy, the term *secondary energy* is used to refer to the energy content in energy carriers. For the most part, consumers like you and I purchase energy carriers, or secondary energy, and not primary energy. Energy carriers are the forms of energy consumed at the point they are converted to useful work (and heat) that provides some desired service (see Fig. 2.1). Example

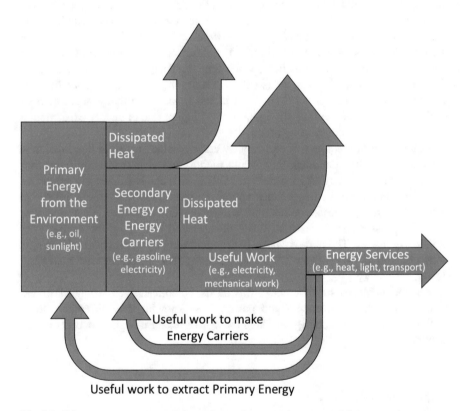

Fig. 2.1 Primary energy extracted from the environment is converted into secondary energy carriers that are then consumed to produce useful work to provide energy services. Some useful work is required to extract primary energy itself, and make energy carriers

---

[17]U.S. energy consumption for 1 year = (100 EJ/year) × (1 year) = 100 EJ. 1 EJ = 1 × 10^{18} J and 1 J is one "joule" of energy.

secondary energy carriers are the gasoline you put in your car, the natural gas that heats your home, and the electricity that turns on your lights and charges your mobile phone. This primary to secondary to useful work concept is relevant for understanding how energy data sets are compiled and used in economic analyses. The concept also gives insight for understanding the efficiency of using energy from the beginning to the end of the supply chain.

For those that want to understand additional details on units of energy and power, as well as differences among how agencies actually count primary energy, see the Appendix. There I summarize the different methods for counting primary energy consumption. You do not have to know the energy accounting methods to comprehend the content in this book, but it provides the background on why there is no one number for total energy consumption.

To obtain a feel for the *cost* of energy, this chapter compares primary energy consumption to GDP, or net economic output. Before we do this, the next section provides a quick summary of the GDP metric itself.

A Brief Description of GDP

In the following sections describing energy and food data, I use gross domestic product, or GDP, data to provide context and a metric for the cost of food and energy. This metric is spending on food and energy divided by GDP, and it proves to be very insightful. Practically, because historical time series estimates of GDP exist, energy spending per GDP is one of the few long-term metrics we can calculate for representing the feedback of the energy system on the economy.

The original concept for GDP developed in the U.S. in the 1930s. GDP is equal to the total monetary value of all final goods and services that have been exchanged within a country, usually specified for a year. GDP increases by exporting goods and services. It decreases by importing. When businesses invest more, consumers purchase more, and the federal government spends more, GDP increases. Because each of us decides to buy any given good or service at the price presented, in theory our consumer purchases include how much we "value" what we purchase. If something is not worth the price, then we can choose not to make the purchase, and GDP is lower than if we did make the purchase. If we do not make enough purchases of certain goods and services, businesses will stop producing and investing in them, and GDP goes down. Thus, much of GDP is supposed to be an aggregate measure of what consumers value and sellers produce. If GDP is higher, then it must be because the economy is producing output that people increasingly want. Right?

However, GDP as a metric itself has several limitations.

While GDP is a pretty good metric for the production of physical stuff, from its beginning GDP was never intended to be a measure of social welfare. That is to say, if GDP or GDP per person is larger in a given country, that does not necessarily mean that there is more social welfare, longer and healthier livelihoods, or more happiness and contentment. Most of what we hear in political and economic discourse in the news is the GDP growth rate, or lack thereof during recessions. Citizens are led to believe that growing GDP is always good, and declining GDP always bad. This is not strictly true, but various well-being indicators (e.g., literacy levels, health outcomes) do correlate well with increased GDP and energy consumption ... up to a point. The literature shows that up to certain per capita levels of GDP and energy consumption, many well-being indicators increase, but after that point there are minimal gains (see various metrics in [28]). These indicators include child mortality, life expectancy, and literacy rates. An approximate threshold seems to be about 100 gigajoules of primary energy consumption per person (GJ/person). Below this number, certain indicators tend to be low, and above it they tend to be high but do not increase much with higher energy consumption. For reference, the U.S. consumes about three times this (arbitrary) threshold, and the European Union consumes about 130 GJ/person.

A second major limitation is that GDP measures a rate of economic output, not the amount of wealth within a country. GDP also measures some amount of produced "bads" as well as "goods." One example is war activity. GDP increases with increased production of "goods" such as missiles and aircraft specifically used to destroy both man-made and natural capital. When you destroy these capitals you remove the services they provide, thus producing bads by subtracting goods. As some say, with war, it is "our" missiles destroying "their" wealth. However, we can do it to ourselves. Advertising that promoted increased smoking and other unhealthy lifestyles increased GDP since more cigarettes were sold, more people were treated for cancer, and more jobs and technology were created to treat cancer. However, smoking reduced the value of our human and natural capital by increasing rates of cancer. Fortunately, in the case of smoking, much of the Western world has limited advertising for smoking as a tactic to reduce the occurrence of cancer. Herman Daly referred to increasing GDP by producing more bads as "uneconomic growth." [2]

Another major limitation of GDP is that while the phrase or acronym stays the same, its mathematical and economic definition has not. An easy example to discuss is computers. Clearly we could not consider the economic output from making computers in the year 1900 because there were no computers in 1900. Today we know there is some economic value from making, and using, computers including mobile devices. The invention of new goods affects the calculation of GDP because at first you do not know about them to include them in the calculation. Thus, there is a delay in counting contributions from new products and services.[18] However, it is even trickier than that. Old goods and services that were not counted as economic output might become included at later dates. One good example of this is prostitution. A few years ago the European Union started counting prostitution (the world's "oldest profession"), and illegal drugs, as part of GDP.[19] Further, U.S. states

---

[18]Robert Gordon discusses this problem at length in [13].

[19]New York Times (2014), Sizing Up Black Markets and Red-Light Districts for G.D.P.: https://www.nytimes.com/2014/07/10/business/international/eu-nations-counting-sex-and-drug-trades-toward-gdp.html UK Daily Mail (2014), Who said crime doesn't pay? Counting prostitution

are increasingly legalizing marijuana to be openly grown and sold for recreational, not only medical, purposes. Again, this then establishes official state accounting of previously uncounted sales of a particular item, in this case marijuana.

From more theoretical and political perspectives, in *The Value of Everything* Mariana Mazzucato explains that before 1993, statistical agencies did not count banking and financial services toward GDP: "... until the 1990s the services it [the banking sector] represented were assumed to be fully consumed by financial and non-financial companies, so none made it through to final output. The 1993 SNA [System of National Accounts] revision, however, began the process of counting FISIM [financial services] as value added, so that it contributed to GDP. This turned what had previously been viewed as a deadweight cost into a source of value added overnight. The change was formally floated at the International Association of Official Statistics conference in 2002, and incorporated into most national accounts just in time for the 2008 financial crisis."[20] For most of history, paying interest on loans, was seen as a non-productive cost of business, not a productive way to make money.

Practically all changes to GDP measures make the metric appear larger, not smaller, than it would otherwise be. Is an economic metric that always grows really what we want? Will we redefine GDP as needed such that it always grows? I will not now digress on this important philosophical topic of whether there is or should be some economy-wide number, such as GDP, that we inherently seek to maximize. Chapter 8 discusses how we might consider the economy as "seeking" to maximize something, but we have a while to go before we get to that topic. Part III summarizes reasons why some pose the use of alternative metrics of "progress." For more thorough discussions of the history and limits of using GDP as a metric, I refer you to other literature that provides context and in-depth discussion of that matter [1, 14, 15].

---

and drugs in the GDP figure has seen the UK's economy overtake France as fifth largest in the   world: http://www.dailymail.co.uk/news/article-2888416/Who-said-crime-doesn-t-pay-Counting-prostitution-drugs-GDP-figure-seen-UK-s-economy-overtake-France-fifth-largest-world.html.

[20]Banking services are encompassed within FISIM. "The cost of 'financial intermediation services, indirectly measured' (FISIM) is calculated by the extent to which banks can mark up their customers' borrowing rates over the lowest available interest rate. National statisticians assume a 'reference rate' of interest that borrowers and lenders would be happy to pay and receive (the 'pure' costs of borrowing). They measure FISIM as the extent to which banks can push lenders' rates below and/or borrowers' rates above this reference rate, multiplied by the outstanding stock of loans.

The persistence of this differential is, according to the economists who invented FISIM, a sign that banks are doing a useful job. If the gap between their lending rates and borrowing rates goes up, they must be getting better at their job. That is especially true given that, since the late 1990s, major banks have succeeded in imposing more direct charges for their services as well as maintaining their 'indirect' charge through the interest-rate gap.

According to this reasoning, banks make a positive contribution to national output, and their ability to raise the cost of borrowing above the cost of lending is a principal measure of that contribution." [23, pp. 107–108].

Energy Consumption: How did We Get to Today?

The rest of this chapter presents three major trends of primary energy consumption for three different geographies. The first trend is gross primary energy consumption, the second is energy cost relative to GDP, and the third is food cost relative to GDP. The three geographies are England and the United Kingdom (U.K.), the United States of America (U.S.), and finally the world overall.

In this section I consider the quantification of absolute trends in energy consumption but not food consumption for all geographies. However, I do describe some food consumption within the U.K. and describe cost data for both energy and food. In doing this, I focus on the most core of Maslow's Hierarchy of Needs: physiological. Essentially, this means that if we do not have air, food, water, and shelter, then our survival is in such jeopardy that we do not have time to worry about social problems. This sentiment underpinned the modeling and worldview of *The Limits to Growth* authors: "Food, resources, and a healthy environment are necessary but not sufficient conditions for growth. Even if they are abundant, growth may be stopped by social problems." [24] Research supports that people do report being happier with more income to be secure in basic necessities, but higher incomes that support additional consumption and more luxurious items do not increase happiness much after the basic needs are met [22].[21] Let us dive into the energy and food data in the context of being necessary, but not sufficient for addressing social goals.

By observing these energy data across time for the three geographic scales, we understand the fossil energy transition within the first industrialized nation (the U.K.), energy trends within the post-World War II global power (the U.S.), and the energy production and costs in the context of the entire world. The world context is extremely important to understand each country's position in the global economy. We must always ask ourselves if each country in the world can develop along similar pathways as those demonstrated for the U.K. and the U.S.

The major purpose of presenting the energy consumption and cost data in such detail is to emphasize that there is one assumption about these energy trends that many analysts and policymakers take for granted, but that it is irresponsible to do so. Sometimes this assumption is a deliberate component to their narrative and worldview, but sometimes the assumption does not represent a conscious choice because individuals are unaware of its importance.

The incorrect assumption I speak of is that energy and food costs will always decline, and because they always decline, energy and food have not constrained socio-economic outcomes. However, the long-term data of this chapter show that:

> in the context of industrialization, *energy plus food costs are no longer declining*.

---

[21] Page 59 of [22] states: "…when money is relatively scarce, money buys happiness; when it is relatively plentiful, it ceases to do so."

This trend is a relatively new, important, and unappreciated indicator of the state of the world. Declining relative energy and foods costs (spending divided by GDP or income) are a defining characteristic of the industrial and fossil fuel era. No matter if you consider the U.K., U.S., or the world, spending on food and energy relative to economic output declined with industrialization until around the year 2000. For over 15 years since that time, we have been unable to continue this declining cost trend that many proponents of the techno-optimistic economic narrative assume continues irreversibly in an infinite world. In addition to looking at total energy costs in this chapter, Chap. 3 looks at the price of oil to further explore whether its cost has a clear ever-declining trend and to discuss price as an indicator of resource scarcity.

Before we can fully decipher the contemporary narratives of energy in Chap. 3, we must first consider the historical energy and economic data.

The United Kingdom

Primary Energy Consumption—U.K.

Perhaps the best way to envision the long-term change in both primary energy consumption and energy costs is by looking at a long historical time series of energy consumption and costs. Thanks to Roger Fouquet at the London School of Economics, we have estimates of England and the United Kingdom energy consumption and spending on energy since the year 1300.[22]

In Figs. 2.2, 2.3, 2.4, and 2.5 we see that pre-industrial England and U.K. were dominated by three types of biomass fuels: wood, fodder, and food for physical labor of humans. In terms of the pre-industrial energy services provided by biomass fuels, wood was primarily used for domestic heating (including cooking). Fodder refers to biomass and silage fed to animals that performed physical work. Fodder is still a sizable percentage of fuel in many developing countries. Fodder and food were the fuels for animals and people (that provided physical labor power), respectively [7]. This power was largely used in the fields for farming to grow the fodder and food itself. In 1700, England's economy consumed about 0.13 exajoules (EJ) of energy. 1 EJ is a billion billion joules, and burning one kitchen match releases about 1000 joules. Thus, the amount of energy that England consumed in 1700 was about *130 billion kitchen matches*. Since 1700, coal and other fossil fuels dominate primary energy consumption for all energy services, primarily the provision of heat and physical power via steam engines during the Industrial Revolution. The peak U.K. primary energy consumption occurred in 1973 at about 10.4 EJ, equivalent to burning *10 trillion kitchen matches*!

---

[22]The United Kingdom came into existence in 1707 by merging the Kingdom of England, including Wales, with the Kingdom of Scotland.

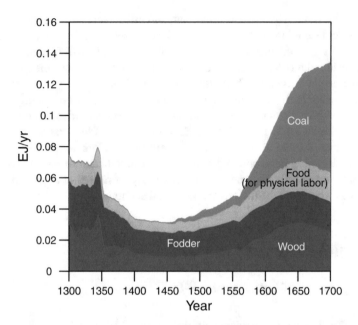

Fig. 2.2 The annual energy consumption (EJ/year) for England and the United Kingdom [9]. The fuels before 1700 were wood (brown), fodder (blue), food for physical labor (orange), and coal (gray). EJ = exajoule: 1 EJ = 1×10^{15} J

The percentage of the England/U.K. energy supply derived from biomass energy was practically 100% before 1450 and still greater than 85% in 1550 [8, 9]. This fraction dropped steadily to nearly 13% by 1830 with the rapid increase use of coal starting around 1600 (see Fig. 2.3). Note that this increase in coal use started well before the invention of the steam engine in 1712 by Thomas Newcomen and James Watt's design in 1769 that was the basis for the Industrial Revolution. This is because coal was already beneficial for domestic heating before engines existed [7].

But make no mistake, the steam engine undoubtedly affected trajectory of human history by spawning our modern economy. Here is a short story of how it all started.

In 1866, William Stanley Jevons stated in *The Coal Question*: "The terms in which the [steam] engine was described, and the way in which it was actually used for nearly two centuries, show that the raising of water out of our [coal] mines was the all important ...purpose."[23] Industrialization accelerated because coal was burned in steam engines, steam engines operated water pumps, the pumps removed water from coal mines, and dry coal mines enabled access to more coal deposits and higher rates of mining that increased the flow of coal output from the production cycle. Thus, there was a motivation to describe how these steam engines actually functioned in order to engineer them to be more powerful and efficient and

---

[23] Jevons [16, p. 98].

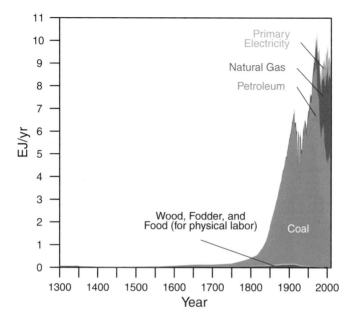

Fig. 2.3 The annual energy consumption (EJ/year) for England and the United Kingdom from 1300 to 2008 [9]. Primary electricity includes hydropower, nuclear, wind power, and solar power. EJ = exajoule: $1 \, EJ = 1 \times 10^{15} \, J$

accelerate the process faster again. As stated earlier, the laws of thermodynamics were largely derived from this need to understand the function of the steam engine. The medium defines the message.[24] Coal is the medium that became synonymous with early industrialization and accelerating economic growth.

Coal combustion was responsible for nearly 80% of primary energy consumption before natural gas, petroleum, or primary electricity (e.g., hydropower) played any role.[25] After World War I, petroleum started to dominate consumption. After World War II, natural gas and primary electricity increased in use to take appreciable shares of total primary energy provision. By the twenty-first century, the share of coal use dropped to approximately 20%, and in 2000 U.K. coal consumption was 3% of its peak that was reached during World War I.

---

[24]"The medium is the message" was a phrase coined by Marshall McLuhan and described in Chapter 1 of *Understanding Media: The Extensions of Man* by Marshall McLuhan, 1964.

[25]Fouquet's primary energy data assume the partial substitution method for translating primary electricity to thermal primary energy units.

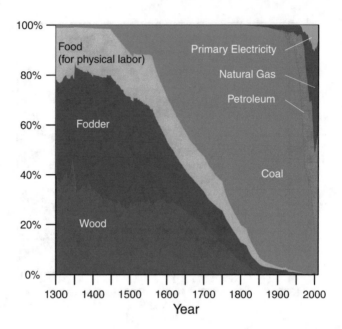

Fig. 2.4 The annual energy consumption by percentage of fuel for England and the United Kingdom from 1300 to 2008 [7, 9, 10]. Primary electricity includes hydropower, nuclear, wind power, and solar power

Spending on Fuels—U.K.

The dramatic rise in U.K. primary energy consumption starting in the 1800s coincides with an equally dramatic decline in the cost of energy. Energy consumption increased because energy became much cheaper and more abundant. I do not mean cheaper by a little bit, but cheaper by a wide margin. In using the word *cheap*, I refer to spending on energy relative to GDP. By spending I refer to expenditures by industry to produce food and energy and/or consumer purchases of energy and food.

Figures 2.5 and 2.6 show England and U.K. spending on energy, and it is well worth discussing the data at some length. I know of no other data set estimating the cost of energy that spans a longer time period. First and foremost, the numbers are much higher on the left side than on the right side of the figures. Energy was relatively expensive before the 1800s, and it has been relatively cheap since the mid-1900s.

The pre-industrial English economy (1300 to about 1800) typically spent between 30% and 40% equivalent of its GDP for what we might today call "energy" in the form of fuels (see Fig. 2.5) that *do not* include food for humans performing physical labor. When we include food as a fuel input for humans to perform physical labor, then this cost of energy jumps to 50–60% relative to GDP. In order to discuss pre-industrial society we must consider food as an energy resource, a fuel, for

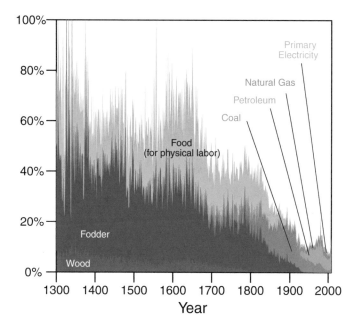

Fig. 2.5 The plot represents expenditures on energy, by type of fuel, divided by GDP for England/U.K. Included is food consumed by humans for performing physical labor. Data from Roger Fouquet [7, 9]. Primary electricity includes hydropower, nuclear, wind power, and solar power

preforming physical labor. In fact, fodder (biomass feed for animals) and food dominate the cost of energy up until around 1800. Food for humans and fodder for animals are the fuel sources, and muscles were the dominant *prime movers* of the pre-industrial era that provided useful work as power over the course of the day.[26]

We can see this dependence upon muscles for power needs in Fig. 2.6. This figure shows the same calculation as in Fig. 2.5 in terms of spending on energy divided by GDP. However, this time, it shows the results in terms of the cost of fuels to provide different *energy services* instead of the cost per type of each fuel itself. Ultimately, we seek energy services, and not necessarily energy itself. The various energy services that we seek are generally categorized as power (stationary useful work), heat (for industry and domestic homes), transportation (moving people and freight from place to place), and light [7, 9, 10]. That is to say, I can spend one hundred dollars to purchase natural gas, but 50 dollars of natural gas can go to provide heat while the other 50 dollars goes to provide power. All $100 would show up as spending on natural gas in Fig. 2.5, but in Fig. 2.6 $50 would show up as spending for power and $50 would show up as spending for domestic heat.

---

[26]Prime movers are devices that convert fuel, consumed at some rate, into force and motion, or power output.

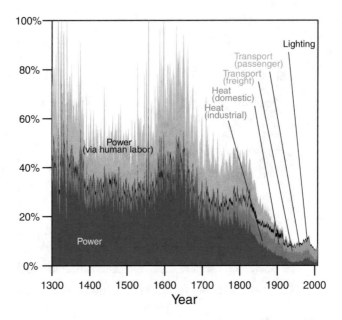

Fig. 2.6 The plot represents expenditures on energy, divided by GDP for England/U.K. from 1300 to 2008, for purchasing energy for services of industrial power, industrial heating, domestic heating, freight transport, passenger transport, and lighting. Included is food consumed by humans for performing physical labor. Data are from Roger Fouquet [7, 9]

The two time periods (the early 1300s and early 1600s) in which spending on energy was more than 60% of England GDP also correspond to times of higher population pressure relative to native food supply [30]. Thus, more people put more pressure on food and biomass resource costs.

During the time in which spending on energy was relatively high, until about 1800, the English economy grew at a slow rate of less than 1%/year for both real GDP/year and real GDP/person/year [7, 10, 12]. In effect, pre-industrial society was "power-limited" because physical power was provided primarily by muscles fed by fodder (animals) and food (people). Further, these biomass fuel sources grew at a rate limited by the sunlight, land area, and existing technologies and practices to grow biomass.

The cost of energy, including food for labor, in the U.K. did not fall below 30% relative to GDP until the 1840s. During this time the absolute energy dissipated from biomass consumption increased considerably, but coal consumption increased at a much more rapid rate, thus taking over the majority of the primary energy mix. After the 1840s the relative cost of energy dropped quickly for 80 years through the 1920s, eventually to below 10% during World War II, as the benefits accumulated from investments associated with the Industrial Revolution and fossil fuel consumption.

While the United Kingdom has the longest string of data on energy consumption and costs, allowing us to track patterns from a pre-industrial to post-industrial economy, we know the Industrial Revolution did not stay within the confines of the British Isles. All developed nations went through similar transitions in using fossil energy and hydropower to release themselves from the burden of energy and power constraints. For example, Sweden's spending for energy (including food and fodder for animate power) relative to its GDP also consistently declined from 90% in the early 1800s to less than 20% after 1925 as Sweden shifted from biomass to coal [17, 29]. I now turn to discuss energy consumption and cost trends for the United States.

The United States: Post-World War II Superpower

Primary Energy Consumption—U.S.

The United States went through a similar, but faster, transition as did the U.K. in terms of increasing use of coal. One major difference is that the colonists (before the U.S. was a country) on the eastern seaboard of North America did not use appreciable amounts of coal. The U.S. did not start using significant quantities of coal until the mid-1800s, over two centuries after coal was of significant use in the U.K. Because of the later use of coal, the U.S.'s transition from a biomass to fossil-dominated economy was faster than that of the U.K.

U.S. total primary energy consumption increased tremendously from the late 1800s until around 2000. Figure 2.7 indicates that major changes in the trends of increasing energy consumption coincided with the Great Depression, the two oil "crises" in the 1970s (discussed in Chap. 3), and the mid-2000s as the time of highest energy consumption through 2018. As the data in Fig. 2.7 come from the EIA, the conversion of electricity from hydroelectric, wind, and solar power assumes the partial substitution content method. (See Appendix for details about different accounting methods for primary energy.)

In broad terms, U.S. primary energy consumption increased at an exponential rate of about 3%/year from 1900 to 1973. It increased approximately linearly at +1 EJ/year/year from 1973 to 2000, and stayed approximately constant since 2000. It is important to understand these changes in trend (first increasing quickly from 1900 to 1973, then increasing more slowly from 1973 to 2000, and then stagnating from 2000 to present) within the context of the dynamics of both U.S. and global economic and demographic factors. This discussion, however, must wait until Chap. 5 after introducing some of these non-energy trends in Chap. 4.

As a share of U.S. primary energy consumption, coal peaked in the first decade of the 1900s near 75% (Fig. 2.8), but in total rate of consumption coal peaked in 2005 at 24 EJ/year. That is to say, even though the *fraction* of coal use peaked at the beginning of the twentieth century, the *rate* of coal consumption peaked at the beginning of the twenty-first century. Thus, a "transition" in share is not the same as "transition" in quantity or rate of use.

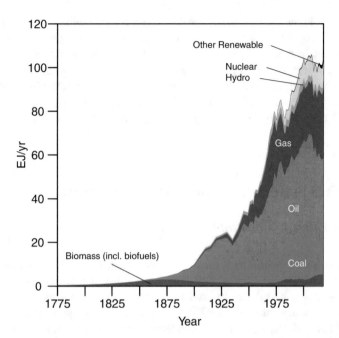

Fig. 2.7 U.S. primary energy consumption (EJ/year) by fuel from 1775 to 2018 [Energy Information Administration, Monthly Energy Review, Table 1.3 and Appendix E]. Biomass includes liquid biofuels. Other renewable includes primary electricity from geothermal, wind, and solar power plants. EJ = exajoule: $1\ EJ = 1 \times 10^{18}\ J$

The highest shares of oil and natural gas consumption occurred in mid-1970s and early 1970s at 48% and 32%, respectively. Through 2018, the highest absolute rate of energy consumption from oil was in 2005 (42 EJ/year) and from natural gas was in 2018 (33 EJ/year). As with coal, the highest shares of use of oil and natural gas are not coincident with the highest absolute rates of their consumption. Thus, a higher share for a primary energy resource does not necessarily mean there was more absolute consumption of that resource.

Since 1981 the U.S. share of consumption from each fossil fuel has remained within a relatively constant range: coal from 13% to 23%, oil from 35% to 42%, and natural gas from 22% to 31%. However, since 2014, the share of coal declined to the lower end of its range, and natural gas to the higher end of its range. For the last decade both natural gas and renewable energy consumption have increased while the total primary energy consumption rate of the U.S. has remained relatively constant, between 99 and 106 EJ/year, since 1996. After 2008, increased horizontal drilling with hydraulic fracturing in tight sand and shale formations extracted increasing quantities of natural gas that displaced significant quantities of coal use for power generation [4].

Since declining below 10% of the total in the 1920s, the share of total renewable power has typically been between 6 and 8 percent. Total hydropower generation

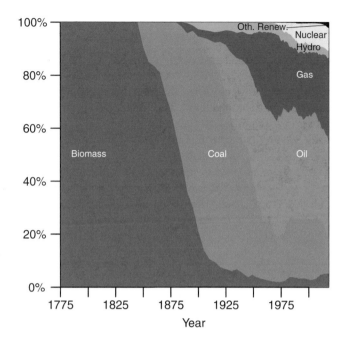

Fig. 2.8 U.S. primary energy consumption by percentage of each fuel from 1775 to 2018 [Energy Information Administration, Monthly Energy Review Table 1.3 and Appendix E]

(e.g., in kWh) increased through the 1970s, but has remained flat since. Liquid biofuels (e.g., ethanol and biodiesel), wind and solar power are responsible for increasing the share of total renewable energy consumption above 8% since 2009. The absolute rate of consumption of renewable energy has seen a slow but steady average increase of less than 0.1 EJ/year/year from 1990 through 2007 and approximately 0.5 EJ/year/year since 2007. In 2017, total renewable energy consumption was above 10 EJ/year for the first time in U.S. history.

Nuclear power in the U.S. rose quickly from the mid-1950s until the 1990s as the only major wave of nuclear construction commenced after World War II. The share of nuclear power peaked at just under 9% in 2009 roughly coincident with its maximum absolute quantity of production of just under 9 EJ/year in 2007–2010. Since 1999, nuclear energy consumption has been greater than 8 EJ/year but it did not increase substantially after that point, and it is expected to decline in the near term due to expectations of power plant retirements along with few to no new reactors coming online. As of the time of this writing, there are only two new reactors in construction (Vogtle power plant reactors 3 and 4 in Georgia), and in 2017 construction was halted for two other reactors that had begun construction at the same time (V.C. Summer reactors 2 and 3). Chapter 3 summarizes the reasons why we are unlikely to see near-term increases in nuclear power in the U.S.

Spending on Fuels—U.S.

The declining energy and food cost trends witnessed for the U.K. are repeated in
the U.S. The United States post-World War II era is characterized by a continuous
decline in relative food costs until 2006, and a decline in combined food and energy
costs from the 1930s until around 2000. This trend holds from two perspectives of
energy spending.

First, consider "consumer expenditures" on food and energy goods and services
relative to GDP (Fig. 2.9a). As a category, consumer expenditures are the largest of
the components that are summed to estimate GDP.[27] Consumer food expenditures
and prices refer to what you and I pay at the grocery, and these prices include
the cost to produce food in addition to transportation, storage, packaging, and
marketing. Consumer energy expenditures refer to our purchase of fuels such as

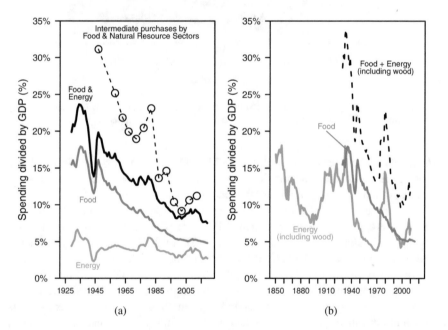

Fig. 2.9 U.S. spending on food and energy using two different energy estimates. (**a**) Energy and
food data are annual personal consumption expenditures from BEA Table 2.3.5. Food and resources
sectors as intermediate purchases (open circles) are from BEA benchmark summary input–output
tables as calculated in [20]. (**b**) Food data are annual personal consumption expenditures from BEA
Table 2.3.5. Energy spending estimate, including for wood, is from Fizaine and Court [6]

---

[27]Gross Domestic Product, or GDP, is equal to the sum of (1) consumer expenditures, (2)
investment by companies, (3) (federal and state) government expenditures and investment, (4)
net exports (= exports − imports) of goods and services to other countries, and (5) change in
inventories.

gasoline and electricity in our homes. However, consumer energy expenditures do not include total U.S. energy spending. For example, government spending on energy is excluded.

It is very important to note that *the majority of the long-term decline in total food and energy spending is due to declining food costs.* Historically, consumer food spending has been higher than for energy, but today is no longer the case. Figure 2.9a shows data from the Bureau of Economic Analysis (BEA) that indicates U.S. consumer spending on food was 18% of GDP in 1935, during the Great Depression, and approximately 5% of GDP for the last two decades. Relative to consumer spending on energy, food spending was two to three times larger from the 1930s to the 1950s, after which time consumer food spending per GDP declined through 2006. Since 2006, food spending per GDP has been approximately level near 6%.

The change in consumer spending for energy has not declined as dramatically as that for food. The consumer cost of energy goods and services has had a slow declining trend since the 1930s while averaging just above 4% of GDP and typically staying between 3% and 5%. The time periods of consumer spending significantly greater than 5% of GDP on energy generally correspond to times of declining or low economic growth (e.g., Great Depression, oil crises of the 1970s) As I will discuss in Chap. 3, seemingly small fluctuations in the cost of energy can have large ramifications depending upon the level from which they start. That is to say, increasing economy-wide energy expenditures 1% from 4% to 5% is largely unnoticeable, but changing 1% from 7% to 8% has been the difference between recession and growth (see the energy spending peaks in the 1970s and 2008 in Fig. 2.9).

A second way to consider food and energy costs is not by how much you and I (as "consumers") spend at grocery stores, restaurants, and gasoline stations, but by how much companies spend to provide the food and energy that we end up purchasing. We can calculate this spending by using data from the BEA that summarizes "intermediate" purchases,[28] or spending by the economic sectors associated with food and energy production. These sectors are those such as farming, oil and gas drilling, and electricity. Their collective spending is represented by the circles in Fig. 2.9a.[29] Spending by the food and energy sectors, relative to GDP, dropped from 31% in 1947 to 9% in 2002. In 2012, the last year with benchmark data, U.S. food and energy sectors spent an amount equal to 11% of GDP. Here again,

> *just as with the data for U.S. personal consumption expenditures, relative to GDP, the low point in U.S. intermediate spending for food and energy occurs in the early or mid-2000s.*

---

[28]These intermediate purchases are those that are used to provide final products to consumers like you and me.

[29]The BEA data presented here are derived from the benchmark input–output tables that are estimated approximately every 5 years.

Figure 2.9b uses a different estimate of energy costs as an additional comparison and verification of energy cost trends in the U.S. The food data are the same as in Fig. 2.9a, but the energy data more closely represent an estimate of the cost of primary energy supplies instead of only secondary energy carriers purchased by consumers. Further, the data estimate begins in 1850 and includes an estimate of the cost of wood used for energy. By including a cost estimate for wood, the pre-World War II energy expenditures increase substantially to typically 12–15%, and over 15% for the 1850s. This combined food and energy cost more closely matches the intermediate spending in Fig. 2.9a, and thus is also shown as a dashed line for easier comparison.

Just as with the combined energy and food estimates in Fig. 2.9a, those in Fig. 2.9b show a distinctive and clear declining trend from the 1930s until around the year 2000.

Thus, no matter how you slice the data for the U.S., we can declare that the cost of food and energy in the U.S. declined for 70 years after the Great Depression until about the year 2000. After that year, energy and food have no longer become less expensive, and on average they have been more expensive than in 2002.

The World

Primary Energy Consumption—The World

Figure 2.10 shows an estimate of global primary energy consumption from 1800 through 2012, and Fig. 2.11 shows the percentage share of each fuel type [6].[30] The conversion of electricity from hydropower, wind, and photovoltaics into primary energy assumes the partial substitution method.[31]

One striking difference between the data for the world and those of both the U.K. and the U.S. is that world primary energy consumption is still rising while that of the U.K. peaked in the 1970s and that of the U.S. has plateaued since the 2000s. Globally, each type of primary energy resource has increased in absolute rate of consumption until very recently. According to the BP Statistical Energy Review, 2014 marked the maximum energy consumption rate from coal worldwide, with the

---

[30] World primary energy data from [6] is stated as: "We retrieved global primary energy productions through the online data portal of The Shift Project (2015) which is built on the original work of Etemad and Luciani (1991) for 1900–1980 and EIA (2014) for 1981–2012. Prior to 1900, we completed the different fossil fuel time series with the original 5-year interval data of Etemad and Luciani (1991) and filled the gaps by linear interpolation. The work of Fernandes et al. (2007) and Smil (2010) was used to retrieve historical global consumption of traditional biomass energy (including wood fuel and crop residues but excluding fodder and traditional windmills and waterwheels)."

[31] See Appendix for summary of different energy accounting methods.

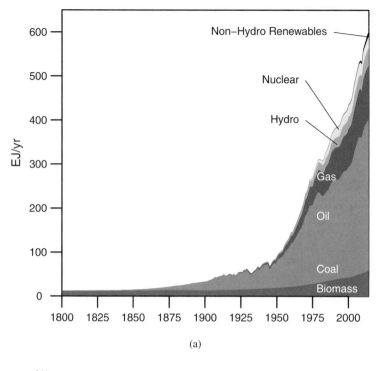

(a)

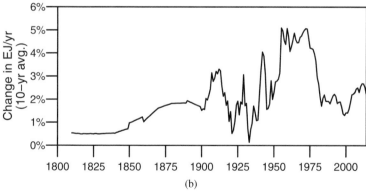

(b)

Fig. 2.10 (**a**) World primary energy consumption (EJ/year) per primary energy source from 1800 to 2014. (**b**) 10-year average growth rate in global energy consumption. Data from 1800 to 1899 as used in [6], and data from 1900 to 2014 from International Institute for Applied Systems Analysis Primary, Final and Useful Energy Database (PFUDB) [3]

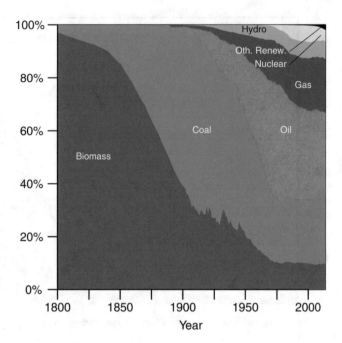

Fig. 2.11 The percentage of world gross primary energy consumption per each supply type from 1800 to 2014. Data from 1800 to 1899 as used in [6], and data from 1900 to 2014 from International Institute for Applied Systems Analysis Primary, Final and Useful Energy Database (PFUDB) [3]

rate of coal consumption declining from 162 EJ in 2014 to 156 EJ in 2017. However, they reported higher coal consumption of 158 EJ in 2018.[32] It remains to be seen if the decade of the 2010s exhibits the peak coal consumption rate as it would be the first time a fossil resource has peaked in the *absolute* rate of consumption since the industrial era. Some argue that the worldwide production rate of "conventional crude oil" peaked in the mid-2000s, and I discuss the disagreements over this matter in Chap. 3.

Another remarkable feature of world primary energy consumption is its rate of increase from 1955 to 1979. This is the only time in history that the 10-year running average growth rate in energy consumption was greater than 4%/year. The rapid increases in production of oil and natural gas drive this statistical trend, and the U.S. was the dominant player for extraction and consumption during this period.

In 1950, the U.K. and the U.S. combined for approximately 45% of total world primary energy consumption. This is amazing. The populations of the two countries represented 8.3% of world population yet commanded 45% of approximately

---

[32] BP Statistical Review of World Energy 2019, downloaded September 22, 2019 from https://www.bp.com/content/dam/bp/business-sites/en/global/corporate/xlsx/energy-economics/statistical-review/bp-stats-review-2019-all-data.xlsx.

100 EJ/year of worldwide energy consumption in 1950.[33] Also, during this time, the U.S. dominated worldwide consumption of oil, accounting for 65% in 1949 declining to 30% by 1979 even though the U.S.'s absolute consumption rate of oil increased through 1978. The U.S. command of oil and gas consumption in the two decades after World War II was enabled by its high domestic extraction. The U.K. did not become a major oil and gas extractor until the 1970s as it responded to the Arab Oil Embargo in 1973 and the OPEC oil price increase in January of 1974 by exploring and extracting oil and gas from the North Sea. Chapter 5 further explores the worldwide shift in energy systems caused by events in the 1970s.

The worldwide shift to different consumption of the different primary energy resources and technologies is qualitatively the same as for the U.K. and the U.S. Because early energy fossil energy consumption was dominated by the U.K. and the U.S. Thus, the U.S. and the U.K. largely determined the initial global shift to fossil fuels. I defer further discussion of the timing of the change in world energy mix until the Summary of this chapter.

Spending on Fuels—The World

Both the U.K. and the U.S. data indicate that energy and food costs declined since industrialization until the 2000s. The same trend also holds for the overall world economy.

Figure 2.12 shows two estimates for world energy expenditures since 1850 [6], and adds these to world food production costs from the Food and Agriculture Organization (FAO). The "no wood" data are estimates of marketed energy, primarily oil, natural gas, coal, and electricity from renewable and nuclear power. The cost estimate "including wood" uses data for wood prices in the U.S. to multiply by an estimate of global wood consumption.

One takeaway from Fig. 2.12 is that the cost of energy, including wood, typically fluctuated between 6% and 8% of global GDP from 1850 until the 1950s. Starting around the end of World War II, the cost of energy declined almost continually until 1970. In these data, the lowest cost energy (including wood) for the world was 4.0% in 1970 (the year of peak oil production rate in the U.S.), matched very closely in 1998 [6]. The post-World War II multi-decadal trends for spending on energy are largely dictated by swings in oil prices.

The FAO data show that, since the 1990s, the world cost of food production has kept declining, but at a much slower rate than before 1980. The FAO food cost estimate in Fig. 2.12 is that of cost of food production by farmers rather than food purchased by consumers like you and me. From 2007 to 2014, world food production costs per global GDP remained about the same at 3.6–3.8%, before

---

[33] United Nations, Department of Economic and Social Affairs, Population Division (2013). World Population Prospects: The 2012 Revision, DVD Edition. File POP/DB/WPP/Rev.2012/POP/F01-1.

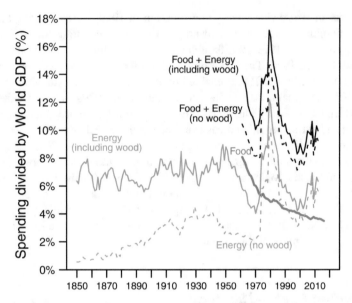

Fig. 2.12 World spending on energy (1850–2012) and food (1961–2016) divided by global gross domestic product (GDP) and expressed as a percentage. Data for energy costs are separated into time series that includes the cost of wood (solid lines) and without the cost of wood (dashed lines) from [6]. Data for food expenditures is from the Food and Agricultural Organization as World "Gross Production Value (constant 2004–2006 million US$)" for Food (item code 2054) divided by GWP from the World Bank. The food cost calculations before 1992 are shifted upward by the difference in the Gross Production Value of food from 1991 to 1992 because there are no FAO data for U.S.S.R (1991 and earlier) but there are FAO data for the states formed from the U.S.S.R starting in 1992. Thus, the shift is an estimate of the Gross Production Value of the U.S.S.R.

declining to 3.5% in 2015. While food has become quite cheap, it is approaching its lower limit. Thus, per Figs. 2.12 and 2.13,

> *the combined world energy and food expenditures data indicate the*
> *worldwide trend of energy and food costs as a share of global gross*
> *domestic product reached its minimum around the year 2000.*

Unfortunately, the data in Figs. 2.12 and 2.13 do not include fodder eaten by animals that perform work on farms as was the case in the UK data of Figs. 2.5 and 2.6. Thus, given the large share of energy costs for fodder (to produce power on the farm) in pre-industrial UK and England, we should expect that the global totals in Fig. 2.12 are likely 10–20% higher in 1850. We should also expect cost for fodder today is not zero due to a non-trivial portion of developing countries' agriculture still dependent on animate power for farming.

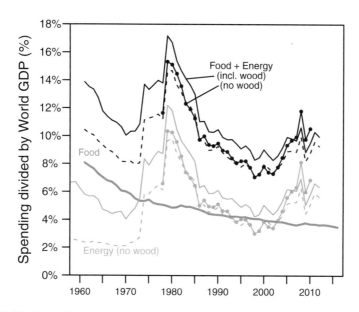

Fig. 2.13 World spending on energy (1961–2012) and food (1961–2016) divided by global gross domestic product (GDP) and expressed as a percentage. Data are the same as in Fig. 2.12 except for one additional time series for comparison of energy expenditures: energy expenditures data (for 1978–2010) represented by the thin lines with filled circles come from [21]

The fundamental shift in global food costs is also evident when considering consumer level prices instead of farm level prices, as in Fig. 2.14. Consumer food costs are represented by the FAO's Real Food Price Index. This index generally declines from the mid-1960s until around the year 2000. During that interval there is both a sharp rise in food costs in 1973 and 1974 and a sharp decline in food costs in 1985 and 1986. The price rise coincided with the Arab Oil Embargo in 1973 and the OPEC oil price increase in January of 1974. The food price decline coincided with declining oil prices that followed a decade of massive investments in both oil drilling and efficiency in use of oil (e.g., fuel economy standards for cars and no longer using oil to fuel significant quantities of power generation in OECD countries). These correlations show how coupled oil is to food production and distribution.

The FAO Real Food Price Index resides at its lowest levels from the mid-1980s until the mid-2000s. After 2005 the food index rises quickly, staying high until 2014 before dropping in 2015 and 2016 but staying above the 2006 value through 2019. *The FAO real food price index data show that world consumer food prices have increased since the early 2000s.* Thus, while producer costs might only be stagnating since the 2000s, consumer costs are on the rise.

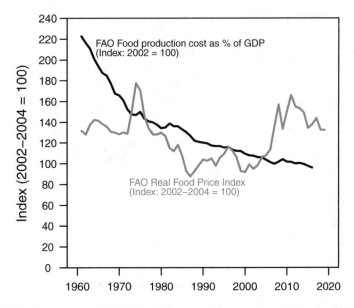

Fig. 2.14 World spending on food (1961–2016) expressed as an index with the year 2002 as index = 100. The black line is world FAO food production costs divided by gross world product (the same data as in Figs. 2.12 and 2.13), and the red line is the FAO Real Food Price Index (1961–2019). The Food Price Index represents *consumer level* spending which is larger than food production costs at the *farm level* due to additional costs of distribution, processing, and other services

Summary

Primary Energy Consumption and Energy and Food Costs

There are a few major takeaways from considering the historical data for both the consumption rates of primary energy and the cost of energy and food within the U.K., U.S., and world overall.

> First, in the history of mankind, the cost of energy plus food has never been cheaper than around the year 2000.

Up until that point in history, food and energy costs generally declined, greatly accelerated during the transition to the use of fossil-fueled machinery, and after that point they have approximately held steady (after a rise to 2008) from the producer perspective and increased from the consumer perspective. This shift away from declining costs holds whether we look at over 700 years of data describing England and the United Kingdom [10, 18], 200 years of data for Sweden [17], almost 100 years of data for the United States [5, 19], or the last 60 years of data for world food and energy costs [19, 21]. Thus, the combined cost of *food*, the fundamental input that allows people to live, and *energy*, the fundamental input that drives our economy, has crossed a major turning point during our current industrial and fossil fuel era.

Second, coal led the transition from a biomass to fossil economy, and this transition occurred first in the U.K., then the U.S., and then the world overall.

Third, while the world has experienced definitive shifts in *the share* of primary energy obtained from different sources, these shifts have typically come with increased *absolute* total primary energy consumption from each supply. Only very recently have we seen evidence for the possible decrease in the worldwide consumption rate of one energy source: coal. Historically, new sources of primary power have simply been additions to the existing sources, not replacements.

Fourth, the first region to reach its maximum total primary energy consumption was the U.K. in 1970, the second was the U.S. in 2005 (but relatively constant over the last 20 years), and the world has not yet experienced a peak in energy consumption (see Fig. 2.10). The fact that both the U.K. and the U.S. no longer consume more energy within their borders is important to consider in the contexts of broader megatrends (Chap. 4) and systems thinking (Chap. 5). That is to ask, if the U.K. (the first industrialized country, small in land mass and population) and the U.S. (a country with abundant energy resources, a large land mass, and the largest economy in the world since World War II) both peaked in energy consumption, then should we expect this to eventually occur for the world also?

Looking Backward to See Forward: Renewable and Fossil Energy Transitions

We simply cannot explain the current state of the world without considering the full context of the increase in the rate of energy consumption, and decline in the cost of this consumption, since industrialization. To consider energy over the course of more than 100 years, we have to consider food, fodder, and biomass as energy resources from which developed countries initially transitioned. In the pre-industrial age, most people were farmers. Food and fodder were the fuels that enabled human and animal (e.g., horses, oxen) muscles to perform physical work, such as plowing, grinding, and harvesting on the farm [28]. Before modern machines, animals and laborers were the major "machines" into which fuels were input to enable force and motion. With the advent of steam and internal combustion engines in combination with coal and oil, the course of history was changed.

While practically all disciplines and perspectives recognize the unprecedented enhancements from fossil fuel-driven machinery, they do not all recognize that these enhancements cannot increase indefinitely. Even though it is a fact that the Earth is finite, some holding the techno-optimistic economic narrative and fossil energy narrative see no limitation in our practical ability to advance technology to extract more technically challenging fossil energy and material resources. For many of those in the renewable energy narrative that also hold to economic techno-optimism, they see a limitation in our fossil-powered society, either due to climate change or declining resource quality, but they do not see any similar limitation in our ability to extract renewable resources.

More often than not there is a belief in continuous technological innovation, but too often only for the energy narrative that one is promoting. Fossil energy narrative: We will always find more fossil energy and never run out, so we do not need renewable energy. Renewable energy narrative: Costs of renewable energy systems will decline indefinitely, just as have the costs for mobile phones, and costs for fossil fuels will eventually increase such that we will eventually and easily substitute renewable for fossil energy.

It is to a comparison of the two energy narratives that we now turn.

References

1. Coyle, D.: GDP: A Brief but Affectionate History. Princeton University Press (2014)
2. Daly, H.E.: uneconomic growth in theory and fact. http://www.feasta.org/documents/feastareview/daly.htm (1999). Online; accessed 1-February-2020
3. De Stercke, S.: Dynamics of energy systems: A useful perspective. Tech. rep., International Institute for Applied Systems Analysis (2014)
4. DOE: Staff report to the secretary on electricity markets and reliability. Tech. rep., U.S. Department of Energy (2017). https://energy.gov/sites/prod/files/2017/08/f36/Staff%20Report%20on%20Electricity%20Markets%20and%20Reliability_0.pdf. Accessed September 3, 2017
5. Bureau of Economic Analysis, U.S.A.: Personal consumption expenditures by major type of product, table 2.3.5 (2017). Accessed data Last Revised: January 4, 2017
6. Fizaine, F., Court, V.: Energy expenditure, economic growth, and the minimum EROI of society. Energy Policy **95**, 172 – 186 (2016). http://dx.doi.org/10.1016/j.enpol.2016.04.039. http://www.sciencedirect.com/science/article/pii/S0301421516302087
7. Fouquet, R.: Heat, Power, and Light: Revolutions in Energy Services. Edward Elgar Publishing Limited, Northampton, Massachusetts (2008)
8. Fouquet, R.: The slow search for solutions: Lessons from historical energy transitions by sector and service. Energy Policy **38**(11), 6586–6596 (2010). https://doi.org/10.1016/j.enpol.2010.06.029
9. Fouquet, R.: Divergences in long-run trends in the prices of energy and energy services. Review of Environmental Economics and Policy **5**(2), 196–218 (2011). https://doi.org/10.1093/reep/rer008
10. Fouquet, R.: Long-run demand for energy services: Income and price elasticities over two hundred years. Review of Environmental Economics and Policy **8**(2), 186–207 (2014). https://doi.org/10.1093/reep/reu002. http://reep.oxfordjournals.org/content/8/2/186.abstract
11. Georgescu-Roegen, N.: Energy and economic myths. Southern Economic Journal **41**(3), 347–381 (1975)
12. Gordon, R.: Is U.S. economic growth over? faltering innovation confronts the six headwinds. NBER working paper no. 18315 (2012)
13. Gordon, R.J.: The Rise and Fall of American Growth: the U.S. Standard of Living since the Civil War. Princeton University Press, Princeton, NJ (2016)
14. Heun, M.K., Carbajales-Dale, M., Haney, B.R.: Beyond GDP: National Accounting in the Age of Resource Depletion. Springer International Publishing (2015)
15. Jackson, T.: Prosperity Without Growth: Foundations for the Economy of Tomorrow, second edition edn. Routledge, Milton, UK and New York, NY, USA (2017)

16. Jevons, W.S.: The Coal Question: An Inquiry Concerning the Progress of the Nation, and the Probable Exhaustion of Our Coal Mines, second edition, revised edn. Macmillan and Co., London (1866). Kessinger Legacy Reprints
17. Kander, A., Stern, D.I.: Economic growth and the transition from traditional to modern energy in Sweden. Energy Economics **46**(0), 56–65 (2014). http://dx.doi.org/10.1016/j.eneco.2014. 08.025. http://www.sciencedirect.com/science/article/pii/S0140988314002072
18. King, C.W.: Comparing world economic and net energy metrics, part 3: Macroeconomic historical and future perspectives. Energies **8**(11), 12,348 (2015). https://doi.org/10.3390/ en81112348. http://www.mdpi.com/1996-1073/8/11/12348
19. King, C.W.: The rising cost of resources and global indicators of change. American Scientist **103**, 6 (2015)
20. King, C.W.: Information theory to assess relations between energy and structure of the u.s. economy over time. BioPhysical Economics and Resource Quality **1**(2), 10 (2016). https://doi. org/10.1007/s41247-016-0011-y.
21. King, C.W., Maxwell, J.P., Donovan, A.: Comparing world economic and net energy metrics, part 2: Total economy expenditure perspective. Energies **8**(11), 12,347 (2015). https://doi.org/ 10.3390/en81112347. http://www.mdpi.com/1996-1073/8/11/12347
22. Lane, R.E.: The Loss of Happiness in Market Democracies. Yale University Press (2000)
23. Mazzucato, M.: The Value of Everything Making and Taking in the Global Economy. Public Affairs, New York (2018)
24. Meadows, D.H., Meadows, D.L., Randers, J., Behrens, W.W.I.: Limits to Growth: A Report for the Club of Rome's Project on the Predicament of Mankind. Universe Books, New York (1972)
25. Mirowski, P.: More Heat than Light: Economics as Social Physics, Physics as Nature's Economics. Cambridge University Press, Cambridge (1989)
26. Sautoy, M.D.: The Great Unknown: Seven Journeys to the Frontiers of Science. Penguin Press, New York (2016)
27. Simon, J.L.: The Ultimate Resource 2, revised edn. Princeton University Press, Princeton, N.J (1996)
28. Smil, V.: Energy in Nature and Society: General Energetics of Complex Systems. The MIT Press, Cambridge, Mass. (2008)
29. Stern, D.I., Kander, A.: The role of energy in the industrial revolution and modern economic growth. The Energy Journal **33** (2012). https://doi.org/10.5547/01956574.33.3.5
30. Turchin, P., Nefedov, S.A.: Secular Cycles. Princeton University Press (2009)

Chapter 3
The Energy Narratives: Fossil Fuels Versus Renewables

Fossil Fuels:

> Coal is pretty much everything wind isn't ...[1]—Alex Fitzsimmons (2017)

> Renewables, when they come on and off, it screws up the whole the physics of the grid ...So when people want to talk about science, they ought to talk about the physics of the grid and know what real science is, and that is how do you keep the lights on? And it is with fossil fuels and nuclear.[2]—Bernard McNamee (2018), President Donald Trump's appointee to the Federal Energy Regulatory Commission

versus Renewables:

> I'd put my money on the sun and solar energy. What a source of power! I hope we don't have to wait till oil and coal run out before we tackle that. I wish I had more years left.[3]—Thomas Edison

> By 2050, we could get all the energy we need from renewable sources. This report shows that such a transition is not only possible but also cost-effective, providing energy that is affordable for all and producing it in ways that can be sustained by the global economy and the planet.[4]—James P. Leape (2011), Director General, WWF International

---

[1]Former spokesman for Fueling U.S. Forward and senior adviser in the U.S. Department of Energy's Office of Energy Efficiency and Renewable Energy as appointed by President Donald Trump. Quote from "Fossil fuel promoter settles into renewable energy office" by Hannah Northey, *E & E News*, July 17, 2011, https://www.eenews.net/greenwire/2017/07/11/stories/1060057185, accessed July 11, 2017.

[2]Gavin Bade, "FERC nominee McNamee slams renewables, green groups in Feb. video," *Utility Dive* November 20, 2018: https://www.utilitydive.com/news/ferc-nominee-mcnamee-slams-renewables-green-groups-in-feb-video/542702/?mc_cid=a5911a78d7&mc_eid=d6e84014c0.

[3]This website http://quoteinvestigator.com/2015/08/09/solar/ says this quote is attributed to Edison via his friend James D. Newton via Newton's 1987 book *Uncommon Friends: Life with Thomas Edison, Henry Ford, Harvey Firestone, Alexis Carrel, & Charles Lindbergh*.

[4]The Energy Report: 100% Renewable Energy by 2050, [38].

© Springer Nature Switzerland AG 2021
C. W. King, *The Economic Superorganism*,
https://doi.org/10.1007/978-3-030-50295-9_3

There are many concepts that proponents and detractors of the fossil and renewable energy narratives use to argue for and against energy technologies and resources. In these arguments, they also invoke the economic narratives. This chapter summarizes the following topics, including statements from articles, books, and interviews, to highlight many of the different arguments for and against different energy extraction and generation options:

1. Expenditures—so important, Chap. 2 highlighted energy spending in detail.
2. Size—my resource is bigger than yours.
3. Price—my prices are lower, and getting more so.
4. Reliability—it's about time.
5. Morality—the poor, health, and "progress."
6. Development—my energy helps poor countries more than yours.
7. Environmental—your environmental impacts are worse than mine (land, water, greenhouse gases).
8. Government Support, or Subsidies—you get too many subsidies, but I don't.

These are not the only topics that are relevant and of concern. However, they are some of the most widely debated concepts over energy. Absolute and definitive statements are hard to justify in the energy narratives, and some of the statements are made with significant hyperbole. Thus, I could neither resist including them, because the book is more entertaining as such, nor comment on them with 100% objectivity.

This is the longest chapter of the book, and its subject could be the basis of an entire book itself. This chapter presents modern-day statements for the energy narratives that provide fodder to dive deeper into the science of energy and economic systems (Chap. 5) and economic theories (Chap. 6) that we can blend into a coherent viewpoint of how the world operates (Chap. 8). This chapter also provides examples that seemingly justify overly simplified economic narratives to support and refute certain ideologies (Chap. 9) for how we should and can organize our social, political, and economic systems.

But enough set up, let's begin.

The Expenditures Discussion

Chapter 2 went to great lengths to discuss total energy expenditures. It is a good idea to refer back to the energy (and food) expenditures data as needed. There are three points that are important for considering energy expenditures in the context of the energy and economic narratives.

First, total expenditures on any particular fuel or energy carrier equals the price of that fuel multiplied times the quantity that you purchase. Expenditures = price × quantity purchased.

Second, what matters most at the macroeconomic level is not the price or amount of expenditures on any one fuel or energy carrier, but the total expenditures *on all fuels and energy carriers.* Total energy expenditures relative to GDP (or income) *is the collective piece of information* that feeds back to economic growth [7, 34, 59].

Of course the price of any given fuel (e.g., dollars per gallon of gasoline) is important. The price is the main piece of information that affects immediate consumer and producer decisions at the individual level. However, by focusing on total energy expenditures one avoids problems associated with disagreements over substitution of one primary energy resource or energy carrier with another. For example, assume I consume two fuels, Fuel A and B. The price of Fuel A doubles in price from 1 to 2 $/unit so that I consume 5 units instead of 10, and the price of Fuel B declines in price from 2 to 1 $/unit so that I consume 10 units instead of 5. In both the before and after situations, I spend $20 on both Fuel A and B. The mix changed. There was substitution, but my expenditures were the same. The data in Chap. 2 show that the major trend of industrialization via the use of fossil fuels was to decrease total energy and food spending relative to GDP, at least through the year 2000, even as the mix of energy sources continues to change.

Third, the focus on price allows one to narrow discussion to whether your technology's price is lower than your competitor's while ignoring consideration of whether one or both prices are too high or sufficiently low. People creating narratives for or against any given energy technology often ignore expenditures and focus only on price or quantity without considering their combined feedback on the economy, discussed later in Chap. 5.

The next two sections discuss the data and narratives of the two components of expenditures: quantity (or size) and price.

The Size Discussion: My Resource Is Bigger Than Yours

Even if you don't often pay attention to energy issues, you may have heard statements similar to those in this section that discuss just how large are energy resources, both renewable and fossil. It is easy for the energy narratives to speak past each other because fossil energy resources are *stocks* of materials stored in the Earth, and renewable energy resources can be both stocks and *flows*, such as wind and sunlight.

Size: We Have More Than Enough Renewable Energy

The sunlight striking the earth's surface in just one hour delivers enough energy to power the world economy for one year.[5]—Lester Brown (2015), President of Earth Policy Institute

---

[5]From Article dated June 17, 2015 at http://www.alternet.org/environment/sunlight-striking-earths-surface-just-one-hour-delivers-enough-energy-power-world, accessed August 3, 2017 and

The solar energy quote attributed to Thomas Edison at the beginning of this chapter is often used for inspiration. If the father of the light bulb, and one of America's greatest inventors, can see the solar future, then why can't we all?

Comments such as Lester Brown's are a common starting point for discussing why solar-powered technologies are often favored from a size, or quantitative, standpoint. The average amount of power in sunlight reaching the Earth's surface is about 175 watts per square meter (W/m^2) [87, 92]. Multiplying this by the surface area of the Earth[6] gives 89,300 trillion Watts. The energy input from one hour at this average power equals 320 exajoules (EJ) [92]. Figure 2.10 from Chap. 2 shows that worldwide primary energy consumption was around 600 EJ in 2014. Thus, today human primary energy consumption is approximately twice that which comes from the sun over 1-h. Close enough for Brown's statement.

However, while the quote is not misleading, it is not informative in three aspects: technology specifications, space, and timing.

The technology and pure quantity of land use are not particularly restrictive. Even assuming conversion of sunlight to electricity using photovoltaic (PV) panels at 10% efficiency (well within capability of existing technology)[7] using 10% of the entire Earth's surface, it would take about 1 week to produce as much solar electricity as all primary energy consumed in 2014. However, given that only 29% of Earth is land, allocating 10% of the Earth's surface to PV panels is not a trivially small area. For example, only 3% U.S. land is urban,[8] and the Food and Agriculture Organization data estimates less than 1% of global land is covered by artificial surfaces.[9] Nonetheless, this solar calculation shows that the constraints are not the pure technology and resource combination, but other factors such as the economic and social concerns (e.g., total cost, aesthetic, environmental, agricultural) of placing solar converting technologies in specific locations and connecting them to load centers.

Of course, we don't consume power for 1 week only, and then cease all activity for the other 51 weeks. It is the timing issue of solar-generated electricity that draws the ire of many detractors and the focus of many engineering and scientific efforts. We can convert sunlight to power every day, and the extent of this engineering challenge is discussed later in this chapter within the section describing *reliability*.

---

stated as an excerpt from book *The Great Transition: Shifting from Fossil Fuels to Solar and Wind Energy*, by Lester R. Brown, with Janet Larsen, J. Matthew Roney, and Emily E. Adams.

[6] Approximating the Earth as a sphere, its surface area = $4\pi r^2 \sim 510,000,000\,km^2$ with the radius of the Earth approximately $r = 6378\,km$ [92].

[7] See the National Renewable Energy Laboratory chart that tracks the historical trend of efficiency for various solar photovoltaic technologies: https://www.nrel.gov/pv/cell-efficiency.html.

[8] USDA Economic Research Service data for 2007, accessed August 17, 2017 at https://www.ers.usda.gov/data-products/major-land-uses.aspx#25988.

[9] FAO Global Land Cover SHARE database, http://www.fao.org/news/story/en/item/216144/icode/ accessed August 17, 2017.

Size: We Have More Than Enough Fossil Fuels

I do not say that "infinite substitutability" is possible now or at any future moment. What I do say is that substitutability is *increasing* with the passage of time; there have been more and cheaper substitutes for each raw material with the passage of time.

...There is no doubt that my assertion of nonfiniteness...regrettably...is not explicit in standard economics, though it is not incompatible with standard received economics. But the critics simply do not come to grips with the matter that the available data are not consistent with the assumption of finiteness.[10]—Julian Simon (1996)

...economically you have a very simple test of whether anything is an exhaustible resource, namely: is its price rising over time? If we look at the price of energy and of oil, its real price has been going down over time.[11]—Milton Friedman (1978)

There is no such thing [as an exhaustible natural resource ...a fixed stock such as oil]. ...The total mineral in the earth is an irrelevant non-binding constraint. If expected finding-development costs exceed the expected net revenues, investment dries up, and the industry disappears. Whatever is left in the ground is unknown, probably unknowable, but surely unimportant; a geological fact of no economic interest.[12]—Morris A. Adelman (1990)

But the issue of exhaustion is resolved. Energy supplies are—for all practical purposes—infinite.[13]—Peter Huber and Mark Mills (2005)

The above quotes from Julian Simon, Milton Friedman, and Morris Adelman cover a vast, shall I say infinite, space between two concepts, both of which say the finite nature of Earth is not relevant. The extended concept by Simon is that we don't even know how to measure the size of the environment in which we live. Over the long-run, he considers the "cosmos," or all of the universe, as the potential domain of human existence and our economy. People like billionaire Elon Musk are perhaps halfway between these concepts of infinity. Musk seeks to colonize Mars, clearly thinking we don't have to restrict ourselves to living on Earth [69]. But one reason Musk seeks to make humans a "multi-planetary species" is as a backup plan in case we fail to achieve a sustainable energy economy, without the use of fossil fuels, on Earth.[14] My discussion here stays within the Earth's confines, accepting

---

[10]Simon [85, pp. 596–597].

[11]The Energy Crisis: A Humane Solution, available September 1, 2017 at https://miltonfriedman. hoover.org/friedman_images/Collections/2016c21/BP_1978_1.pdf.

[12]The full unedited passage from [2] is: "The assumption dropped is that there exists "an exhaustible natural resource ...a fixed stock of oil to divide between two [or more] periods" (Stiglitz, 1976). There is no such thing. The total mineral in the earth is an irrelevant non-binding constraint. If expected finding-development costs exceed the expected net revenues, investment dries up, and the industry disappears. Whatever is left in the ground is unknown, probably unknowable, but surely unimportant; a geological fact of no economic interest."

[13]Huber and Mills [48, p. 181].

[14]"By definition, we must at some point achieve a sustainable energy economy or we will run out of fossil fuels to burn and civilization will collapse. Given that we must get off fossil fuels anyway and that virtually all scientists agree that dramatically increasing atmospheric and oceanic

that the more existential discussion of human inhabitation beyond Earth can be both productive and distractive.

The economist Adelman tells us not to concern ourselves with the "geologic fact" that stocks of mineral resources are finite. While accepting that the Earth, and the stock of minerals, is finite, he more precisely states this fact is of "no economic interest." One point he is explaining is the very important difference between mineral *resources* and mineral *reserves*.   The word resource refers to a quantity of the amount of a mineral stock (e.g., oil) estimated to exist in total. The word reserve refers to only the portion of resources that one estimates can be extracted profitably. "The well's proved reserves are the forecast cumulative profitable output, not the total amount of oil that is believed to be in the ground."[3] That is to say, if an oil driller knows there is more oil in a certain location, but it costs more money to pull it out of the ground than he can receive by selling it, then he leaves the oil in the ground. That oil left in the ground counts as part of the total resource, but none of it counts as part of total reserves.

Because reserve estimates are based on profitability, and profits are based on price, then reserves estimates are also dependent upon (the current) price. Further, because our human-constructed economic rules affect prices via cost inputs, market dynamics, treaties, and regulations, the rules inherently affect profits and estimates of reserves. There is a continuous feedback loop among these factors, and both producers and consumers make adjustments using price as the major piece of information.

In short, *there is no pure technological description of fossil fuel reserves*, and certainly not only from the perspective of extraction technology.

The Organization of Petroleum Exporting Countries (OPEC) provides a good example of how oil reserves have been affected by rules, in this case outside of market forces. OPEC is a group of countries that are net exporters of oil and gas. As a group it has some level of influence on the price of oil by using any "spare" production capacity to ramp up or down extraction and influence oil price downward or upward, respectively. In the 1980s, facing low oil prices from low demand and higher oil production capacity, OPEC changed the rules by which it allocated production among its members. Driven by Saudi Arabia, the new rules stated that the allocation of production would be proportional to country reserves. Thus, countries with higher oil reserves would be allowed to produce more oil, and countries with lower reserves were to produce less. Miraculously, the reserves estimates of OPEC countries rose significantly.

Julian Simon had a good analogy for thinking of mineral reserves. They are really an estimate of a current inventory, like the amount of food in your home or even in

---

carbon levels is insane, the faster we achieve sustainability, the better." [70] "I think there are really two fundamental paths. History is going to bifurcate along two directions. One path is we stay on Earth forever, and then there will be some eventual extinction event. I do not have an immediate doomsday prophecy, but eventually, history suggests, there will be some doomsday event. The alternative is to become a space-bearing civilization and a multi-planetary species, which I hope you would agree is the right way to go." [69].

the grocery down the street.[15] We don't assume we will run out of food in our lifetime if our shelves at home or local grocer don't have all of the food we will need over our entire lifetime. But when our home or store shelves get bare, it is a signal to acquire more. Thus, if mineral reserves relative to consumption (or more commonly the reserves divided by production rate) become smaller than usual, it is a signal to raise the price, which can in principle increase reserves without any new physical exploration. If reserves relative to consumption increases (e.g., due to lower consumption), it can be a signal to decrease price and reserves can move lower.

Adelman additionally says that estimating the total *resource* size is unimportant because it is impossible: "To predict ultimate reserves, we need an accurate prediction of future science and technology. To know ultimate reserves, we must first have ultimate knowledge. Nobody knows this, and nobody should pretend to know."[3] Thus, taking "ultimate reserves" to be the amount of the resource that will be produced over all of time, it is easy to imagine how hard it is to estimate that number. Estimating "ultimate reserves" requires future estimates, for all of time, of the social and technological factors that dictate the extraction cost and consumer incomes that in turn affect sales price, revenues, and quantity sold. Thus, the argument goes, even if you accept that the Earth has limited mass, you don't have a way of knowing how much of this mass we will ever be able to economically use, and, thus, you should not try to estimate this unknowable number.

Adelman is telling us that because we will assuredly extract less than 100% of the amount of any finite mineral from the Earth over all time (or say practically over the next several hundred years), there will always be a quantity remaining in the ground due to economic reasons. Thus, we might as well think of the quantity of ultimate reserves as infinite in size. You can decide for yourself if you can simultaneously believe that resources are finite and reserves (a subset of resources) are infinite.

Nonetheless, many interested parties are compelled to continuously update estimates of finite quantities of reserves and resources. One major reason is that an updated estimate of oil and gas reserves (not estimated ultimately recoverable quantities) is a major piece of information for valuing oil and gas companies because it is the basis for estimating future revenues over the next years to decades.

In summary, *reserve* is an economic term, *resource* is a geologic term. The conflation of the terms resource and reserve is often at the heart of energy narratives, or at minimum statements of confusion.

Even the experts can lose focus. In a 2016 press release, Rystad Energy estimated "...total global recoverable oil reserves at 2092 billion barrels ..." [72]. However, in a 2017 press release, they estimated "...total global oil recoverable oil resources at 2.2 trillion [2200 billion] barrels." [28] Notice the subtle difference? In 2016 they use the word reserves, and in 2017 they use the word resources to report numbers derived from the same methodology for estimating a quantity of oil in the ground.

---

[15]Simon [85, p. 172].

Size: No, Renewable Resources Are Not Big Enough (and They Aren't Even Renewable!)

> The hydrocarbons in the earth's crust amount to more than 500,000 exajoules of energy. (This includes methane clathrates—gas on the ocean floor in solid, ice-like form—which may or may not be accessible as fuel someday.) The whole planet uses about 500 exajoules a year, so there may be a millennium's worth of hydrocarbons left at current rates.
>
> Contrast that with blue whales, cod and passenger pigeons, all of which plainly renew themselves by breeding. But exploiting them caused their populations to collapse or disappear in just a few short decades. It's a startling fact that such "renewable" resources keep running short, while no non-renewable resource has yet run out: not oil, gold, uranium or phosphate. [79]—Matt Ridley (2011)

Matt Ridley, journalist, author, and self-proclaimed *rational optimist*, says we "may or may not" have access to nearly 1000 years in known quantity of fossil energy consumption at current rates and that renewable resources aren't even renewable.[16] He does not refer to the word resources, but his stated quantity is an estimate of fossil resources, not reserves. The major data sources stating estimates (including government agency reported values) of fossil reserves are an order of magnitude lower, translating to the 10,000s of EJ [11, 98].[17]

Ridley's discussion of depleting renewable resources references what we can call *renewable stocks* such as animals (e.g., schools of cod) and trees (e.g., forests) that "renew themselves by breeding." These stocks are different from *renewable flows*, such as the wind and sunlight, that we associate with other renewable energy generation technologies such as wind turbines and photovoltaic panels. His definition includes human choice, or "exploitation," of these renewable stocks as part of the definition of whether they are renewable. Thus, he effectively includes characteristics of both the renewable stocks themselves—their size, regeneration rate, and death rate—and humans as part of the definition of renewable. This combination of concepts is useful, but not in the way in which Ridley uses them.

Consider these definitions of *renewable*:

- **renewable:** "capable of being replaced by natural ecological cycles or sound management practices"[18]—Merriam-Webster, and
- **renewable:** "able to be renewed" with *renew* as being able "to restore to a former state" or "to make effective for an additional period"[19]—Dictionary.com (Random House Unabridged Dictionary)

---

[16]Matt Ridley: http://www.rationaloptimist.com and author of *The Rational Optimist: How Prosperity Evolves*.

[17]Also Energy Information Administration International Energy Statistics for "Crude Oil Proved Reserves" accessed August 17, 2017 at https://www.eia.gov/beta/international/data/browser/.

[18](Adjective) Merriam-Webster dictionary, accessed August 9, 2017 at http://www.merriam-webster.com/dictionary/renewable.

[19](Adjective) Dictionary.com accessed August 9, 2017 at http://www.dictionary.com/browse/renewable and http://www.dictionary.com/browse/renew.

The definitions refer to the renewable resource having the characteristic that it *can be, if managed properly*, replaced or brought back to an original state. Thus, we might "exploit" a renewable stock resource such that it becomes fully depleted, such as the passenger pigeon example by Ridley, but that doesn't change the fact that the renewable resource has the capability of being renewed as long as we don't deplete it faster than it regenerates.

The Ecological Footprint is one calculation performed in the context of the rate of renewable resource regeneration in comparison to human consumption.[20] However, the Global Footprint Network calculates this number not to convince people that renewable resources are not renewable, but to incentivize, in their minds, appropriate levels of consumption (generally lower for developed countries) as part of the "sound management practices" in the definitions above.

Here is my rephrase of Ridley's statement with the goal of increased objectivity and informativeness that includes Morris Adelman's concepts on estimating fossil mineral quantities from the previous section:

> (My rephrase): Renewable resource stocks such as fish and trees have the capability of replacing themselves in quantity as long as they are not over-exploited by human activities or natural predators. A renewable resource stock does not become "nonrenewable" just because we can or did deplete the stock. Further, the physical and economic inability to extract 100% of any fossil mineral from all of the Earth's crust does not make it renewable even though the resource, though finite by definition, can never be fully exploited.

Just as humans can decimate stocks of animals, like passenger pigeons and cod, we can deplete forests too:

> Haiti meets about 60% of its energy needs with charcoal produced from forests. ...Full marks to renewable Haiti, the harbinger of a sustainable future! Or maybe not: Haiti has felled 98% of its tree cover and counting; it's an ecological disaster compared with its fossil-fuel burning neighbor, the Dominican Republic, whose forest cover is 41% and stable. [79]—Matt Ridley (2011)

Ridley's reference to Haiti is one often used. You can view satellite imagery of the border between Haiti and the Dominican Republic to observe stark contrasts in vegetation cover on one side versus another. And as noted by one of the most detailed analyses to-date of Haiti's land cover, there are many peer-reviewed publications indicating less than 3% of forest cover for Haiti, as implied by Ridley [15]. However, a 2014 study indicates that using higher resolution satellite imagery increases the estimated forest cover to 29% of Haiti, significantly higher than the 2005 value of 4% reported by the Food and Agricultural Organization [15].

We can't blame Ridley for (likely) using numbers reported in established papers and reports of academics and the United Nations' Food and Agricultural Organization. But we can attribute the drastic change in forest cover estimate to an important idea: resolution. The higher spatial resolution of land cover data, the more

---

[20]See Global Footprint Network: http://www.footprintnetwork.org/our-work/ecological-footprint/.

accurate is the estimate of tree cover,[21] and previous estimates were based on coarse resolution data [15]. Just as higher resolution can affect one's estimate of land use cover, higher temporal resolution can affect one's estimate of how much electricity is generated from different technologies.

Size: No, Fossil Fuel Resources Are Not Big Enough

A ... type of transient growth is ... represented by ... the quantity [that] grows exponentially for a while. Then the growth rate diminishes until the quantity reaches one or more maxima, and then undergoes a negative-exponential decline back to zero. This is the type of growth curve that must be followed in the exploitation of any exhaustible resource such as coal or oil, or deposits of metallic ores."[22]—M. King Hubbert (1974)

M. King Hubbert was not the first person to be concerned about the inability to forever extract more of a given fossil fuel. Chapter 2 mentioned William Stanley Jevons and his 1866 book *The Coal Question* that focused on understanding extraction of coal in the United Kingdom [58]. Jevons understood that the U.K.'s use of its domestic coal was a primary reason for its economic prominence at the time. Thus, he was concerned about how long the U.K. could produce coal. In 2018 U.K. coal extraction was almost zero, and the peak rate occurred at the eve of World War I [83].

If not the first person to concern himself with limits in fossil fuel extraction rates, M. King Hubbert was the dominant twentieth century figure discussing peak oil extraction, the notion that any well's or region's extraction rate eventually declines after its initial rise. He was the most prominent modern-day geologist to explain that finite fossil resources cannot be extracted at continuously higher rates and that ultimately this fact poses a limit to physical growth on Earth. This thinking led Hubbert to promote the use of science and engineering for public policy, including via the Technocracy movement of the 1930s [52].

Having worked for both Shell Oil and the United States Geological Survey, Hubbert is known most for his methods for estimating future production rates of U.S., world, and other regions' oil, natural gas, and coal using what became known as "Hubbert curves." These curves were extrapolations based upon historical

---

[21]"Plotting the recoded land cover statistics calculated for this study (Table 10) and data product resolution (Table 2) suggests a correlation between spatial resolution and land cover percentages (Fig. 8). The best fit trend-line for tree cover suggests a linear relationship with an R2= 0.996. For the shrub cover/herbaceous class the correlation with a linear trend is not as good, R2= 0.728. The trend suggests that coarse resolution imagery will tend to underestimate the amount of tree cover and potentially overestimate the amount of shrub cover/herbaceous cover." [15].

[22]The full quote from [47] is "A third type of transient growth is that represented by Curve III in Figure 1. Here, the quantity grows exponentially for a while. Then the growth rate diminishes until the quantity reaches one or more maxima, and then undergoes a negative-exponential decline back to zero. This is the type of growth curve that must be followed in the exploitation of any exhaustible resource such as coal or oil, or deposits of metallic ores."

production and an estimate of ultimate recoverable reserves, or the amount of the fossil resource estimated to be produced over all future years. While Hubbert production curves are often portrayed with the rate of increase and decrease being symmetric, or portraying a bell curve,[23] Hubbert did not fundamentally place this restriction on his thinking. As in his quote above, the important part was that there was an increase, and there would be a decrease approximately as steep as the increase. This held not only for individual fossil reservoirs and regions but also for the world overall. It was because of his unflinching repetition of this concept that Hubbert has been both reviled and praised.

Critics of Hubbert claim that his methods are focused only on geology and ignore economic drivers and technological innovation. However, historical oil extraction rates were certainly influenced by the economic and technological factors of the time. Thus, when Hubbert based his analysis on the historical data, he had no choice but to make some inherent assumption about how people within companies assess the state of the economy and technology to decide how much to invest in drilling and exploring for oil. Extrapolating into the future is a different story. Estimating future production requires assumptions that Adelman tells us will always be wrong because we can't predict the future due to our lack of "ultimate knowledge." However, even though we are ignorant, is it still useful to compare assumptions about the future to gauge a range of possibilities?

Consider Hubbert's extrapolations for U.S. oil, coal, and natural gas production from his 1956 paper presented to a meeting of the American Petroleum Institute in San Antonio, Texas [45]. In this paper Hubbert made two assumptions about ultimate oil reserves in the U.S. for "... oil capable of being extracted by present techniques" [45]. Extrapolating from historical production and assuming 150 billion barrels (BBLs) of ultimate reserves, Hubbert drew a curve showing a peak production rate of oil of approximately 2.7 billion BBLs/year in 1965.

Assuming 200 billion BBLs of ultimate reserves, his peak rate was 3.0 billion BBLs/year in 1970. It turned out that the peak crude oil production rate of the continental U.S. *did* occur in 1970, but at 3.4 billion BBL.[24]

Because the U.S. oil extraction rate did reach a peak in 1970, Hubbert's 1956 extrapolations have been touted as accurate enough for many planning and strategic purposes. Because the actual peak rate was a little higher than his extrapolations, some have said Hubbert was inaccurate. In retrospect, Hubbert's prediction was astounding in its accuracy. At a time when practically no one was discussing any peak in U.S. oil extraction, he ended up correctly estimated, to within a few years, a peak 9–14 years out from the year 1956 (for the continental U.S. using only primary oil recovery methods).

---

[23] A bell curve is more formally described as the normal distribution that is symmetric about the middle point.

[24] The peak oil extraction rate in 1970 when not including extraction from Alaska was 3.4 billion BBL. The peak extraction rate in 1970 when including extraction from Alaska was 3.5 billion BBL. Data for crude oil from EIA Monthly Energy Review, Table 3.1 Petroleum Overview, release date July 28, 2015.

But was he lucky or good with his U.S. oil production rate estimations? In 2014, U.S. oil production surpassed the 1970 peak in oil production rate. If you are a Hubbert fan, you would say he considered only conventional oil (i.e., "present techniques" in the 1950s). If you are a Hubbert skeptic, you would say he missed the mark precisely because an analysis assuming constant technology will always be wrong. Predicting the future is difficult. In making future projections we can't assume zero technological change, but we can't assume infinite capability either.

In the same 1956 paper, Hubbert also projected world oil extraction. He assumed 1250 billion BBLs for world ultimate oil recovery with a peak in extraction near 12.5 billion BBLs/year around the year 2000. In 1970 the world extracted approximately 16 billion BBL/year [11],[25] and in 2015 it produced 29 billion BBLs/year of crude oil.[26] These are well above Hubbert's 1956 estimates.

While Hubbert's 1956 estimate of 2015 world oil extraction was quite a bit too low relative to the actual rate, his estimate of future U.S. coal extraction was too high. He projected U.S. extraction of nearly 1.5 billion metric tonnes (Gt/year) in 2015 whereas U.S. extraction was below 0.9 Gt/year in 2015 and only topped 1 Gt/year in the mid-2000s [11]. Similar to his world oil projection, his estimate of world coal extraction also underestimated 2015 production by approximately 3 Gt/year as Hubbert approximated only 5 Gt/year in 2015.

Thus, for the U.S., Hubbert was relatively close to the mark, or slightly overpredicting of fossil extraction. For the world, he significantly underpredicted.

Ultimately, Hubbert and Adelman recognized the same fact of finite mineral resources, but they held widely different perspectives for how to deal with that fact. Their two perspectives are examples of divergent viewpoints on how to consider utilization and dependence upon finite resource stocks, with Adelman and Hubbert representative of the techno-optimistic and techno-realistic economic narratives, respectively.

Hubbert focused discussion on the long term. The techno-realism and finite Earth narrative usually does. Many of his fossil fuel extraction extrapolations project over 100 years into the future. Hubbert thought that because extraction rates (of any given resource) can never continue to increase forever, we need to continuously track how much we are extracting relative to our best estimates of ultimate extraction such that we can plan for the inevitable "...transition between the period of increase and the period of decline." [46]

---

[25] BP Statistical Review of World Energy (2016) by reducing the reported oil extraction rate of 48,056 MMBBL/day to 45,000 MMBBL/day to subtract approximately 3000 MMBBL/day of natural gas liquids from the total.

[26] "Crude   oil   and   lease   condensate"   as   reported   by   the   Energy   Information Administration   International   Energy   Statistics   beta   website: https://www.eia.gov/ beta/international/data/browser/#/?pa=000gfs00000000000000000000000000000vg&c= 4100000002000060000000000000g0002000000000000000001&tl_id=5-A&vs=~~~~~INTL. 58-1-WORL-TBPD.A~~~~~~~~~~~~~~~~INTL.57-1-WORL-TBPD.A&cy=2014&vo=0&v= T&start=1980&end=2016&s=INTL.57-1-WORL-TBPD.A.

Adelman, Simon, Friedman, and others focused on time scales shorter than a few decades. The techno-optimism economic narrative usually does. The rate of exploration and extraction of fossil fuels has been quite dynamic from decade to decade, and much more so after 1970 when the U.S. reached an initial peak in the rate of oil extraction. These dynamics are affected by many social and technological factors, such that given our ignorance of the future, Adelman stated there is no use in estimating fossil mineral extraction rates a hundred years or more into the future.

Can both the Hubberts and Adelmans of the world be right? Yes.

Both of their perspectives are valid and useful. We need to think across time scales to understand how shorter-term dynamics affect longer-term outcomes and how the finite nature of the Earth is affecting our immediate decisions. If the concept of the finite Earth is not within your worldview or your economic model, then it is impossible to associate any change as the result of a feedback from a finite Earth.

Size: Fossil Fuels Are Renewable Too

> Even fossil fuels are renewable in the sense that they are still being laid down somewhere in the world—not nearly as fast as we use them . . . [79]—Matt Ridley (2011)

If you are interested in time scales longer than the millions of years required to form new fossil fuels, then you can take this quote seriously. I'll move on.

Size: The Stone Age (or Substitution) Argument

> Thirty years from now there will be a huge amount of oil—and no buyers. Oil will be left in the ground. *The Stone Age came to an end, not because we had a lack of stones*, and the oil age will come to an end not because we have a lack of oil. [33] (emphasis not in original)—attributed to Sheikh Zaki Yamani (2000), Saudi Arabia's oil minister from 1962 to 1986

> Our ability to find and extract fossil fuels continues to improve, and economically recoverable reservoirs around the world are likely to keep pace with the rising demand for decades. As the saying goes, *the Stone Age did not end because we ran out of stones*; we transitioned to better solutions. [14] (emphasis not in original)—Steven Chu (2013), 1997 Nobel Prize in Physics and U.S. Secretary of Energy (2009–2013)

The Stone Age analogy is attributed to the quote above from Sheikh Zaki Yamani, Saudi Arabia's oil minister from 1962 to 1986. The concept is so catchy that not even Nobel Prize winner and former U.S. Secretary of Energy Steven Chu was immune to letting it creep into his speeches.

The analogy is both apt and unfitting. Considering both stones and oil generically as natural resources that we appropriate for various end means (e.g., hunting, grinding, transportation), then the analogy is OK. On the unfitting side, unlike oil, the stones for which the Stone Age is named are not burned and converted into heat and work in the thermodynamic sense. Stones are shaped into tools using other tools held by prime movers (human muscles) fed by fuels (food). They are ineffective unless attached to a prime mover, such as muscles in a human arm, that applies energy to the tool like when killing an animal. Oil is mainly a fuel, but it can also be converted into plastic tools and other products such as bottles and clothing. In our modern economy, the primary role of oil is as the fuel fed into modern prime movers such as internal combustion engines and turbines.

In short, a stone is not fuel, but oil is.

The implication of the Stone Age argument is that if we no longer use oil it will be because we've moved on to something better. But, as we saw in Chap. 2, as new energy resources have come into use (e.g., coal, oil, hydropower, solar), we haven't yet stopped using the existing resources. As we transitioned to use oil, we haven't yet stopped using coal or even declined global coal consumption. As we have generated electricity powered by the sun and the wind, we haven't slowed our use of oil and natural gas.

While for all practical purposes we no longer directly use raw or hand-shaped stones as tools (e.g., stone arrow and spear points), we have not absolutely stopped using stones in tools—diamonds are used not only for jewelry but also in cutting tools. More generally, raw shaped stones are used as building materials (e.g., limestone walls of homes and buildings) and functional surfaces (e.g., granite counter tops). Finally, perhaps in ultimate circularity, large amounts of very fine-grained stones, or sand, are now used to extract oil and gas itself in the horizontal drilling and hydraulic fracturing process. *It seems the latest revolution of the Oil Age is still dependent on stone-based technology.* In short, we have not stopped using stones.

This next quote from Matt Ridley, referencing "limits" and the Stone Age argument is for direct contrast to the quote starting the next section.

> How many times have you heard that we humans are "using up" the world's resources, "running out" of oil, "reaching the limits" of the atmosphere's capacity to cope with pollution or "approaching the carrying capacity" of the land's ability to support a greater population? The assumption behind all such statements is that there is a fixed amount of stuff—metals, oil, clean air, land—and that we risk exhausting it through our consumption. ...But here's a peculiar feature of human history: We burst through such limits again and again. After all, as a Saudi oil minister once said, *the Stone Age didn't end for lack of stone.* [80, 81] (emphasis in original)—Matt Ridley (2014)

Size: Technological Change Is Not Enough—The Earth Is Still Finite

> Can this physical [population and physical capital] growth realistically continue forever? Our answer is no! ...There is no question about whether growth in the ecological footprint

will stop; the only questions are when and by what means.[27]—Meadows, Randers, and Meadows (2004)

We did NOT prove that there are limits to physical growth on a finite planet. We assumed it.[64][28]—Dennis Meadows (2012)

Perhaps nothing contrasts more with the Stone Age argument than *The Limits to Growth* books. While growth proponents and techno-optimists discuss bursting through (apparent) limits again and again, Hubbert, Meadows *et al.*, and other techno-skeptics point out how these breeches only represent temporary overshoot of the limits.

Overshoot is the concept that a long-term limit in the rate of use of either a natural resource or a sink (e.g., for pollutants) can be surpassed, but only temporarily. For example, if a forest regenerates itself at 5% per year, then you could harvest 5% of the total forest each year to be sustainable in the long term. However, you could harvest 10% of the forest for a few years, overshooting the sustainable harvest rate before eventually depleting the forest if maintaining that high harvest rate.

As simply stated by Dennis Meadows' quote above, while we can physically and mathematically prove there are limits to physical growth on a finite planet, each person still has to make the choice of whether to include the finite Earth into his worldview. For those that don't, they have not assumed a limit, and their analyses, models, and reasoning, therefore, cannot anticipate any affect of a limit that they assume does not exist. The earlier quote from Julian Simon is a relatively direct reply to Meadow's statement.[29] If the finite Earth is affecting socio-economic outcomes, Simon could never use the finite Earth as an explanation since he assumed it away from the beginning. Simon even replied to his critics that the universe, not the Earth, was the relevant system for human exploitation.[30]

Size—Summary

Size is relative.

An elephant is large compared to a mouse, but the same elephant is small compared to a blue whale. A fish in an aquarium has a much smaller world in which to live than one living within an ocean reef. We don't study fish without considering the context of their habitat. We shouldn't study humans outside of that context either.

---

[27] Meadows et al. [65, p. 48].

[28] Presentation at Smithsonian Institution in honor of the 40th Anniversary of the book *The Limits to Growth*, Washington DC, February 29, 2012.

[29] "...critics are reduced to saying that all the evidence of history is merely "temporary" and must reverse course "sometime," which is the sort of statement that is outside the canon of ordinary science." Julian Simon (1996) [85].

[30] Simon uses the word "cosmos" instead of universe [85, pp. 596–597].

The size of the Earth might not be an explanation for many trends we see, but then again, it might, and we'll continue to explore this question throughout the book. If you don't put a size of Earth in your model of global economics, then you inherently assume it is infinite, and infinite is clearly an incorrect size.

For many of those that do include the finite Earth into their worldview, the ultimate answer is clear: it matters. They only have questions for how to estimate and plan for the timing and magnitude of the feedback effects from the finite Earth. Social constructs and economic markets cannot overcome physical reality, and most disagreements about finite limits center on our perceptions of finite reality. As the authors of *The Limits to Growth* stated, they were not "... anti-market. We understand and respect the capacities of the market. ... We count on improvements in market signals, as well as in technologies, to bring about a productive and prosperous sustainable society. But we do not have faith, and we have no objective basis for expecting, that technological advance or markets, by themselves, unchanged, unguided by understanding, respect, or commitment to sustainability, can create a sustainable society."[31]

By including the major factors driving physical and economic activity into conceptual and mathematical models we can approach holistic understanding. Thus, if we want to understand the opportunities and limits of technology and finite resources, then we must represent and couple both in one way or another. Chapter 5 discusses scientific ideas for this representation and Chap. 6 discusses the competing economic theories, using scientific approaches or not, used to explain energy and economic patterns we observe. But now we turn to the concept that economists definitely do not ignore. We might even say economists focus on it too much: price.

The Price Discussion

Consequently if you have an exhaustible resource, unless people expect the future price to be higher than the present price, they have an incentive to bring it out now. That tends to drive down the present price; it tends to drive up the future price. Therefore economically you have a very simple test of whether anything is an exhaustible resource, namely: is its price rising over time? If we look at the price of energy and of oil, its real price has been going down over time.

and

So the alternative interpretation, and the only one that makes sense, is that in any economic sense oil, far from being an exhaustible resource, is a producible resource at more or less constant or indeed declining cost because of the improvements in the technology of drilling and exploring and so on. You can find more oil, and therefore the future price could not rise above the present because if it tended to do so it would give somebody an incentive to go

---

[31] Meadows et al. [65, pp. 205–206].

out and find more and add to the supply. Of course that situation may change.[32]—Milton Friedman (1978)

Transmutation of the elements,—unlimited power, ability to investigate the working of living cells by tracer atoms, the secret of photosynthesis about to be uncovered,—these and a host of other results all in 15 short years. It is not too much to expect that our children will enjoy in their homes electrical energy too cheap to meter . . . [33]—Lewis L. Strauss (1954), Chairman of the U.S. Atomic Energy Commission

Price A piece of information, or metric, often attributed with magical or potentially sinister influence.

On the magical attribute of price, there are market prices. A textbook market price, say of an energy commodity, is supposed to simultaneously include all influential information on supply from multiple producers and demand from multiple consumers. In the real world a market price neglects many *external* costs related to common goods such as the environment, sharing of information among all actors, disproportionate influence of actors (e.g., due to size), and other facets that are necessary for a textbook *free market* to exist. Even though all of the conditions for a perfect market never exist, existing markets work very well for many purposes.

On the neutral-to-sinister attribute of price, monopolies or cartels can set restrictions on the flow of commodities and specify prices to achieve various goals including the balance of supply and demand, the maximization of profit, geopolitical influence, and the elimination of competitors. The U.S. has been no stranger to these practices, with oil trade front and center.

Oil extraction from the early 1930s East Texas oil boom was so prolific that the 1933 National Industrial Recovery Act (NIRA) specified state-level oil extraction quotas. After the U.S. Supreme Court struck down the NIRA, Texas senator Thomas Connally sponsored the Connally Hot Oil Act that became law in 1935.[34] This law made it a federal offense to transport oil produced in excess of state quotas across state lines.

In addition to federal regulation of oil trade, the state of Texas ran what was arguably the world's first energy cartel via the Texas Railroad Commission. Before 1970, the U.S., largely because of Texas, was the world's *marginal* oil producer. If you are the marginal producer of something, then you are the one that reduces output if demand goes down a little bit or increases output if demand goes up a little bit. In theory, in economic markets where companies compete to sell products at the lowest price, the costs of the marginal producer are closest to the market price. But before the 1980s, there was no short-term market for trading (buying and selling) oil. During the few decades leading up to 1970, the Texas Railroad Commission

---

[32]"The Energy Crisis: A Humane Solution" https://miltonfriedman.hoover.org/friedman_images/Collections/2016c21/BP_1978_1.pdf.

[33]Remarks Prepared by. Lewis L. Strauss, Chairman, United States Atomic Energy Commission, For Delivery At the Founders' Day Dinner, National Association of Science Writers. September 16, 1954, New York, New York.

[34]"Connally Hot Oil Act of 1935, Texas State Historical Association website accessed December 12, 2019 at: https://tshaonline.org/handbook/online/articles/mlc03.

maintained oil price stability by throttling Texas oil production to prevent a price collapse due to overproduction [99]. Once Texas oil wells were producing at 100% of capability in 1970, the Commission lost the power of price control because it lost the control on power in the form of the flow rate of oil.

Today the Organization of Petroleum Exporting Countries (OPEC) is an energy cartel for which many in the West have views ranging from a neutral world actor (trying to maximize their economic returns) to a sinister arm of foreign governments punishing the West (for its religious, political, and/or economic viewpoints and power) by controlling oil exports and thus world oil prices.[35] However, OPEC members and other energy exporters have several concerns that govern their actions. Many OPEC countries rely on oil and gas export revenues for the majority of their government funding and social programs. Thus, each exporting country needs a certain oil price to prevent a budget deficit [4]. In many countries, this "fiscal break-even" price for the government is often higher than the market price of oil.

In addition to markets and cartels, there is the concept of cost-plus, or markup, pricing. Using this type of pricing a producer sets a price equal to costs plus an added percentage of costs to target a known profit ratio. For example, markup pricing is used to set electricity rates ($/kWh) that consumers are charged by regulated electric utilities as well as transmission and distribution utilities where wholesale electricity markets exist. Upon public utility commissions and utilities agreeing to both the cost and need for a given capital investment in electricity infrastructure, the utility's owners (e.g., stock and bondholders) are then guaranteed a rate of return on that investment above the capital and operating costs.

These and other pricing concepts are and have been used to influence and determine prices of energy commodities. People do incorporate additional non-price information into their purchase decisions (e.g., company social and environmental investments), but the stated prices of goods and services are the pieces of information that consumers must immediately take into account. Let's now look into oil price history in some detail.

Price: Oil

For the last 100 years, oil, by most accounts, has been the most productive and valuable energy resource. Its high energy density (energy per mass) and ease of storage (e.g., gasoline in your car tank, jet fuel in a plane) enabled it to power internal combustion engines and turbines well beyond the capabilities of coal-powered steam engines. Because it dominates as a fuel for transportation, oil has

---

[35]Here I use the "West" as the Western Civilization in the context as discussed by Samuel Huntington's "clash of civilizations" anticipated for the post-Cold War era where "The conflicts of the future will occur along the cultural fault lines separating civilizations." [49, 50]. The West is largely the United States, Canada, Western Europe, and Australia with Western Christian, as opposed to Orthodox Christian, history.

been the single dominant factor in total energy expenditures, consumption, and useful work since World War II (see Figures in Chap. 2).

Thus, it is worth investigating the oil price trends. There are several ways to consider the price of oil. Depending on the narrative you might want to tell, you might focus on one price metric and ignore others. The goal of this discussion of oil price is to look at the same data in multiple ways to gain valuable perspective to interpret the various oil-economic narratives. Julian Simon had the following statement on how to think of the price of natural resource stocks, such as oil:

> So price, together with related measures such as cost of production and share of income, is the appropriate operational test of scarcity at any given moment. What matters to us as consumers is how much we have to pay to obtain goods that give us particular services; from our standpoint, it couldn't matter less how much iron or oil there "really" is in the natural "stockpile." Therefore, to understand the economics of natural resources, it is crucial to understand the most appropriate *economic* measure of scarcity is the price of a natural resource compared to some relevant benchmark.[36]—Julian Simon (1996)

As Julian Simon suggests, prices of something relative to a benchmark are the most appropriate economic measure of scarcity. If you want to know if we are running out of any given resource, check the price relative to your pocket book. This is effectively what Chap. 2 presents as energy and food expenditures relative to GDP. Gross domestic product is the proxy for our collective budget.

In concert with Simon's statement, Fig. 3.1 has four charts that explore different perspectives for viewing the price of oil. Three charts (Fig. 3.1a–c) explain the relationship of changes in oil price and expenditures over time. The fourth chart (Fig. 3.1d) shows U.S. oil price relative to U.S. oil consumption.

Figure 3.1a, the real price of oil since 1861 [11], shows three time periods that distinguish fundamentally different oil prices.[37] The average oil price during these time periods is indicated by the dotted horizontal lines. Initially (1861–1879) the price averaged 54 $/BBL. It is common for new products to have relatively high prices as there are few familiar customers or uses for the new product. The first major use of oil in Europe and the U.S. in the mid-1800s was for lighting via kerosene lamps.[38]

The second time period (1880–1973) is one of a low average oil price at 19 $/BBL. In 1880 oil extraction accounted for 2% of the quantity of global energy consumption from coal. The 1880s saw the first designs for internal combustion engines in Germany (Gottlieb Daimler and Wilhelm Maybach; Karl Benz) that were both designed to burn petroleum-based fuels and light enough for transportation rather than only stationary power.[39] In 1908, Henry Ford started producing the Model-T. Both oil price and consumption were low; thus so were world oil expenditures at less than 0.5% of the value of gross world product through World War II when coal and wood dominated as fuels [34].

---

[36] Simon [85, p. 26].

[37] All prices stated and in Fig. 3.1 are in real 2015 U.S. dollars.

[38] Yergin [99, pp. 24–25].

[39] Smil [87, pp. 230–232].

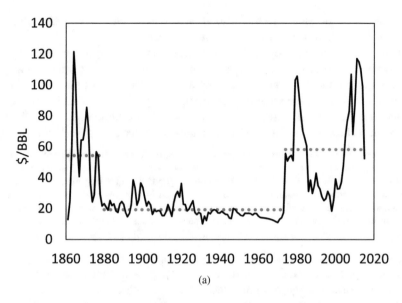

(a)

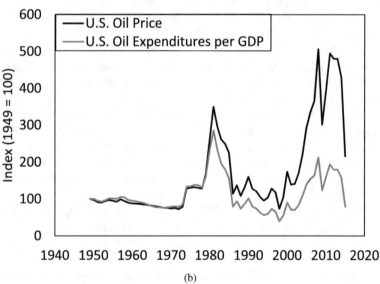

(b)

Fig. 3.1 (a) Over the long term, the oil price (in real 2015 U.S. dollars per barrel) can be described by three major time periods: before 1880 at an average of 54 $/BBL, between 1881 and 1973 averaging 19 $2015/BBL, and after 1973 (1974–2016) averaging 58 $/BBL ["Oil-Crude prices since 1861" from BP Statistical Review (2016) [11]]. (b) The indices of U.S. real oil price ($/BBL) [nominal price from EIA scaled to $2015 using U.S. GDP deflator from U.S. BEA] and U.S. oil expenditures divided by U.S. GDP [Petroleum consumption: EIA MER Table 1.3. Nominal oil price: EIA MER Table 9.1. Nominal GDP: St. Louis Federal Reserve, BEA series "GDPA"].

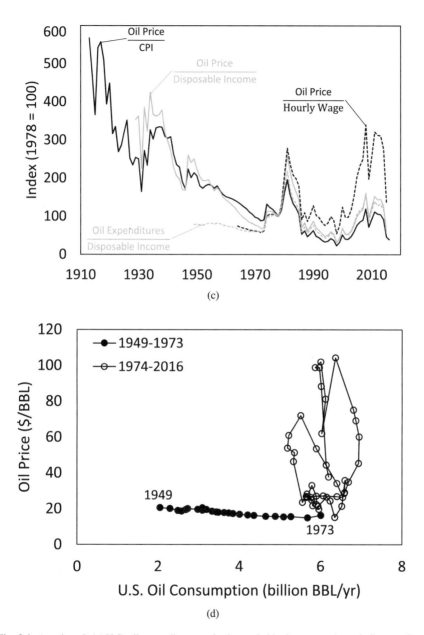

Fig. 3.1 (continued) (**c**) U.S. oil expenditures and price scaled by income and cost indicators. Pre-1949 oil price is from BP Statistical Review (2016) [11], and oil price from 1949 to 2016 is from EIA Monthly Energy Review [27]. CPI = consumer price index, a measure of the cost of consumer goods and services. Hourly wage = Average Hourly Earnings of Production and Nonsupervisory Employees: Total Private, Dollars per Hour, Annual, Seasonally Adjusted (data series: AHETPI from St. Louis Federal Reserve). (**d**) U.S. oil price (in real 2015 U.S. dollars per barrel) versus U.S. oil consumption (EIA reported energy from oil [27] converted at 5.8 MMBtu/BBL)

The rate of worldwide oil extraction increased at about 7%/year from 1880 to World War II and almost at 8%/year afterwards to 1973.[40] However, the finite Earth economic narrative states, exponential growth on Earth cannot continue forever. The data indicate that after 1973, it didn't.

The U.S. peak in oil extraction rate in 1970 took away the price-regulating power of the Texas Railroad Commission because the oil flow from Texas wells could no longer be increased at whim. The 1973 Arab Oil Embargo and OPEC's subsequent 127% increase in the posted price of oil starting January 1, 1974 signaled a new era when the West lost significant control over world oil production and price.[41]

During the 1970s oil-exporting countries sought to gain significant geopolitical power by using the "oil weapon" against Western nations. The West responded by finding ways to reduce oil consumption (e.g., in power generation), targeting oil and energy efficiency, and significantly increasing the rate of oil exploration and drilling. Via a major effort to shape the discussion, or *narrative* of the day, the oil industry and U.S. President Nixon deflected attention from oil specifically by describing the Middle East-driven oil crisis as a broader "energy crisis."[42]

The 1979 Iranian Revolution and strike by oil workers was the second major oil shock of the 1970s to trigger a significant recessionary impact in the West. By the early 1980s world oil production capacity was significantly larger than oil demand, causing oil prices to collapse. In 1983, the New York Mercantile Exchange began trading crude oil futures, and the era of the global oil market began in earnest. Despite the 1980s oil price collapse, the average price of oil from 1974 to 2018 was 58 $2015/BBL, about three times the price of the pre-1973 era.

Figure 3.1a shows that at broad approximation, the oil price has exhibited three phases over its history: an initial high price (before 1880) followed by a long period of low prices (1880–1973) to the current third phase of the last 45 years again at a relatively high price. A transcript of a speech by Milton Friedman in 1978 indicates at the time he found it nonsensical to think of an oil price breakpoint at 1973 [36].[43] However, with over 40 years of data since 1973, the real oil price has been unquestionably higher than the previous 90 years. Some explain this post-1970s higher oil price, and its significant volatility (the real oil price has fluctuated in a range from 20 to over 100 $2015/BBL) as effects from oil trading in financial

---

[40]Using data from [34], power from oil extraction increased at 7.0%/year from 1880 to 1945 and 7.7%/year from 1946 to 1973.

[41]The OPEC Gulf Six nations (Iran, Iraq, Abu Dhabi, Kuwait, Saudi Arabia, and Qatar) raise their posted prices for oil from $5.12 to $11.65 per barrel [57]. Prices had been rising during the previous several years, but this was the largest single increase.

[42]Mitchell [66, pp. 178–179].

[43]"But tell me, is it sensible to talk the way so many people talk about the oil industry. The way I interpret them is they say the whole history of oil is divided into two periods: from 1859 when oil was first discovered in Titusville, Pennsylvania to 1973 and from 1973 to 1978. Conceivably that could be true, but it seems to show a very shortsighted point of view and a lack of perspective to divide all of 120 years of history into two parts, one 115 years long and one 5 years long, and say "We know that there has been a fundamental change." Maybe, but if we allowed the market to work, if there were such a fundamental change that would show up in the form of a change in the price pattern and in a change in the incentives to people to find, exploit, and use oil." [36].

markets, such that there is no consistent oil price signal to consider. However, as noted earlier in this chapter, if not for the pre-1970 restriction in oil extraction rates at both the U.S. and Texas state levels, oil prices would have most certainly been even lower between 1930 and 1970.

You can call the post-1973 oil price regime a "new era" or just a time of higher prices, but there were many unprecedented responses to oil-related events of the 1970s. As already mentioned the concept of energy efficiency, such as for cars, largely did not exist before 1973, but now it is a household concept. A significantly debated energy narrative surrounds the fundamental reasons associated with pursuing energy efficiency. Some believe more energy efficient devices decrease total energy consumption, and some believe the exact opposite. Chapter 5 explains this disagreement, and after completing this book, the reader will understand why the post-1973 focus on energy end-use efficiency is a natural response to high prices and energy constraints.

Before considering the major post-1973 responses to higher oil prices (e.g., efficiency and increased exploration), we can see their economic effects over the following 40 years. One way to contextualize the post-1973 higher oil price is by comparing it to total oil expenditures in the same way we considered energy expenditures in Chap. 2 (see Fig. 3.1b). If oil expenditures per income or per GDP increase less than the price of oil, this indicates that adjustments and growth in the economy have adapted to consume less oil per unit of monetary output. From 1949 to 1979 oil price and U.S. oil expenditures divided by GDP moved in lock-step. After a roller coaster ride up and down since 1980, U.S. oil expenditures relative to GDP in 2015 were near the same level as in 1973 (about 8% lower in 2015), whereas the oil price was 170% higher. Amazing. U.S. annual energy consumption from oil was practically the same in 2015 as in 1973, about 6 billion barrels (see Fig. 3.1d), but more total dollars were spent on oil in 2015. Even though total dollars for U.S. oil expenditures in 2015 were about 170% higher than in 1973, real U.S. GDP was 200% higher. Thus, oil expenditures relative to GDP were similar in 2015 as they were in 1973.

How is this possible to have the same oil expenditures, relative to GDP, today as in 1973? How can we afford a higher real oil price today? The answer is energy efficiency. Starting in 1980, policy-driven technology changes, such as more fuel-efficient cars, started to take hold. Thus, if the price of energy goes up too high and too fast, consumption must come down. Consumption can come down by doing less (e.g., drive fewer miles) and by using less fuel input to do the same amount of activity (e.g., efficiency). The former happens in the short term (weeks to years), the latter over longer periods of time (years to decades).

As suggested by Julian Simon, another way to contextualize oil price and expenditures is to compare them to some benchmark. He preferred to consider cost (or price) relative to wages and consumer price index (CPI) [85], and Fig. 3.1c shows oil price relative to both of those as well as total U.S. disposable income.[44] The

---

[44]U.S. Disposable income (defined for all Americans) is equal to total income minus personal taxes.

overall trend is that oil price fell relative to after-tax income (i.e., disposable income) as well as overall goods and services (i.e., per CPI) throughout the twentieth century but with notable periods of increases during the world wars, Great Depression, and 1970s oil shocks. Of course, these periods of higher prices were not independent of oil or other resource access. For example, Japan's bombing of Pearl Harbor, that ultimately compelled the U.S. to enter World War II, was driven by U.S. action to slow Japan's military advances in Asia by limiting Japan's access to oil via U.S. exports (by freezing Japanese financial assets) and the oil fields of the East Indies [99].[45]

The trend for oil price relative to *hourly wages* contrasts to those relative to CPI and income, but it is the same if dividing oil price by *median income*. Oil price divided by the average hourly wage for "Production and Nonsupervisory Employees" is higher after 1974 and lower before 1974. Notably, oil expenditures divided by disposable income follows the oil price trend relative to wages before 1974 while it then follows the price trends relative to disposable income and CPI after 1974. What this means is that before the 1974, disposable income of the U.S. overall grew at about the same rate as hourly wages of workers. After 1974, the income of Americans *overall* grew faster than hourly wages.

Before 1974 Americans consumed increasing quantities of oil at roughly constant prices.[46] After 1974, Americans consumed approximately a constant amount of oil at prices that varied from 15 to 140 $2015/BBL.[47] Figure 3.1d more clearly shows this structural change after 1974 in terms of how the U.S. economy consumed oil. Before 1974 the trend is approximately a horizontal line (constant price, increasing consumption). After 1974, the trend is approximately a vertical line (constant consumption, varying price). I'd like to think that today, as opposed to 1978, Milton Friedman would clearly acknowledge the 1973 breakpoint in the trends of U.S. oil consumption rate and price. It is difficult to show data with a more stark change, and it is important to consider the ramifications. Chapters 4–7 discuss various ways we can view the linkage between the unequal distribution of income and money within the U.S. economy as compared to the rate and cost of energy consumption.

As of this writing, the full context of the post-2000 oil price and expenditures trends are unclear. U.S. oil prices from 2005–2014 were all above 60 $2015/BBL, but the price in 2015 and 2016 was 44 and 38 $2015/BBL, respectively. Oil prices since 2016 have stayed above 45 $2015/BBL. However, it is not yet obvious if oil expenditures (relative to GDP or disposable income) can decrease any further going forward. A good bet for the next few decades is to assume that U.S. and Western economy oil spending relative to GDP will average approximately 2–3% of GDP with excursions above that range relating to poor economic growth or even recession. This approximately constant spending share of GDP occurs due to

---

[45] Yergin [99, pp. 309–314; 316–319].

[46] Oil consumption was 2 billion BBL/year in 1949 increasing to 6 billion BBL/year in 1973. Oil prices varied from 15 to 21 $2015/BBL.

[47] Annual oil consumption varied from 5 to 7 billion BBL/year.

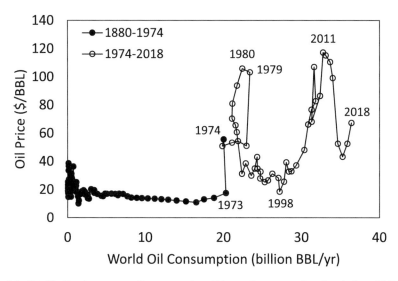

Fig. 3.2 World oil price versus oil consumption. Price and consumption data before 1965 are from Fizaine and Court (2016) [34] as taken from Etemad and Luciani (1991) with oil production assumed as consumption. Data starting in 1965 are from BP Statistical Review [12]. Prices are expressed in 2015 U.S. dollars per barrel

responses that increase oil efficiency/substitution and reduce services traditionally fueled by oil. Of course, what we'd really like to know is just how much of each of these responses will occur. That is a much harder question.

Is the U.S. economy no longer increasing oil consumption because Americans don't want to, can't afford to, or already feel rich enough?

I briefly address this question by showing global data of oil price versus consumption in Fig. 3.2 to compare to the U.S.-specific version in Fig. 3.1d. The main difference is that while the U.S. has not consumed oil at a higher rate since 1973, the world overall has while exposed to essentially the same oil prices. The reason this is the case is that the vast majority of the world population in developing countries did not have as much oil-consuming capital before 1973. Since they were starting at a low economic level, they were able to grow their economies by consuming increasing quantities of higher-priced oil, yet still reside below the level of consumption and economic development in the U.S. and other OECD countries. The "rich countries" already had a large number of low-efficiency cars, trucks, and aircraft to replace with higher-efficiency versions. But that switch to focusing on efficiency takes time, and it seems to have possibly had a side effect on wages. As hinted by the trend of oil price relative to wages in Fig. 3.1c, and as we'll learn further in Chaps. 4 and 6, the 1970s signaled a stagnation in wages to most Americans such that they just didn't have the money to buy as much oil as they used to.

Let's now switch to an example price trend for renewable energy, that of solar photovoltaic panels.

Price: Solar Photovoltaics

[O]ne of the big differences between today and a decade ago is that we do have the solutions now. It is remarkable now that solar electricity and wind electricity have followed the pattern that we have seen with computer chips, and mobile phones, and flat screen TVs. Some areas of technology come down in cost and then when production scales up they come down even faster in cost. And it's wonderful that that pattern is being seen in renewable energy.[48]—Al Gore, National Public Radio interview (July 24, 2017)

The cost and functionality of PV panels have certainly come a long way since the discovery of the photovoltaic effect in 1839 and the construction of the first PV cell in 1877, with a conversion efficiency of only 1–2%. This is similar to the efficiency that plants and trees convert sunlight into biomass. The practical reality of solar PV cells only emerged in 1954 when researchers at Bell Laboratories produced silicon solar cells at 6% efficiency.[49] Today, several PV cell designs convert over 20% of sunlight into electricity.

With higher efficiency comes lower cost. In his quote above, Al Gore invokes a common concept regarding the declining price of wind turbines and photovoltaic (PV) panels. In 2017 the U.S. Department of Energy declared victory 3 years before the target date for its SunShot goal of enabling utility-scale photovoltaic installations at less than 1 \$/W and 0.06 \$/kWh [37]. Figure 3.3 shows how rapidly both the PV module price and the installed cost of PV systems have decreased over the last 40 years. Thanks China and globalization!

While Japan and the U.S. were the main pioneers in the development of PV technology, China and other southeast Asian nations are the dominant manufacturers of PV today. In 2015, China made 65% of PV cells, Taiwan 14%, and Malaysia 6%, accounting for 85% of worldwide PV cell manufacturing. Japan, Germany, and the U.S. accounted for 4%, 2%, and 2%, respectively. On top of PV *cell* manufacturing, China, Japan, and South Korea manufacture 69%, 5%, and 5% of the PV *modules* that house individual cells in a protected unit [51].

The decline in PV installation cost has occurred across the supply chain from the cells, modules, other equipment (such as inverters), and business costs such as labor and profits. And because the supply chain extends across the globe, globalized trade has enabled these cost reductions. Germany's policies with lucrative incentives for generating renewable power, such as the feed-in tariff started in the 2000s, were meant to incent both local solar manufacturing and electricity generation from local installations of solar panels. However, almost all of the PV manufacturing that launched in Germany has shut down due to lack of cost-competitiveness. To date, the roads leading to cheap solar panels are trans-Pacific shipping lanes.

To pure proponents of the renewable energy narrative, cost curves such as these in Fig. 3.3 for solar PV are all the evidence they need to say "...we do have

---

[48] http://www.npr.org/2017/07/24/538391386/despite-climate-change-setbacks-al-gore-comes-down-on-the-side-of-hope, accessed September 3, 2017.

[49] Data in this paragraph from Smil [87, pp. 255–256].

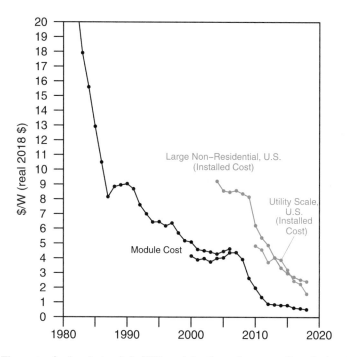

Fig. 3.3 The costs of solar photovoltaic (PV) modules themselves, as well as the installed costs for PV systems, have dropped rapidly over the last several decades. (black lines) PV module prices: data 1975–2006 from Earth Policy Institute (2007) [25], and 2000–2018 data from Lawrence Berkeley National Laboratory Tracking The Sun, 2019 Edition [6]. (red line) The U.S. installed cost of large non-residential PV in $/W DC (direct-current). (gray line) The U.S. installed cost of utility-scale PV in $/W AC (alternating-current) from Lawrence Berkeley National Utility-Scale Solar, 2019 Edition [10]

the solutions now." To many pro-renewable and/or agnostic system-wide thinkers, affordable PV and wind power are no doubt crucial components in the suite of solutions, but not the entire story, as I discuss in the last section of this chapter.

Price—Summary

Whether it be dollars per gallon of gasoline in your car ($/gal) or cents per kilowatt-hour of electricity consumed at your house (¢/kWh), these energy prices provide important information to consumers. While they do not tell the full story, they give many hints as to the activities of energy consumers and producers.

The oil price history shows that, before the 1970s, it was easy to argue that fossil prices were low and, aside from hydropower generation, renewable electricity prices were high. After all, practical solar photovoltaic designs didn't exist until the 1950s. Since 1974 the oil price has averaged about three times higher than the previous

90 years, and it became much more volatile. The start of this post-1973 trend was influenced by the peaking (at the time) of U.S. oil extraction, the doubling of oil prices by OPEC in 1974, and the reduction in oil supply from the 1979 Iranian Revolution.

Not until the 2000s did renewable electricity costs from wind and solar systems become low enough to make an impact, even with government price support. Today major wind turbine manufacturers exist in the U.S., Europe, and Asia and the lifetime cost of wind electricity is cost-competitive without government support in regions with good wind resources. But critically, over the last decade, due to mass manufacturing in China and Southeast Asia, solar photovoltaic panels now also produce cost-competitive electricity over the life of the panel.

Timing matters. The costs during the time when a company invests in a new energy project affect which project it chooses. Twenty years ago, wind turbines, solar panels, and horizontal drilling with hydraulic fracturing were just showing up on the appetizer menu. Today, they are all part of the main course (although for fracking, so far only in the U.S.)

But timing matters in another important sense. For how long does the oil or gas come out of a well? How much electricity is driven by the wind and sun over the course of the next year, month, day, hour, or even the next few minutes and seconds? This type of timing is a major point of contention between the energy narratives. How much do we need to know about when energy is extracted from the environment in order to call it *reliable*?

The Reliability Discussion: It's About Time

Reliability A word thrown around discussions of energy much too loosely. Reliability does not mean the same thing to all people. To a lobbyist advocating for the fossil fuel narrative, reliability means one thing. To a lobbyist advocating for the renewable energy narrative, it means another. To operators responsible for maintaining the day-to-day technical operation of the electric grid, reliability has additional meaning that few people contemplate. Depending upon the point someone wants to make, he might bolster his argument by describing only the reliability aspects that promote one energy narrative. Very rarely does a single person describe reliability in the context of both narratives.

One can take a philosophical view of reliability:

What Is Reliable?

Prediction is always a leap of faith; there is no scientific guarantee that the sun will come up tomorrow.[50]—Julian Simon (1996)

One can take a practical view of reliability:

---

[50]Simon [85, p. 30].

We Need Reliable Energy

Wherever access to reliable, affordable energy goes up, so does the quality of life. ... It is hard to overstate the impact that clean, affordable, reliable energy will have. It will make most countries energy-independent, stabilize prices, and provide low- and middle-income countries the resources they need to develop their economies and help more people escape poverty.[51]—Bill Gates (2015)

One can use reliability to support the fossil fuel narrative:

Fossil Fuel Energy Is Reliable

Energy is the basis of modern life and the engine of American innovation. Fossil fuels— our most abundant, reliable, and affordable energy sources—are key to ensuring Americans continue to prosper and thrive.[52]—website of Fueling U.S. Forward (2017)

One can use reliability to support the renewable energy narrative:

Renewable Energy Is Reliable

People talk about fusion and all that, but the sun is a giant fusion reactor in the sky. It's really reliable. It comes up every day. If it doesn't, we've got bigger problems.[53]—Elon Musk (2017)

One can use reliability to undermine the renewable energy narrative:

Renewable Energy Is Not Reliable

In fact, *modern solar and wind technology do not produce reliable energy, period.*[54]—Alex Epstein (2014) (emphasis in the original)

And last, but not least, one can use reliability to undermine the fossil fuel narrative:

Fossil Fuel Energy Is Not Reliable

We "progressed" from [the sun] an external, reliable, and constant source of energy to one [fossil fuels] that is internal, unreliable, and variable.[55]—Geoffrey West (2017)

From the philosophical to the practical, the concept of reliability provides much fodder for the energy narratives. While little orphan Annie tells you to "bet your

---

[51]"Energy Innovation Why We Need It and How to Get It", white paper downloaded from https://www.gatesnotes.com/Energy/Investing-in-Energy-Innovation on September 3, 2017. Direct url is https://www.gatesnotes.com/-/media/Files/Energy/Energy_Innovation_Nov_30_2015.pdf?la=en&hash=6EAE95501FB01629D4817599F0636B0CFB12378B.

[52]https://fuelingusforward.com/ accessed August 2, 2017.

[53]Per Elon Musk's address at the 2017 Summer meeting of the National Governors Association. "Here's Elon Musk's Plan to Power the U.S. on Solar Energy", https://www.inverse.com/article/34239-how-many-solar-panels-to-power-the-usa accessed February 3, 2020.

[54]Epstein [31, p. 48].

[55]West [96, p. 236].

bottom dollar that tomorrow there'll be sun," Julian Simon says there is no scientific guarantee the sun will come up tomorrow. Simon's statement is that of the example from philosopher David Hume who claimed that, using deductive reasoning alone, we are not justified to believe the sun would come up tomorrow. To Hume this lack of belief in tomorrow's sunrise is warranted because it is based on observing only part of the past (since we can't observe everything) and the assumption that future events obey the same laws as past events.

But science is not based solely on logic and deductive reasoning. Science is based on observing the past and using these observations as a basis for predicting future events, such as that the sun will come up tomorrow (or rather that the sun will keep shining and Earth keep rotating). This is inductive reasoning. "We rely on inductive reasoning in arriving at beliefs about what we have not observed, including, most obviously, our beliefs about what will happen in future. ... Science is heavily dependent on induction. Scientific theories are supposed to hold for all times and places, including those we have not observed. ... the only evidence we have for their truth is what we have observed."[56]

While the energy narratives sometimes philosophize for their view of the future, most discussions of energy reliability are more practical. This section considers the practical reliability of the generation of electricity from fossil and renewable energy technologies and their integration within system-wide operation of the electric grid. This is a common context in which people argue over the reliability of fossil, nuclear, and renewable energy technologies, and it also important because many analysts see increased electrification of the energy system, for example, electric vehicles, as a trend for the foreseeable future.

Merriam-Webster defines reliability in two contexts which are both useful for understanding issues regarding the function and operation of the electric grid:[57]

1. **reliability:** "the quality or state of being *reliable*," where *reliable* is "suitable or fit to be *relied* on," and where *rely* is "to be dependent."
2. **reliability:** "the extent to which an experiment, test, or measuring procedure yields the same results on repeated trials."

Thus, *reliability* in the first definition describes an object, person, or system on which one can be dependent that a certain function or action will occur as we expect. This definition relates to the purpose and function of the electric grid in our society, regardless of the elements that make up the grid itself. Thus, if the purpose of the grid is to provide electricity to homes and businesses 99.9% of the time, then fundamentally we shouldn't care if the power generation mix is 100% fossil or 100% renewable as long as either system suits that purpose with all other factors (e.g., cost, environmental impact) the same.

In the second definition, *reliability* describes whether or not we can actually design a system such that some measured quantity is the same each time we run a

---

[56] Blog of Stephen Law, "Problem of Induction explained simply ... (from my book The Philosophy Gym)," October 12, 2012 at: http://stephenlaw.blogspot.com/2012/10/problem-of-induction-explained-simply.html.

[57] https://www.merriam-webster.com/dictionary, accessed July 18, 2017.

test. This definition relates to how the electric grid actually operates, without regard to its purpose to society. In effect, the electric grid is a complex machine in which we continuously run an experiment at each instant to generate electricity equal to the demand from all of us turning on our lights, air conditioners, and toaster ovens.

Let's see how proponents of the energy narratives talk circles around each other. Consider the first definition of reliability. Why might we rely on electricity? I might rely on the electric grid for the purpose of delivering electricity to my home to power an air conditioner that keeps my house cool, a refrigerator that keeps my food fresh, and a television or computer that I use to receive information and watch movies for entertainment. While not having electricity to watch a movie is an inconvenience, other consumers of electricity have health and safety priorities that most of us would consider more important than movies. For example, hospitals use electricity to power devices that keep people alive (e.g., artificial lung breathing machines) and run time-critical tests that enable doctors to diagnose and treat patients before their injuries or diseases overcome them. Other electricity consumers are companies that provide salaries for employees and operate stores where we shop, restaurants where we eat, and factories that make intermediate and final products from raw materials. There is even the need for electricity to power the control rooms of the electric grid operators such that they can manage the electric grid itself.

Each of us can think of the priority of these purposeful uses of electricity. To me, the highest priority would be electricity for hospitals and the electric grid operators, and the lowest would be for electricity in my home, even though there are health concerns if a person lives in a home that is too hot or too cold. In fact, because maintaining the function of facilities such as hospitals and grid control rooms is such a high priority, they usually have significant backup systems, directly at their premises, that provide electricity that allows them to continue functioning even if the electric grid stops operating.

That brings us back to the second definition of reliability. This definition relates to how the electric grid actually operates and is managed on time scales that range from milliseconds to decades. Each of several time scales has different implications for planning and operating the grid, and, thus, each time scale has direct relevance to what the fossil or renewable narrative proponents have to say about technology reliability.

In short, *it's about time*.

When it comes to reliability, the energy narratives often clash because they discuss different times spans.

To explore this idea, I'll ask the following question: "What is the *time scale or time span* at which an experiment, test, or measuring procedure yields the same results on repeated trials?" Let's consider the context of the electric grid and go from shortest to longest time scale (see Fig. 3.4).

Less Than Seconds: Physics
At time scales less than 1 s, changes occur so fast that it is unreasonable to expect human decisions and markets to intervene. The dynamics and solutions here are the domain of physics and automatic engineering controls, not politics, environment, or markets.

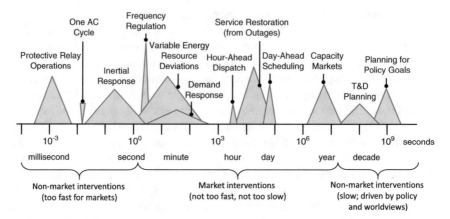

Fig. 3.4 Depending upon the time scale, there are different operations that are needed to maintain reliability of the electric grid. Figure adapted from [23] and [22]

Seconds to Minutes: Regulation

At the time scale of seconds, there are adjustments on the grid to maintain the *frequency* of the grid [29]. In North America, the electrons flowing in the grid alternate direction at a frequency of sixty times per second, or 60 Hertz (Hz).

There is some deviation allowed from the targeted frequency, but not very much. If the grid frequency is a little high, it means there is a little bit too much power generation than there is demand for power, and vice versa. If the grid frequency is a little high, the grid operator tells one or more electric generators to generate less. If the grid frequency is a little low, the grid operator asks for increased power generation. The control of these small up and down commands is called *regulation*. Because humans can't act fast enough to observe the state of the grid and tell power generators what to do every second, grid operators have an automatic control system that provides these regulation commands to power plants every one or few seconds in order to maintain the grid frequency. Only generators and energy storage devices that are currently operating can provide regulation services (i.e., the generator has to be "on"), and their owners can be paid for these services.

Minutes to Hour: Load Following

At the time scale of minutes to perhaps a little over 1 h, there are adjustments on the grid to perform what is called *load following*. Load following is required when the actual electricity demand "right now" is different than what was projected during planning and scheduling the day or hours before.

Grid operators have many rules and processes for performing load following. They hold what are called *spinning* and *non-spinning reserves* of power generation that can ramp up to deal with unexpected increases in power demand. Here, the term *ramp* refers to the action of an electricity generating unit to relatively quickly (e.g., in 5–30 min) increase (ramp up) or decrease (ramp down) from its current power output.

Several factors explain why the actual electricity demand is different from the demand projected the day before. A major factor is the weather. Cold fronts, clouds, and changes in wind patterns are difficult to predict every 30 min to a few hours. Not only does the weather affect variable renewable energy generation (e.g., wind and solar), but it also affects load. A hotter day than projected can drive more air-conditioning load, and a colder day can drive more heating.

The more uncertainty there is in projecting both power demand and power generation from variable renewable energy technologies, the more operational reserves are needed to enable adjustment to real-time conditions. These same operational reserves are needed to handle uncertain outages from dispatchable power plants such as nuclear, coal, natural gas, and hydropower. For example, an unexpected event can cause a large capacity nuclear or coal power plant to go offline in less than 1 s. These events can be natural disasters, terrorist acts, or some unforeseen internal breakage that causes the power plant to automatically "trip" offline and begin a shutdown process. Grid operators plan for this type of contingency, and one approach specifies a *non-spinning reserve* that ensures there is enough excess generation capacity available, but not yet generating power, to come online quickly (e.g., in less than 30 min) in case the largest power plant on the grid trips off for any reason. The largest power plant on the grid is usually a coal or nuclear facility, often rated at over 2 gigawatts (GW) of capacity. Thus, grid operators maintain a spare capacity equal to the capacity of the largest generator. This spare capacity waits, while generating no power, just in case it is needed.

These additional reserves that provide the ability to respond to uncertain conditions require more investment that increases the total cost of the grid. Hence *the costs of reserves for handling increased quantities variable renewable generation from wind and solar plants is one of the major points of contention for fossil fuel proponents. These same proponents rarely bring up the need to back up large fossil and nuclear generators as well* using non-spinning types of reserves.

With the increasing use of electric meters and appliances that are connected to the Internet, or controlled otherwise, grid operators are increasingly making use of *demand response* to address minute-to-hour mismatches in electricity supply and demand. From the operational engineering standpoint, there is equal benefit to increasing power generation as there is decreasing demand. Both can help maintain grid reliability.

Hours to Days: Scheduling

At the time scale of hours to one or more days, grid operators perform what is called *unit commitment*. This unit commitment is the *scheduling of electric generation units* in terms of which ones turn on, turn off, ramp up, and ramp down over the course of one or more days. Depending upon the regulatory and market structure that governs the operation of the electric grid, the unit commitment schedule is defined by 5–30 min increments. The market-based priority to unit commitment is based upon operational costs only. Those with the lowest operational costs are scheduled first. Operational costs are driven by items such as fuel cost and employees that work at the power plant.

Included in this unit commitment scheduling is information on normal diurnal patterns of electricity demand and generation from renewable energy technologies such as wind and solar power. For example, in Texas the peak electricity demand is in the summer afternoons when air-conditioning demand makes the afternoon power draw approximately twice as high as the minimum power demand for the day. Thus, the grid operator must schedule more power capacity for the afternoon than the early morning. In more northern climates, peak demand for electricity occurs due to heating demand in winter months. In addition to scheduling for load, information from weather forecasts is used to estimate day-ahead generation from both wind and solar generation, and of course the position of the sun in the sky is used to inform an estimate of solar power generation for the next day.

Days to Year: Maintenance and Revenue
At the time scale of multiple days to a year, grid operators and owners of electric generation units consider when they expect generation units to be available for operation. Power plants need regular maintenance to operate efficiently and safely, and this maintenance is scheduled on a regular basis. Power plant owners typically perform this maintenance during times of low demand such as in the spring when weather is neither hot nor cold. The exchange of fuel rods in nuclear power facilities also occurs during such times. In addition to maintenance, power plant owners with high operational costs might schedule to have their power plants unavailable during times of the year when power demand is low because the power plant would not be dispatched enough to have revenues exceed operational costs.

Years to Decades: Investment
At the longest time scale of years to decades, electric utilities, regulators, power plant investors, environmental advocates, and many other possible stakeholders consider *investment* decisions. These stakeholders must consider what power plants and other infrastructure to build and retire such that consumers have electricity for the social purposes (our first definition of reliability) and infrastructure and power plant owners are financially sound. Whatever uncertainties exist at grid operational times scales of seconds to days, these are influenced by investment decisions that have a lag time of several years. That is to say, once a utility or investor commits to invest in a power plant, it might take several years until the power plant is actually in operation. It takes decades to pay for it.

The difficulties in deciding what and how much investment to put into any type of technology are enhanced because the profitability of one investment can be affected by the investments, consumption choices, and regulatory changes of the various actors. Further, all stakeholders rarely agree on what power plants should be built. Because of this, political goals drive power plant investments as much as economic criteria. If the cost for a given technology is not low enough, then monetary incentives, or subsidies, of one type or another are often used. Many electricity stakeholders complain that political, or "out-of-market" policies are ruining the power markets. But power plant investment decisions are not driven by pure-market criteria. They never have been, and likely they never will.

Arguments abound on the regulatory structures that should govern grid investment and operation. There are two main type of relationships between consumers and power generators: regulated utilities and deregulated, or restructured, markets. First, regulated utilities. In the case of investor-owned utilities (IOUs), they are guaranteed a rate of return on the power plants they own, and each investment is approved by state public utility commissions. Electricity customers within these IOU territories cannot choose to buy electricity from another company. These *monopolistic* arrangements are the same business model set up at the beginning of the electricity industry in the U.S., and they continue in some regions of the U.S. today [93]. Regulators adjust the electricity rate ($/kWh) for electricity sold by the utilities to meet the regulated rate of return on capital investments, usually 8–10%.

However, there are limits as to how much regulators are willing to guarantee that consumers ultimately pay for all investment. A good example is the Kemper power plant owned by Southern Company's Mississippi Power Company subsidiary. This power plant was to be a state-of-the-art facility that captured the carbon dioxide emissions emitted from burning the coal. After severe cost overruns and years of delays in getting the power plant fully functional, in 2017 the Mississippi Public Service Commission said the power plant should run as a natural gas plant instead of as intended [91]. This meant Southern Company and its shareholders had to absorb several billion dollars of losses, and customers are not on the hook for the vast majority of the cost overruns [8].

An even more extreme case relates to cost overruns on new nuclear power plants. Staying in the southeastern U.S., the 2-reactor expansion of the V.C. Summer nuclear power plant in South Carolina was canceled halfway through construction after the main contractor, Westinghouse Electric Co., was forced into bankruptcy in March of 2017. The projected cost to finish the expansion rose from the original $14 billion to almost $26 billion. As stated by the chairman of one of the utilities that commissioned the project, "The costs of these units are simply too much for our customers to bear."[58] [8] As of late 2019, Southern Company was still planning to finish the half-completed Vogtle nuclear plant expansion of two reactors in Georgia that use the same Westinghouse design as meant for V.C. Summer. In doing so, the company became the project manager.

At the other extreme of the investing spectrum, there are restructured, or deregulated markets. These markets are set up to provide a price signal for power plant investment decisions. The investors get rewards from profits and any government incentives. They also bear the risk if the power plant does not make enough money. Just as regulated utilities can invest too much, so can private companies competing to sell electricity in the open market. Examples of bankruptcy from investing in the electricity market range from a single power plant to an entire company, and I describe an example of the latter in the penultimate section of this chapter [63].

---

[58]"Scana stops work on partially built V.C. Summer Nuclear Station as estimates of cost to finish it rise to $25 billion," *Wall Street Journal* https://www.wsj.com/articles/scana-halts-south-carolina-nuclear-power-project-1501524763 (accessed July 31, 2107 4:02 pm ET).

Table 3.1 Operational and/or forecasting reliability of the quantity of electricity generation from a given type of power plant at different time scales of interest

Technology	Time scale					
	Seconds	Minutes	Hours	Days	Years	Decades
Coal	High	High	High	High	Medium	Medium
Nuclear	High	High	High	High	Medium	Medium
Natural gas	High	High	High	High	Medium	Medium
Biomass	High	High	High	High	Medium	Medium
Hydropower	High	High	High	High	Medium	Medium
Wind	High	Medium	Medium	Low	High	High
Solar	High	Low	Medium	Medium	High	High
Geothermal	High	High	High	High	Medium	Medium
Electricity demand	High	High	High	Medium	Medium	Low

Now that we understand the various time scales that are of interest to maintain reliable operation of the electric grid, and the reasons for considering these time scales, we can consider how well each type of power generation helps or hinders electric grid operational reliability in the context of the narratives we are told.

To sift through the narratives, let's take a system-wide perspective. Table 3.1 summarizes my comparison of how each major power generation technology addresses grid operational reliability (the second definition of reliability) at each time scale of interest. Thus, my "test" for filling out the table is to ask this question: "In the next second/minute/hour/day/year/decade, can I accurately predict the amount of power generation from a given electricity technology?" To stay at a high conceptual level, I only use general categories of low, medium, and high reliability.

For example, consider solar electricity. We can be highly confident as to the next second of solar power with decreasing confidence for the next several minutes (e.g., is a cloud nearby or not) but fairly predictable the next few hours (e.g., will clouds form later today). Over the course of decades, we expect the sun to shine at approximately the same intensity, the Earth to spin at the same rate such that diurnal patterns of sunlight are consistent, as the Sun, Earth, and Moon continue their celestial dance as predicted by astronomers (e.g., we can predict solar eclipses). This is the gist of Elon Musk's quote at the beginning of this section. It is the medium time scales of hours to days that are the least predictable for a collective set of wind and solar power plants. Their electricity output is not completely unpredictable at medium time scales, just less so than over the course of the next few minutes and the next 12 months.

Next consider dispatchable fossil, nuclear, hydropower, and geothermal generation. We have high certainty in knowing if we can generate power from them at short time scales, but the longer time scales are less certain. At short time scales we know if they have fuel and if they are ready to operate. At long time scales we do

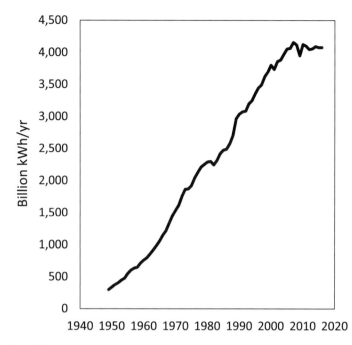

Fig. 3.5 Total U.S. annual electricity generation increased at an approximate constant linear rate until 2007, after which it remained relatively constant for 10 years [EIA Monthly Energy Review Table 7.1]

not know if they will be economically viable in the future due to increased fuel costs or new regulations.

Finally, we cannot discuss electric grid reliability without considering the electricity demand. At short time scales there is high certainty in the daily patterns of electricity usage. Over the course of days and weeks the uncertainty increases due to lack of predictability of the weather. Finally, at the longest time scales, we do not know what will be the demand for electricity. The data in Fig. 3.5 are evidence for the high uncertainty in predicting long-term electricity demand. For practically the entire history of the electricity industry, electricity generation increased at a nearly constant rate. Pivoting about 2007 with the onset of the Great Recession, this trend was broken. Since 2007, U.S. electricity generation (due to flat demand) has remained nearly constant at about 4000 billion kilowatt-hours per year.

Note that even if we can reliably predict electricity output from a given technology, that reliable prediction might not be an outcome of high power generation. For example, hydropower reservoirs store water for multiple purposes such as drinking water and irrigation supplies. Thus, a low water storage situation might necessitate that very little water is released for power generation, and hence we could reliably predict low or no power generation for days to months. Similarly, when power plants are off during regular maintenance, we can reliably predict no power generation.

Let's now consider some quotes from both pro-fossil and pro-renewable advocates that promote their narrative or trash the other.

Reliability: The Intermittency Argument (Narratives at Short Time Scales)

Intermittency The word describes a characteristic of alternately stopping and starting again.[59] As such, it is a word often part of narratives explaining how wind and solar power technologies are unreliable at short times scales of seconds to days. Consider the following quotes:

> We know from experience that the sun doesn't shine all the time, let alone with the same intensity all the time, and the wind doesn't blow all the time—and leaving aside the assurance that the sun will be "off" at night, they can be extremely unpredictable. ... That's the real problem—the intermittency problem, or more colloquially, the unreliability problem.[60]—Alex Epstein (2014)

> Wind energy whether it's being produced in the U.S., the U.K., or China is a scarce, expensive, unreliable, resource. And the more wind you have the more problems you have.[61]—Alex Fitzsimmons (2014)

Given that I spent some considerable text relating the definitions of the word *reliability* to the context of the electric grid, you can imagine that the above quotes are not quite nuanced enough for me. Here Epstein and Fitzsimmons essentially equate intermittency to unreliability, and this is too simplified for a reasonable discussion.

We know from experience that the sun does not turn on and off. It is the spinning of the Earth that Epstein wants us to think about, but instead he tells us that the sun turns on and "off" (yes, he does put on quotes so as not to imply that Sun literally turns off) as if we live on a flat Earth. In fact, if the sun stopped shining, we wouldn't have to worry about how we want to produce energy for our economy because life on Earth would cease as we know it! And while there is variation in the solar irradiance, it typically varies less than 0.1% and has been estimated to vary

---

[59]*Intermittent*: (1) coming and going at intervals: not continuous, (2) appearing and disappearing seasonally (Merriam-Webster.com). *intermittent*: stopping or ceasing for a time; alternately ceasing and beginning again (Dictionary.com).

[60]Epstein [31, p. 50].

[61] Appointed by President Trump in 2017 as a senior adviser appointed to the U.S. Department of Energy's Office of Energy Efficiency and Renewable Energy, former employee of Fueling U.S. Forward (https://fuelingusforward.com/ and policy director for the Institute for Energy Research from 2013 to 2016 [71]. Quotes from BBC Interview January 8, 2014 accessed July 11, 2017 at https://www.youtube.com/watch?v=OlX5ognB9sA.

by up to 0.5% (over centuries) as inferred from tree ring data.[62] This quantity of variation is unimportant for estimating power generation from photovoltaic panels, but it can be relevant for understanding the temperature on Earth.

These properties of the sun are what motivates quotes, such as those of Geoffrey West and Elon Musk at the beginning of this book section, that state the exact opposite of Epstein and Fitzsimmons in terms of whether solar or fossil energy is more reliable and constant.

Those that promote renewable energy tend to use the term *variable* instead of intermittent because it provides the connotation that we know much about how wind and sunlight change over time for any given location on Earth.[63] Most literature that seriously discusses grid management and integration of wind and solar power also uses the term variable (for example, see [18]).

In his book, Epstein shows example wind output, solar PV output, and demand data for Germany on time scales of 15-min, days, and each month of a year [31]. He uses these to show the patterns are not uniform and that the wind and PV output do not exactly match the demand. For anyone trying to seriously consider the increased use of wind and solar power, Epstein's revelations are essentially givens, not showstoppers. That is to say:

> to proponents of the fossil fuel narrative, the characteristics of wind and solar generation are insurmountable annoyances. To proponents of the renewable energy narrative they are the ultimate engineering and policy challenge worthy of their full effort.

Importantly, intermittency arguments usually avoid one fundamental issue related to assessing the reliability of wind and solar compared to fossil and nuclear power: time scales. Think about my test for reliability and Table 3.1. The times scales referenced by Epstein are the most difficult to forecast accurately and thus pose challenges for integrating wind and solar power within the electric grid. These are also the time scales at which generation from dispatchable generators (thermal power plants) is most reliably predictable. To date, at low penetrations of wind and solar, the challenges have been manageable, and we can expect the need to augment power grid management, markets, and investment (e.g., electricity storage) as variable renewable power increases as a fraction of total generation.

However, time scales of minutes to months are not the only time scales of interest for society. One advantage of wind and solar technologies is that we're very confident that the fuel source, which is ultimately the sun, will be shining for centuries into the future. Further, for any given location, we expect the amount of

---

[62]"The intensity of the Sun varies along with the 11-year sunspot cycle. When sunspots are numerous the solar constant is high (about 1367 W/m^2); when sunspots are scarce the value is low (about 1365 W/m^2). Eleven years isn't the only "beat," however. The solar constant can fluctuate by 0.1% over days and weeks as sunspots grow and dissipate. The solar constant also drifts by 0.2–0.6% over many centuries, according to scientists who study tree rings." https://science.nasa.gov/science-news/science-at-nasa/2003/17jan_solcon, accessed September 3, 2017.

[63]*Variable*: able or apt to vary; subject to variation or changes "variable winds" (Merriam-Webster.com). *variable*: apt or liable to vary or change; changeable (Dictionary.com).

sunlight and the amount of wind to be about the same year after year. They won't be exactly the same every year, but we know that for decades windy locations will stay windy and sunny locations will stay sunny.

Epstein notes that "... the wind doesn't always blow ..." Practically, the wind is always blowing somewhere on Earth. However, barring a single worldwide electric grid connecting all people to windy locations, we practically consider continental scale electric grids. For example, researchers have shown that, due to low wind conditions, any one wind farm in the U.S. might have two or three thousand hours per year (out of 8760 h in 1 year) when the power output is less than 15% of its maximum capacity. However, if you integrated wind farms across nine locations that span the continental U.S., you would have about one thousand of hours per year generating less than 15% of capacity [43]. Using thresholds of 5% and 1% of generating capacity instead of 15%, the nine wind locations analyzed in [43] would generate below those thresholds for only 36 and practically 0 h per year, respectively. Thus, the wind is always blowing somewhere, but to take advantage of that you have to connect these disparate locations with transmission lines, which increases costs. Keep in mind that over the course of any given year, a typical wind farm generates electricity at about 30–40% *capacity factor*. The capacity factor of a power plant is the percentage of electricity actually generated relative to the amount of electricity generation if the power plant operated 100% of the time at its maximum power output.

Wind and solar power output variability is an important factor for power grid design and operation, but the goal of fossil fuel proponents is to promote fossil fuels rather than be objective. Over periods of time more than 1 year, we have confidence in predicting the *amount of energy* produced from these technologies. Over periods of time from minutes to months, we have less confidence in predicting the *amount of power* generated at any given instant. The distinction between power and energy is key for understanding the impact of wind and solar electricity variability at time scales less than 1 year. The difference between energy is time, and the purpose of energy storage is to shift power across time.

Reliability: The Storage Argument (Narratives at Intermediate Time Scales)

Energy Storage The concept of storing energy at one time for discharge at a later time is embedded in both energy narratives. Energy storage is a practical necessity to power our industrialized world entirely by renewable energy flows of wind, water, and sunlight. However, what is often missed, is that energy storage has already been critical for fossil, biomass, and nuclear power plants. The truth is that our electric grid has historically operated with complete dependence on energy storage. The challenge for operating a 100% renewable energy grid is to see how little, not how much, storage we must have.

Energy storage describes a *stock* of *potential energy* at our beck and call. These stocks grow over some period of time by accumulating a *flow* of inputs. This time period of accumulation ranges from seconds to millions of years. An electrochemical battery or capacitor in the transmission and distribution grid, or even your home, might charge and discharge over periods of seconds to hours. Fossil fuels are stocks of energy that accumulated flows of biomass millions of years ago that were buried and cooked into coal and hydrocarbons.

Historically, when we have operated the electric grid we do so by converting energy stocks, as stored potential energy, into a flow of energy as electricity. Some energy storages are the piles of coal sitting beside power plants, the natural gas compressed in a network of pipelines and underground caverns, and the water sitting behind a dam. For example, to handle fluctuations in demand between winter and summer, the natural gas system in the U.S. stores a quantity of gas equal to about one-quarter to one-third of annual consumption.[64] Even the uranium fuel rods in nuclear power plants are highly refined stocks of enormous quantities of potential energy.

Thus, a power grid operated solely by coal, natural gas, nuclear fission, and hydropower is a grid that operates with 100% of its electricity derived by inputting fuel stored at or near the power plant. This energy storage is relatively cheap, and it doesn't involve exotic technology (aside from uranium fuel refining). The reason fossil fuel storage is cheap is because nature has done all the work for us. Fossil fuels, stored as gases, solids, and liquids are free in the natural state. We only have to pay to get them to the Earth's surface where we use them. Further, we convert these energy stocks into electric power at our command. If we need more power, we input coal and natural gas into the power plant at a faster rate. If we need less power, we slow the rate of fuel input.

Now consider what happens if we install one wind turbine on the grid. When that wind turbine generates power it composes a very small fraction of total grid power that is no longer generated by converting a stock of energy into power. The wind turbine is converting the instantaneous flow of the wind into electricity. No energy storage required. The same concept applies to solar photovoltaic panels.

When working to enable the 100% renewable grid of the renewable energy narrative, the engineering challenges are thus to determine how *little* storage we can get away with, what storage characteristics are needed, what technologies have those characteristics, and at what cost can this system be constructed and operated.

Both energy narratives recognize the value and necessity of energy storage. They are simply promoting different *forms* of energy storage. The renewable narrative wants to remove all grid storage in the form of coal, natural gas, and oil, and sometimes nuclear fuel as well. Thus, other forms of energy storage must substitute.

---

[64]See Table 4.4 "Natural Gas in Underground Storage" in the Monthly Energy Review of the Energy Information Administration.

In the mean time, to date the existing fossil energy storage on the grid has enabled us to integrate increasing quantities of renewable electricity without much difficulty. Given the zero-sum game of constant U.S. electricity demand since 2007 as shown in Fig. 3.5, more renewable and natural gas-fired electricity has meant less coal-fired electricity. Coal proponents took notice.

Because of the inherent energy storage of the coal and nuclear power supply chains, some believe this provides an enhanced value for which current electricity markets do not provide compensation. Included in this group was President Trump's first Secretary of Energy Rick Perry: the same Rick Perry who ran for U.S. president in the previous election cycle, who in a 2011 Republican debate, could not remember the third executive department that he claimed he would eliminate if elected. He became head of that third department: the Department of Energy.[65]

Following the 2017 delivery of his requested Department of Energy (DOE) Staff Report on the topic of electric grid reliability, Perry sent a "Notice of Proposed Rulemaking" that directed the Federal Energy Regulatory Commission to come up with a rule that would allow power plants with 90 days of fuel supply to acquire additional compensation for this fuel storage.[66] Perry's notion was that more fuel storage at the site of power plants makes the overall grid more resilient to disruption from accidents and weather extremes. Practically, only coal and nuclear power plants would qualify for this compensation. Unfortunately for Perry, this fossil fuel narrative solution for on-site fuel storage was not high on the list of recommendations in the DOE Staff Report that was requested to specifically summarize the major reliability issues. In addition, most energy industry analysts and companies opposed additional compensation for on-site fuel storage.[67] Thus, many interpreted Perry's action as an anti-market repayment to coal companies that supported Trump's 2016 presidential campaign promise.[68]

For electricity generation, the fossil fuel narrative against renewable variability focuses on short time scales, and the battle over energy storage requirements focuses on intermediate time scales. It is when we reach the longest time scales that the renewable energy advocates believe they have the strongest argument.

---

[65] The other two executive departments that Rick Perry stated he would eliminate if elected preside were those of Commerce and Education.

[66] Staff Report to the Secretary on Electricity Markets and Reliability, accessible December 31, 2017 at https://energy.gov/downloads/download-staff-report-secretary-electricity-markets-and-reliability. Grid Resiliency Pricing Rule, Docket No. RM17-3-000, Agency: Department of Energy. Action: Notice of Proposed Rulemaking. September 28, 2017.

[67] Naureen S Malik Mark Chediak, and Jim Polson, "Fate of $700 Billion Power Trade Hinges on Trump Buzzword," *Bloomberg*, December 6, 2017 at https://www.bloomberg.com/news/articles/2017-12-06/fate-of-700-billion-power-trade-hinges-on-one-trump-buzzword.

[68] Jody Freeman and Joseph Goffman, "Rick Perry's Anti-Market Plan to Help Coal," Op-Ed in *The New York Times*, October 25, 2017.

Reliability: The Depletion Argument (Narratives at Long Time Scales)

Consider the following quote:

> Our main fossil fuel sources—oil, coal, and gas—are finite natural resources, and we are depleting them at a rapid rate. Furthermore they are the main contributors to climate change, and the race to the last 'cheap' fossil resources evokes disasters for the natural environment as seen recently in the case of the BP oil spill in the Gulf of Mexico.[38]—World Wildlife Fund (2011)

The World Wildlife Fund (WWF), along with most proponents of renewable energy are usually taking the long view of decades, and sometimes more. Anthropogenic impacts on the climate are anticipated to last for centuries. In this quote, WWF is effectively saying there are three main arguments to stop using fossil fuels. First, they are finite, and, thus, we can't use fossil fuels forever (the long view of decades to centuries). This argument generally follows that of M. King Hubbert and the earlier discussion on fossil resource size.

Second, fossil fuels are getting more physically and technologically difficult to obtain (e.g., deepwater offshore drilling), and, thus, our activities to extract them have increasingly broad impacts on the environment because we pursue fossil extraction in more remote locations. In the quote, WWF also appeals to shorter time scales (months and years to perhaps decades) by referring to the oil spill related to the tragic April 2010 sinking of the *Deepwater Horizon* offshore drilling rig while drilling the Macondo Prospect in the Gulf of Mexico. This event is credited as the largest marine oil spill in the history of the petroleum industry at nearly 4.9 million barrels of oil.[69] It caused many observable and unobservable impacts to marine wildlife along with the deaths of 11 individuals working on the drilling rig.

Third, we emit carbon dioxide when we burn fossil fuels, and we are doing so at a rate beyond which the Earth's carbon sinks can absorb. Thus, WWF and other renewable advocates anticipate the related environmental damages from atmospheric CO_2 accumulation to last for centuries—time scales longer than the useful life of any individual investment decision in the energy system.

Reliability—Summary

Time is the concept that relates power to energy. It is also the key to understand how energy narratives discuss reliability. Each energy resource and technology is more or less predictable, or reliable, in its capability at some time scales more than others. At short time scales of seconds to a few minutes, we generally assume predictable output from all electricity technologies. At medium time scales of hours

---

[69]https://www.dco.uscg.mil/Portals/9/OCSNCOE/OCS%20Investigation%20Reports/Macondo %20-%20DWH%20Reports/DWH%20ROI%20USCG%20Vol%20I%20Redacted%20Final.pdf? ver=2017-10-05-072821-053

to a few days, wind and solar output become less predictable, but not completely uncertain. At long time scales of decades, renewable technologies produce relatively predictable electricity output compared to fossil fuels. If we throw in the uncertainty of policy to limit climate damages from carbon dioxide emissions, then fossil fuel energy consumption is quite unknown several decades out. Because these "climate damages" are not fully known and might last for centuries to millennia, the transition from fossil fuels often includes a moral argument that renewable energy is the best way to care for the future. The fossil narrative does not concede this point. To morality we now turn.

The Morality Argument

Morality is about right and wrong, not correct and incorrect. A person's worldview, including religious belief, is the basis for the interpretation of data to form opinions on morality. A quote from the 30-year update to the *Limits to Growth* sums it up well:

> It is crucial to remember that every book, every computer model, every public statement is shaped at least as much by the worldview of the authors as by any "objective" data or analysis."[70]—Meadows, Randers, and Meadows (2004)

The domain of energy is not immune to imposition of morality. Energy is highly correlated to economic wealth, education, and both political and human rights [87]. Energy also operates infrastructure, such as municipal water supply and wastewater treatment, that keeps dense city populations healthy.

There is no more obvious choice for summarizing the moral argument for fossil fuels than from the man who dedicated an entire book to topic, *The Moral Case for Fossil Fuels*, and who now gets paid to spread that gospel:

> Here, in a sentence, is the moral case for fossil fuels, the single thought that can empower us to empower the world: Mankind's use of fossil fuels is supremely virtuous—because human life is the standard of value, and because using fossil fuels transforms our environment to make it wonderful for human life.[71]—Alex Epstein (2014)

Epstein's imposition of "human life [as] the standard of value" can't be more anthropocentric, and as I discuss next, completely opposite to Pope Francis. Epstein readily acknowledges, but rejects, other standards of value such as those based on religion or "pristine nature."[72] You either think that humans are part of the environment, or separate from the environment. The fossil fuel narrative is usually about the latter. If we are separate, then it is up to us to control and manipulate the environment as we wish. If we are part of the environment, then we are completely

---

[70]Meadows et al. [65, p. 4].

[71]Epstein [31, p. 209].

[72]p. 30 of [31].

coupled such that our actions cause changes in the environment, and changes in the environment in turn cause us to act.

The fossil fuel narrative often focuses on short-term day-to-day improvements in living conditions. It states that if you want more refrigerated food, climate controlled environments within buildings, medicines, jobs, and income now and in the future, then *the answer is to put more fossil fuel in the fire today.* By concerning oneself only with planning horizons shorter than about one decade (e.g., planning and construction of a nuclear power plant or offshore oil project representing perhaps the longest span of concern), the thought is that even incremental setbacks in short-term human conditions ultimately lead to large improvements in the long term:

> More people, and higher income, cause problems of increased demand for and consumption of resources in the short run. Heightened demand causes prices to rise for awhile. The higher prices present opportunity for businesses to make money and for investors to gain satisfaction and glory with new inventions, prompting investors and entrepreneurs to search for solutions. many fail, at cost to themselves. But in a free society, solutions are eventually found. And in the long run the new developments leave us *better off than if the problems had not arisen.* That is, prices end up lower than before the increased scarcity occurred.[73]—
> Julian Simon (1996) (emphasis in the original)

In this quote Julian Simon does not necessarily prioritize fossil fuels over renewable energy, but his faith lies in assuming solutions to resource constraints, via substitution and technology change one way or another, will always arise in the long-run. Because Simon did not see evidence for fossil energy resource constraints, he saw no reason to remove fossil fuels from the set of options.

The extreme end of the fossil fuel argument can be linked to the techno-optimism economic narrative and the idea that there is no conscious planning needed for environmental preservation. The argument for not explicitly making regulations for environmental preservation is that as society progresses, humans will maintain the environment in a satisfactory state because we will, by necessity, make markets that send price signals to incentivize us to invent practices and technologies as much as needed to continue growth. (Chapters 8 and 9 explore thinkers and ideologies that promote markets as the mechanism by which we provide the necessary price signals to spur technological innovations for socio-economic problems.) At the extreme of techno-optimism might be technologies to remove carbon dioxide directly from the atmosphere if that becomes necessary: we can have our carbon dioxide emissions and our low-carbon atmosphere too.

Simon is saying, because of our ignorance, it is immoral and nonsensical to pre-specify the state of the environment that we will need in the long term. How can we plan for a distant future that we can't predict? Epstein must be a big fan of Julian Simon. He states "...a productive civilization buys us time to think and discover, and then use that knowledge to become more productive, and buy more time to think and discover." and "The production of energy increases the production of knowledge, and it is knowledge that enables one generation to begin where the last

---

[73] Simon [85, pp. 382–383].

left off." But can't we generate more energy from renewable technology just as from using fossil fuels? No, say fossil proponents such as Epstein: "If we slow down our progress … by using inferior energy, we deserve nothing but contempt from future generations …" Of course, for Epstein renewable energy is "inferior energy." We are told that fossil fuels were the most productive energy resource of the past, and they will be for the future. Thus, they are what will enable us more time to increase our productivity, knowledge, and ultimately our quality of life ongoing.[74]

It is difficult to pose a more dichotomous moral viewpoint than that expressed by Pope Francis. Pope Francis is not your traditional leader of the Catholic Church. He was anointed because of his exceptional dedication to the poor. I post several short quotes from Francis' Encyclical Letter *On Care for our Common Home* that attempt to summarize one of his major arguments: we must treat our environment well as that is a reflection of how we treat ourselves.

> 23 …The problem [of global warming] is aggravated by a model of development based on the intensive use of fossil fuels, which is at the heart of the worldwide energy system.

> 68. This [mutual responsibility between human beings and nature] for God's earth means that human beings, endowed with intelligence, must respect the laws of nature and the delicate equilibria existing between the creatures of this world, …The laws found in the Bible dwell on relationships, not only among individuals but also with other living beings. …Clearly, the Bible has no place for a tyrannical anthropocentrism unconcerned for other creatures.

> 122. A misguided anthropocentrism leads to a misguided lifestyle. …When human beings place themselves at the centre, they give absolute priority to immediate convenience and all else becomes relative. Hence we should not be surprised to find, in conjunction with the omnipresent technocratic paradigm and the cult of unlimited human power, the rise of a relativism which sees everything as irrelevant unless it serves one's own immediate interests. There is a logic in all this whereby different attitudes can feed on one another, leading to environmental degradation and social decay.—Pope Francis (2015) [35]

Are you kidding me? TheGlobalization, quotes from Francis' Encyclical Letter might as well have come from speakers at conferences on ecology or environmental justice. These statements come from the leader of the Catholic Church, the same organization that took 359 years to apologize for the persecution of Galileo for being correct that the Earth was not the center of the universe.[75]

Just as the fossil fuel narrative argues it provides the best route to long-term prosperity, so does the renewable energy narrative. If you want more refrigerated food, climate controlled environments within buildings, medicines, jobs, and income now and in the future, then *the renewable narrative answer is to use more renewable energy today*. Because biomass-based energy requires significant land and water input, most renewable proponents now focus primarily on non-biomass renewable systems. While it is easy for fossil fuel advocates to oppose liquid biofuel programs,

---

[74]Quotes from [31] in this paragraph are on p. 184.

[75]Alan Cowell, "After 350 Years, Vatican Says Galileo Was Right: It Moves," *New York Times*, October 31, 1992, available December 31, 2017 at http://www.nytimes.com/1992/10/31/world/after-350-years-vatican-says-galileo-was-right-it-moves.html.

a decade after the biofuels policies of Europe and the United States started in the mid-2000s, even pro-renewable advocates largely acknowledge the energetic and environmental limits of industrial scale liquid biofuel production.[76]

WWF International's 2011 report on how to reach 100% renewable energy stated "Bioenergy (liquid biofuels and solid biomass) is used as a last resort where other renewable energy sources are not viable—primarily in providing fuels for aeroplanes, ships, and trucks, and in industrial processes that require very high temperatures."[38] Thus, to avoid too much land use for energy and retain habitat for biodiversity, the renewable energy movement now largely focuses on a *renewable electricity* future, where that electricity might be used to make liquid or gaseous fuels in the form of hydrogen itself or as a building block to synthetic diesel and gasoline.

In addition to land and water use, both energy narratives use a health-based argument against burning biomass, in the form of wood, crop residues, and dung, for heating and cooking in homes. The World Health Organization states that "[a]round 3 billion people still cook using solid fuels (such as wood, crop wastes, charcoal, coal, and dung) and kerosene in open fires and inefficient stoves," leading to almost four million premature deaths per year.[77]

Perhaps *the* major argument for a transition to renewable energy is to provide energy services while limiting the anticipated negative impacts from climate change. This is done by reducing the rate of greenhouse gas (GHG) emissions, primarily carbon dioxide that is emitted from burning fossil fuels, such that the GHG concentration in the atmosphere becomes stabilized. The most recent global agreement on climate change is the Paris Agreement forged in 2015. This agreement sought to limit the rise of the global average temperature to less than 2 °C. Most nations signed the agreement, including the United States under President Obama.

Even though each country's commitments are non-binding, in November 2019 President Trump notified the United Nations that the U.S. was withdrawing from the Paris climate agreement.[78] This act requires no effort, so Trump largely fulfills a campaign pledge by doing nothing.

President Trump most assuredly campaigned on fossil fuel narrative, most pointedly with his slogan "Trump digs coal." He appointed former Texas Governor Rick Perry as his first Secretary of Energy. The same Rick Perry who proposed a regulation to provide additional compensation to coal-fired power plants if they can store enough coal on-site. The same Rick Perry who attempted to fast-track the permitting of over one dozen new coal-fired power plants in Texas in 2007, just

---

[76]For example, the Renewable Fuels Standard enacted in the Energy Policy Act of 2005.

[77]World Health Organization, "Household air pollution and health," May 8, 2018 at: https://www.who.int/news-room/fact-sheets/detail/household-air-pollution-and-health.

[78]"U.S. Formally Begins To Leave The Paris Climate Agreement," All Things Considered, National Public Radio November 4, 2019 at: https://www.npr.org/2019/11/04/773474657/u-s-formally-begins-to-leave-the-paris-climate-agreement.

before the Global Financial crisis and the hydraulic fracturing boom combined to place significant headwinds on coal generation.[79]

One interesting point is that when Perry was Texas governor, neither he nor Texas state legislators needed climate change as a moral argument to promote renewable energy, and in particular wind energy. They used the *economic* development argument. In 2005 Perry signed Texas Senate Bill 20 (SB 20) that established what were called Competitive Renewable Energy Zones (CREZ). The CREZs were designated areas of high quality wind resource in west Texas and the Texas panhandle. SB 20 mandated that transmission lines be constructed to connect those CREZs to the main regions of electricity demand in the central part of the state: Dallas-Fort Worth, Austin, and San Antonio. Bolstered by analysis indicating that 11.5 GW of new wind power capacity would save money for Texas consumers (by offsetting anticipated high-cost natural gas-fired electricity),[80] provide royalties for land owners, create jobs, and establish a new tax base in rural counties, Texas created, signed, and successfully implemented a centrally coordinated seven billion dollar transmission project to allow consumers access to renewable energy [42].[81]

Morality—Summary

In this section the views on morality of energy use centered on how to best benefit humans, whether or not to explicitly consider human dependence on the environment, and whether humanity's primary role is to exploit the environment or preserve a symbiotic relationship. Thus, the morality discussion of the energy narratives can occur without regard to economics. But in our industrial economy, decisions usually come down to dollars and cents. From this standpoint, even the former governor of Texas supported both the fossil (via coal) and renewable (via wind) energy narratives. If renewable energy can work in Texas, the U.S. state with the highest coal, natural gas, and oil consumption, then can it work anywhere? In particular, what about developing countries?

---

[79] Story on fast-tracking coal power plant permits is by Kelly Shannon, "Perry's fast-track order derailed," February 20, 2007, *Plainview Daily Herald*, accessed December 31, 2017 at http://www.myplainview.com/news/article/Perry-s-fast-track-order-derailed-8660632.php.

[80] ERCOT (2008) Competitive Renewable Energy Zones (CREZ) Transmission Optimization Study, April 2, 2008, available on September 22 at https://www.nrc.gov/docs/ML0914/ML091420467.pdf.

[81] The transmission line costs are included in the bills of all ERCOT customers at an anticipated cost of 4–5 \$/month [42].

The Developing Country Argument: Follow or Leapfrog the Rich Countries?

> The fossil fuel-based economy is not delivering the economic potential of sub-Saharan Africa. There is very poor energy security in the region, with instability in both the supply and the price of fossil fuels, and a huge draw on foreign export earnings to deliver energy demands. Conventional fuels are not delivering energy for the poor or meeting global climate change objectives.
>
> …
>
> Africa has a big opportunity to leapfrog and transition to a low-carbon development path and at the same time still expand access to energy services. It is possible to lift Africa out of energy poverty without increasing emissions. But it will need financial and technological support to rise to the occasion. Africa, in other words, can be a low-carbon leader. [24]—Christian Aid (2011)

> Finite and increasingly expensive fossil fuels are not the answer for developing countries. But renewable energy sources offer the potential to transform the quality of life and improve the economic prospects of billions.[82]—World Wildlife Fund (2011)

Since most developed economies are no longer increasing energy or electricity consumption (see U.K. and U.S. data in Chap. 2 and Fig. 3.5), energy companies look to developing countries for future increases in energy consumption and infrastructure investment. Large developing countries, such as India and China, can effectively pursue many options and manufacture or trade for much of the needed infrastructure. However, the energy narratives differ on which electricity technologies best suit much of Africa that is underserved by electricity, primarily the sub-Saharan region.

As mentioned in the previous "Morality" section, burning biomass (or coal) in homes leads to severe health problems due to poor indoor air quality. By operating homes on electricity we avoid this health problem. You might not be surprised that both fossil fuel and renewable energy narratives claim they are best suited for helping developing countries make this biomass-to-electricity transition.

Again, Alex Epstein fosters the fossil narrative on why developing countries need fossil-fueled power:

> …this book is focused on *fossil energy*—but only, as you'll see, because I believe that it is the most essential technology for producing energy for 7 billion people to improve their lives, at least over the next several decades. If there was a better form of energy and it was under attack in a way that wildly exaggerated its negatives and undervalued its positives, I'd be writing the moral case for *that* form of energy.[83]—Alex Epstein (2014)

Before the above quoted passage from Epstein's book in which he states his moral case for fossil fuels, he recites "Kathryn's Story" from the non-profit organization Power Up Gambia. This story recounts how the author, while in The

---

[82]WWF (2011) *The Energy Report: 100% Renewable Energy by 2050*, p. 13.

[83]Epstein [31, p. 39].

Gambia, witnessed an emergency C-section of what turned out to be a stillborn underweight baby. "The surgeon later explained that the baby had suffocated in utero. If only they had had enough power to use the ultrasound machine for each pregnancy, he would have detected the problem earlier and been able to plan the C-section. Without early detection, the C-section became an emergency, moreover, the surgery had to wait for the generator to be powered on. The loss of precious minutes meant the loss of a precious life."[84]

During a trip to Africa in 2017, then U.S. Secretary of Energy Rick Perry had a similar thought of how fossil fuels can help protect African women in remote villages:

> When the lights are on, when you have light, it shines the righteousness, if you will, on those [sexual assault] types of acts ...From the standpoint of how you really affect people's lives [in African villages], fossil fuel is going to play a role in that.[85]—Rick Perry (2017)

To advocates of fossil fuel narrative, coal, oil, and natural gas are always the answer to energy shortages. However, the disturbing thing about Epstein's use of Kathryn's story about medical issues in a developing country, is that while he uses it to argue the moral choice for more fossil fuels, Power Up Gambia was actually advocating for renewable solar power to address the same exact issue:

> Our mission is to improve healthcare delivery in The Gambia by providing proven, reliable, and sustainable electricity through solar energy. Hospitals and clinics in The Gambia are still without access to electricity....Solar power can save lives![86]—Power Up Gambia (2017)

In fact, the tagline of Power Up Gambia is "Transforming Healthcare through Solar Energy." It is hard to spin a narrative more than Epstein's quoting an organization as justification for a policy that is the opposite policy of that same organization.

So what gives? Should developing African nations, such as The Gambia, invest in fossil or renewable electricity? Should they follow the path trailed by the developed world that built large-scale fossil and hydropower plants that are connected via a transmission grid, or should they install small-scale wind and solar systems that are not connected to a larger country-wide or regional electric grid?

Some advocate that developing countries *leapfrog* the technologies by which the developed world provided universal access to electricity. Because urban areas dissipate so much power that they need to be connected via a grid to multiple power plants (renewable or fossil), the leapfrog concept largely (but not only) applies to

---

[84]Website of Power Up Gambia, http://powerupgambia.org/about/story/, accessed September 14, 2017.

[85]Secretary of Energy in response to a story that a young woman from a[n African] village told him that electricity was important to her not only because it would free her from having to read by the light of a fire with choking fumes, but also from the standpoint of sexual assault [78].

[86]Website of Power Up Gambia, http://powerupgambia.org/about/story/, accessed September 14, 2017.

distributed renewable electricity in rural areas. It is important to keep in mind that even in the United States rural areas obtained universal access to electricity well after it was ubiquitous in cities. Before the 1930s, less than 10% of farms had access to the electric grid as private investors did not consider rural electrification a significantly profitable investment [94]. Programs, such as those implemented by the Rural Electrification Administration (REA) created after the Great Depression, led to universal electricity access across the U.S. "Rather than simply build power systems, the REA made loans to electric cooperatives that were repaid over 30 years. Country folk came together, organized cooperatives, and provided labor to build the systems that they ultimately came to own."[87] While there were clear economic benefits of U.S. rural electrification, the justification for investment was just as much related to a moral, or political, push for equal electricity access for all citizens.

While the U.S. rural electrification programs were financed with internal federal money, there is push for African rural electrification to be financed or simply paid for by developed countries:

> Developed countries need to commit to deliver sufficient financing to developing countries that will be delivered through a special dedicated window under the Green Climate Fund, with democratic and equitable governance, that will enable African countries and other developing countries to pursue energy access and sustainable development through a clean development model. This should be a leapfrog fund for low-carbon energy access."[24]— Christian Aid (2011)

This presents a question: does a developing country leapfrog, or skip a class of supposedly inferior technology, if it has to use money and products from other countries in order to establish rural (or even urban) electrification? Rural electrification infrastructure of the U.S. was largely manufactured within the U.S. using U.S. companies and intellect. Thus, while serving a social need, private businesses also made money installing, operating, and selling equipment. Further, in addition to creating the physical capital of the electric grid, the U.S. as a whole developed the human capital, or know-how, to maintain and improve it.

Wind turbines, solar panels, and battery systems are largely designed and manufactured outside of Africa. Thus, the faster African nations want to use new renewable energy systems, the more they must rely on the intellectual and financial capacities of other countries. Some leapfrog advocates recognize this issue: "Limited technical expertise in operating and maintaining technology further heightens costs for investors. One potential option to mitigate these high costs would be to invest in research and development, with the aim of reducing the need for and associated cost of importing technical expertise from abroad." [24]

This point is key. If developing nations don't now have the governmental stability, technical knowledge, manufacturing capability, and legal and regulatory structures to provide universal access to an electricity system similar to that of the developed countries, how are they going to leapfrog those electricity systems to

---

[87]Wallace Jr., Howard D., Smithsonian (2016), http://americanhistory.si.edu/blog/rural-electrification, accessed September 16, 2017.

something new? More than research and development, they need to develop their own manufacturing and intellectual capabilities that include a population educated in engineering and technical training to maintain both large- and small-scale energy systems—no short-term task. Some see the manufacturing of renewable energy as exactly the way to develop a new industrialized economy:

> Apart from obvious energy security and environmental security issues, renewables offer industrialising countries their best chance of breaking into manufacturing value chains. Whereas the global fossil fuel economy offers countries like India a marginal role at best, the world of manufacturing renewables and deploying them in solar and wind farms offers real economic benefits—local employment, exports and integration into value chains. These are substantial benefits that have little to do with global climate change and everything to do with building wealth and incomes through industrialisation.[88]—John Mathews (2015)

Arguments such as Mathews' substantially expand the concepts of how energy benefits people, and stay squarely within the techno-optimistic economic narrative. Is energy only beneficial via its direct use to provide energy services, such as heating, cooling, light, power (e.g., substituting for physical labor), and transportation? Or is there something larger, that he who controls the means to produce and distribute energy is thus in control of other aspects of political and economic power? Yes, this larger context is very important. Give a man a fish, and you feed him for a day; teach a man to fish, and you feed him for a lifetime. Does that proverb also hold for both manufacturing solar panels and operating oil drilling rigs? Should it hold for countries as much as individuals?

The answer to these questions are not so straightforward. In pre-industrial solar-powered agrarian societies there were practically no democracies or constitutional republics comparable to those 100 years into a coal-fired industrialized economy [66]. Countries need more than natural resources to be economically successful [1], but an economy cannot be fully balanced if too dependent on others for energy. Flows of money and energy are interdependent.

Advocates for renewable energy in developing countries don't necessarily want to wait for local industrial development. At the small scale of installing solar in developing countries was Mobisol, a German company that installed solar systems in Africa ranging from 40 to 200 W each, or one to a few solar panels. Mobisol, started in 2011, was one of the most promising new companies, selling 12 megawatts (MW) of solar systems in Africa by 2019. Unfortunately, it went insolvent in 2019 and was sold to Engie, a major French utility holding company with an established presence in Africa, already selling solar home systems in six countries.[89] One challenge for Mobisol was that their investors wanted competitive monetary returns.

---

[88]John Mathews, professor at the Macquarie Graduate School of Management at Macquarie University, December 23, 2015, chinadialogue.net, https://www.chinadialogue.net/article/show/single/en/8509-Let-them-eat-solar-panels.

[89]John Dizard, "Mobisol's rescue does not assure success in Africa," *Financial Times*, September 12, 2019 at: https://www.ft.com/content/d5d667b3-b45b-3870-8443-fe21d669f6d6. The 12 MW of solar installed by Mobisol come from its website, https://plugintheworld.com/, accessed December 7, 2019. An Engie press release on September 3, 2019 summarizes its acquisition of Mobisol,

This challenge holds for any company with the same goals and funders who are effectively trying to make a buck from people living at or near the poverty line. In the end, Mobisol's "...10-year subscription micro-finance model was always going to struggle in sub-Saharan Africa, where incomes are so reliant on seasons, healthy crops and other factors far removed from local people's controls."[90] Since so many people are farmers in sub-Saharan Africa, when the crop yields are too low, incomes dry up and food becomes the priority. Referring to impacts from a cyclone in 2019, Gwen Parry, manager of Challenges Zambia stated "...in Malawi and Mozambique, where farms have been completely wiped out, well, I don't think those people will be caring about their solar kits repayments when they've no food and nowhere to live."[91]

At the level of international aid from governments, are developing countries going to give money, or lend on terms favorable to African nations, at a level allowing African energy system to leapfrog those in OECD countries?[92] So far, the answer to this last question seems to be no. At the 15th Conference of Parties (COP) meeting of the UN Framework Convention on Climate Change (UNFCCC), developed countries pledged US$100 billion per year (by 2020) to aid developing countries for adaptation and mitigation related to climate change. Yet the actual progress toward that commitment is in serious doubt as there is not even agreement on what monetary flows actually can count toward the commitment (e.g., a loan versus a pure grant of money). A 2015 report by the OECD estimated nearly $62 billion was committed in 2014, but the assumptions behind that number were not transparent and developing countries simply balked at the claim [73, 82]. Development banks and aid agencies are trying to help. For example, for large solar projects, the World Bank's Scaling Solar campaign seeks to help developing countries overcome, among other things, their lack of institutional capacity, or business experience.

However, the developed countries' (and China) economies and governments are not necessarily organized to help "develop" developing countries more than themselves, and populism is on the rise in the U.S. and Europe in the last decade. In the past, multilateral development banks, such as the World Bank and International Monetary Fund that are funded by developed countries, often simply shuffled "donated" money within the developed countries themselves. China has

---

available December 7, 2019 at: https://www.engie.com/sites/default/files/assets/documents/2020-01/engie-mobisol-v4-en.pdf.

[90] The Challenges Group, "Time for solar PV sector to find alternative ways to power remote communities," on *Medium* June 21, 2019 at: https://medium.com/the-challenges-group/after-mobisol-its-time-for-africa-s-solar-pv-sector-to-find-alternative-ways-to-power-communities-2c80db308dae.

[91] The Challenges Group, "Time for solar PV sector to find alternative ways to power remote communities," on *Medium* June 21, 2019 at: https://medium.com/the-challenges-group/after-mobisol-its-time-for-africa-s-solar-pv-sector-to-find-alternative-ways-to-power-communities-2c80db308dae.

[92] OECD: Organization for Economic Cooperation and Development.

now established similar institutions independent of the Western countries. The development bank money-shuffling occurs via loans for infrastructure in developing nations where the principal goes immediately to pay rich-world contractors, and the loan is repaid via proceeds from the installed infrastructure. Aside from the construction and engineering work primarily going to the rich-world contractors, the assumed benefits from the infrastructure were often purposefully overstated such that loan cannot feasibly be repaid. Thus, the developing country must make concessions (e.g., of resource access, privatization of public assets) to the developed countries that didn't directly help build in-country intellectual and institutional capacity in the first place. The benefits end up in rich-world private contractors, and the burden falls onto developing populations as increased debt. This tactic is described fully by John Perkins' *Confessions of an Economic Hit Man* [75].

Developing Countries and Energy—Summary

Can Africa leapfrog developed economies' energy infrastructure? I don't know, but there are serious challenges if developing countries deploy state-of-the-art energy systems without sufficient in-country intellectual, political, and legal support structures. These institutional capabilities, along with the legal and regulatory system that ensures their stability, are key to economic development [1]. The capabilities and universality of energy infrastructure took many decades to establish in developed countries, and we should expect many decades also for developing countries. Making energy system investments that place too much of a debt-burden on low-income citizens to pay back foreign investors is also not sustainable. It will be a long path. The fruits will come much later, but they will taste much sweeter.

The Environmental Discussion

In addition to the issue of climate change and greenhouse gas emissions, the energy narratives battle on more traditional environmental grounds. While both narratives generally agree we want to avoid the detrimental health impacts from indoor burning of solid fuels, such as wood, dung, and coal, there is still plenty to argue about in terms of materials and land use.

Environmental: Renewable Energy Isn't Renewable

> The wind may never stop blowing, but the wind industry depends on steel, concrete and rare-earth metals (for the turbine magnets), none of which are renewable.[79]—Matt Ridley (2011)

> The basic problem is that the *process* for solar and wind to generate reliable electricity requires so many resources that it has never been cheap and plentiful. ...The diluteness problem is that the sun and the wind don't deliver concentrated energy, which means you need a lot of materials per unit of energy produced.[93]—Alex Epstein (2014)

Matt Ridley's above statement is generally correct, but in an engineering sense, we don't have to use rare-earth metals in an electrical generator connected to a wind turbine. It also holds true for industrial solar photovoltaic and concentrating mirror-based systems. For that matter, his statement holds true for fossil and nuclear power plants. The statement also holds generally for all modern industrial technologies and systems that are power-consuming, as opposed to power-producing, such as the computer on which I'm now typing.

However, if we take Ridley seriously regarding his earlier quote that "The hydrocarbons in the earth's crust amount to more than 500,000 exajoules of energy" that can provide close to a millennium of present primary energy consumption rates, then should we have concern with wind and solar technologies being made of fossil minerals? [79] If we believe we can ultimately extract the quantity of fossil energy *resource* he states, then we can also conclude we can access all of the minerals needed to make wind and solar farms. In other words, at one extreme we could directly burn this abundant fossil energy directly in power plants to generate electricity. At another extreme we could burn fossil fuels in machines that specifically extract the other fossil minerals required to make and install renewable energy power plants, and the rest of the economy consumes electricity from the renewables. Of course, in reality, we're currently combining both approaches.

The point of the Ridley and Epstein quotes is that the amount of power plant materials (concrete, steel, etc.) per megawatt of capacity or megawatt-hour (MWh) generated from wind and solar systems is not zero. So let's take a look at some numbers in Table 3.2. Solar photovoltaic (PV) systems use up to 3 kg of iron and 2 kg of cement per MWh of electricity generated [44]. Onshore wind turbines use up to 9 kg of iron and 3 kg of cement per MWh. The amount of cement and iron to generate a MWh from coal and natural gas power plants is about ten times less. The fossil energy narrative often stops here, but the renewable narrative says this is not the complete story.

The renewable narrative counter argument is that fossil fuel systems are *more dependent* on fossil minerals and materials per MWh. By considering the mass of the fossil fuels themselves, coal and natural gas power plants require an order of magnitude more fossil material per MWh over their lifetime than do renewable electricity technologies. The total fossil fuel mass used over the life cycles of natural gas combined cycle and coal electricity are around 150 and 350 kg per MWh. On a life cycle material basis, natural gas combined cycle and coal power use nearly four and ten times more material than solar PV, and ten to forty times more than wind.

So there you have it. This simple example of counting the mass needed for renewable and fossil electricity can support either narrative. If you neglect the

---

[93]Epstein [31, pp. 48–49].

Table 3.2 The mass of material per unit of output (kg/MWh) required over the estimated lifetime of the power generation technology

Technology	Cement (kg/MWh)	Iron (kg/MWh)	Fossil (nonrenewable) energy demand (GJ/MWh)	Fossil (nonrenewable) material demand (kg/MWh)
Solar PV (ground-mounted)	0.5–2	2–3	<1	<40[a]
Wind (onshore)	1–3	3–9	<0.3	<10[a]
Coal	0.3	0.5	8–10	310–360[b]
Natural gas combined cycle	0.1	0.3	8	150[c]

Data from Figure 1 of [44]

[a]Assumes (worst case scenario) that all fossil energy to build and operate comes from coal (26.2 MJ/kg)

[b]Considers only the coal transported for combustion at the power plant (26.2 MJ/kg)

[c]Considers only the natural gas transported for combustion at the power plant (0.187 m^3/kWh at 0.8 kg/m^3 density of natural gas)

material aspect of the fuels for natural gas and coal-fired electricity, then fossil electricity appears less material intensive than renewable electricity. If you include the mass of the coal and natural gas, you come to the opposite conclusion.

But it takes energy to move mass from place to place. Using renewable electricity requires us to *move less mass during operation* because we don't have to physically transport coal, oil, and natural gas within networks. Chapter 5 expands on how the concept of distribution networks helps understand relationships we observe between energy consumption and the size of the economy.

Strictly speaking, renewable energy technologies are labeled as such because the fuel or resource flow that they take as an input during *operation* is renewable, not because they are composed of only renewable materials. However, the fact that all industrial technologies, energy-related or otherwise, require fossil minerals is more profound than immediately implied by the Epstein and Ridley quotes. The modern economy has been built and is currently enabled by extracting fossil energy, which allows us to extract other fossil minerals that we shape and combine into most of the man-made stuff we have around us today. This recognition is one reason that the computer model used in *The Limits to Growth* did not explicitly represent renewable energy akin to that generated from solar panels or wind turbines. It is possible to conceive of the extraction of fossil minerals without the use of fossil minerals and fuels, but it is practically quite difficult to take to the extreme. For example, as discussed in Chap. 2, even early coal mining was performed with simple (fossil) metal hand tools and human animate (renewable) power, but mining only increased rapidly with the steam engine.

If the cost of extracting fossil minerals, energy or otherwise, becomes too high to perpetuate further growth with the existing economic structure, then in addition to efforts to improve extraction technology, we can choose to use these minerals more judiciously in at least two ways. First, we can choose to put the minerals into devices that extract power from renewable flows because, while they are composed of fossil minerals, they don't require a continuous flow of them to keep operating.

Secondly, we can choose to recycle materials from our industrial products such that we no longer have to extract raw materials from the Earth's crust. There is already substantial recycling of aluminum, steel, and plastics (made from hydrocarbons). This recycling, however, requires energy itself as an input. It remains to be seen whether or not we can operate and live within a *circular economy*, one that no longer extracts raw materials but only reuses materials from existing products and structures, at current developed world lifestyles. Further discussion on modeling and relationships between energy and materials consumption for economic activity occurs in Chaps. 6 and 9.

A third way to use Earth's fossil minerals is to make spacecraft that either take humans to live on other planets or somehow return materials to Earth. The 2009 movie *Avatar* portrayed a techno-optimistic dream scenario where humans go to another planet, Pandora, to extract *unobtanium* and return it to Earth.[94] In movies we can have Unobtanium which "...is not only the key to Earth's energy needs in the twenty-second century, but it is the enabler of interstellar travel and the establishment of a truly spacefaring civilization. This makes a feedback loop; the more unobtanium is mined, the more ships can be built and the more mining equipment can be sent to Pandora."[95] This feedback is the same as that of coal at the beginning of the Industrial Revolution. However, in engineering schools, the idea of unobtanium is a joke. If you design the perfect machine, but there are no known materials with the properties needed to make the machine, then your colleagues say "great, now all we need is some unobtainium."[96] While techno-optimists and Julian Simon might have been proud of *Avatar*'s technological vision, by showing humans exploit another planet in the "cosmos," the movie also implies that humans still can't figure out how to continue growth without destroying others' cultures and environments for the sake of "our progress."

Environmental: Power and Energy Density

> ...the average wind turbine has a power density of about 1.2 watts per square meter [of land area] ...[97]—Robert Bryce (2010)

---

[94]Unobtainium is pronounced un-ub-tain-ee-um.

[95]http://james-camerons-avatar.wikia.com/wiki/Unobtanium, accessed September 13, 2017.

[96]The unobtainium joke is also referenced at http://james-camerons-avatar.wikia.com/wiki/Unobtanium.

[97]Bryce [13, p. 235].

The grim land-use numbers behind all-renewable proposals aren't speculation. Arriving at them requires only a bit of investigation, and yes, that we do the math.[98]—Robert Bryce (2018)

The concept of power and energy density is well-established in energy circles. The prolific energy writer Vaclav Smil has done a great job at forcing this into discussion via writings in many forums [86–89]. There are three main ways to measure power and energy density: area, mass, and volume. All are relevant, as each density concept provides a unique understanding.

The discussion of the "used" or disturbed land area of renewable and fossil energy systems is one of the most confusing. We can measure the amount of the Earth's land surface residing above primary energy stocks such as oil, gas, and coal deposits. The units are energy per area such as joules per hectare or British Thermal units per acre. We can also measure the amount of wind and solar energy flowing across the land. The units are power per area such as watts per hectare. Thus, one reason why energy narratives talk past each other with regard to land use is because fossil energy resources are stocks and renewable resources (sunlight and wind) are flows. A further reason is the specification of the area of interest.

Considering wind farms per the Bryce quote above, Bryce's number refers to the "total project area" of a wind farm which is approximately all of the land circumscribed by all of the individual wind turbines.[99] Considering the entire project area of a wind farm, the installed capacity per area is typically $1–11\,\text{W/m}^2$ [21]. Assuming wind farms operate at a typical 30–40% capacity factor, then the average power output density *of a wind farm* per total project area is $0.3–4\,\text{W/m}^2$.

A proponent of the renewable narrative might not like to focus on "total project area" but instead focus on the "permanent direct impact area" because this area is much smaller [84]. There is a lot of space in between wind turbines in a wind farm. Figure 3.6 illustrates one example. Wind turbines typically have several hundreds of meters between each of them. A landowner can utilize this space for ranching and farming. When considering only the direct impact area of wind turbines in a wind farm, the installed capacity per area is typically between 100 and $1000\,\text{W/m}^2$ [21]. This makes sense because the wind resource itself (e.g., the average power in the blowing wind at the height of the turbine blades) is typically in the hundreds of W/m^2. Assuming again that wind turbines operate at 30–40% capacity factor, then the power output density *of a wind turbine* per permanent direct impact area is $30–400\,\text{W/m}^2$.

---

[98]Robert Bryce. All-renewable energy in California? Sorry, land-use calculations say it's not going to happen, *LA Times*, August 21, 2018 at http://www.latimes.com/opinion/op-ed/la-oe-bryce-renewables-california-20180821-story.html.

[99]By circumscribed, I mean the area that would be enclosed by a polygon if one were to draw a line from one turbine to another such that there are no wind turbines outside of the polygon. Due to varied and unique layouts of wind turbines within wind farms, the actual total project area is not fundamentally derived using the purely geometrical circumscribed boundary, but I have used the concept as an approximative idea.

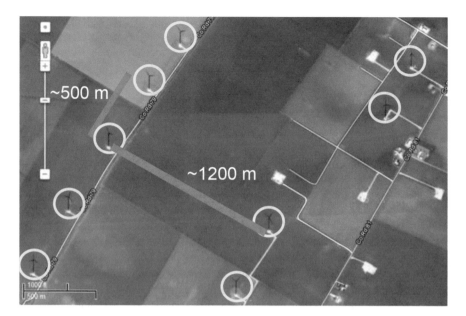

Fig. 3.6 A satellite image of a wind farm situated among cropland along the coast of Texas. The circles highlight where each wind turbine resides, several hundred to over one thousand meters apart

The capacity and power output densities using permanent direct impact area are about 100 times larger than when considering the total project area. Thus, using the same data for capacity of wind turbines and their actual electricity output, one can calculate two numbers that are two orders of magnitude apart. Both numbers are relevant and depend on your reason for understanding land use. If your goal is to understand how many wind turbines you can put in a given area, then the power density per total project area is the relevant calculation. If your goal is to understand how much area is removed from ranching and agriculture by installing wind turbines, then the power density per permanent direct impact areas is the relevant calculation. If your goal is to understand how land use of wind turbines affects wildlife, then both power density numbers can be relevant depending upon the species of interest.

Fossil Fuel Narrative: The Renewable Energy Footprint Is Too Large

The land-focused energy and power density arguments are powerful. If our infrastructure, energy or otherwise, takes up more land, then there is less pristine land for both biodiversity and sheer numbers of other species that reside on our planet. Peter Huber and Mark Mills' 2005 book *The Bottomless Well* is a techno-optimistic economic narrative perspective promoting that our fossil-fueled world is generally better for the environment and economy than a renewably powered world. Particularly they point out the energetic benefits of high energy density fossil over biomass energy sources:

...the carbohydrate fueled stomach [of a person, fueled by carbohydrates, pedaling a bike] is a whole lot worse for the atmosphere than the hydrocarbon-fueled motor that has replaced it.[100]

...such that...

No conceivable mix of solar, biomass, or wind technology could meet even half our current energy demand without doubling the human footprint on the surface of the continent.[101]— Peter Huber and Mark Mills (2005)

The second statement is particularly relevant even for purveyors that hold both the renewable energy and techno-realism narratives, roughly the opposite of the spectrum from Huber and Mills. This group recognizes that more land and fixed capital are needed for renewable energy, whether including biomass or not, and, thus, we probably cannot continue with our current energy demand in a near 100% renewable energy future. They see this curbed demand as acceptable because the finite Earth economic narrative says you can't grow energy demand forever. Thus, there is no point in pretending otherwise. Purveyors that hold both the renewable energy and techno-optimism narratives believe that we don't have to reduce energy consumption in a 100% renewable future. Thus, everyone pushing for a renewable energy and/or low-carbon future does not agree with each other. I discuss an example in the last section of this chapter where academics, most within the renewable energy camp (though that was debated) waged war over a computer model.

Even many leaning to the side of fossil fuel narrative don't completely discount renewable energy. For example, Huber and Mills' *The Bottomless Well* is not strictly anti-renewable energy, and their book somewhat promotes a blend of the energy narratives. One reason why the book subtitle states "...we will never run out of energy" is because of their techno-optimistic belief in human ability to continuously find ways to produce and consume energy at higher rates. One of these ways might be via renewable technologies since "...engineers will undoubtedly, in time, find ways to incorporate cheap, high-efficiency semiconductor junctions [photovoltaic cells] in roofs, walls, and widely used construction materials."[102] However, they are skeptical, as the passage continues "Less clear, however, is whether any of these renewable energy technologies will improve faster than conventional ones, as they must, to catch up." (See footnote 101). Mills' multiple writings and speeches are generally skeptical of a transition to a renewably powered economy.

Since the writing of their 2005 book, both fossil and renewable energy technologies have indeed improved at a high rate. Fossil technology advancement largely came from hydraulic fracturing and horizontal drilling for oil and gas. Renewable technology advancement largely came from larger wind turbines and mass solar photovoltaic manufacturing in Asia. It is coal and nuclear power technologies that

---

[100]Huber and Mills [48, p. 165].

[101]Huber and Mills [48, p. 167].

[102]Huber and Mills [48, p. 180].

have stagnated, and are thus losing market share today in Western countries. Part of the reason is that coal power plants produce cheaper power when they are large. But in a countries no longer growing fast in either population or electricity consumption (recall Fig. 3.5), building a large power plant is a big problem, not a benefit.

As more wind turbines and solar panels have become more effective and prevalent, their impacts have become more scrutinized. A common critique of wind turbines is their affect on bats and birds, including birds that normally reside on the ground. If you are a bird that lives on the ground, a lot of animals want to eat you. Some of the predators, such as foxes and coyotes, have four legs. Some, such as snakes, have no legs. But some are other birds, with two legs and two wings, such as hawks and eagles. Thus, a set of blades rotating in the air and causing shadows can make some ground birds such as grouse and prairie chickens avoid areas near turbines. The ground birds also tend to like wide open areas that don't have perching areas for their avian predators. However, agriculture and oil and gas development, due to historical development, can be primary causes of loss of ground bird habitat.[103]

Birds can be directly struck by the spinning wind turbine blades, and it is not hard to find videos on the Internet showing this. However, bats can be killed even without direct contact with wind turbine blades because they often fly behind the blades in a low pressure zone that causes their blood vessels to burst (See footnote 102). While wind turbines have a measurable impact on wildlife, it is important to understand our overall impact if we want to maintain biodiversity. One study of bird deaths in the U.S. estimated 370,000 bird deaths from wind turbines per year for the wind capacity of the early 2010s [32]. To put this into perspective, annual bird deaths from communications towers are about seven million, buildings nearly a billion, and household cats (raised by humans) about four million.[104] If you want to preserve and increase bird populations, energy infrastructure is not the only concern.

Renewable Energy Narrative: The Fossil Energy Footprint Is Too Large
Of course fossil fuels are the subject of many discussions of energy-related environmental impacts. On any given day we can find several news articles discussing the long-term impacts related to carbon dioxide emissions and climate change. Coal is the most common villain. Even independent of climate change impacts, many environmental advocates claim there is no such thing as "clean coal."[105] There are many reasons, but two of the most prevalent are due to landscape destruction

---

[103] Big Wind, Big Questions, Texas Parks and Wildlife television series, Program 2513. August 4, 2010: https://youtu.be/N6sx-dmQlnU; https://www.youtube.com/watch?v=ceCwBTXFuC8.

[104] Are Songbirds the Forgotten Wind Power Victim? *Conservation*, Conservation This Week, September 23, 2014 viewed at: http://www.conservationmagazine.org/2014/09/are-songbirds-the-forgotten-wind-power-victim/.

[105] "Clean coal" is sometimes used as a term for coal-fired power plants that are designed to capture a large percentage of the carbon dioxide emissions before they reach the atmosphere, and other criteria pollutants also tend to be removed from the exhaust gases.

(e.g., coal mining via mountaintop removal in Appalachia destroys the landscape and local streams) and an increase in air pollutants that impair human health (e.g., asthma). Most of the detrimental health outcomes due to low air quality occur in cities where large numbers of people are exposed to poorer air quality that is primarily caused from burning petroleum fuels for transportation than from coal or natural gas for power generation. However, in most Western countries pollution controls on vehicles, power plants, and other industrial plants prevent the majority of pollutants from reaching our air. The impacts are not zero, but they are far below what would be the case without pollution control. In the U.S. the economic damages from air emissions have been declining in recent decades [68]. Anyone that has visited some major world cities, such as Beijing, China, and Delhi, India, in the past decade can attest to the poor air quality related to few pollution controls. Of course, achieving good air quality is a challenge when concentrating ten or twenty million people in one metropolitan area with millions of car, truck, and moped engines.

Back to the issue of greenhouse gas emissions. Due to the recent prevalence of hydraulic fracturing and the set of steps related to completing these "unconventional" oil and gas wells, a significant amount of study has focused on the methane emissions from these steps. Depending upon the time frame you want to consider the global warming impact of methane emissions, a policy question as much as scientific one, they have twenty to nearly one hundred times more warming effect than carbon dioxide. Because methane reacts with and decays into other molecules, the longer the time scale of interest, the less global warming impact due to methane emitted today. Since 1–2% of extracted methane ends up leaking along the supply chain, the global warming impact is not trivial [61]. Environmental Defense Fund estimates that as long as natural gas supply chain leakage rates stay below 3.2%, then natural gas power has less global warming impact than new coal power plants [5].[106] A simple estimate of carbon dioxide emissions from new natural gas power plants is that they emit about half the quantity as a coal power plant (half a tonne of carbon dioxide per megawatt-hour versus about one for coal). Thus, any leaking methane only makes total greenhouse gas emissions from the natural gas supply chain even closer to that for coal.

Occasionally, we hear about short-term environmental consequences of fossil energy supply chain. When something goes wrong extracting, handling, or storing fossil fuels or their waste products, it can go really wrong. Renewable narrative proponents don't let the fossil fuel narrative dominate the discussion of impacts to nature, biodiversity, and ultimately our communities and economy:

> Many [oil and gas] reserves are located in some of the world's most pristine places—such as tropical rainforests and the Arctic—that are vital for biodiversity and the ecosystem services that we all depend on, from freshwater to a healthy atmosphere. Extracting them is difficult and dangerous, and costly to businesses, communities and economies when things go wrong.[38]—World Wildlife Fund (2011)

---

[106]The climate impacts of methane emissions, April 2012 at: https://www.edf.org/climate-impacts-methane-emissions.

Concentrated energy dense fuels and wastes can have concentrated impacts. Coal ash spills can bury sections of rivers and homes. For example, in 2008 the dike used to contain coal ash failed at the dewatering area of the Tennessee Valley Authority Kingston coal power plant spreading over 300 acres an into the local Emory river [30]. Large oil spills become front page news and are given household names: the 1989 Exxon *Valdez* oil tanker spill in Alaska, the 2010 *Deepwater Horizon* blowout and sinking in the Gulf of Mexico. Because coal is a solid and oil a liquid, when they come out of their containers and pipelines, we can easily see them.

If methane (or natural gas) leaks, we don't see it with our eyes, but we can see it using different types of cameras. In 2015 the Aliso Canyon gas storage facility in California experienced a pipe failure that resulted in the worst natural gas leak to that date in U.S. history.[107] The scope of the leak was brought to public attention by using infrared cameras that clearly showed the gas plume that is invisible to the unaided human eye.[108]

Three years later there was an even larger methane leak. In 2018 in Ohio, there was an "extreme" 20-day methane blowout from a natural gas well during drilling operations [74]. The leak was detected by sensors on satellites, and the leak amounted to "...leaking more methane in 20 days than all but three European nations emit over an entire year."[109] While some local areas need to be evacuated during these leaks, the long-term consequences of these greenhouse gas leaks have global impacts.

The Subsidies Discussion

> There is no industry in this country that speaks more loudly about the virtues of free enterprise and does more to undermine it than the oil industry.[110]—Milton Friedman (1978)

Subsidies Believe it or not, the discussion of subsidies is one area of agreement between the energy narratives. Each side tells you the other side's subsidies are unfair and need to go, and each side has a good reason for keeping their own.

---

[107]Matt McGrath, California methane leak "largest in US history," BBC News, February 25, 2016, http://www.bbc.com/news/science-environment-35659947.

[108]See website of the Environmental Defense Fund: http://blogs.edf.org/energyexchange/2015/12/10/infrared-camera-reveals-huge-wafting-cloud-of-methane-over-californias-aliso-canyon/. Also see Environmental Defense Fund, December 23, 2015, New Footage Reveals First Aerial View of Methane Leak Polluting Los Angeles County, https://www.edf.org/media/new-footage-reveals-first-aerial-view-methane-leak-polluting-los-angeles-county and https://www.youtube.com/watch?v=exfJ8VPQDTY&feature=youtu.be.

[109]Steven Mufson, "A blowout turned an Ohio natural gas well into a methane 'super-emitter'," *The Washington Examiner*, December 16, 2019 available at: https://www.washingtonpost.com/climate-environment/a-blowout-turned-an-ohio-gas-well-into-a-methane-super-emitter/2019/12/16/fcbdf622-1f9e-11ea-bed5-880264cc91a9_story.html.

[110]"The Energy Crisis: A Humane Solution" https://miltonfriedman.hoover.org/friedman_images/Collections/2016c21/BP_1978_1.pdf.

In the U.S. every energy generation technology and resource extraction industry has one or more financial support mechanisms from one or more government authorities across federal, state, and local governments that include taxing authorities such as school districts, cities, and counties. I will not describe the details of all energy-related subsidies, but it is useful to highlight different types of subsidies.

The subsidy concept sounds simple by its definition:

1. **subsidy:** "A sum of money granted by the government or a public body to assist an industry or business so that the price of a commodity or service may remain low or competitive."[111]
2. **subsidy:** "a grant by a government to a private person or company to assist an enterprise deemed advantageous to the public."[112]

As stated by the U.S. Department of Energy (DOE) in a 2017 report investigating electricity reliability: "Federal and state governments use subsidies, mandates, and prohibitions to affect how public and private entities behave. Subsidies make the favored behavior or product more appealing relative to other competing products by accelerating its development (as with R&D and direct construction expenditures), lowering its ultimate cost to the consumer (as with tax incentives, low lease payments or grants), or making the product better known and more appealing (customer education, ratings, and marketing)." [23]

It has been said that there is nothing certain but death and taxes.[113] Since, as stated by the DOE, energy subsidies are often defined by taxes, they can be difficult to remove once in place. A good summary of the politicization of energy subsidies is the paraphrase attributed to Senator Grassley of Iowa, a state with significant wind power and ethanol production that historically benefited from the relevant energy tax incentives. Responding in August of 2016 whether then presidential candidate Trump could repeal support for wind power via the Production Tax Credit (PTC):

If he wants to do away with it [PTC], he'll have to get a bill through Congress, and he'll do it over my dead body.[114]—Senator Chuck Grassley (R-IA)

---

[111]Oxford English Dictionary, https://en.oxforddictionaries.com/definition/subsidy, accessed September 4, 2017.

[112]Merriam-Webster Dictionary, https://www.merriam-webster.com/dictionary/subsidy, accessed September 4, 2017.

[113]"This is usually attributed to Benjamin Franklin, who wrote in a 1789 letter that 'Our new Constitution is now established, and has an appearance that promises permanency; but in this world nothing can be said to be certain, except death and taxes.' However, *The Yale Book of Quotations* quotes 'Tis impossible to be sure of any thing but Death and Taxes,' from Christopher Bullock, *The Cobler of Preston* (1716). The YBQ also quotes 'Death and Taxes, they are certain,' from Edward Ward, *The Dancing Devils* (1724)." from http://freakonomics.com/2011/02/17/quotes-uncovered-death-and-taxes/ on October 28, 2017.

[114]Devin Henry, Grassley: Trump will attack wind energy "over my dead body," *The Hill*, August 31, 2016 at http://thehill.com/policy/energy-environment/293924-grassley-trump-will-attack-wind-energy-over-my-dead-body.

Politicians attempting to remove energy-related subsidies might very well expedite their own death, politically or physically. Hyperbole aside, the U.S. Congress allowed the 30-year ethanol production tax credit to expire in 2012.[115] No politicians dead in the aisles, but after 30 years, natural attrition played its part.

The second subsidy definition above assumes a subsidy is "advantageous to the public," but it is hard to know if any individual subsidy confers a net cost or benefit to the public over a long period of time. A person's worldview affects how she might analyze the subsidy's impact. In practice, even under strict economic interpretations, assessing subsidies and their net impacts are notoriously difficult. One reason is that while upon creation of the subsidy, economic analyses might estimate a net benefit or cost, after implemented, a net benefit might turn into a net cost, or vice versa. For example, California estimates the cost of its Renewable Portfolio Standard (which sets targets for the percent of all electricity obtained from renewable technologies) by comparing the cost of renewable electricity to the cost of a natural gas combined cycle (NGCC) power plant.   Calculating the cost of natural gas-fired electricity "today" necessitates an assumed natural gas price over the life of the project. If (really when!) the price of natural gas changes from that which was assumed, then the originally calculated cost and benefit no longer represent reality. This is not to suggest we should never estimate costs and benefits of energy subsidies and investment alternatives. While we can't predict the future, we can make educated decisions and keep their uncertainty in context of larger trends. Part of this context means not getting too caught up in pitting one side versus another (e.g., fossil versus renewables) without considering whether our policies are likely to achieve social, economic, and environmental goals that are "advantageous to the public."

By arguing over subsidies that go to businesses, it takes time away from discussions about how our economic and political systems affect people and the environment. This distractive effect is another reason to not define a subsidy as necessarily having a net economic benefit or cost. A subsidy is not necessarily "good" or "bad" for all people. Not all values are able to be expressed in economic calculations. The concept of fairness, or determining if there is a "level playing field," among companies competing to sell energy to consumers can distract from the bigger picture of understanding how humans collectively operate in the economy. As discussed in Chap. 8, a significant quantity of evidence indicates that our collective human decisions seek to expand the human economy. Thus, these quibbles over subsidies might merely be part of a larger process to find the energy resources and technologies that continuously grow the economy [39, 40, 56]. In other words, it is possible that government subsidies help the economy grow by filling an otherwise unmet "need." In the context of the economy, by speeding up the process of technological development, subsidies could be taking the equivalent

---

[115]"After Three Decades, Tax Credit for Ethanol Expires," *The New York Times*, January 1, 2012, http://www.nytimes.com/2012/01/02/business/energy-environment/after-three-decades-federal-tax-credit-for-ethanol-expires.html?mcubz=0 accessed August 20, 2017.

role of accelerating genetic mutations in biological evolution (Chap. 8 expands on this analogy).

There is a question as to what it means for a government to give a subsidy. A government can give some amount of money that it previously did not. For example, a grant is an outright gift, usually for research purposes. A government can also use a tax credit or tax rate reduction to take less money, via taxes, than what was historically taken. Grants, tax credits, and rate reductions are often considered subsidies. However, some might think otherwise. For example, governments often charge a *severance tax* on the revenues from mineral extraction within its jurisdiction. If the government normally taxes mineral extraction (e.g., oil, natural gas, coal) at 7% of its value, but changes the tax to 3%, did it subsidize that industry with a 4% tax break, or does the government still tax at 3% with no subsidy?

There is also significant disagreement regarding not only the categories to include as a subsidy but also on the set of assumptions needed to estimate the magnitude of each category. Consider the diversity within the United States. Some states have a severance tax for some fossil fuels, but not others. For example, Texas has a severance tax on oil and natural gas extraction (7.5% for natural gas, 4.6% for oil and condensate), but not its lignite (coal).[116] Because practically all of Texas' coal is lignite, a low-quality variety usually burned at power plants adjacent to the surface mines, it is generally not sold on the open market and thus also has no *ad valorem tax*.[117] Wyoming on the other hand, which extracts a significant portion of U.S. coal for sale to companies in other states, has a severance tax on all three fossil fuels (6.0% for oil and natural gas, 7.0% for surface coal).[118] Each government makes a subjective choice on whether to tax energy (and other) minerals based on their extraction, sales, or neither. ˙

Governments must have some amount of revenue to function, and taxing mineral extraction or sales is one of many options (e.g., governments tax incomes, property, sales of goods and services). A government decides both how much revenue it needs and how to obtain that revenue. In deciding the appropriate amount of tax on energy mineral extraction, a person's worldview dictates whether they think the tax has filled the government's coffers half full, or whether the coffers are still half empty.

---

[116]A severance tax is one imposed on the removal of natural resources. The severance tax rates for Texas are listed by the Texas Railroad Commission, http://www.rrc.state.tx.us/oil-gas/publications-and-notices/texas-severance-tax-incentives-past-and-present/, accessed September 6, 2017.

[117]An ad valorem tax is a tax whose amount is based on the value of a transaction, typically applied at the time of the transaction.

[118]Wyoming Department of Revenue, http://revenue.wyo.gov/mineral-tax-division/severance-tax-filing-information and https://0ebaeb71-a-84cef9ff-s-sites.googlegroups.com/a/wyo.gov/ wy-dor/SeveranceTaxRates.pdf?atta chauth=ANoY7crI4hXuewCP1XQDgmlaa2z6QO2GkGqRo3\penalty-\@MoeaO7BFMLgJBtBavl0Geb5KZlx35\penalty-\@MJndr5tGK6nocLxjNaf0ljYZuuJALX2\penalty-\@M_ntyXMqYmug1E5yCEhvph1UnzRLO8VFQYUIZsoay0bJnJpo4xpTB59ixpsbaRm884JCeO6j-PsZYk8mUybrwuZ3rv_YeoKHl75Fm7ZkhQ3u9P3t0Qgisj3wQtUtapBQGPQ%3D%3D&attredirects=0, accessed September 6, 2017.

If a country or state has considerable mineral extraction within its borders, then by taxing that extraction, it has the opportunity to lower other taxes. A state with no mineral extraction does not have that option.

Consider Oklahoma, one of the top states for both hydrocarbon extraction and wind power generation. Its legislators' decision in 2017 is a great example of "my subsidy is OK, but yours isn't." In this case, the fossil energy narrative won its argument.

In 2017 the state faced an $870 million dollar revenue shortfall. As part of the package of budget balancing solutions, Oklahoma state elected officials ended a tax credit of 0.5 cents per kilowatt-hour for wind power generation 3 years earlier than it was scheduled to expire. This action was estimated to save $500 million over a decade [95]. Thus, they ended a wind tax credit worth about 6% of the 2017 budget shortfall, but it was projected to increase in future years due to increases in wind power generation. As Sen. Bryce Marlatt (R-Woodward), an Oklahoma legislator, put it, "The tax credit was put in place to try to help the [wind] industry get off the ground and get going . . . Obviously, with the success we've had and the amount of development we've had, it's done its job and the tax credit's not needed anymore." [90] The 2017 legislation only applied to new wind farms. So existing wind farms continued to receive the tax credit. Oklahoma's opponents to the wind industry were not content with simply letting the wind tax credit fade with time. They established a non-profit, *WindWaste*, ". . . dedicated to educating Oklahomans about the harmful effects of Industrial Wind."[119]

Contrast Oklahoma's wind tax credit repeal with maintaining a tax reduction for its oil and gas industry. In 2014 the government extended a tax break, set to expire in 2015, for horizontal drilling that was originally put in place to assist what was in the recent past a nascent technology in the well-established oil and gas industry. This tax break lowers the severance tax on oil and gas extraction to 2% for the first 3 years of extraction from new horizontal wells before then returning to the normal 7% rate [9]. This length of time for the lower severance tax is extremely relevant since a majority of extraction from horizontal and hydraulically fractured wells usually comes during the first 3 years. Thus, most of Oklahoma's oil and gas is taxed at a low rate. In 2017, the same year Oklahoma legislators prematurely ended the $50 million per year tax break for wind developers, the lowered severance tax on oil and gas cost the state approximately $450 million, or about half of the state budget shortfall [9].

If the Oklahoma wind tax credit was removed because the wind industry was considered mature enough, then by leaving the severance tax reduction on horizontal wells, it appears as though Oklahoma legislators believe the oil and gas industry is still not mature enough to afford its historical 7% tax rate. Modern wind turbines of size greater than 1 MW have been around since the late 1990s, about 30 years. The current concept of hydraulic fracturing was first used in the 1940s, over 60 years

---

[119]WindWaste: https://windwaste.com/.

ago.[120] What else should we expect from a state with a governor who proclaimed a day in 2016 as "Oilfield Prayer Day?"[121]

Horizontal drilling and hydraulic fracturing of tight sand formations has been touted as a "revolution" in technology for hydrocarbon extraction. There is no doubt these combined technologies allow access to significant quantities of resources that were previously uneconomic (i.e., it turned many *resources* into *reserves*). But if these technologies are truly revolutionary, why do companies using them need a tax break? It is this type of question that we cannot answer without the systems context of Chap. 5 that considers the individual energy and economic trends of Chaps. 2 and 4.

For now, consider that Texas seems to agree with Oklahoma. At any given time Texas houses approximately 40–50% of all operating oil and gas drilling rigs within the U.S.[122] In 1989 the Texas Legislature codified a severance tax reduction for "high-cost gas" that now also applies to natural gas from horizontal and hydraulically fractured wells.[123] The Texas Comptroller of Public Accounts states that the tax incentive originally applied to 3% of gas produced in 1997 but from 2004 to 2014 applied to over 40% of gas production, and over 60% from 2009 to 2012 [76]. The average tax rate for all high-cost gas wells was 1–2% (as opposed to the normal 7.5%), and from 2005 to 2014, the annual state-wide tax reduction for all high-cost wells ranged $0.8–$2.0 billion equating to 0.19–0.48 $/Mcf (Mcf = thousand cubic feet), or 4–8% of the Henry Hub market price of natural gas.[124] Ironically, high-cost gas wells (as defined by the Texas Legislature) are largely responsible for keeping natural gas prices below the historic highs of the mid-2000s. Never judge a tax break by its name.

To date, most discussions of subsidies for fossil fuel extraction largely focus on what tax rate, greater than zero, should be applied to their extraction. Thus, the government is increasing the cost (as seen by the consumer) of fossil energy to some degree in order to obtain revenue for providing social services. Most discussions

---

[120]FracFocus website, "Hydraulic fracturing is not new. The first commercial application of hydraulic fracturing as a well treatment technology designed to stimulate the production of oil or gas likely occurred in either the Hugoton field of Kansas in 1946 or near Duncan Oklahoma in 1949," visited April 28, 2018: https://fracfocus.org/hydraulic-fracturing-how-it-works/history-hydraulic-fracturing.

[121] AP (2016) "Oklahoma governor urges Christians to pray for oil industry," http://www.foxnews.com/us/2016/10/10/oklahoma-governor-urges-christians-to-pray-for-oil-industry.html.

[122]Baker Hughes North American Rig count, accessed September 2017 at http://phx.corporate-ir.net/phoenix.zhtml?c=79687&p=irol-reportsother.

[123]The tax reduction applies to the "...first 120 consecutive calendar months beginning on the first day of production, or until the cumulative value of the tax reduction equals 50 percent of the drilling and completion costs incurred for the well, whichever occurs first." Texas Tax Code Chapter 201, Section 201.057. Accessed July 17, 2017 at http://www.statutes.legis.state.tx.us/Docs/TX/htm/TX.201.htm.

[124]Mcf = thousand cubic feet of natural gas. Natural gas prices at Henry Hub (benchmark price for North America) typically range from 2 to 6 $/Mcf, but much of the time from 2004 to 2008 prices were greater than $6/Mcf.

of subsidies for renewable energy largely focus on what amount of money will be given, as a tax credit or grant, that effectively lowers the cost (as seen by the consumer) of renewable energy. The most prominent U.S. federal subsidies for wind and solar electricity have been the Investment Tax Credit (ITC) and Production Tax Credit (PTC).

The ITC reimburses the capital cost of a power plant by some stated percentage. The subsidy is independent of the amount of electricity generated. The PTC provides a tax credit at some value for every megawatt-hour of electricity generated (e.g., some $/MWh of subsidy) independent of capital cost. For most of its history, the ITC reimbursed capital costs by 30%, and the PTC for renewable electricity was 23 dollars per megawatt-hour. In multiple instances over the last two decades these highly influential subsidies have been temporarily renewed as they were typically designed to expire every 2–4 years. The drumbeat to remove them is summarized in the following quote:

> The PTC [Production Tax Credit] is a massive waste of taxpayer dollars. Perhaps worse, the PTC distorts energy markets by allowing wind producers to actually pay the grid to take their electricity so they can continue to collect federal largesse. This harms reliable energy sources like natural gas that Americans depend on keep the lights on. It is long past time for Congress to let the PTC permanently expire.[125]—Travis Fisher and Alex Fitzsimmons, Institute for Energy Research (2013)

Alex Fitzsimmons, one of the authors, later worked for Fueling U.S. Forward, an organization backed by Charles and David Koch that promoted fossil fuels, as well as the Department of Energy for the Trump administration.[126]

As of this writing, the ITC and PTC were set to gradually expire in the early 2020s as outlined in The Consolidated Appropriations Act of 2016 (under President Obama). This act was a compromise with fossil fuel interests: the renewable tax credits were extended and oil producers were allowed to export crude oil from the United States for the first time in 40 years.

The ITC and PTC concepts are not only applicable to renewable energy technologies. For example, the Energy Policy Act of 2005 established a federal PTC for up to 6000 MW of new nuclear power plants coming online before 2021. Because now new reactors were going to fit that timeframe, Congress extended the nuclear PTC of $18/MWh for the next 6000 MW of nuclear power built any time after 2020.[127] As noted earlier in this chapter, due to cost overruns during

---

[125]Travis Fisher and Alex Fitzsimmons, Institute for Energy Research (September 19, 2013), accessed July 20, 2017 at https://www.instituteforenergyresearch.org/analysis/wind-ptc-threatens-grid-reliability/.

[126]https://www.eenews.net/greenwire/2017/07/11/stories/1060057185.

[127]Nuclear Energy Institute, "Congress Passes Nuclear Tax Credit in Big Boost for New Construction," February 9, 2018, at: https://www.nei.org/news/2018/congress-passes-nuclear-production-tax-credit.

construction of the last four nuclear reactors in the U.S., two were canceled halfway into construction, but as of 2019 two continue in construction.[128]

Tax credits like the ITC and PTC are interesting subsidies. By definition, they reduce the tax burden of a profitable business. All business operations, including energy industries, are taxed on profits. Should the energy project owner generate more tax credit than its own tax burden, the money associated with the ITC and PTC can be sold to other companies. The tax credit can even be sold, for profit, if the electricity facility owner did not otherwise make a profit on selling the electricity. This design provides some flexibility in that small project owners with low tax burdens can sell "excess" tax credits to larger companies that have tax burdens but have no relation to the energy sector. While tax credits such as the ITC and PTC clearly lower the cost of developing power plants, their structure raises the question of whether they are meant to benefit only the energy sector versus increase the profitability of businesses in general. One downside of tax credits is that during an overall economic recession, because there are fewer companies with profits, and thus lower taxes across the economy, the value of the credits can be carried forward to future years.

Among all of this discussion of giving subsidies to renewable electricity generation, one state policy stands out with regard to renewable energy taxation— Wyoming. While taxes have been historically applied to fossil fuel extraction, it is not common to specifically tax renewable energy production. While Wyoming doesn't directly tax the wind or sun as an input for power generation, the state has applied a 1 dollar per MWh tax on output electricity generation from wind turbines, and in 2017 legislators debated moving it to 5 $/MWh.[129] This Wyoming wind production tax is the closest concept to a severance tax on renewable power. Many people see the Wyoming wind production tax as a political effort to deter wind development and promote Wyoming's coal extraction industry with claims for the need to preserve viewscapes and wildlife.[130] But there is another viable viewpoint. Wyoming has exceptional wind resources and a low human population density. Thus, taxing wind production (exported to other states) could possibly substitute for what has, over the last few years, become a declining severance tax from declining coal extraction. The discussion in 2017 in Wyoming was not over the size of a tax subsidy for the wind industry, but the size of the tax. In short, if you have coal extraction, tax it. If you have wind power, then perhaps you need to tax that too. In a strange way, by taxing wind electricity Wyoming is respecting the maturity of the wind industry and the quality of its wind resource.

---

[128]"Scana stops work on partially built V.C. Summer Nuclear Station as estimates of cost to finish it rise to $25 billion," *Wall Street Journal* https://www.wsj.com/articles/scana-halts-south-carolina-nuclear-power-project-1501524763 (accessed July 31, 2107 4:02 pm ET).

[129]Hancock, Laura (2017), Legislators kill wind tax bill, *Casper Star Tribune*, available on September 26, 2017 at: http://trib.com/news/state-and-regional/govt-and-politics/legislators-kill-wind-tax-bill/article_2cce36b6-b8e7-51e2-b4cf-3add20f8719a.html.

[130]Such as the sage brush grouse, a ground bird that avoids wind turbines, is prominent in Wyoming, and a species that the U.S. Fish and Wildlife Service decided in 2015 not to list as endangered.

Subsidies: Should We Care?

At a high level, the discussions of different tax types at various levels of government divert us from broader more important questions. Should we have taxes on the extraction of minerals such as fossil fuels, or production of electricity from wind turbines and solar panels? Does the tax code, in aggregate, favor any energy type over others? If it does, should we care?

A few years ago I performed a study with colleagues at The University of Texas at Austin. We calculated *only a subset* of subsidies, or government financial support, that relate to the electricity supply chain. Other studies consider much broader boundaries of analysis for categorizing a subsidy. For example, one can count military spending on aircraft carrier fleets patrolling the Persian/Arabian Gulf in the Middle East as a subsidy to consumers and oil companies investing in that area.

In our relatively narrow focus on electricity, the subsidies for the U.S. in 2016 amount to approximately 0.1% of gross domestic product (GDP) [41]. For California in 2016 the electricity-related government support as a percent of gross state product was 0.2%, and for Texas it was 0.1% or 0.2% (depending on if counting the cost of the Competitive Renewable Energy Zones transmission lines, mentioned earlier in this chapter) [42]. At a high level the two states contribute approximately the same relative amount to electricity-related subsidies as the federal government. Is this a large amount?

The 2016 U.S. federal budget was approximately 21% of GDP, and consumer spending on energy goods and services since 1929 has typically been between 3% and 5% of GDP with a slight downward trend over time.[131] Thus, federal plus state electricity-related subsidies amount to a few percent of total spending on energy. Not extremely small. Not extremely large.

But if energy is already relatively expensive it doesn't take much of an increase in cost to be the difference between economic growth versus contraction. Since World War II, high energy expenditures were usually driven by sharp increases in oil prices. There is a nonlinear negative feedback. If oil expenditures increased from 2% to 3% of GDP, oil prices make the news, but we all keep about our daily routines. But if oil expenditures increase from 3.5% to 4.5%, this has historically triggered recession [60]. We'll further discuss this concept of an energy spending threshold in Chap. 5.

Subsidies: Taking a Step Back

It is simply a matter of fact that some people work for companies associated with one set of energy options over another. Thus, considerable effort goes into analyses of subsidies supporting or refuting the renewable or fossil energy narratives. By

---

[131]This fraction is taking "Energy Goods and Services" spending from the Bureau of Economic Analysis Annual Personal Consumption Expenditures Table 2.3.5 and dividing by U.S. GDP.

stepping out from both the fossil and renewable perspectives, we can view subsidies for what they are: efforts to grow the economy. By making energy appear cheaper to consumers than what it otherwise is, consumers make choices that are more compatible with increased consumption. This notion has historically been much more pronounced in some oil and gas exporting countries that used sales from petroleum exports to enable their citizens to purchase gasoline below the market price. This is a major reason why Venezuela has had such economic difficulty since 2014. The country kicked out the international oil companies and subsequently was unable to maintain its (relatively high cost) oil production and provide subsidized consumption to its citizens. Once the Venezuelan population became exposed to world market prices for energy, food, and other goods, they could not afford their levels of consumption, and the economy went into deep recession.

Subsidies that lower the prices of consumer energy commodities, like electricity and gasoline, are different from those that affect the profitability and decision making of investors and energy companies. First, let's consider subsidizing energy commodities. For example, any oil-exporting country that charges its citizens less for petroleum fuels (or any number of consumer goods) than the going world prices is exposing itself to a potential problem if it no longer exports enough oil at a high enough price. If you sell oil at a global price of $100/BBL but charge your citizens as if oil costs $60/BBL, then as long as you sell enough oil on the global market to pay for this difference of $40/BBL, you are OK. However, if you no longer have any excess oil to sell, every barrel you buy on the market is $100/BBL and you must start charging your citizens more for oil products or have the oil company, and perhaps the country itself, go into a budget deficit and accumulate debt. Alternatively, the same problem can occur even if you still sell the same amount of oil, but instead the global oil price drops from 100 to 50 $/BBL. This declining revenue from oil exports roughly explains the situation in the Venezuelan example.

Now let's consider subsidizing energy extraction, electricity generation, or capital investments in projects like wind and solar farms. These types of subsidies are used in market economies. They do affect consumer energy prices, but only indirectly. They primarily affect the profitability of companies, and many companies can take advantage of them. For example, in the cases of the Investment Tax Credit and Production Tax Credit, companies that have no role in the energy industry can take advantage of them by either buying the tax credits directly from the project developer or by investing in the project while factoring in shared benefits from the tax credits. Large investment funds, such as pension funds, can take this latter approach to have relatively stable returns over decades.

Because subsidies are one factor that affects the ability of energy companies to attract investors, the narratives for and against energy technologies persist. But the battles aren't only pitting fossil fuels versus renewables. Disagreement about the benefits of fossil and renewable energy resources are not limited to arguments *between* fossil fuel and renewable proponents. Disagreements also occur *within* each energy narrative.

With Friends Like This, Who Needs Enemies?—Fossil Energy Narrative

Fossil fuels is not a catch all phrase for all companies that extract, transport, or burn oil, natural gas, and coal. Different groups of fossil fuel advocates can compete vehemently with one another.

In 2006, TXU, a power generation company in Texas, announced they wanted to build 11 new coal-fired power plants in Texas. Other power companies stated they wanted to build an additional eight. The quantity of suggested new power capacity, around 11 GW installed over approximately 5–10 years, was significantly more than what was likely needed to serve increased demand for electricity. But hey, this was 2006, before the 2008 financial crisis rocked the world economy and before *fracking* for natural gas and oil was a common word. Worldwide demand for commodities, largely stemming from Chinese demand, was pushing up prices for oil and many other commodities. The increasing oil prices increased the cost of drilling rigs and thus the cost of drilling for natural gas as well. Because natural gas was the marginal fuel for power generation in Texas, it set the price of electricity. Higher natural gas prices meant higher prices for electricity, and larger profits for coal generators. The solution, according to many power companies: build more coal plants.

As part of its push for more coal power, TXU ran a series of television and newspaper ads targeting primal human instincts. One ad featured a toddler in a dark room afraid of monsters. The ads stated "Our new, advanced coal plants could generate energy for over 5 million new homes and enough power to keep the monsters away." [26, 62] Are your kids afraid of monsters? Burn more coal. TXU stated that the ad campaign was "... an opportunity to get our message across in an unbiased way."[62] Unbiased? Judge for yourself in Fig. 3.7.

These advertisements and the stated plans for the additional coal plants prompted what became known as the "Texas coal wars." Robert Redford's Redford Center even co-sponsored a film about the struggle: *Fighting Goliath: Texas Coal Wars*. But there weren't only your expected anti-coal environmental groups fighting in this war. Oil and gas producers, coal mining companies, electric utilities, and independent power producers do not speak with a monolithic voice. In regulated or restructured electric markets, fossil fuel producers and consumers can be in conflict with each other.

Some natural gas companies were part of an organization called the Texas Clean Sky Coalition that sprung up to fight the coal plant expansion. While natural gas-fired power plants produce less air pollution, than those of coal, the issue wasn't only about air quality. This group, whose funders included Chesapeake Energy Corporation, wanted to keep natural gas as the dominant fuel for Texas' power generation [62]. The Texas Clean Sky Coalition countered TXU's ads with their own showing actors with "studiocreated" black makeup smudges on the faces of men, women, and children [62]. Does Fig. 3.8 show the face of a coal miner? Instead of conjuring up monsters going after your children, these anti-coal ads asked questions such as "Would you bathe your child in coal? Sprinkle arsenic, mercury, and lead

Fig. 3.7 The "monsters" advertisement run by TXU in 2007 to persuade Texans to support them building eleven new coal-fired power plants

on your husband's cereal? Treat your friends to a big dose of radiation?" [26] Other ads were more simple: "Face it. Coal is filthy." and "Live Better. Live Longer. No more coal plants." [62].

What was the response of TXU to the ads from Texas Clean Sky Coalition? "Because the facts are not on their side, it seems these groups have resorted to staged, misleading images as tactics to scare the public," [77]. Misleading images?

Fig. 3.8 The "Coal is Filthy" advertisement run by Texas Clean Sky Coalition in 2007 to persuade Texans to oppose TXU and other companies' plans to build up to 19 new coal-fired power plants

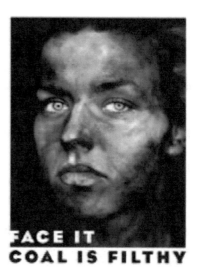

Yes, but on both sides. Apparently Texans' choice was between monsters going after your sleeping child holding a Teddy bear or arsenic on your cereal. Take your pick.

How did this all turn out? The excitement over Texas power generation led an investment group buying TXU in what was the most expensive private equity investment to date. In 2007, the private equity firms Kohlberg, Kravis, Roberts (or KKR), and the Texas Pacific Group, along with four investment banks including Goldman Sachs, paid $32 billion in cash along with assuming $13 billion in debt to take over TXU [20]. Luminant, the power generation arm of the newly formed private company Energy Future Holdings, only ended up building three new coal-fired power plants. It turns out that was probably three too many.

Anticipating that natural gas prices would stay high for a long time, the investment anticipated making high margins from owning lower-cost coal plants operating in Texas' high-cost gas power market. It didn't turn out that way. The boom in horizontal drilling and hydraulic fracturing, along with significant increases in wind-power electricity, took away market share and lowered wholesale electricity prices too much for the private equity owners to make money or pay back their debt. In 2014, Energy Future Holdings filed for bankruptcy with $40 billion in debt [16].

With Friends Like This, Who Needs Enemies?—Renewable Energy Narrative

Just as different fossil advocates can argue against each other, so can advocates for renewable energy. Here my example is that of academics in the arena of simulation and analysis rather than natural gas and coal companies battling within electricity markets. However, this was not simply an academic battle among ivory towers, there was a lawsuit and policies that hung in the balance.

This *academic* controversy centers on studies led by Mark Jacobson of Stanford University and Mark Delucchi of University of California at Berkeley. Jacobson and Delucchi have published several peer-reviewed studies [19, 53, 54] on how the U.S. can transition to 100% renewable energy, not only for electricity generation, but for all primary energy needs. In particular, it was their 2015 paper published in the *Proceedings of the National Academy of Sciences*, or *PNAS*, that led to considerable discussion and controversy [54]. This paper led to a critique by 21 authors that was published in *PNAS* in 2017 [17], and *PNAS* also allowed Jacobson et al. to write a response to the critique [55]. This was so enthralling to the energy nerd (shall I say) community, that Greentech Media even dedicated a webpage to the controversy.[132]

To give a feel for the expression of each group of authors, I post and discuss quotes from the critique of the Jacobson research, which I refer to as "Clack et al. (2017)."[17] I refer to Jacobson's rebuttal to Clack et al. (2017) as "Jacobson et al. (2017)."[55] For those interested in reading a summary on one of the major points of contention in this saga, that of the treatment of hydropower in Jacobson's model, see the Appendix that discusses this specific mathematical modeling point of controversy as illustrative of the challenges of projecting an energy future far-removed from ours of today.

From the Clack et al. (2017) critique of Jacobson et al. (2015):

> The scenarios of [Jacobson et al. (2015)] can, at best, be described as a poorly executed exploration of an interesting hypothesis. The study's numerous shortcomings and errors render it unreliable as a guide about the likely cost, technical reliability, or feasibility of a 100% wind, solar, and hydroelectric power system. It is one thing to explore the potential use of technologies in a clearly caveated hypothetical analysis; it is quite another to claim that a model using these technologies at an unprecedented scale conclusively shows the feasibility and reliability of the modeled energy system implemented by midcentury. [17]

Two quotes from the response by Jacobson et al. (2017) to the critique of their work are:

> The premise and all error claims by Clack et al. [(2017)] in *PNAS*, about Jacobson et al.'s [(2015)] report, are demonstrably false. We reaffirm Jacobson et al.'s conclusions. [55]

and,

> In sum, Clack et al.'s [(2017)] analysis is riddled with errors and has no impact on Jacobson et al.'s [(2015)] conclusions. [55]

It is hard to be more starkly adamant and of opposite opinion than these competing statements. Clack et al. (2017) state there are errors. Jacobson et al. (2017) say "no there aren't."

It is important to keep in mind that *practically all authors involved, in both papers, are generally interested in understanding how to use more renewable and/or*

---

[132] Your Guide to the Bitter Debate Over 100% Renewable Energy: https://www.greentechmedia.com/articles/read/100-renewable-energy-debate.

low-carbon energy! That is to say they are largely in the same camp in terms of performing engineering, scientific, and economic analyses that discuss how to wean off of fossil fuels, or at least reduce CO_2 emissions. However, Jacobson rejects the use of nuclear power or technologies to capture and store CO_2 emitted from fossil power, whereas many of his critics consider one or both of these as viable options.

Renewable energy proponents disagree on whether or not this scientific discussion, centering on mathematical modeling, should be out in the public. There are two general positions on this issue of openly discussing disagreements on models of future renewable energy scenarios.

Position 1 is that all open discussion of science, engineering, and economic analyses of how to transition to more use of renewable energy is a good thing. That is to say, more discussion leads to more accurate analyses in the long-run, and we want the most accurate analyses as possible. If this discussion involves a lot of both good and bad information exchange during the process, then that is OK because that is just the scientific process going through its motions.

Position 2 is that open discussion of disagreement on the technical and/or feasibility of increasing the use of renewable energy is bad thing because it gives ammunition to fossil fuel advocates.[133] Quite simply, fossil fuel advocates can point to peer-reviewed publications that say how going to 100% renewable energy is not possible because studies claiming that possibility are flawed.

I clearly fall into the group advocating Position 1. If I didn't, I wouldn't have written this book.

In the June 22, 2017 Greentech Media podcast of The Energy Gang, Jigar Shah and Katherine Hamilton argue that the Jacobson paper is flawed but proposes a vision of 100% renewables.[134] They see the journal paper as benign because policymakers don't take it seriously. Stephen Lacey, the reporter and host of The Energy Gang, disagreed noting that publishing in peer-reviewed literature is a serious matter. I agree with Lacey's view.

Peer-reviewed articles are not the place for visions. They are the place for analyses with all assumptions and caveats explained for interpretations by other researchers. However, we cannot kid ourselves. When writing legislation, policymakers do in fact feel more justified in basing legislation on results from peer-reviewed literature than other published formats, such as white papers or articles in lay magazines. I know because I have specifically asked this question to congressional staff. In discussing my and my colleagues' research with persons at White House offices as well as staff members of House and Senate committees, these staff members confirm that they prefer peer-reviewed literature to inform energy and environmentally related legislation. It is also true that a congressional staff member can usually find a peer-reviewed journal article, or a white paper from a think tank, to support a predetermined position.

---

[133] Joe Romm, Dear scientists: Stop bickering about a 100% renewable power grid and start making it happen: https://thinkprogress.org/a-carbon-free-grid-is-unstoppable-so-why-did-a-nasty-debate-about-it-just-erupt-fa2bf7a6827a/, accessed June 21, 2017.

[134] June 22, 2017: https://www.greentechmedia.com/articles/read/100-renewable-energy-debate.

Jacobson also points how his concerns are more than greenhouse gases and current economics. His scenarios target such ideas as the improvement of local air quality, increasing jobs, and minimizing land use for energy by avoiding bioenergy. In an interview with Greentech Media he criticizes a few of the critiquing authors by name for being pro-nuclear, and thus in Jacobson's view, having a vested interest against his idea of using *only* wind, water, and solar (WWS) energy technologies.[135] Others are criticized, but not by name, for being part an academic department of a university that gets money from natural gas companies that in Jacobson's opinion influence the view of that author. He also points out that *PNAS* typically requires co-authors to have provided a substantial contribution, which has to be listed, to the written article. Of the 21 authors of Clack et al., only 3 are listed as designing the research, performing the research, or analyzing the data. The other 18 co-authors are listed only as writing the paper.

Jacobson also points out that the Clack et al. (2017) paper is very uncommon because while it presents a critique of original research rather than original research itself, it is published as a regular "research report." Usually, scientists discuss disagreements over methodology or results of a research article via back-and-forth letters to the editor of the journal. Why allow a research paper simply as a critique of another research paper? Clack et al.'s response is that they became frustrated when they could not constructively engage with Jacobson regarding his research, and thus they felt they had no choice but to write a full rebuttal article. The *PNAS* editors agreed with this argument and decided to allow the Clack et al. (2017) critique in the more formal full length manner.[136]

At this point I'll end this section by summarizing the end of the saga itself. This drama peaked in severity in November 2017 when Jacobson filed a lawsuit, seeking damages of $10 million, against both Christopher Clack and the U.S. National Academy of Sciences, the latter being the organization that publishes the journal *PNAS* [67, 97]. Jacobson claimed that he notified *PNAS* there were multiple falsehoods and several "materially misleading statements" in the pre-published version of the Clack et al. (2017) paper. However, the *PNAS* editors considered Jacobson's concerns, sent the paper through the peer-review process, and as Clack noted, found Jacobson's criticisms to be "without merit." [97]

---

[135]July 12, 2017: https://www.greentechmedia.com/articles/read/an-interview-with-mark-jacobson-about-100-percent-renewable-energy.

[136]From PNAS website, http://www.pnas.org/site/authors/authorfaq.xhtml, accessed August 13, 2017:

"Submissions must be:

- original scientific research of exceptional importance,
- work that appears to an NAS member to be of particular importance, and
- intelligible to a broad scientific audience."

In February 2018, Jacobson dropped his lawsuit and posted an FAQ document on his university-based website in which he provided a lengthy explanation of his argument and reasoning for dropping the lawsuit.[137]

Summary

There are three kinds of lies: lies, damned lies and statistics.—unknown

The above quote is in Mark Twain's *Chapters from My Autobiography*, but the originator of the quote seems to be unknown. In this chapter I have not so much shown statistics, but some historical and calculated numbers that we can consider as relatively uncontroversial facts. I have also quoted many statements of words that are used to argue points that, in the context of energy, are often better left to the numerical domain. Thus, sometimes people agree on the numbers and the facts yet use them to argue for opposing narratives.

Reliability is a good example. In the context of electricity generation, the concept of reliability is much better left to the mathematical and computational analyses within engineering and economics. A conceptual debate on whether the sun, wind, your spouse, your mother, or your dog are reliable is a fun discussion for philosophers and friends over a nice drink, but that debate does not help keep the lights on.

If you have a fossil or renewable energy argument you want to make, you can usually find a number to support it. If you can't find a number, then as this chapter has shown, you can simply move to another argument. There are many characteristics of energy systems that are important and relevant to consider, and practically impossible to digest them all in one conversation. How are we to weigh them all? Is there any point to nit-picking over details of how one energy resource or technology is better or worse on one metric versus another? We will continually explore these questions in the rest of the book. But now, we need to explore several important socio-economic megatrends that are critical to consider in the context of energy consumption. These are the bases for the economic narratives, and they set the stage for whether the economic narratives allow for energy to play any role in explaining the trends.

References

1. Acemoglu, D., Robinson, J.A.: Why Nations Fail: The Origins of Power, Prosperity, and Poverty. Crown Business, New York (2012)
2. Adelman, M.A.: Mineral depletion, with special reference to petroleum. The Review of Economics and Statistics **72**, 1–10 (1990). http://www.jstor.org/stable/2109733
3. Adelman, M.A.: The real oil problem. Regulation pp. 17–21 (2004)

---

[137]Jacobson (2018) FAQs and Withdrawal of Lawsuit against Clack," downloaded Feb. 23, 2018 from http://web.stanford.edu/group/efmh/jacobson/Articles/I/CombiningRenew/18-02-FAQs.pdf.

4. Aissaoui, A.: Fiscal break-even prices revisited: What more could they tell us about OPEC policy intent? Tech. Rep. 8–9 (2012)
5. Alvarez, R.A., Pacala, S.W., Winebrake, J.J., Chameides, W.L., Hamburg, S.P.: Greater focus needed on methane leakage from natural gas infrastructure. Proceedings of the National Academy of Sciences 109(17), 6435–6440 (2012). https://doi.org/10.1073/pnas.1202407109. http://www.pnas.org/content/109/17/6435
6. Barbose, G., Darghouth, N.: Tracking the sun pricing and design trends for distributed photovoltaic systems in the united states, 2019 edition. Tech. rep., Lawrence Berkeley National Laboratory (2019). https://emp.lbl.gov/tracking-the-sun
7. Bashmakov, I.: Three laws of energy transitions. Energy Policy 35, 3583–3594 (2007)
8. Bauerlein, V., Gold, R.: South Carolina seeks ways to salvage nuclear project (2017). https://www.wsj.com/articles/south-carolina-seeks-ways-to-salvage-nuclear-project-1502125431
9. Blatt, D.: We must end oil and gas tax breaks to save Oklahoma communities (2017). http://okpolicy.org/must-end-oil-gas-tax-breaks-save-oklahoma-communities/. Accessed September 3, 2017
10. Bolinger, M., Seel, J., Robson, D.: Utility-scale solar: Empirical trends in project technology, cost, performance, and PPA pricing in the United States – 2019 edition. Tech. rep., Lawrence Berkeley National Laboratory (2019). https://emp.lbl.gov/utility-scale-solar
11. BP: Bp statistical review of world energy, 2016 (2016). http://www.bp.com/en/global/corporate/energy-economics/statistical-review-of-world-energy/downloads.htm. Accessed August, 2016
12. BP: Bp statistical review of world energy, 2019 (2019). http://www.bp.com/en/global/corporate/energy-economics/statistical-review-of-world-energy/downloads.htm. Accessed August, 2019
13. Bryce, R.: Power Hungry: The Myths of "Green" Energy and the Real Fuels of the Future. Perseus Books (2010)
14. Chu, S.: Letter from secretary steven chu to energy department employees. Tech. rep. (2013). https://energy.gov/articles/letter-secretary-steven-chu-energy-department-employees. Accessed August 3, 2017
15. Churches, C.E., Wampler, P.J., Sun, W., Smith, A.J.: Evaluation of forest cover estimates for Haiti using supervised classification of Landsat data. International Journal of Applied Earth Observation and Geoinformation 30, 203–216 (2014). http://dx.doi.org/10.1016/j.jag.2014.01.020. http://www.sciencedirect.com/science/article/pii/S0303243414000300
16. Chutchian, M.: Texas utility giant EFH poised to exit bankruptcy after three years. Forbes (2017). https://www.forbes.com/sites/debtwire/2017/02/16/texas-utility-giant-efh-poised-to-exit-bankruptcy-after-three-years/#37f4817f3a6b. Accessed July 20, 2017
17. Clack, C.T.M., Qvist, S.A., Apt, J., Bazilian, M., Brandt, A.R., Caldeira, K., Davis, S.J., Diakov, V., Handschy, M.A., Hines, P.D.H., Jaramillo, P., Kammen, D.M., Long, J.C.S., Morgan, M.G., Reed, A., Sivaram, V., Sweeney, J., Tynan, G.R., Victor, D.G., Weyant, J.P., Whitacre, J.F.: Evaluation of a proposal for reliable low-cost grid power with 100% wind, water, and solar. Proceedings of the National Academy of Sciences 114(26), 6722–6727 (2017). https://doi.org/10.1073/pnas.1610381114. http://www.pnas.org/content/114/26/6722.abstract
18. NERC (North American Electric Reliability Corporation): Special report: Accommodating high levels of variable generation (2009). www.nerc.com/files/ivgtf_report_041609.pdf. Accessed July 21, 2017
19. Delucchi, M.A., Jacobson, M.Z.: Providing all global energy with wind, water, and solar power, Part II: Reliability, system and transmission costs, and policies. ENERGY POLICY 39(3), 1170–1190 (2011). https://doi.org/10.1016/j.enpol.2010.11.045
20. Demos, T.: Don't mess with texas: State politicians could derail the carefully constructed TXU deal, says Fortune's Telis demos. Fortune (2007). http://archive.fortune.com/2007/03/08/news/companies/txu_lawmakers.fortune/index.htm?section=money_lates
21. Denholm, P., Hand, M., Jackson, M., Ong, S.: Land-use requirements of modern wind power plants in the united states. Tech. rep., National Renewable Energy Laboratory (2009). http://www.nrel.gov/docs/fy09osti/45834.pdf. Accessed August 6, 2017

22. DOE: Quadrennial energy review, transforming the nation's electricity system: The second installment of the QER. Tech. rep., U.S. Department of Energy (2017). https://www.energy. gov/sites/prod/files/2017/02/f34/Quadrennial%20Energy%20Review--Second%20Installment %20%28Full%20Report%29.pdf. Accessed February 10, 2017
23. DOE: Staff report to the secretary on electricity markets and reliability. Tech. rep., U.S. Department of Energy (2017). https://energy.gov/sites/prod/files/2017/08/f36/Staff%20Report%20on %20Electricity%20Markets%20and%20Reliability_0.pdf. Accessed September 3, 2017
24. Doig, A., Adow, M.: Low-carbon Africa: Leapfrogging to a green future. Tech. rep., Christian Aid (2011). http://www.christianaid.org.uk/resources/policy/climate/low-carbon-africa.aspx. Online; accessed September 16, 2017
25. Dorn, J.G.: Eco-economy indicators, solar power, solar cell production jumps 50 percent in 2007. Tech. rep., Earth Policy Institute (2007). http://www.earth-policy.org/datacenter/ xls/indicator12_2007_7.xls. Main website: http://www.earth-policy.org/indicators/C47/solar_ power_2007
26. EENews: Ad war putting face on texas power plant debate. E&E News (February 16) (2007). https://www.eenews.net/greenwire/stories/51697. Accessed July 11, 2017
27. EIA: Monthly energy review (2012)
28. Energy, R.: Rystad energy annual review of global recoverable oil resources: Saudi Arabia adds oil resources ahead of IPO. Rystad Energy website (2017). https://www.rystadenergy.com/ NewsEvents/PressReleases/2017-annual-oil-recoverable-resource-review. Accessed August 4, 2017
29. DOE (Department of Energy): The importance of flexible electricity supply, solar integration series. 1 of 3. Tech. rep., DOE (2011)
30. EPA: "U.S. Environmental Protection Agency" and "Tennessee Valley Authority" "Kingston" coal ash release site, project completion fact sheet. https://semspub.epa.gov/work/04/ 11015836.pdf (2014). Online; accessed 22-April-2018
31. Epstein, A.: The Moral Case for Fossil Fuels. Portfolio (2014)
32. Erickson, W.P., Wolfe, M.M., Bay, K.J., Johnson, D.H., Gehring, J.L.: A comprehensive analysis of small-passerine fatalities from collision with turbines at wind energy facilities. PLoS ONE **9**(9), e107,491 (2014). https://doi.org/10.1371/journal.pone.0107491. http://dx.doi. org/10.13712Fjournal.pone.0107491
33. Fagan, M.: Sheikh Yamani predicts price crash as age of oil ends. UK Telegraph (2000). http://www.telegraph.co.uk/news/uknews/1344832/Sheikh-Yamani-predicts- price-crash-as-age-of-oil-ends.html. Accessed August 3, 2017
34. Fizaine, F., Court, V.: Energy expenditure, economic growth, and the minimum EROI of society. Energy Policy **95**, 172 – 186 (2016). http://dx.doi.org/10.1016/j.enpol.2016.04.039. http://www.sciencedirect.com/science/article/pii/S0301421516302087
35. Francis, P.: Laudato si' – encyclical letter on care for our common home, Francis I (2015). http://w2.vatican.va/content/francesco/en/encyclicals/documents/papa-francesco_ 20150524_enciclica-laudato-si.html. Accessed August 14, 2017
36. Friedman, M.: The energy crisis: A humane solution (1978). https://miltonfriedman.hoover. org/objects/57283/the-energy-crisis-a-humane-solution?ctx=ee796946-b436-4b49-a567- 762daa0fcf71&idx=3. Milton Friedman Speaks. Lecture sponsored by Bank of America in San Francisco, CA, February 10, 1978. The transcript of this speech was also published in *The Economics of Freedom*. Milton Friedman Speaks was a series of 15 lectures, including question and answer sessions, given by Friedman between 1977 and 1978 that formed the basis for the Free to Choose television series. URL for transcript accessed August 22, 2017.
37. Fu, R., Feldman, D., Margolis, R., Woodhouse, M., Ardani, K.: U.S. solar photovoltaic system cost benchmark: Q1 2017, nrel/tp-6a20-68925. Tech. rep., National Renewable ENergy Laboratory (2017). https://www.nrel.gov/docs/fy17osti/68925.pdf. Online; accessed September 16, 2017
38. WWF (World Wildlife Fund): The energy report: 100% renewable energy by 2050 (2011). https://www.worldwildlife.org/publications/the-energy-report. Accessed July 21, 2017

39. Garrett, T.J.: Are there basic physical constraints on future anthropogenic emissions of carbon dioxide? Climatic Change **104**(3), 437–455 (2011). https://doi.org/10.1007/s10584-009-9717-9
40. Garrett, T.J.: Long-run evolution of the global economy: 1. physical basis. Earth's Future **2**(3), 127–151 (2014). http://dx.doi.org/10.1002/2013EF000171
41. Griffiths, B.W., Gülen, G., Dyer, J.S., Spence, D.B., King, C.W.: Federal financial support for electricity generation technologies, white paper utei/2016-11-2. http://energy.utexas.edu/the-full-cost-of-electricity-fce/ (2017). White Paper UTEI/2016-11-2, one in a series of white papers for the Full Costs of Electricity Project of the Energy Institute of the University of Texas at Austin
42. Griffiths, B.W., Gülen, G., Dyer, J.S., Spence, D.B., King, C.W.: State level financial support for electricity generation technologies: An analysis of texas & California, white paper utei/2017-xx-yy. http://energy.utexas.edu/the-full-cost-of-electricity-fce/ (2017). White Paper UTEI/2017-XX-YY, one in a series of white papers for the Full Costs of Electricity Project of the Energy Institute of the University of Texas at Austin
43. Handschy, M.A., Rose, S., Apt, J.: Is it always windy somewhere? occurrence of low-wind-power events over large areas. Renewable Energy **101**, 1124–1130 (2017). http://dx.doi.org/10.1016/j.renene.2016.10.004. http://www.sciencedirect.com/science/article/pii/S0960148116308680
44. Hertwich, E.G., Gibon, T., Bouman, E.A., Arvesen, A., Suh, S., Heath, G.A., Bergesen, J.D., Ramirez, A., Vega, M.I., Shi, L.: Integrated life-cycle assessment of electricity-supply scenarios confirms global environmental benefit of low-carbon technologies. Proceedings of the National Academy of Sciences (2014). https://doi.org/10.1073/pnas.1312753111. http://www.pnas.org/content/early/2014/10/02/1312753111.abstract
45. Hubbert, M.K.: Nuclear energy and the fossil fuels. Tech. rep., Shell Development Company, Exploration and Research Division (1956). Presented before the Spring Meeting of the Southern District Division of Petroleum, American Petroleum Institute, Plaza Hotel, San Antonio, Texas, March 7-8-9, 1956
46. Hubbert, M.K.: Degree of advancement of petroleum exploration in united states. American Association of Petroleum Geologists Bulletin **51**(11), 2207–2227 (1967)
47. Hubbert, M.K.: Testimony for the national energy conservation policy act of 1974, hearings at the hearings before the subcommittee on the environment of the committee on interior and insular affairs house of representatives. Tech. rep. (1974)
48. Huber, P.W., Mills, M.P.: The Bottomless Well: The Twilight of Fuel, the Virtue of Waste, and Why We Will Never Run Out of Energy. Basic Books, New York (2005)
49. Huntington, S.P.: The clash of civilizations? Foreign Affairs **72**(3), 22–49 (1993). http://www.jstor.org/stable/20045621
50. Huntington, S.P.: The Clash of Civilizations and the Remaking of World Order. Simon & Schuster, New York, NY USA (2011). ISBN-13: 978-1451628975
51. IEA: Trends 2016 in photovoltaic applications, 21st edition. Tech. Rep. IEA PVPS T1-30:2016, International Energy Agency of the Organisation for Economic Co-operation and Development (2016)
52. Inman, M.: The Oracle of Oil: A Maverick Geologist's QUest for a Sustainable Future. W. W. Norton & Company, New York, NY (2016)
53. Jacobson, M.Z., Delucchi, M.A.: Providing all global energy with wind, water, and solar power, Part I: Technologies, energy resources, quantities and areas of infrastructure, and materials. ENERGY POLICY **39**(3), 1154–1169 (2011). https://doi.org/10.1016/j.enpol.2010.11.040
54. Jacobson, M.Z., Delucchi, M.A., Cameron, M.A., Frew, B.A.: Low-cost solution to the grid reliability problem with 100% penetration of intermittent wind, water, and solar for all purposes. Proceedings of the National Academy of Sciences **112**(49), 15,060–15,065 (2015). https://doi.org/10.1073/pnas.1510028112. http://www.pnas.org/content/112/49/15060.abstract
55. Jacobson, M.Z., Delucchi, M.A., Cameron, M.A., Frew, B.A.: The united states can keep the grid stable at low cost with 100% clean, renewable energy in all sectors despite inaccurate claims. Proceedings of the National Academy of Sciences **114**(26), E5021–E5023 (2017). https://doi.org/10.1073/pnas.1708069114. http://www.pnas.org/content/114/26/E5021.short

56. Jarvis, A.J., Jarvis, S.J., Hewitt, C.N.: Resource acquisition, distribution and end-use efficiencies and the growth of industrial society. Earth System Dynamics **6**(2), 689–702 (2015). https://doi.org/10.5194/esd-6-689-2015. http://www.earth-syst-dynam.net/6/689/2015/

57. Jefferson, M.: Shell scenarios: What really happened in the 1970s and what may be learned for current world prospects. Technological Forecasting and Social Change **79**(1), 186–197 (2012)

58. Jevons, W.S.: The Coal Question: An Inquiry Concerning the Progress of the Nation, and the Probable Exhaustion of Our Coal Mines, second edition, revised edn. Macmillan and Co., London (1866). Kessinger Legacy Reprints

59. King, C.W.: Comparing world economic and net energy metrics, part 3: Macroeconomic historical and future perspectives. Energies **8**(11), 12,348 (2015). https://doi.org/10.3390/en81112348. http://www.mdpi.com/1996-1073/8/11/12348

60. Kopits, S.: Oil: What price can America afford? (research note). Tech. rep., Douglass-Westwood (2009)

61. Littlefield, J.A., Marriott, J., Schivley, G.A., Skone, T.J.: Synthesis of recent ground-level methane emission measurements from the U.S. natural gas supply chain. Journal of Cleaner Production **148**, 118–126 (2017). https://doi.org/10.1016/j.jclepro.2017.01.101. http://www.sciencedirect.com/science/article/pii/S0959652617301166

62. Loftis, R.L., Souder, E.: Airing out a fight on coal power. The Dallas Morning News (2007)

63. Maloney, P.: Panda temple bankruptcy could chill new gas plant buildout in ERCOT market. Utility Dive (2017). http://www.utilitydive.com/news/panda-temple-bankruptcy-could-chill-new-gas-plant-buildout-in-ercot-market/442582/. Accessed May 15, 2017

64. Meadows, D.: It is too late for sustainable development. In: Perspectives on *Limits to Growth*: Challenges to Building a Sustainable Planet. a symposium hosted by the Smithsonian Institution and the Club of Rome, Washington, DC (2012)

65. Meadows, D.H., Randers, J., Meadows, D.L.: Limits to Growth: The 30-Year Update. Chelsea Green Publishing, White River Junction, Vermont (2004)

66. Mitchell, T.: Carbon Democracy: Political Power in the Age of Oil. Verso, London and New York (2013)

67. Mooney, C.: Stanford professor files $10 million lawsuit against scientific journal over clean energy claims (2017)

68. Muller, N.Z.: Boosting GDP growth by accounting for the environment. Science **345**(6199), 873–874 (2014). https://doi.org/10.1126/science.1253506. http://science.sciencemag.org/content/345/6199/873

69. Musk, E.: Making humans a multi-planetary species. New Space **5**(2), 46–61 (2016). https://doi.org/10.1089/space.2017.29009.emu

70. Musk, E.: Master plan, part deux (2016). https://www.tesla.com/blog/masterplanpartdeux. Accessed July 22, 2016

71. Northey, H.: Fossil fuel promoter settles into renewable energy office. E&E News (2017). https://www.eenews.net/greenwire/stories/1060057185. Accessed July 11, 2017

72. Nysveen, P.M.: United states now holds more recoverable oil than Saudi Arabia. Rystad Energy website (2016). https://www.rystadenergy.com/NewsEvents/PressReleases/united-states-now-holds-more-oil-reserves-than-saudi-arabia. Accessed August 4, 2017

73. {OECD}, {Climate Policy Initiative}: Climate finance in 2013–14 and the USD 100 billion goal: A report by the OECD in collaboration with climate policy initiative. http://www.oecd.org/env/climate-finance-in-2013-14-and-the-usd-100-billion-goal-9789264249424-en.htm (2015). Online; accessed September 16, 2017

74. Pandey, S., Gautam, R., Houweling, S., van der Gon, H.D., Sadavarte, P., Borsdorff, T., Hasekamp, O., Landgraf, J., Tol, P., van Kempen, T., Hoogeveen, R., van Hees, R., Hamburg, S.P., Maasakkers, J.D., Aben, I.: Satellite observations reveal extreme methane leakage from a natural gas well blowout. Proceedings of the National Academy of Sciences **116**(52), 26,376–26,381 (2019). https://doi.org/10.1073/pnas.1908712116. https://www.pnas.org/content/116/52/26376

75. Perkinis, J.: Confessions of an Economic Hit Man. Plume Press, New York (2004)

76. TCPA: High-cost natural gas tax rate incentive study. Tech. rep., Texas Comptroller of Public Accounts (2014). https://www.comptroller.texas.gov/transparency/whitepapers/high-cost-natural-gas/high-cost-natural-gas.pdf. Accessed March 10, 2017

77. Ratcliffe, R., Babineck, M.: Group launches 'coal is filthy' ads against TXU. Houston Chronicle (2007). http://www.chron.com/business/energy/article/Group-launches-Coal-is-Filthy-ads-against-TXU-1801939.php

78. Reuters: U.S. energy chief says fossil fuels could help prevent sexual assaults in Africa. Reuters (2017). http://www.reuters.com/article/us-usa-energy-assaults/u-s-energy-chief-says-fossil-fuels-could-help-prevent-sexual-assaults-in-africa-idUSKBN1D22U7. newblock Accessed November 2, 2017

79. Ridley, M.: Why renewables keep running out. The Rational Optimist - blog (2011). http://www.rationaloptimist.com/blog/why-renewables-keep-running-out/. Accessed August 4, 2017

80. Ridley, M.: Why most resources don't run out. The Rational Optimist blog (2014). http://www.rationaloptimist.com/blog/why-most-resources-dont-run-out/. Accessed August 3, 2017

81. Ridley, M.: The world's resources aren't running out. Wall Street Journal (2014). https://www.wsj.com/articles/the-worlds-resources-arent-running-out-1398469459

82. Roberts, T., Weikmans, R.: Roadmap to where? is the '$100 billion by 2020' pledge from Copenhagen still realistic? https://www.brookings.edu/blog/planetpolicy/2016/10/20/roadmap-to-where-is-the-100-billion-by-2020-pledge-from-copenhagen-still-realistic/ (2016). Online; accessed September 17, 2017

83. Rutledge, D.: Estimating long-term world coal production with logit and probit transforms. International Journal of Coal Geology $85(1)$, $23 - 33$ (2011). http://dx.doi.org/10.1016/j.coal.2010.10.012. http://www.sciencedirect.com/science/article/pii/S0166516210002144

84. Siegel, A.: Energy bookshelf: A power hungry gushing of lies (2010). http://getenergysmartnow.com/2010/08/16/energy-bookshelf-a-power-hungry-gushing-of-lies/

85. Simon, J.L.: The Ultimate Resource 2, revised edn. Princeton University Press, Princeton, N.J (1996)

86. Smil, V.: Energy at the Crossroads: Global Perspectives and Uncertainties. MIT Press, Cambridge, Mass. (2003)

87. Smil, V.: Energy in Nature and Society: General Energetics of Complex Systems. The MIT Press, Cambridge, Mass. (2008)

88. Smil, V.: Power density primer: Understanding the spatial dimension of the unfolding transition to renewable electricity generation (part i - definitions). http://vaclavsmil.com/wp-content/uploads/docs/smil-article-power-density-primer.pdf (2010). Online; accessed August 5, 2015

89. Smil, V.: Power Density: A Key to Understanding Energy Sources and Uses. The MIT Press, Cambridge, Mass. (2015)

90. Stecklein, J.: State senate votes to sunset wind subsidy. EnidNews.com (2017). http://www.enidnews.com/news/state/state-senate-votes-to-sunset-wind-subsidy/article_bb6fcd83-b66e-5d32-87e7-4ed04915790d.html. Accessed September 3, 2017

91. Swartz, K.E.: Southern co. suspends $7.5b Kemper plant. E&E News (2017). https://www.eenews.net/greenwire/stories/1060056757/. Accessed June 28, 2017

92. Tsao, J., Lewis, N., Crabtree, G.: Solar FAQs, working draft version 2006 Apr 20. Tech. rep. (2006). http://www.sandia.gov/~jytsao/Solar%20FAQs.pdf. Accessed August 3, 2017

93. Tuttle, D.P., Gülen, G., Hebner, R., King, C.W., Spence, D.B., Andrade, J., Wible, J.A., Baldick, R., Duncan, R.: The history and evolution of the U.S. electricity industry (2016). http://energy.utexas.edu/the-full-cost-of-electricity-fce/.

94. Tuttle, D.P., Gülen, G., Hebner, R., King, C.W., Spence, D.B., Andrade, J., Wible, J.A., Baldwick, R., Duncan, R.: The history and evolution of the U.S. electricity industry, white paper utei/2016-05-2. http://energy.utexas.edu/the-full-cost-of-electricity-fce/ (2016). White Paper UTEI/2016-05-2, one in a series of white papers for the Full Costs of Electricity Project of the Energy Institute of the University of Texas at Austin

95. Wertz, J.: Fallin signs bill to end tax credit that helped fuel Oklahoma's wind-energy boom. State Impact (2017). https://stateimpact.npr.org/oklahoma/2017/04/18/fallin-signs-bill-to-end-tax-credit-that-helped-fuel-oklahomas-wind-energy-boom/. Accessed September 3, 2017

96. West, G.: Scale The Universal Laws of Growth, Innovation, Sustainability, and the Pace of Life in Organisms, Cities, Economies, and Companies. Penguin Press, New York, NY (2017)

97. Woolston, C.: Energy researcher sues the us national academy of sciences for millions of dollars (2017). http://www.nature.com/news/energy-researcher-sues-the-us-national-academy-of-sciences-for-millions-of-dollars-1.22944

98. Xu, C., Bell, L.: Worldwide crude oil reserves down, production holds steady. Oil & Gas Journal **December 5** (2016). 2442 words

99. Yergin, D.: The Prize: The Epic Quest for Oil, Money, & Power. Free Press, New York (1991)

Chapter 4
Other Megatrends

If something cannot go on forever, it will stop.[1]—Herbert Stein

The introductory quote is often referred to as Stein's Law. Pundits have likely generalized the statement more than was intended by the author, Herbert Stein, in the original article. Nonetheless, the concept is simple and relevant to the purpose of this book and particularly useful to consider in this chapter. Chapter 1 defined the techno-realism narrative via a similar statement attributed to Kenneth Boulding that only madmen or economists think there can be exponential growth on a finite planet. Stein's Law represents a more general rephrasing of the concept. If a certain trend cannot continue, then it won't. But how do we know if a trend cannot continue? Before we address this question in Part II, we need to consider the trends.

This chapter presents many economic and demographic data and trends for which we can decide if they can "go on forever" or not. We can contemplate which economic narrative best explains these trends: techno-optimism and infinite substitutability or techno-realism and the finite Earth. The answer could be neither. Keep in mind that the characteristics of the energy system, and the energy trends of the previous chapter, inherently influence and are influenced by the "non-energy" trends of this chapter. Every viable energy and economic narrative must be consistent with the data, and not for just a short time period within the history.

Think about Stein's Law as follows. If for one reason or another you know that some statistical trend cannot continue on its present or long-term trend, then you

---

[1]Stein, H. (1989) Problems and Not-Problems of the American Economy, *The AEI Economist*, June, 1989. This citation is noted per column by Robert J. Samuelson (May 30, 2013), http://www.washingtonpost.com/opinions/robert-j-samuelson-is-steins-law-real/2013/05/30/716942f2-c942-11e2-8da7-d274bc611a47_story.html, "...an intern who waded through a decade's worth of "AEI Economists." Stein's Law appeared on Page One of the June 1989 issue under the headline "Problems and Not-Problems of the American Economy." The reference was inspired by America's trade and budget deficits, which have probably lasted longer than Stein imagined likely."

© Springer Nature Switzerland AG 2021
C. W. King, *The Economic Superorganism*,
https://doi.org/10.1007/978-3-030-50295-9_4

cannot use that trend as a basis for thinking about the future. For example, recall the food and energy cost trends of Chap. 2. We witnessed that until the end of the twentieth century, the cost of energy and food relative to economic output had been generally declining since the start of the Industrial Revolution. But since the turn of the century, energy and food have no longer continued getting cheaper. Stein's Law tells us to expect, at some point, that this cost trend would in fact cease to decline as it had done for over 100 years. There are only two ways that the trend of energy and food expenditures divided by gross world product (GWP) could eternally decline. Energy and food costs would have to decline to zero cost, or GWP would have to grow to infinity. Neither of these are likely on a finite planet, at least one that has a human population and economy that resemble anything what we have today. One could pontificate that machines take over the human population, and thus food expenditure would go to zero if our population dies out. However, that world would no longer resemble ours today.

In addition to the energy and food trends of Chap. 2, there are many other important long-term megatrends that are important to understand the state of the world. This chapter presents these trends, and they reinforce the unlikely event that the world will reverse its recent course and pay less for food and energy than that already achieved around the turn of the century. These trends also make the case that the Finite Earth Narrative more plausibly explains how we have reached the state of the world today. I start with population.

Population and Age Demographics

Population is one of the most important metrics informing the state of society. To the techno-optimistic/infinite substitutability economic narrative, an increasing population is a sign of progress, an indicator that the human condition is improving for more people. After all, if the human condition is worsening, then why is population still increasing? Why is total energy consumption per person still increasing (Fig. 4.1)? Why are people living longer with reduced infant mortality? To the techno-realistic/finite Earth economic narrative, human population cannot increase indefinitely. Thus, there is no reason to postulate, promote, or praise an ever-increasing population that would seemingly rise only for the reason that it can, until it ultimately can't. There is no intelligence in that pursuit.

Regardless of the costs and benefits of increased population, we can look at the data and at least understand how population is growing. Analysts and pundits often discuss population growth as "exponential." Strictly speaking, exponential growth means that the growth of some stock, such as people, depends on the amount of that stock already in existence. While many times we hear someone exclaim that population is growing exponentially, they often imply (without acknowledgment or clarification) that population growth is "rapid" such that we need to take some action to ameliorate or mitigate potential impacts from too many people on the planet. However, just stating that population grows exponentially is no more than

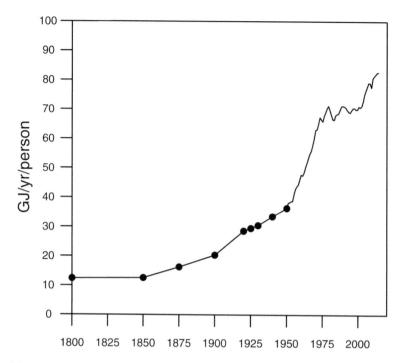

Fig. 4.1 Global energy consumption per person (1800–2014) is still increasing despite a stagnation in this trend from the 1970s–2000s. Energy data from 1800–1899 are as used in [10], and data from 1900–2014 are from International Institute for Applied Systems Analysis Primary, Final and Useful Energy Database (PFUDB) [6]. Population data up to 1950 are from Kremer [17] and from 1950 to 2014 are from the United Nations *World Population Prospects, the 2017 Revision*

a definition. It is also true that population can decline exponentially. That is to say, the more people there are, the more deaths occur as well as more births. If there are more births than deaths, population rises, and if the reverse, population declines. In both cases the growth or decline is exponential.

Instead of inferring the meaning of exponential population growth, let's just look at the data in Fig. 4.2. Until the early 1970s, world population had been growing faster and faster every year since the 1700s. That is to say, not only was the *absolute* number of people increasing each year (seen in Fig. 4.2a), the *growth rate* was also increasing each year (seen in Fig. 4.2b).

Consider the increasing growth rate. Before 1940, rate of global population growth was less than 1.0%/year. Then by 1950, population grew at 1.4%/year, and by 1970 the growth rate was 2.2%/year. Thus, the population growth rate itself grew from less than 1 to 2.2%. This is the concept of *acceleration* that you experience when driving a car, and why the pedal you push to make the car go is called the accelerator pedal. The more you push the accelerator pedal, the faster the growth rate of the speed of the car. For over a century leading up to 1970, the accelerator pedal of population growth was getting pressed more and more. In addition, Fig. 4.1

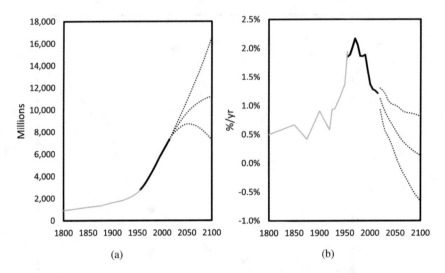

Fig. 4.2 Historical (solid) and projected (dotted) (**a**) world population (in millions of people) and (**b**) population growth rates (%/year) indicate that while world population is still growing, it is growing at a slower rate since approximately 1970. Pre-1950 data from Kremer [17]. 1950–2015 data and post-2015 projections (low, medium, and high variants) from the United Nations *World Population Prospects, the 2015 Revision*

shows that per capita energy consumption was increasing faster than population, also at an exponential increasing rate up to the early 1970s.

By witnessing the accelerating rate of population growth during the 1960s, an adherent to the finite Earth narrative might have easily seen a ticking "population bomb" ready to explode and outstrip the ecological limits of the planet to support human life:

> Sometime in the next fifteen years the end will come. And by the end I mean an utter breakdown of the capacity of the planet to support humanity.[2]—Paul Ehrlich (1970)

While the rate of population growth, not the absolute growth, has declined since Paul Erlich wrote his 1968 book *The Population Bomb* and made the above statement in 1970, 45 years later, he stuck to his original premise:

> I do not think my language is too apocalyptic in *The Population Bomb*. My language would be even more apocalyptic today.[3]—Paul Erlich (2015)

---

[2]Video footage from 1970 as part of Retro Report video (time 6:50) "The Population Bomb?" by Clyde Haberman, available at https://www.nytimes.com/2015/06/01/us/the-unrealized-horrors-of-population-explosion.html.

[3]Interview as part of Retro Report video (time 11:20) "The Population Bomb?" by Clyde Haberman, available at https://www.nytimes.com/2015/06/01/us/the-unrealized-horrors-of-population-explosion.html.

While the 2015 population of over 7.3 billion is a number that many see as already too large, barring large-scale warfare, we are likely going to add at least two billion before global population stops growing. I will not speculate on the timing or the quantity of the peak global population as I need not do that to know that the finite Earth is already providing a feedback signal to mitigate population growth.

This signal is in the population data: *the population growth rate has been decreasing* for over 40 years. Further, an annual rate of global population increase above 1%/year is the exception rather than the rule. While the deaths during World Wars I and II certainly took their toll on population growth (mostly in Europe and Asia), they did not halt the drive to growth rates above 1% starting in the 1940s. For the entire history of mankind leading to the 1930s, global population grew at slower than 1%/year. It is likely (but not for certain) that by 2050 the global population growth rate will again be less than 1%/year. Thus, there might be only one single span of 100 years in which humans experience population growth greater than 1%/year, and we are living during that time.

Just what might be the causal mechanism for the declining population growth rate? A common answer is that as people get richer, they decide to have fewer children. This is more of an observation than an explanation. By plotting data on net births (birth rates minus death rates) versus income per person, one can draw this conclusion (see Fig. 4.3).

If population growth declines as we get richer, then we should just work to ensure everyone becomes as rich as possible. Right? Eventually, we might become so rich as to no longer increase population. Of course, the correlation among data does not mean one variable is the cause of the other, but it could be. While the correlation of per capita income and to net birth rate is undeniable, I classify this "income" explanation within the techno-optimism economic narrative because the answer itself is usually devoid of any hint of the consumption of physical resources that drive the increase in income. If you believe the Earth doesn't constrain economic or population growth, then perhaps you can believe that socio-economic growth will self-limit population. No biological or physical explanation is required. For the finite Earth and techno-realism economic narrative, there is no need to resort to socio-economic data: a limit to population growth is a biophysical constraint, pure and simple.[4]

Importantly, both the fossil and renewable energy narrative proponents typically take some sort of socio-economic choice as the causal reason for declining population growth rates. The narrative is that birth rates have declined, by choice, because we don't need as much farm labor as we did in agrarian and early industrial times. It is true we don't need a large family of farm hands any longer. Fossil-fueled tractors, fertilizers, and irrigation removed the need for a large fraction of population

---

[4]The biophysical constraint could range from the basic Malthusian idea of population outgrowing food supply (with food supply limited from inputs such as energy and fertilizers or distribution costs) to an ultimate far-reaching scenario of running out of physical space for people live because there are too many people right next to them (you can only put an infinite number of angels on the head of a pin).

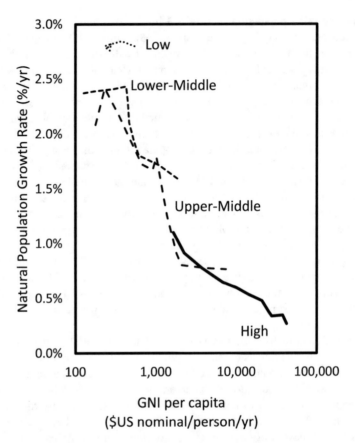

Fig. 4.3 Natural population growth of countries, grouped by income category, versus gross national income (GNI) per capita (on logarithmic scale). This growth rate represents the portion of population growth (or decline) determined exclusively by births and deaths (i.e., no immigration or emigration). The 1950–2015 population change data are "Rate of natural increase by region, subregion, and country, 1950–2100 (per 1000 population)" from the United Nations *World Population Prospects, the 2017 Revision.* Data for Gross National Income per person are from World Bank as "GNI per capita, Atlas method (current US$)"

to perform physical work in agriculture.[5] Some facets of developed economy agriculture still require significant labor today, such as harvesting vegetables typically performed by immigrants in the U.S., but engineering designs continue to mechanize more farming tasks.

---

[5]In the grand scheme of agriculture, the use of mechanized tractors is recent, starting in 1905. Tractors and other agricultural productivity gains lowered U.S. farm labor from 7 h per ton of wheat grain in 1900 to about 90 min per ton in 2000 [24, p. 307]. Modern agriculture technologies in the U.S. lowered the percentage of the total workforce in farming from 40% in 1900 to 15% in 1950 and less than 1.5% in 2015 [24, p. 307].

As fewer people could earn a living farming, more people were forced to move to cities in search of work as selling even a larger quantity of crops didn't make up for the drop in crop prices. Productivity increases on farms lowered the cost of food, and farm revenues no longer supported large farming communities. Agriculture shifted from relatively large farming communities selling relatively high-cost food to small farming communities selling low-cost food. While there seems to be a growth-enhancing effect of cities that drives their formation as a city's economic output increases faster than its population [3, 30], the energetic gains on the farm first drive people to leave farms and increase the population of cities. Thus, urban area formation is a socio-economic adaptation to integrate the portion of the population that is no longer needed to feed itself. This is one reason why data show increased urbanization accompanies declines in population growth rates for countries spanning vastly different cultures and histories. Chapter 9 revisits this farm-to-city migration in the context of economic thought, private land ownership and capitalism, and the early industrialization of England.

There are important economic improvements that do not require as much of a "physical work" explanation as explained for farming. Foremost is that scientists eventually developed basic biological knowledge (e.g., of bacterial infections and diseases) leading to practical medical care (e.g., antibiotics and vaccinations) and provision of clean drinking water and wastewater treatment. With this knowledge, infant mortality and overall death rates declined. Fewer deaths means that fewer births are needed for any given growth rate. The term "developing countries" largely refers to those countries that still do not have the levels of health and water services that exist in the "developed," or relatively rich, countries. Of course, municipal water supply systems do require infrastructure and energy inputs to operate reliably at city scales. This is one direction of the *energy–water nexus* [16, 25, 29]. While the energy inputs for municipal water and wastewater services are critical for cities to function, these services require a relatively small proportion of the total energy supply. Further, because water storage is easy, water treatment is one of the few core health services that does not have to be performed continuously each minute of the year. Thus, it is entirely viable to power water services with variable renewable energy (e.g., from wind and solar power).

This story of scientific and technological progress of developed countries still doesn't address basic questions regarding population. Why have the relatively rich countries, in mass, chosen to have fewer births when they became richer? Were the choices independent of finite Earth effects?

Certainly there were policy choices and health improvements that are highly influential. Two examples are the choices to develop and distribute contraceptives (i.e., birth control) and the one-child policy in China. The Western use of contraceptives is a bottom-up use of technology to enable people to have sex with minimal chance of pregnancy, and it is indicative of the economic and cultural technological solutions promoted by the infinite substitution economic narrative. China's one-child policy was a top-down economic doctrine, associated with economic penalty, applied to reduce population growth without specification of the methods. Other top-down programs have enforced sterility on some poor and underprivileged populations, such as in the 1970s in India, where issues with targeted sterilizations remain.[6] Given past policies and technologies that deliberately reduce birth rates, why has this occurred in countries that span many cultural and religious doctrines?

The human societal responses to reduce birth rates are consistent with the feedbacks from a finite Earth. If you assume human population is confined to the physical space of the surface of the Earth, which to date it has been, then the human population cannot grow forever. Some proponents of the economic techno-optimism narrative do not make this Earth-limiting assumption. For example, due to possible space travel, humans could leave Earth and increase human population on other planets. Aside from speculation on interplanetary colonies, are there experiments we can perform that inform us about population growth in a finite space? Yes. These are well-known growth experiments on bacteria colonies.

Bacteria colonies confined to a closed medium experience four phases of growth: lag phase (no appreciable growth), exponential growth phase, stationary phase where growth stops due to running out of nutrients or space, and death phase when the live population of bacteria declines exponentially as the reverse of the growth phase.[7] There is no policy, decision, or technology development required to curb the bacteria population growth.

To investigate the time of transition from the bacterial growth to stationary phase we can track the slowing of the growth rate of bacteria. To go from the growth to stagnation phase, the growth rate must go from a positive number to zero. This transition from positive to zero growth does not happen instantaneously. Thus, by looking at the growth rate over time, you can determine the "beginning of the end" of the growth phase as the time when growth rates start to decline. This investigation of growth rate is a way to interpret the declining human population growth rate in Fig. 4.2.

Unlike us humans, as far as we can tell there are no mother and father bacteria taking care of baby bacteria or grandma bacteria. In our society not everyone works, and only those working in the economy can provide for those that are too young or old to work. This point provides a mechanism to think about population *age demographics* in relation to population size, economic growth, and the structure of society.

A calculation called the *dependency ratio* estimates the number of "dependents," or non-working people, relative to the number of working people. If there are more working people than dependents, then the dependency ratio is less than 100. In this case it is relatively easy for workers to support the dependents, via direct care of a child or indirect contributions via redistributive taxes, assuming each worker can support himself or herself plus one dependent.

---

[6]Soutik Biswas, BBC News, November 14, 2014, "India's dark history of sterilisation," accessed January 20, 2018 at: http://www.bbc.com/news/world-asia-india-30040790.

[7]Todar's Online Textbook of Bacteriology, "The Growth of Bacterial Populations (p. 3)," accessed January 10, 2018 at http://textbookofbacteriology.net/growth_3.html.

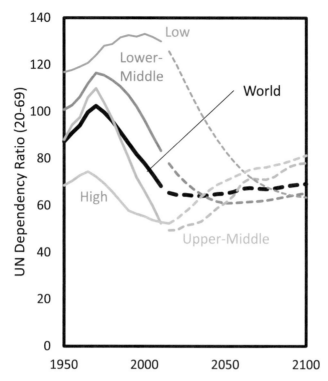

Fig. 4.4 The dependency ratio (scaled to 100) is the percentage of people theoretically too old (≥70) or too young (<20) to work divided by the "working" population aged 20–69. A ratio below 100 indicates more working people than dependents. Labels ("Low," "Upper-Middle," etc.) refer to subsets of countries by income level. Dependency ratio data from United Nations *World Population Prospects, the 2017 Revision*

A simple way to estimate the number of workers and dependents is to assume ages to enter and exit the workforce. Figure 4.4 assumes working age starts at 20 and retirement ends at 69. With these assumptions, over the last 40 years the world working population has become increasingly capable of supporting its dependents primarily because the middle income countries experienced a "demographic dividend" with many young coming into their working ages. The data indicate the low-income countries experienced an increasing fraction of global births. As of the last decade, the high income countries have approached a time of transition in which they shift from an increasingly young to an increasingly aged population.

Thus, the second decade of the twenty-first century might mark a fundamental turning point in age demographics for the world and the high income countries such as the United States. Starting about now, the United Nations population projection expects high income countries will experience increasing average age because birth rates are below replacement rates. *This aging demographic trend is a natural consequence of a slower growing or declining population.* This trend is not something to avoid or fret about. It is something to expect. Declining population growth is a natural consequence of living within a finite space—thus, support for the finite Earth economic narrative. Later in this chapter we will see how the same aging concept applies to U.S. energy infrastructure.

Population: Education and Growing Up

We've learned that population growth is slowing, and it is likely a consequence of the population approaching its confine. Slower population growth in turn translates to an aging population. At the same time a relatively large population today creates pressure to invent new technologies to overcome real and perceived constraints to higher population, higher resource consumption levels, or both. The challenges we face today are fundamentally different than economic challenges of the 1950s just like the challenges of the 1930s were different than those before then.

We cannot overcome new societal challenges for free. There are costs of money, resources, and time. Joseph Tainter uses the phrase the "energy-complexity spiral" to discuss the costs of increasing societal complexity [26, 27]. Generally a more complex society has a larger and a more diverse number of roles in society. Tainter states that "...most of the time complexity increases to solve problems.", and that societies "...subsequently must produce more energy and other resources to pay for the increased complexity." [28].

We can again explore population demographics and the need for increasingly educated workers to find ways to continually grow the economy (assuming for now that is a goal). Over time, society collectively acquires more knowledge. Thus, to make additional intellectual contributions, it takes more time for each person to come up to speed with the present level of knowledge. This increased time is spent in education, including in university that many see as a route to the middle-class. However, we should not be complacent that our educational system will continuously educate all people to a basic level of understanding. Recall Chap. 1 mentioned that 34% of American 18–24-year-olds are not sure that the Earth is round.[8]

Consider the changes in the correlation of education and income. In the U.S. in the 1950s a male worker with a high school diploma could obtain a job with middle-class income and a defined-benefit pension. Very little knowledge of science or mathematics was required for these jobs. Today, a college degree is required for most middle-class jobs that usually have defined-contribution pensions (e.g., 401K plans) that are less onerous to the employer and less secure for the employee. Over the last several decades there has been a steady increase in the number of Americans with

---

[8]Hoang Nguyen, YouGov (April 2, 2018), "Most flat earthers consider themselves very religious," accessed April 7, 2018 at https://today.yougov.com/news/2018/04/02/most-flat-earthers-consider-themselves-religious/.

college degrees since they are increasingly required to obtain middle and upper class incomes.[9] Recent research has suggested that the need for increased education time is one factor in determining that today's adolescence practically extends through age 24 instead of only ages 19 or 20 [22]. In a more complex society, more education is required to make a new contribution, and this education time delays the starting age for working and earning an income.

Consider the world in 1950. If people could start work at age 15, with relatively little education, and retire at age 64, then there were 1.8 workers for every young person and 12 workers for each old person. This is indicated by the "15–64" working age line in Fig. 4.5. In 2015, if you could start working at 15 and retire at 64, then there would still be plenty of workers relative to those too young or old.

But the world in 2015 is much different than that immediately after World War II. Consider the concept that adolescence extends through age 24 because young people have to go to college to learn enough before working to make a good living. If people still stop working at age 64, then there are more dependents than workers because the "25–64" working age line in Fig. 4.5 is to the left of the dashed curve representing a dependency ratio of one. Thus, it is easy to imagine pressure on people to work longer or for governments to delay pension benefits such as Social Security in the U.S.[10] By assuming people work through age 69, this extends the number of workers to again be greater than the total number of dependents in 2015 (see the "25–69" working age line), but just barely.

Population: Summary

A finite world places limits on population growth. The world population growth rate peaked in the 1970s, and it has been declining since (Fig. 4.2). A slower growing population leads to an aging population. A finite world also increasingly constrains physical and economic growth, leading to a need for increased complexity to solve new social problems. A more complex world in turn drives the young to become more educated and places increased pressure on people to start work later in youth and end work at an older age.

Just as we can track the growth rate in the stock of people, we can track the growth rate in the stock of money. One way to do this is to count how much money is borrowed. If we are easily paying for what we want and repaying money that we've borrowed, then our stock of debt would not accumulate. Are we paying back the money we have already borrowed? To further explore if there are signals from a finite world emerging from economic data, it is into the world of debt that we now turn.

---

[9]"Census: More Americans have college degrees than ever before": http://thehill.com/homenews/state-watch/326995-census-more-americans-have-college-degrees-than-ever-before.

[10]Two-thirds of Americans have retired by age 65, and full Social Security retirement age for Americans born after 1960 is 67: https://money.com/ages-people-retire-probably-too-young-early-retirement/.

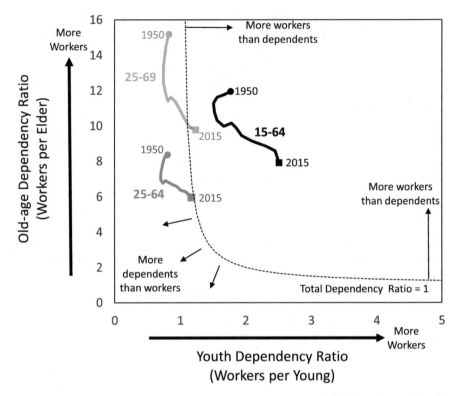

Fig. 4.5 The world dependency ratio disaggregated between its old and young components using three definitions of "working age" (15–64, 25–64, and 25–69). The thick solid lines represent historical estimates (1950–2015). The thin dashed curve represents the threshold of dependency ratio (equal to one) where working population equals the non-working population. A value residing below the threshold indicates the world has a larger old than working population, and a value the left of the threshold indicates the world has a larger young than working population. The population age demographic data are from the United Nations *World Population Prospects, the 2017 Revision*

Debt and Interest

Debt and Interest: Debt

The consequences arising from the continual accumulation of public debts in other countries ought to admonish us to be careful to prevent their growth in our own.[11]—John Adams (1797), First Address to Congress, Nov. 23, 1797.

---

[11] Attributed at John Adams Historical Society, The Official Website at http://www.john-adams-heritage.com/quotes/ accessed on November 12, 2017.

And I sincerely believe with you, that banking establishments are more dangerous than standing armies . . . [12]—Thomas Jefferson (1816)

Let us control the money of a country and we care not who makes its laws.[13]—T.C. Daniel (1913)

Debt The concept can be a confusing one. The 2008 Global Financial Crisis forced us to think more critically about what debt actually is in our modern society. We need to consider whether the amount of debt in the economy, say relative to GDP, is affected by the rate of consumption and cost of energy and other natural resources.

After the 1970s, the rich economies experienced rapid increases in debt. Another rapid increase occurred in the overall worldwide economy after the year 2000. Is this rise in the amount of debt in the economy, say relative to GDP, affected by the rate of consumption and cost of energy and other natural resources? In this chapter we look at the data before Chap. 6 discusses some theoretical foundations and calculations for linking energy consumption to debt levels. This look at debt is critical for explaining the low economic growth rates of the world economy, primarily in the developed countries, since the 2008 financial crisis.

David Graeber's following passage, from his book *Debt: The first 5000 Years*, provides context for thinking about debt in our modern financial economy. He states, we can think of both moral obligations and debts:

> What does it mean when we educe moral obligations to debts? . . . On one level, the difference between and obligation and a debt is simple and obvious. A debt is the obligation to pay a certain sum of money. As a result, a debt, unlike any other form of obligation, can be precisely quantified. This allows debts to become simple, cold, and impersonal—which, in turn, allows them to be transferable. If one owes a favor, or one's life, to another human being, it is owned to that person specifically. But if one owes forty thousand dollars at 12-percent interest, it doesn't really matter who the creditor is; neither does either of the two parties have to think much about what the other party needs, wants, is capable of doing—as they certainly would if what was owed was a favor, or respect, or gratitude.[14]—David Graeber (2014)

We might feel obligated to help our family members in times of medical or other crises. We also might loan them money, but not with the same detailed terms and interest as a bank. Further, when we speak of how much money one country's government and citizens owe to governments and citizens of other countries, it is not some sort of moral obligation, such as a promise to show up at your best friend's wedding.

---

[12]In a letter to John Taylor, May 28, 1816, transcription at https://founders.archives.gov/documents/Jefferson/03-10-02-0053 viewed November 12, 2017.

[13]Attributed to T.C. Daniel, 1857–1923 in letter to President W. Wilson, May 8, 1913; reported in his statement for the joint hearings before the subcommittees of the Committees on Banking and Currency of the Senate and of the House of Representatives, charged with the investigation of rural credits, Sixty-third Congress, second session, part 1, p. 764, February 16, 1914. See https://archive.org/stream/ruralcreditsjoin01unit#page/764/mode/2up pages 764 (transcript of letter to Woodrow Wilson) and 771 (quote during Congressional hearing).

[14]Graeber [13, p. 13].

Both of the words *loan* and *debt* describe money owed by some entity or person to another. However, there are important differences. A loan is a quantity of borrowed money that is owed to the lender—a specific person or company, such as a bank. The borrower can also be a person or a company. Loans are usually repaid via regular payments (e.g., a monthly mortgage or car payment) where a portion of the repayment reduces the principle, or amount of money borrowed, and the rest is the interest payment to the lender.

In our modern banking economy, debt is not the same as a loan. Debt usually involves companies or governments borrowing money rather than an individual person. However, this borrowed money is lent by entities within the general investing community that is composed of individual people, banks, and other investment entities such as pension funds. To do this the borrowing entity issues bonds at a certain price, an interest rate or yield, and a maturation time period. For example, on behalf of the federal government, the U.S. Treasury issues bonds with maturities from a few months to a few decades. Over the life of the bond the price and yield can fluctuate, and the bonds can be bought and sold in a similar manner as company stock. Bondholders are essentially a group of lenders. Similarly to lending for a loan, they receive regular interest payments equal to the yield times the price. But unlike a loan, the repayment of the price of the bond comes at the end time of bond maturity, whereas the loan principle (the amount lent up front) is repaid throughout the time period of the loan.

It is useful to understand how the difference between loans and debt has become blurred in over the last several decades. Consider the time depicted in Frank Capra's 1946 film *It's a Wonderful Life*. Jimmy Stewart's character, George Bailey, convinces the citizens of Bedford Falls not to start a "run" on his savings and loan bank by removing their deposits. He explains that their deposits are *not in the bank*. They exist in the form of loans given to their neighbors, *people they know*.[15]

George pleaded, "We've got to stick together. We've got to have faith in each other." The citizens of Bedford Falls didn't want to force harm on their friends, so they didn't demand 100% of their deposits. In essence, the monetary loans still had hints of moral obligation because the people loaning the money personally knew the people borrowing the money. Today debt is largely a transaction between people that don't know each other.

Can you convert loans into debt? Yes. This type of conversion was at the heart of the 2008 Global Financial Crisis. First, banks lent money as *mortgage loans*. Second, the banks converted the loans into debt. Third, the banks sold this debt to the general public.

Banks lent money in the form of mortgages to U.S. citizens to enable these citizens to "buy" and "own" houses. The words buy and own are in quotes because

---

[15]"You're thinking of this place all wrong As if I had the money back in a safe. The money's not here. Your money's in Joe's house ... right next to yours. And in the Kennedy house, and Mrs. Macklin's house, and a hundred others. Why, you're lending them the money to build, and then, they're going to pay it back to you as best they can. Now what are you going to do? Foreclose on them?"

in many cases the families who were lent the money and lived in the homes did not provide any down payment. Thus, they did not own any equity, and financially speaking, they were not homeowners. This loaned money did not previously exist. It was created by banks that lent the money for the mortgages. Yes, *banks created money from nothing when they made the loans.* If you don't believe me, then ask the Bank of England: "…the majority of money in the modern economy is created by commercial banks making loans." [20].

While the bankers did not necessarily live in the same communities and personally know the borrowers, there was a known relationship between lender (the bank) and borrower (mortgage owner). However, the mortgage loan was clearly not a *moral obligation* since that the banks originating these loans didn't have a personal or community connection to the new mortgage borrowers. In short order, banks converted their mortgage loans into debt. They turned loans into sellable debt by aggregating multiple loans into a group that they could sell to others as an investment much like a bond. This is what was meant when people spoke of packaging, bundling, or securitizing loans. These aggregated mortgage loans were called mortgage-backed securities. These securities repaid owners like a loan, but could be bought and sold like a bond, or debt.

As we learned in the aftermath of the 2008 Financial Crisis, the banks sold the mortgage-backed securities to investment funds and citizens in the U.S. and other countries. These mortgage-backed securities, which included many high risk sub-prime mortgages, were sold to investors as if they were as good as money in the bank. However, the banks selling these contracts knew they were *not* as good as money in the bank. The buyers of mortgage-backed securities did not know the securities were more risky than they were rated, and they certainly didn't know the credit worthiness of the indebted individuals on the other end of the contract.

Why would a bank want to convert its loans into debt-like concepts? One answer is to sell the risk of loan repayment to someone else. Since the banks knew the incomes of the debtors, they knew many of the loans were unlikely to be repaid, and they did not plan on receiving the interest and repayment of principle from the borrowers. However, they charged a fee to create this newly loaned money out of thin air. Thus, the banks could make a fee on the transaction and let someone else worry about collecting the loan payments.

In 2007 and 2008 when the mortgage debtors finally could no longer afford to make their mortgage payments, sometimes only the interest payments, they began to default on their mortgages. Some other people somewhere else in the world, who did not know these mortgage debtors, were losing the money that they traded to the originating banks in order to own the right to repayment of the mortgages.[16]

So let's summarize. Banks created money from thin air to loan to people to "buy" homes, and many of these people did not have the income to pay back the loans. For this, banks charged a fee. They then put many loans into a bundle, and sold the

---

[16]Many were not mortgage holders were not "homeowners" in the sense that they did not outright own any portion of equity in the home in which they lived.

bundle to someone else, pocketing that sale. Creating money is a good way to make money. Selling money you created is even better!

So how much debt and loans are there? A common format to state the level of debt is in relation to income (for people) and net output, or gross domestic product (for countries). For the world, this metric is debt relative to gross world product. The Institute of International Finance estimates total global debt was near 250% of GWP in 2002 and near 320% of GWP in 2016 and 2018.[17] The McKinsey Global Institute estimates global debt in the same ballpark at 250% in 2000 and 290% in 2014 [7].

The United States total quantity debt and loans resides at slightly higher levels than the global estimates (see Fig. 4.6). The peak level of total U.S. credit (debt and loans) relative to GDP was 380% in 2009 during the midst of the Great Recession. As of 2016, the U.S. total credit to GDP ratio was still 350%.

In aggregate, the time period before and after 1980 stand out as starkly different. Before 1980, one can see that total U.S. credit (debt and loans) grew only slightly from the 1950s near 140% of GDP to near 160% of GDP in 1981. The U.S. experienced significant post-war prosperity and productivity driven by abundant and cheap energy, largely oil. During the three and a half decades after World War II, the federal government generally paid off debt (from 90% of GDP in 1947 to 26% in 1974), and the private and consumer sectors of the economy accumulated debt and loans (e.g., mortgages) in an offsetting manner. Before the 1970s, the accumulation of mortgage loans was a sign of increasing confidence of a growing middle class investing in homes and education. The middle class was confident because their material lives had been improving for decades. Confidence comes after observing a trend of improvement, not before. Chapter 9 returns to this concept of consumer and investor confidence as drivers of economic growth.

U.S. credit changes primarily via the federal debt, mortgage loans, and private financial debt. Other major categories do not change as much. For example, state and local government debt increases slightly from the 1950s through 2016, peaking in 2009 at 21% of GDP, but usually residing at 10–20% of GDP. The mandate for states to have balanced budgets is a major driver of this stability. In addition, corporate (non-financial institutions) debt increases slowly but steadily from about 10% of GDP in 1947 to 30% of GDP in 2016.

However, just as so many trends change their direction in the 1970s so do a couple of major categories of debt. First, the federal government debt-to-GDP rose after 1980 through 1993 (to 55%) before declining to around 40% in 2000.

The second category of debt that changes its trend in the 1970s is the debt taken on by financial institutions. Relative to GDP it was almost non-existent coming out of World War II. Financial institution debt steadily increases from about 1% of

---

[17]Data from Institute of International Finance reported by Chibuike Oguh and Alexandre Tanzi, *Bloomberg*, January 15, 2019, "Global Debt of $244 Trillion Nears Record Despite Faster Growth" https://www.bloomberg.com/news/articles/2019-01-15/global-debt-of-244-trillion-nears-record-despite-faster-growth.

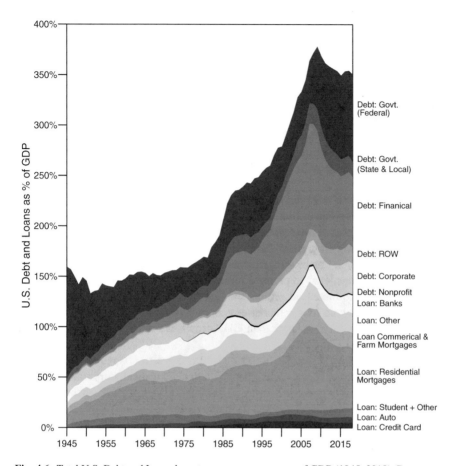

Fig. 4.6 Total U.S. Debt and Loans by category as a percentage of GDP (1945–2018). Data come from U.S. Federal Reserve Bank z.1 Financial Accounts of the United States, tables L.208 (Debt, listed as liabilities by sector), L.214 (Loans, listed by instrument), and L.222 (Consumer Credit, as four categories as a subset of Loans: credit card balances, automobile loans, student loans, and "other.") GDP from St. Louis FRED data series GDPA

GDP in 1947, to 17% in 1980, to its peak of 104% in 2008. This demonstrates the "financialization" of the economy that accelerated after the 1970s.[18]

The turn of the twenty-first century saw new swings in debt and loans. At the beginning of this century, the financial and household sectors increased debt the most rapidly. A rapid rise of loans provided for residential mortgages started around the year 2000 and increased until 2007. The quantities of these loans dropped off substantially after 2009 as mortgage owners defaulted and paid off loans.

---

[18]See Nicholas Shaxson's *The Finance Curse: How Global Finance Is Making Us All Poorer* for an extended discussion on financialization of the economy [21].

After the financial crisis ensued, central governments bailed out banks by taking on their debt, and thus government debt and central bank assets rose substantially for several years after 2007. Starting in 2009 the federal government debt grew from 40 to 83% of GDP by 2014, remaining at 86% in 2016. Financial institution debt decreased to 72% of GDP by 2015 because the Federal Reserve purchased much of the bad mortgage securities. Consider that the Federal Reserve Bank of the U.S. had assets of $960 billion in 2007 rising to $4.5 trillion in 2014.[19]

The post-2008 jumps in U.S. government debt and Federal Reserve Bank assets is the government bailing out commercial banks by taking their debt and putting that burden onto U.S. citizens. It is a perfect example of socializing losses from private companies.

But why would private companies and banks need a bailout? Think of the relative growth rate of economic output, or GDP, to the growth rate of debt. This metric is called the *marginal debt productivity* of the economy. This metric helps answer the following question: "How much additional economic output, or GDP, is added for each additional dollar borrowed?" Businesses borrow money to invest in new capital (machines, buildings, etc.) to produce more products and hopefully increase profit. If companies make increased profits, this translates to growing economic output. If borrowing money no longer increases economic output, then that is important to know. If debt accumulates faster than GDP grows, we must understand the social and economic ramifications.

Figure 4.7 shows the change in U.S. GDP relative to change in debt of state and federal governments as well as financial and non-financial corporations. If this number is greater than one, the debt-to-GDP ratio declines. If it is less than one, the debt-to-GDP ratio increases. It is just another way to look at the data in Fig. 4.6, but in this case only the debt and not the loans.

From the end of World War II through 1980, marginal debt productivity is greater than one. For every dollar borrowed by U.S. companies and governments, there was more than one dollar of GDP during that time. The economy was "productive" because it generated more annual value for every borrowed dollar. Since 1982 marginal debt productivity is below 1, and it has hovered near 0.5 with a noticeable dip in the few years leading up to the Great Recession of 2008. After 1980, every extra dollar borrowed only paid back an extra 50 cents. A practically identical "loan productivity" trend occurs when considering loans instead of debt. In the case of loans, individuals borrow money for home mortgages, cars, and education. Since the early 1980s, loans accumulated faster than the growth in GDP.

Companies choose how much of their profit to invest. After paying workers, taxes, and interest, profit is money companies use to invest and provide dividends to investors. Part of investment replaces capital that is outdated or is no longer functional. The rest creates new capital. Historically U.S. companies invest more than their profit, 50–150% more as shown in Fig. 4.8. How do companies invest

---

[19]Central Bank total assets as reported in Table s.61.a of the Federal Reserve Statistical Release, Z.1 Financial Accounts of the United States.

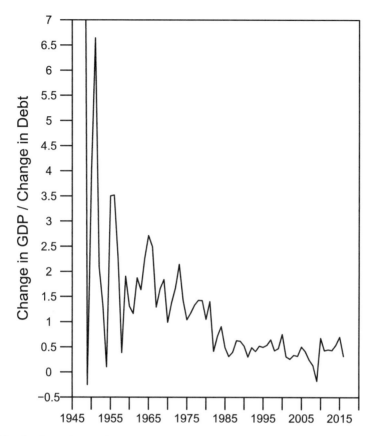

Fig. 4.7 The change in U.S. GDP per the change in total U.S. debt. Debt data are from St. Louis Federal Reserve data series ASTDSL

more than the money they have for investment? One way is to borrow money from banks, and this creates debt. Companies borrow money just like people do for mortgages. The nature of capitalism rewards those that are optimistic and take investment risks. If you believe future economic growth will enable you to pay back your debt, you are more willing to take the risk of borrowing. Chapter 6 discusses how most economic growth models ignore the role of debt, but by including the concept of debt and investment behavior as observed in Fig. 4.8, we can mimic the overall debt-to-GDP trends in Figs. 4.6 and 4.7.

Why should we care about debt productivity and debt-to-GDP ratios? Just what does it mean for U.S. and global debt ratios to be over 300% of their respective economic output? Debt is a stock of money, say in units of dollars. GWP is global "net output," in aggregate equal to "value added," say in units of dollars per year. Value added is essentially the flow of money that is split among paying wages (to workers), profits (to owners of businesses and stocks), rents (to those that own

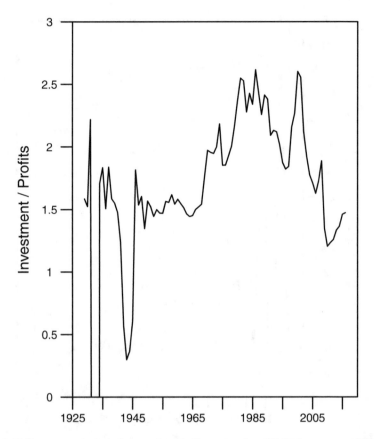

Fig. 4.8 U.S. gross corporate investment is typically greater than 150% of corporate profits. Data are gross private domestic investment (BEA Table 1.1.5, series A006RC), and corporate profits with inventory value adjustment, IVA, and capital consumption adjustment, CCAdj (BEA Table 1.1.12, series A051RC)

property), government taxes, and interest payments on loans and debt (to those that lend money).

By dividing debt by GWP, you get units of time, or years. Thus, a debt-to-GWP ratio of 330% means that if all GWP was allocated to paying off debt, it would take about 3.3 years to pay off the debt. During that time no one would receive a paycheck for over 3 years.

Of course we don't have to pay down all of our debt right away at the expense of paying wages and profits. As previously stated, we only repay some partial amount each month, the interest payments that depend on the *interest rate*, and interest rates change over time.

Debt and Interest: Interest and Interest Rates

Interest payments equal the amount of debt owed times the interest rate on that debt. The previous subsection discussed the first half of that equation (the level of debt), and this subsection summarizes the second half (the interest rate). If there is zero debt, there are no interest payments. Likewise, if interest rates are 0%, there are also no interest payments. People, largely central bankers, set interest rates in response to observed trends in the economy. The purpose of this subsection is to provide the data that allow us to consider whether historical changes to interest rates were responses to the effects of energy costs and consumption.

People and companies that lend money receive income as the interest payments on that loan. Those that borrow money pay the interest. A new company might need a loan to purchase machinery or pay wages before it has time to make enough revenue to be profitable. An existing company might obtain a loan in anticipation that future revenue increases more than the amount of the interest payment on the new loan. Thus, companies obtain loans with the anticipation of obtaining future profits.

Broadly speaking, the profits of a company are reduced by its interest payments. If there is not enough economic activity, lower interest rates on existing or future loans should entice more lending. This is why central banks lowered interest rates during the recession after the 2008 Financial Crisis. With the economies in recession, they wanted to entice businesses to obtain loans with which to invest in business ventures that would in turn hire workers.

Figure 4.9 shows the interest rates as set by the central banks of four countries: England, Japan, Canada, and the United States. These are *nominal* interest rates. One striking aspect of the data is that, while interest rates change frequently within a year and from year-to-year, they have historically resided in the range of 2–8%.

We see three regimes of interest rate change over the last couple hundred years. The first is witnessed by the relatively constant rate of the Bank of England since its inception until the Great Depression. England has a long history of banking and investing because of its colonial and sea trading history. For around 100 years in the 1700s the interest rate for the Bank of England was 5%. The principals of the Bank of England would loan money to England and receive 5%/year rate of return on that loan. In the 1800s the rate fluctuated between 2 and 7%, but usually near 3–4%/year. During that century, British companies could borrow money at these relatively low rates because they were effectively backed by the soundness of the global British Empire, including its navy and military, upon which "the sun would never set."[20]

The second regime for interest rates is from the end of World War II until the 1980s as interest rates rose for more than 30 years. As shown in Chap. 2, global and developed economy energy consumption grew at its fastest rate ever during this time as the middle classes became established in the U.S. and Western Europe. The

---

[20]Galbraith [11, p. 101].

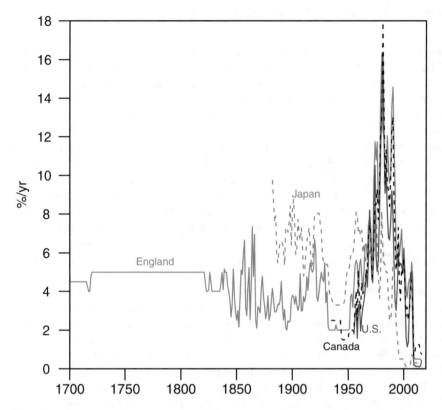

Fig. 4.9 Central bank interest rates (to 2016) of the U.S. (Federal Funds Effective Rate: H15/H15/RIFSPFF_N.A), England (Bank of England), Japan (Basic Discount Rate and Basic Loan Rate), and Canada (annual average of monthly bank rate v122530)

1970s and 1980s are the only time in which interest rates were *above 10%*. These high interest rates were set as a reaction to the increasing energy costs and wages, and the next major section expands upon the change in the growth of wages before and after the 1970s.

The final regime is one of declining interest rates from the mid-1980s until the financial crisis in 2008. The decade after 2008 is the only time in which central bank rates for many Western countries resided *below 1%* for any extended period of time. Only during the Great Depression and immediately after World War II did central banks set interest rates almost as low to spur economic activity, such as rebuilding after the war. Before the 1970s, the central bank rates for Japan are notably higher than the Western countries, but after the 1970s Japanese rates declined ahead of the others. Japan's central bank rate has been less than 1% since 1996.

Economists have stated that the post-2008 Western economies were seemingly stuck in a "new normal" situation of "secular stagnation" characterized by low economic growth, low interest rates, and stagnant wages [11]. As witnessed via the data in Fig. 4.9, there has not been a "normal" direction of change in interest rates since before the Great Depression. There were rising interest rates from World War II until the 1980s, and then there were falling interest rates through 2008. This up-down reaction by central banks was a response to changes in costs to businesses, namely wages and energy costs, or inflation.

Debt and Interest: Get Real

What I've discussed so far are *nominal* interest rates, specifically those that central banks adjust in response to the overall economic conditions such as employment and inflation. The *real* interest rate is (approximately) the nominal interest rate minus inflation, and it ultimately determines whether a person is earning more on an investment relative to how the prices of goods and services are changing. If you earned 5% nominal interest last year on the money in your savings account, but the prices of the things you buy also increased by 5%, then your real interest rate for your bank account is really 0%.

Fast rising prices are the same as high inflation. If inflation is high, central banks tend to raise nominal interest rates to make it more expensive to borrow money. More expensive money raises the cost to run a business and obtain a mortgage. If employment is too low or prices are not rising, they tend to lower interest rates to incentivize borrowing that increases economic activity.

Figure 4.10 shows both nominal and real interest rates for the U.K., U.S., Japan, and Canada. From the 1960s through the 1980s, real interest rates were below nominal rates. The difference is most stark for the U.K. and Japan. During this time prices were rising at a slower rate than the central bank interest rate. After 1990, real interest rates more closely match the nominal interest rates.

During the 1990s and 2000s Japan experienced negative inflation, or deflation. Prices were declining such that real interest rates were positive, while nominal rates were near zero. Japan's real interest rate declined from 3.2% in 2011 to −1% in 2015. What happened in 2011? In March 2011 an earthquake off the coast caused a tsunami that flooded and destroyed a significant amount of Japanese coastal infrastructure, including the 4.7 GW Fukushima Daiichi nuclear power station. Japan's government reacted by shutting down almost all nuclear facilities, removing about 13% of Japan's energy supply.[21] Japan experienced higher energy prices because of a declining supply of electricity, and by replacing some of the lost nuclear electricity with natural gas power plants. Japan imports all of its natural gas from

---

[21] The primary energy from Japan's nuclear fleet from 2010 to 2014 was 66.2, 36.9, 4.1, 3.3, and 0 million tonnes of oil equivalent (Mtoe) out of a total primary energy consumption of 503.8, 477.8, 475.0, 471.3, 456.7 Mtoe. Thus, in 2010, nuclear served 13.1% of Japan's primary energy, and in 2014, it served 0%. Data from the BP Statistical Review of World Energy, 2018.

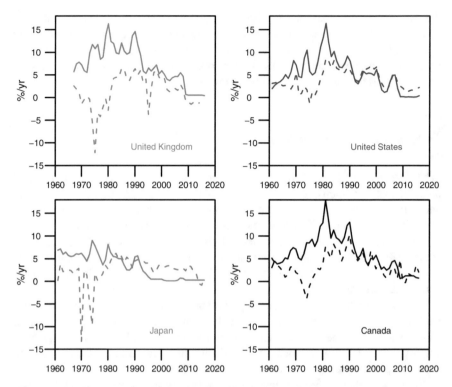

Fig. 4.10 Nominal (solid line) and real (dashed line) central bank interest rates of the U.S. (Federal Funds Effective Rate: H15/H15/RIFSPFF_N.A), United Kingdom (Bank of England), Japan (Basic Discount Rate and Basic Loan Rate), and Canada (annual average of monthly bank rate v122530). Real interest rates are from the World Bank (indicator FR.INR.RINR) defined as "Real interest rate is the lending interest rate adjusted for inflation as measured by the GDP deflator"

liquified natural gas (LNG) tankers, and LNG prices were very high during this time.

Increasing energy prices cause inflation and thus can be one reason that real interest rates decline. The OPEC increase in oil prices at the beginning of 1974 coincides exactly with the sharp decline in real interest rates shown in Fig. 4.10. The year 1974 experienced the lowest real interest rates in the U.K. and Canada, and second lowest in Japan. Since 1960, 1975 is the only year in which the U.S. experienced a negative real interest rate.

When economists speak of the economic rate of inflation, or the rate at which prices are increasing, it is common to discard "volatile food and energy prices" from the calculation. It is thought that energy prices are always relatively constant over the long term, but can fluctuate more wildly in the short term. Thus, we are told we can ignore the short-term variability in energy prices. Economist James Galbraith, in his book *The End of Normal*, says we should not so quickly dismiss the idea that significant changes in our energy system are crucial factors behind the major shifts in inflation, and thus interest rate responses:

So far as I'm aware, no study of the [2008] *financial* crisis has yet suggested that resource costs lie at the heart—or near the heart—of it. But it remains equally true that resource costs have moved from the shadows, and are now understood by all informed, practical people to play a central role in economic performance–even though formal economics continues to neglect them. They are the simplest, clearest way to understand the crisis of the 1970s, and why inflation emerged then but disappeared in the 1980s and 1990s. They can also help explain why the energy-using world fell into troubles again after 2000, just as resource costs roughly doubled in relation to the prices of goods and services produced in the resource-using lands. And why, meanwhile, the energy-producing world, in the Middle East and in Latin America, experienced no financial crisis at all. No one suggests that resource costs alone are the full story of the Great Crisis–only that they are one underlying part of it. For now, that is enough.[22]—James Galbraith (2014)

Galbraith is right. Look back at Fig. 2.13. When energy costs rise, sparking inflation and a recession, central banks tend to respond by increasing nominal interest rates to prevent negative real rates. This sequence happened in the 1970s and with more muted dynamics during the 2000s leading up to 2008. Following each major recessionary period was a time of cheaper energy that enabled a downward adjustment in nominal interest rates. Historically, moderate nominal interest rates have been associated with the trend of steadily decreasing food and energy costs with increasing energy consumption. The last 10–15 years are unique in that rich countries have not increased their energy consumption despite relatively moderate energy costs and a low interest rate regime.

Debt and Interest: Is That Your Negative Bond Yield Showing?

Not only are the post-2008 near-zero central bank interest rates unprecedented, but investors buying government debt are also betting on zero, and even negative yields. As of mid-2019, several major countries such as Germany, Japan, and Switzerland have issued bonds for government debt that returns a *negative yield*. Normally you might lend to a government, and for the privilege of borrowing your money today (by you buying their bond), each year they would pay you a few percent of what you lent them. But globally in mid-2019 about 15–17 trillion dollars worth of government bonds had negative yields, near a quarter of the global bond market.[23] This means that people are effectively *paying governments to borrow money*. Along

---

[22]Galbraith [11, pp. 110–111].

[23]Bloomberg, August 20, 2019, "What trillions of dollars in negative-yielding debt means for markets," https://www.bnnbloomberg.ca/economics/video/what-trillions-of-dollars-in-negative-yielding-debt-means-for-markets~1757903. CNBC, August 13, 2018, "Negative bond yields are not reflecting economic reality, Fitch warns," https://www.cnbc.com/2019/08/13/negative-government-yields-dont-support-credit-rating-fitch-warns.html. CNBC, "How bonds with negative yields work and why this growing phenomenon is so bad for the economy," August 7, 2019, https://www.cnbc.com/2019/08/07/how-bonds-with-negative-yields-work-and-why-this-growing-phenomenon-is-so-bad-for-the-economy.html.

with near-zero interest rates, these negative bond yields for 10-year bonds (and even 30-year bonds for Germany) are unprecedented in history.

But who would do this? Who would lend a government a dollar, euro, or yen only to expect to actually lose money? One answer is that when bond yields go down, bond prices go up. It is largely the increasing demand to buy bonds that drives down the yields. If you think you can buy a bond for one dollar today at −0.5% interest, but then the price increase to 1.05 dollars at −1.0% interest next week, you can then sell the bond and make a quick profit. So some investors are not looking to earn money from bond yield payments, they are looking to buy bonds at "low" prices and sell at higher prices later. This flipping of bonds sounds just like flipping houses before the financial crisis—who cares if you borrowed too much money for too much house because, hey, housing prices always rise!

These negative bond yields are good for the government because people are actually paying off the government's debt. Perhaps investors are just patriotic, or nationalistic, or they ran out of charities to which to donate all of the money they have. Why might investors have so much money they can invest in bonds with negative yields? Many say it's because too much of total national income, or GDP, has been going to too few people. For the last 50 years, the average worker hasn't received much of a real pay raise, and this is why the topic of income inequality has come to the forefront of political discussion since the 2008 financial crisis. We now discuss income distribution.

Wages and Income Distribution

How, if at all, are wages and the distribution of income related to energy consumption and/or costs? To answer this question, we first have to look at the wage and income data.

The topic of wage and income inequality has been at the forefront of economic discussions since the 2008 financial crisis, particularly in Western countries. However, it is useful to consider income distribution both globally and within any given country. In the context of the global population, there are data to estimate income distribution among countries and individual citizens. The trends are different whether considering countries or individuals as the decision making "agent" of interest. We can also consider the level and direction of change in income inequality among countries.

We cannot translate the trends in any one country or group of similar countries to be representative of the world trend. In terms of income distribution, the data indicate that since the 1970s the U.S. and many Western economies are becoming less equal. However, since around the year 2000, the distribution of income across the entire world has become more equal. Thus, much of the economic angst in the U.S. is in part due to "shifting" economic gains from the West to poorer and developing countries (e.g., China, India). In addition, there have been policy changes within developed countries that influenced income distributions within their borders. All of these changes can be considered in the context of access and cost to energy and other resources. First, a look at the United States trends.

Wages and Income Distribution: U.S. Economy

Figure 4.11 indicates one driver of U.S. workers' economic angst that was brewing in the 1970s and that has been revitalized since the Global Financial Crisis. The plot compares U.S. economic "net productivity" to the average hourly wage of private production and nonsupervisory workers [4]. The net productivity is the net

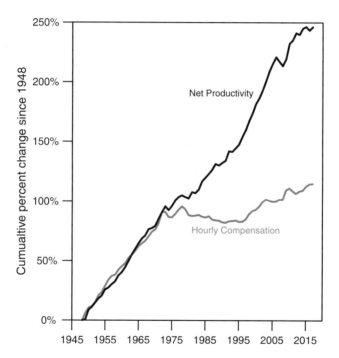

Fig. 4.11 From World War II until 1973, U.S. worker hourly compensation tracked the productivity of the economy. After 1973, net productivity continued to increase, while real (inflation adjusted) hourly compensation stayed the same. Calculations from https://www.epi. org/productivity-pay-gap/ per method of [4]. Note: Data are for average hourly compensation of production/nonsupervisory workers in the private sector and net productivity of the total economy. "Net productivity" is the growth of output of goods and services minus depreciation per hour worked

domestic product of the U.S. divided by all hours worked.[24] Higher productivity means fewer workers are needed to produce the same quantity of economic output. Economically speaking people generally see increased productivity as a good thing. It is often viewed as a measure of "technological progress," but this type of label is vague and misleading. Productivity is usually a word used to describe a statistical trend of "progress" but without explanation of what is driving that trend. For now, just keep in mind that "productivity" and "technological progress" are often poorly defined terms yet used interchangeably. Chapter 6 goes into some detail on economic interpretations of technology, and how energy-focused interpretations provide significant insight.

Hourly compensation is real (inflation adjusted) average compensation for about 80% of the U.S. workers that are neither managers nor executives. This compensation is primarily composed of wages and salaries but also includes supplemental pay (such as paid leave) and employer contributions to health insurance and retirement benefits. There is a stark break in the trend between productivity and hourly compensation starting in 1973. From 1948 to 1973, both productivity and hourly compensation went up together. After 1973, productivity continued to increase, but hourly compensation stayed approximately the same for 40 years.

Other economic data confirm the same breakpoint in U.S. wage trends in the 1970s, as well as one around the year 2000, and Chap. 7 summarizes all in one place. Instead of considering hourly compensation, these data indicate the share of total income going to Americans via different income streams. At the simplest level, there is income going to labor (hourly and salaried workers, as in the top two lines in Fig. 4.12) and income going to capitalists (owners of capital and their profits, as in the bottom line in Fig. 4.12). The top line represents total worker compensation from wages and salaries as well as other employer-provided benefits like health care and retirement pension contributions. The middle line represents only wages and salaries. A 2015 study by the International Labour Organization and the OECD shows this same declining wage share, starting at the same time, also holds for a group of nine rich countries [14].[25]

One takeaway from Fig. 4.12 is that for almost a century there has been a clear tradeoff between worker and capital shares of U.S. national income: one goes up when the other goes down, and vice versa.

There are three phases of change since World War II. Phase 1 is from the end of World War II until the early 1970s. Total workers' compensation share increased by 6% (from 60 to 66%) and capitalists' share decreased by 6% (from 31 to 25%). Phase 2 is from the early 1970s to around 2000. During this phase, both the workers' and capitalists' shares remained constant. Phase 3 is from around 2000 until the

---

[24]Net Domestic Product is equal to gross domestic product minus depreciation of capital, and is thus a metric of economic value added each year after subtracting the costs of maintaining machines, factories, and infrastructure.

[25]From Figure 1 of the report [14], the nine countries are Australia, Canada, Germany, France, Italy, Japan, Spain, the United Kingdom, and the United States.

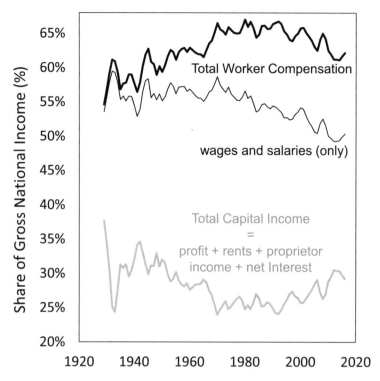

Fig. 4.12 The share of U.S. Gross National Income (GNI) going to workers (upper curve) as total compensation (equal to wages and salaries plus supplements as health insurance and pensions) is a mirror image of the share of GNI going to owners of capital (lower curve) in the combined forms of property (rents), businesses (proprietor income), corporate profit, and net interest to banks (banks earning income by lending money at a higher interest rate than at which they borrow). When one curve goes up, the other one goes down. The middle curve is worker compensation only as wages and salaries. Data are from 1929 to 2016 using U.S. Bureau of Economic Analysis Table 1.12. "National Income by Type of Income"

present. Workers' share decreased by 4.7% from 2001 to 2104 (from 65.8 to 61.1%) and capitalists' share increased by 4.7% (from 25.7 to 30.4%). Thus, while the total U.S. worker compensation share increased to the 1970s and declined after 2000, the opposite occurred for total share of gross national income to owners of capital in the combined forms of property (rents), businesses (proprietor income), corporate profit, and net interest to banks.

Of course in the context of profit and wage shares, any given person can be both a worker and a capitalist earning profits on investments. A person working for a company earns a salary or hourly wage, but she might also separately invest in the stock market and have her employer contribute money to a retirement investment account. This money invested in personal and retirement accounts represents capital that can earn profits (and losses) based on ownership of stock in companies.

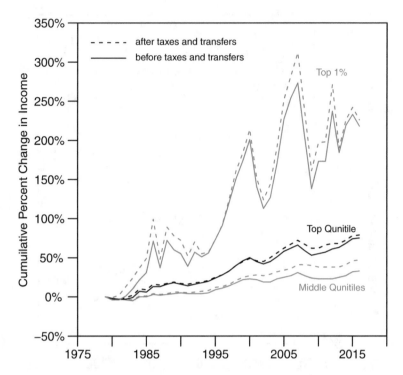

Fig. 4.13 The cumulative percentage change in income both before (solid lines) and after (dashed lines) taxes and transfers for the top 1%, top quintile (top 20%), and middle three quintiles (top 20–80%) of income earners in the United States. Data from U.S. Congressional Budget Office [5]

However, for the vast majority of U.S. workers, the vast majority of their income comes from salaries, wages, and supplements, not capital gains.

U.S. workers also receive transfers of tax revenue via benefits from the federal government. Figure 4.13 shows the cumulative growth in income (for example, wages, capital gains, social security, and Medicare), from 1979 to 2016, both before and after taxes and receiving benefit transfers from the U.S. government.[26] Some argue that income inequality is not too extreme because there is a net transfer of tax revenue to lower incomes, and one should measure inequality based on these after-tax income. The data indeed show higher income for the lower (not shown) and middle income quintiles. However, all income brackets have a higher income after paying taxes and receiving transfers. Further, the Top 1% of income earners

---

[26]**Income before transfers and taxes** consists of market income plus social insurance benefits (including benefits from Social Security, Medicare, unemployment insurance, and workers' compensation). **Means-tested transfers** are cash payments and in-kind services provided through federal, state, and local government assistance programs. Eligibility to receive such transfers is determined primarily on the basis on income, which must be below certain thresholds. **Federal taxes** consist of individual income taxes, payroll taxes, corporate income taxes, and excise taxes.

Fig. 4.14 The political polarization of the U.S. Congress was lowest (less than 0.6) from the Great Depression though the 1970s. Higher values represent higher polarization and less cooperation. Data are from Voteview website: https://www.voteview.com/articles/party_polarization

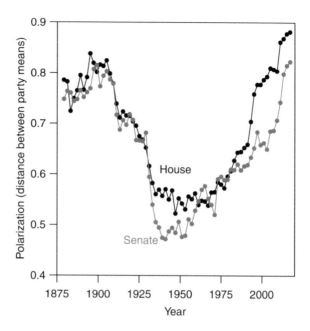

have clearly increased their incomes much more than the rest, and they still receive a net positive transfer of income from the government. After taxes and transfers, relative to 1979, the top 20% of income earners had 79% higher income in 2016, the middle quintiles earned 47% more, and the lowest quintile earned 85% more. Before taxes and transfers the percentages are 75%, 33%, and 33%. Clearly net transfers from the government have a greater "growth effect" for the lower incomes, but that is because their incomes are, well, lower. The transfers do not overcome the increasing allocation of national income to the top 1% of income earners.

Studies of political polarization in the U.S. Congress show a correlated pattern to labor-capital distribution. When more money was going to workers, our elected officials were less polarized. Figure 4.14 shows that the lowest political polarization occurred from the 1930s through the 1970s.[27] This is the same period when the share of national income going to workers was either increasing or at its highest levels, with the opposite trend for the share going to owners of capital.

In addition to the changes in the worker and capitalist share of national income, the distribution of total wages among workers has become less equal since the 1970s. The paychecks of high income workers grew much faster than the paychecks of low-income workers. Common explanations given for this stagnant growth in low-income wages are automation (e.g., factory mechanization and robots),

---

[27] Jeffrey B. Lewis. UCLA Department of Political Science, Voteview website, accessed March 19, 2019: https://www.voteview.com/articles/party_polarization. Plotted data are variable "party.mean.diff.d1."

globalization, and a loss of bargaining power of workers. For example, the authors of the hourly compensation calculation shown in Fig. 4.11 note, "Finally, it also seems worth noting that this decoupling [of net productivity from hourly compensation] coincided with the passage of many policies that explicitly aimed to erode the bargaining power of low- and moderate-wage workers in the labor market." [4] This statement is correct, but there is an additional, perhaps more fundamental question. Why were 1970s workers unable to prevent the loss of bargaining power that they gained in the previous three decades?

In a capitalist economy, the overall goal is to increase, or at least maintain, profits. One way to increase the chance to earn profits is to minimize costs. Business costs fall into three general categories: capital spending on physical and monetary assets (machines, infrastructure, property rents, interest payments on debt), wages for skilled and unskilled labor, and natural resources (energy, environmental regulations). If one or more of these costs are higher at home, then it might make sense to move company activities to another country. This act in turn impacts the global distribution of income.

Wages and Income Distribution: Global Economy

Many of the same trends in inequality within the United States are prevalent in Western Europe and other countries in the "rich club" of the Organisation for Economic Cooperation and Development (OECD). The World Inequality Database (WID) indicates that from the early 1900s to the 1980s the top 10% of population of each of the following large economic regions had a *decreasing* share of income: U.S. plus Canada, Europe, Russia, China, and India [18]. By contrast, from 1980 to 2000 the richest 10% of these population regions increased their share of income, leaving the vast majority of citizens behind (see Fig. 4.15). This trend is corroborated by data from the University of Texas Inequality Project that uses a different method to calculate measures of income inequality. Gini coefficients for several countries are shown in Fig. 4.16 (Gini = 0 means all people earn the same income; Gini = 1 means one person earns 100% of all income).[28] Thus, the higher the Gini coefficient, the more unequal is income distribution. The Gini coefficient calculations and fraction of income going to the top 10% show similar broad trends even though they derive from different data sources, with some differences between metrics for specific countries. Nonetheless, we can discuss general differences in income distribution among OECD countries.

For example, the Scandinavian countries have a lower share of income going to the top 10% than the Western Europeans, Japanese, Australians, and Americans. Cultural differences that influence governmental structure play a large role, and

---

[28]The "UTIP-UNIDO" data of the University of Texas Inequality Project. See https://utip.lbj. utexas.edu/datasets.html.

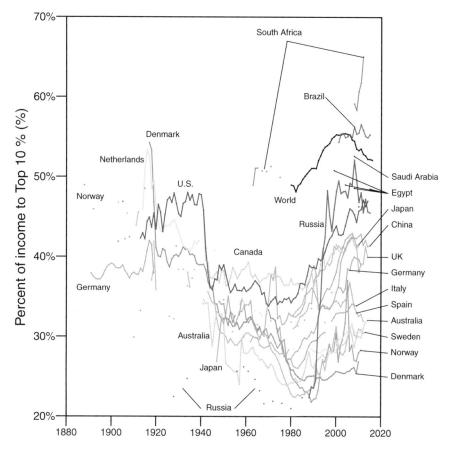

Fig. 4.15 The share of total country income going to the top 10% income earners of the country's population [1]

thus there are choices that affect income and livelihoods.[29] However, practically all OECD countries have steady or declining income inequality during the few decades before 1980 versus increasing inequality after 1980. China also shows an increase in income inequality from the 1980s (when citizens were more equal, but poorer) to the 2000s. However, since 2000 Chinese income inequality is stagnant to declining as it increasingly opened up its economy to the globe and experienced high economic growth rates. The general pattern of the U.S. trend of income to the top 10% and Gini coefficient match that of the share of income to capitalists in Fig. 4.12. There

---

[29]For example, Scandinavian countries are largely socialist democracies with high taxes, large welfare states, and high labor union membership. In contrast, the U.S. since the 1980s has seen several declines in tax rates, lower union membership, and reluctance to move toward universal health care.

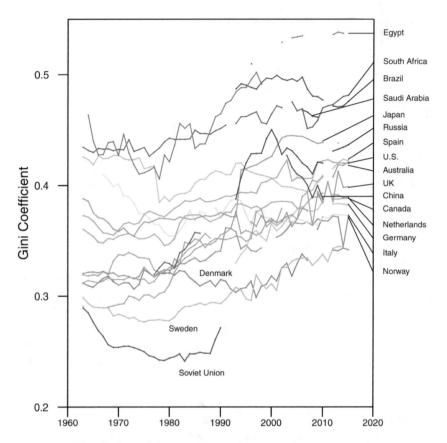

Fig. 4.16 The Gini coefficient is a measure of income inequality. The higher the Gini coefficient, the more unequal is income distribution. Source: The "UTIP-UNIDO" data of the University of Texas Inequality Project

were three phases for the richest income share after World War II—down (to the 1970s), flat for a decade, then up (after the 1980s).

The World Inequality Database (WID) also estimates a value for world income distribution. The top 10% of global income earners capture just over half of all income. From 1980 until the mid-2000s, the global top 10% earned an increasing share, and since that time a decreasing share. Thus, the income data provide some evidence that the recent globalizing economy enables more equal income distribution, particularly due to significant increases in international trade after China joined the World Trade Organization in 2001. However, this decline in the top 10%'s income share is far from some people's vision of a more equal society. The top 10% income earners, largely the middle incomes of OECD countries, still take about half of all global income, and the super elite top 1% of global population increased their share of income more than the bottom 99%.[30] While the global data for low incomes are quite sketchy, the WID data suggest the poorest 50% of global individuals received 10% of global income.[31] Also, the gains of the middle class and wealthy in developing countries (who compose a large portion of the "middle 40%" of global income earners in the income bracket just below the top 10%) have come at the "expense" of low-income growth for the middle class of wealthy countries (part of the global top 10% of income earners).

These trends in income distribution show that middle income groups in the U.S. and Europe are justified in their frustration from experiencing stagnant wage growth. The gains they made in *les trente glorieuses*, the French term for the immediate three post-World War II decades, have for the past four decades gone to poorer individuals in developing countries as well as the richest few percent in rich countries.[32]

Policies emphasizing globalized trade forced a shift in income gains from workers in rich countries to workers in developing countries. As stated by Nobel Prize economist Joseph Stiglitz: "Trade in goods is a substitute for the movement of people. Importing goods from China—goods that require a lot of unskilled workers to produce—reduces the demand for unskilled workers in Europe and the U.S. This force is so strong that if there were no transportation costs, and if the U.S. and Europe had no other source of competitive advantage, such as in technology, eventually it would be as if Chinese workers continued to migrate to the U.S. and Europe until wage differences had been eliminated entirely. Not surprisingly, the neoliberals never advertised this consequence of trade liberalization, as they claimed—one could say lied—that all would benefit."[33]

Economist James Galbraith, who has spent much of his career studying income inequality, agrees with Stiglitz. He states:

> ... the rise in global inequality from 1980 to 2000 was the by-product of a reactionary global financial regime, directed largely from Washington, New York and London [neoliberal policies, or "Washington Consensus"]. ... And the modest reduction in global inequality and poverty, particularly after 2000, can be traced first and foremost to those countries that defied the regime, ... This progress has now ended [by 2018]; we are back to the conditions that generate rising inequality, and the need for comprehensive stabilizing control over global finance is as urgent as it ever was.[12]

In short, Galbraith posits that economic growth in the 2000s, leading up to the 2008 Financial Crisis, was not driven by lending from the rich countries to developing countries, such as from U.S.-led Western development agencies and banks. Another important point is that the reduction in global inequality in the 2000s

---

[30]Table 2.1.1 of [1].

[31]See Galbraith [12] for a discussion of the quality of data and methods within the World Inequality Database that call into question the accuracy of data for low-income countries.

[32]Project Syndicate (2016), Globalization RIP?, https://www.project-syndicate.org/onpoint/globalization-rip-2016-08.

[33]Stiglitz (August 5, 2016), Globalization and its New Discontents, Project Syndicate, accessed August 10, 2017 at https://www.project-syndicate.org/commentary/globalization-new-discontents-by-joseph-e--stiglitz-2016-08.

corresponded with the fastest growth in energy consumption since the 1960s (refer back to Figs. 2.10 and 4.1), largely in China. Higher wages in China occurred with high growth rates of per capita energy consumption in the 2000s just as occurred in the U.S. during the 1950s and 1960s, and Chap. 6 discusses this relationship. Since U.S. wages remained stagnant in the 2000s (and wages as a share of national income dropped), to keep up spending, consumers borrowed money (against rising housing prices) to consume what China was producing, leading to the 2008 Financial Crisis.

But just how much control do we have in affecting the distribution of income? Tax policies are highly influential, but how much? To explore this idea further, I turn to physicists.

Wages and Income Distribution: A Physics-Based Explanation

Discussions of income distribution usually focus on policy choices. How much should we tax the rich, who own many assets and earn high incomes, versus the poor, who own few assets and earn low incomes? Is enough of the population sufficiently educated? How much do historical inequities, such as slavery and inability to own property, translate to outcomes today? These and other questions are perfectly valid and important questions. However, even if we don't contemplate how historical human relations and government policies affect the current state of economic affairs, we can still say something about why we *should not* expect equal incomes for everyone. To understand this we enter the world of *econophysics*.

Econophysics is a word coined by Gene Stanley, scientist and professor at Boston University [31]. It means what it sounds like.[34] For example, some of econophysics has been based on understanding short-term commodity and stock prices for trading purposes. "More precisely, statistical mechanics links the macroscopic, thermodynamic properties of a system to its microscopic constituents. In financial applications envisioned by econophysicists, the market is the macroscopic system while the individual financial agents are the microscopic constituents. Understanding how the principal features of financial markets arise from the microscopic interactions is the main task of econophysics."[35] Econophysicists were some of the "quants" that integrated into Wall Street over the last couple of decades. Some founded consulting firms using econophysics principles to govern stock trading algorithms.

---

[34]Eugene Stanley, The Back Page: Econophysics and the Current Economic Turmoil, *APS News*, 17 (11), 2008. Accessed March 30, 2018 at: https://www.aps.org/publications/apsnews/200812/backpage.cfm.

[35]WorldQuant (2017) Perspectives: Wall Street on a Lattice: Finance Meets Physics, https://www.weareworldquant.com/media/1455/063017_wq-perspectives_wall-st-finance-meets-physics-v2.pdf.

In the late 1990s and 2000s, econophysicists rediscovered the idea of applying statistical physics to monetary transactions and income distribution that sociologist John Angle pioneered in the 1980s [2, 31]. In effect the physicists wondered if, in aggregate, all of us economic "agents" act just like gas molecules as described by statistical and thermodynamic laws. It turns out that most of us do.

The concept is as follows. Each molecule of gas in a room has a temperature. The average of all of these temperatures is what we call the temperature of the room. However, some molecules have temperature below the average, and some above. It is the *distribution* of temperatures that is described by physical and mathematical principles in what is known as the Boltzmann–Gibbs formula.[36] The distribution describes what fraction of all molecules reside at a certain temperature. Each molecule is floating around and randomly bumping into other molecules. In doing this they exchange energy. Some molecules gain energy, and their temperature increases, and vice versa. However, after a long period of time, at thermal "equilibrium," the proportion of molecules at a given temperature no longer changes as long as the total amount of energy of all molecules is the same.

The translation from physics to economics is to compare the temperature of a molecule to the income and wealth of a person or business (or even the GDP of countries). Instead of molecules bumping into each other exchanging energy, people are bumping into each other exchanging money. You might go to the grocery and pay $100 for groceries. Thus, you gave up $100 and the grocery store owner received that same $100. You now have less money, and the grocery store owner has more. Further, the grocery store owner has employees and pays them. In paying wages, the grocery store owner reduces his money supply and the workers increase theirs. Everyone in the economy is both gaining and giving money, and these transactions, big and small, occur billions of times every day.

Physicist and econophysicist Victor Yakovenko and his past students have put the theory to the test using income data from various countries [8, 9]. For the example of the United States, over 97% of people have their incomes distributed as would be expected from the statistical physics, or thermal equilibrium, perspective [23]. This vast majority of the population earns income primarily from wages and salaries, and this "additive" process of getting a paycheck every 2 or 4 weeks is characteristic of the Boltzmann–Gibbs formula. However, the upper 1–3% of the U.S. population with the highest individual incomes cannot be described using the same mathematical pattern. They are "superthermal." They have higher incomes than would be expected using the Boltzmann–Gibbs formula. One of the explanations is that their incomes come primarily via the "multiplicative" process from investments and capital gains that are based on earning some percentage of a quantity of money invested. In 1983 this top 1–3% of income earners took home 4% of all income. This percentage increased such that in 2000 the top 3% earned 18%

---

[36]The fundamental law of equilibrium statistical mechanics is the Boltzmann–Gibbs distribution. Yakovenko (2009) describes some of the historical translation of the Boltzmann–Gibbs distribution to economics via a variety of independent investigations of social scientists and physicists [31].

of all U.S. individual income [23]. When the top 3% of the population takes home 18% of the income, only 82% is left for the other 97%.

Some might doubt the value of the statistical mechanics viewpoint of wealth and income distribution because they think it ignores the influence of our choice of policy. But this is not the case, and the next few paragraphs explain why.

This statistical mechanics viewpoint is tremendously informative given its ridiculously simplifying assumptions. Consider its underlying assumptions for wealth distribution: all persons start with the same amount of money, the minimum wealth and income for any person is zero, the maximum wealth and income is infinity, no person knows anything more than any other person (and there actually isn't anything to know), each person exchanges money (positive or negative) with another randomly chosen person each time step (e.g., each day, year, etc.), and the amount of money (or GDP) in the economy does not increase during the analysis.

Clearly these conditions do not fully describe many important real-world details. However, despite this simplified view, the statistical mechanics concept accurately describes the income distribution for about 97% of U.S. income earners. In doing so it provides valuable insight for thinking about what we mean by words such as equality and fairness. It says that even if everyone were exactly equal in capability, exactly equal in knowledge, and had the same initial amount of money, due purely to random exchange over time the distribution of wealth and incomes would not be equal *if* we allowed the potential for the highest paid person to take an infinite amount of income and prevent the lowest income from going below some threshold, say zero.

This last sentence provides the opening for policy because it can affect the maximum and minimum incomes. In fact, the historical data show that *policy did impact income distribution* in a way we can interpret from the econophysics viewpoint. Because 18% of U.S. income went to the top 3% in 2000 after only 4% did in 1983, this means that the wealthy must have had some additional information or ability to exchange money than did the bottom 97%. Yakovenko's explanation that the superthermal top 3% of earners used multiplicative means of acquiring income through investing represents this additional ability. Quite simply, the ability to make money based on investing money you already have is an additional ability over those that don't have any savings to invest. Air molecules don't have investment accounts!

Since the 1980s and particularly more so in the late 1990s, stock market valuations increased faster than inflation and wages. Thus, those with money were able to acquire more disparate incomes not by working on an hourly basis, but by investing in the stock market, selling stock for gain, and acquiring dividends from stocks and other investments. Further, the early 1980s still had relatively high marginal tax rates on the highest incomes. The top marginal income tax rate fell from 70% in 1980 to 50% in 1983 and has resided between 28 and 40% since. For those with high incomes, starting in the 1980s wealth accumulated at a higher rate than the previous three decades. This is because high income earners both reaped the benefits of compounding wealth accumulation from keeping a larger share of pre-tax income and also reaped the benefits of investing that accumulated income. This is like some "rich" molecules having more information than some "poor" molecules, and we would expect more deviation from the statistical mechanics viewpoint. That is exactly what happened as a lower fraction of total U.S. income could be described by the statistical mechanics approach in 2000 as compared to 1983 [23].

As already noted in this chapter, while income inequality increased in developed countries since the 1970s, it has generally been decreasing when considering the world population overall. Much of the reason for increased global equality in the last one to two decades is because of a decrease in *between-country* average income equality [1]. That is to say if we treat each country as a single entity characterized by its total income divided by population and compare countries on this metric, then the world is becoming more equal.

What happens if you make the same comparison for *energy consumption* per person? Amazingly, the result is strikingly similar [19]. This is because at the country scale, average income and energy consumption go hand in hand. Chapter 6 discusses this again using both theory and U.S. data.

Using data on the average energy consumption per person for each country of the world, Victor Yakovenko and his students determined that global energy consumption is approaching the same equilibrium distribution (line labeled as "Exponential" in Fig. 4.17), as calculated using statistical physics and that describes incomes in the United States and other countries [19]. To construct Fig. 4.17 you gather data on energy consumption and population for each country. You then sort the data from the lowest to highest per capita energy consumption, and calculate the fraction of energy consumption and population in each country. Starting at the lower left corner with the first country, which is the one with the lowest energy consumption per person, move to the right for the fraction of population in that country and up for the fraction of energy consumption in that country and put a point. For the second point, start at the first point, use data from the country with the second lowest per capita energy consumption, and again move to the right for the fraction of population in that country and up for the fraction of energy consumption in that country and put a point. You do this over and over until you run out of countries. Figure 4.17 repeats this procedure for data in 1980, 1990, 2000, and 2010.

The further the curve is from the diagonal line running from the bottom left to upper right, the less equally distributed is energy across the world and the higher the Gini coefficient. In 1980, the distribution of energy consumption was highly skewed to a small number of rich countries. The distribution was characterized by a Gini coefficient of 0.66. By 2010, energy was distributed much more equally around the world with a Gini coefficient of 0.55. Thus, as global trade increased its pace after 1980, income became more equally distributed across the world because developing countries began manufacturing more products. Because more industrial output means more energy consumption, global energy consumption became more equally distributed as well.

Interestingly, the "exponential" distribution predicted by statistical mechanics via the Boltzmann–Gibbs formula has a Gini coefficient of 0.5. This means that a Gini

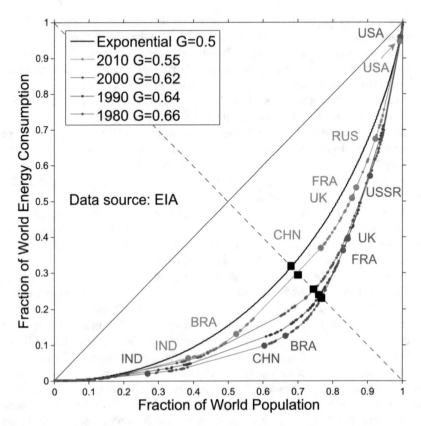

Fig. 4.17 A Lorenz plot of country energy consumption versus population. Over time (from 1980 to 2010) the distribution of each country's average energy consumption per person increasingly approximates that which would be expected by a sequence of random energy exchanges (per Boltzmann–Gibbs equation) among countries as represented by the black line labeled "Exponential" (Figure unmodified from [19] under Creative Commons License 3.0: https://creativecommons.org/licenses/by/3.0/)

coefficient of 0.5 is representative of the distribution in temperatures among a bunch of molecules. Because these molecules are all exactly the same, and none have any advantage over another, an income distribution with a Gini coefficient of 0.5 implies that while *it is an unequal distribution, it represents a fair distribution.*

Wage and Income Distribution: Summary

There are several key takeaways from the wage and income distribution data. First, from the end of World War II until the 1970s, the U.S. and other rich economies increased income equality, but since that time equality has decreased. Second, as income became more concentrated in rich countries, income and energy actually became more equally distributed across the world. From 1980 to the mid-2000s, energy and income were generally shared more equally *between* countries but less so *within* most countries, but income equality actually decreased in some developing countries such as Brazil and China.

Broadly income and energy consumption come together, and thus gains in energy and income that went to Western economies in the 1940s–1970s have since gone to developing countries. From this standpoint, the recent political populism (e.g., the 2016 Brexit vote, the 2016 U.S. presidential election of Donald Trump and popularity of candidate Bernie Sanders) is understandable as a multi-decadal accumulation of citizen resentment. Over the last 100 years, U.S. political polarization has changed in lock step with changes in income inequality: when one went up, so did the other.

But who, if anyone, are the foes of workers in rich countries? Are they low-income workers in developing countries? Are they corporate executives in rich countries who allocate investment to developing countries? Directly and indirectly, people are generally paid to extract resources and turn them into products and services by consuming energy. If we are making things that extract and consume energy, we need the requisite physical machinery, transport systems, and industrial facilities. In short, we need infrastructure.

Infrastructure

Infrastructure As comedian and talk show host John Oliver joked, infrastructure is basically anything that can be destroyed in an action movie.[37]

But on a more serious note, it is important to understand the quantity and age of infrastructure for the same reasons we considered the quantity and age of human population. Just like people, infrastructure accumulates and gets old, and if we slow down the rate of investment in infrastructure, then a higher percentage of infrastructure becomes relatively old. And just like people, if the infrastructure becomes too aged, worn, and feeble, then it can't support economic activities and functions that keep people safe and comfortable in their homes.

It is hard to appreciate all of the economic services that our roads, bridges, railways, ports, pipelines, and electricity grid provide. Even our communications infrastructure—the wires, fiber optic cables, and wireless relay towers—is becoming more critical by the day as it connects more people and devices. This information connectivity is both good (increased access to knowledge) and bad (increased hacking of secure data such a credit card accounts).

---

[37]Last Week Tonight with John Oliver, March 2, 2015. Video available April 15, 2018 at https://www.youtube.com/watch?v=Wpzvaqypav8.

If infrastructure works properly, we don't notice it. We tend to only hear about infrastructure when it stops working. When very important infrastructure fails, it is nationwide or global news, and people make movies about it. Think hurricane Katrina and the failed levies of New Orleans,[38] the sinking of the *Titanic*,[39] the explosion and sinking of the offshore drilling platform *Deepwater Horizon*,[40] and the destruction of the electric grid of Puerto Rico from hurricane Maria in 2017.[41]

The American Society of Civil Engineers (ASCE) tallies an "infrastructure report card" with the state of the U.S. infrastructure expressed in letter grades such as B+ and D−.[42] The ASCE considers roads, railways, airports, pipelines, water and wastewater, energy, and other infrastructures. The latest grades are generally poor, and the grades have declined since ASCE started providing them in 1988. The overall grade for 2017 was D+. While some are alarmed and see this as a call to invest in our nation, others see a narrative from a group whose members' jobs depend upon building infrastructure itself. That brings up an engineering joke. What is the difference between mechanical and aerospace engineers versus civil engineers? Mechanical and aerospace engineers build weapons; civil engineers build targets.

But seriously, all engineers work for non-destructive purposes when building and operating major types of energy infrastructure such as power plants. Just like we thought of the age of the population, we can think of the age of power plants, and they are getting older.

Figure 4.18 shows the fraction of U.S. power plant capacity that is older than a certain age. The power plant age is the number of years from the date the generator began operation. If the lines in Fig. 4.18 are increasing, it means that power plants are aging faster than we are building new ones. If the lines are decreasing, it means there is an investment boom in power capacity. No physical infrastructure lasts forever, and power plants are no exception. They require maintenance, including the replacement of major components.

The post-World War II U.S. economic boom was characterized by the rapid increase in energy consumption, a "baby boom" in population, and a continuing decline in the cost of energy and food (as in Chap. 2). It was also characterized by the decreasing age of power plants. In 1948, 36% of power capacity was older than 20 years. Due to rapid power plant construction after World War II, only 10% of power capacity was older than 20 years in 1971. The types of power plants built in the 1950s and 60s were hydropower, coal, and natural gas with some also consuming oil-derived fuel. Because the electric grid infrastructure was building

---

[38]Cinema Katrina: The Top 10 films inspired by the 2005 storm: http://www.nola.com/movies/index.ssf/2015/07/cinema_katrina_10_years_later.html.

[39]Wikipedia list of films about RMS *Titanic*: https://en.wikipedia.org/wiki/List_of_films_about_the_RMS_Titanic.

[40]Deepwater Horizon (2016): http://www.imdb.com/title/tt1860357/.

[41]*After Maria* (2019), https://www.imdb.com/title/tt10136680/.

[42]https://www.infrastructurereportcard.org.

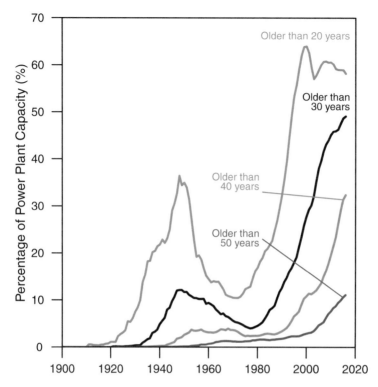

Fig. 4.18 Percentage of U.S. power plant capacity that is older than a certain age. Data are from EIA form 860

from a small base before the war, most of the grid was new. Many home appliances were converted to electricity and purchased for the first time.

The post-World War II wave of power plant installations did not last. As we have seen in so many other economic and energy data, the 1970s are a significant turning point. One U.S. policy response to the oil crises was the Power Plant Fuel Use Act of 1978 that effectively outlawed the use of oil and natural gas as fuels for electric power. This law was repealed in 1987, but the impact on the rate of installed natural gas-fired power plants is evident in Fig. 4.19. This law is one reason for the dominance of coal and nuclear capacity construction in the late 1970s and 1980s.

The U.S. has never had an older fleet of power generation assets than today [15]. In 2016, about 50% of all power capacity was older than 30 years, a higher percentage of power plants of that age than ever before. If we want to have more total generation capacity, we have to install new capacity faster than the existing capacity retires. Increasingly, maintaining and replacing power plants just to keep total capacity at the same level takes resources that have historically been allocated to accumulating more capacity in total [15]. As the total quantity of infrastructure accumulates within the electric grid, at some point, the operation and maintenance

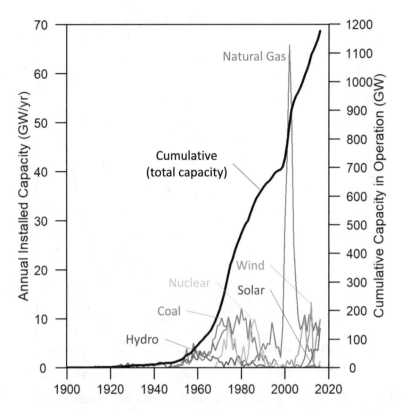

Fig. 4.19 (Left axis) The amount of U.S. power plant capacity installed per year and primary fuel. Solar includes photovoltaics and concentrating solar power. (Right axis) The cumulative amount of U.S. power plant capacity in operation over time. Data are from EIA form 860

costs might become large enough to prevent expansion. We could interpret a non-expanding grid as a response to physical and economic constraints where the costs of maintaining what we have are overwhelming the ability to expand further. Chapter 5 revisits this concept of energy consumption within expanding networks, such as the electric grid.

In addition to having an older power generation fleet, the U.S. no longer consumes more electricity. U.S. electricity generation increased almost continuously from the beginning of the industry until 2007, when it reached about 4000 terawatt-hours just as the Global Financial Crisis hit. Since 2007, annual electricity generation has remained approximately constant (recall Fig. 3.5). Some regions of the country have declining electricity consumption, and some have increasing consumption. Recall from the energy trends in Chap. 2 that U.S. total primary energy consumption has also been flat for over 15 years. Neither of these stagnant consumption trends *have ever been previously experienced in U.S. history.*

Because overall U.S. demand for electricity has been constant for over 15 years, any new power plant must largely be seen as either *replacing* a retiring plant or *displacing* the generation from a power plant with higher operation and maintenance costs. There is little incentive to increase the total installed capacity of power plants. Why make new power investments when people aren't consuming enough power from the existing power plants? In this environment investors lower risk by focusing on smaller rather than larger power plants. Hence recent investments have focused on energy efficiency, demand response, and smaller capacity natural gas, wind, and photovoltaic plants instead of larger coal-fired and nuclear power plants [15]. Figure 4.19 shows this trend in power plant installations. Before the 1990s new power plants were fueled by coal, uranium (nuclear power), natural gas, and water (hydro power). Coal and nuclear plants are more energy efficient, and thus have lower operational costs, when they use relatively large generation units, say more than 600 megawatts (MW) of capacity. Since the late 1990s power plant installations have been dominated by natural gas, wind, and solar that can be installed from 10s to a few 100s of MW at a time.

There are several policy and technological reasons for the post 1990s shift. First, there was a push toward deregulation, or restructuring, of the electricity system in many parts of the country (e.g., Texas, New England, New York). This effort split apart electric monopolies into two separate types of companies. First, the electric poles and wires were still owned by regulated utilities that received a guaranteed rate of return on their investments. Second, the newly classified *independent power producers* now owned the power plants and compete against each other to sell electricity on the grid. These new companies could build power plants to compete with the existing fleet. In the early 2000s, natural gas turbines, derived from the jet engines used in aviation, became affordable and could be combined with traditional steam turbines to make power plants with power efficiency beyond what a coal, nuclear, and existing natural gas plants could achieve. Plus, natural gas was cheap, and these "combined cycle" natural gas power plants had relatively low capital cost. These factors led to the tremendous boom in natural gas-fired generation in the early 2000s. The 6 years from 2000 to 2005 are the only ones in U.S. history with annual installations greater than 15 GW per year for any given type of power plant.

The pulse of new natural gas capacity put pressure on older inefficient power plants. The U.S. witnessed its first major wave of power plant retirements starting in 2001. Amazingly, there were practically no power plant retirements through the year 2000. By 2000 the U.S. had only retired less than 3 GW of capacity. By 2010, 42 GW had retired, and by 2016 it was 133 GW.

Just as discussed with population, the components of the electric grid can grow in number. Just as with population, if we slow the expansion of the grid, then the average age of the grid increases. Just as with population, if the grid is no longer expanding, it is not necessarily representative of a failure in decision making or a lack of investment. Just as we cannot escape the limitations of a finite Earth, the aging grid is also not a problem from which it is possible to escape. It is a reality to embrace, to understand, and within which to adapt.

The U.S. is still feeling the ramifications of the increasing rate of power plant retirements since 2010. Recall from Chap. 3 the bankruptcy of Luminant power. It was representative of companies that owned too much coal power during a time when natural gas became cheap (again after 2008) and wind and solar photovoltaics both benefited from policy and mass manufacturing to make them increasingly affordable. Nuclear power is also struggling to stay economic because the operating cost is often higher than that for natural gas, wind, and solar.

As of 2018, owners of coal and nuclear power plants within the U.S. continue to search for ways to keep them economically viable. A prime example is the lobbying effort of FirstEnergy Solutions, the unregulated arm of FirstEnergy Corporation based in Akron, Ohio. Ohio restructured its electricity system during an extended phase-in period from the mid-2000s to mid-2010s, creating a competitive market to sell wholesale electricity. FirstEnergy Solutions was the unregulated subsidiary created to operate 10 GW of formerly regulated coal and nuclear power plants. In 2018 the company filed for bankruptcy because it was unable to make a profit, and it asked then Secretary of Energy Rick Perry and President Trump to declare a grid emergency for Ohio and provide some type of subsidy that could maintain company profitability.[43]

The reasons cited by FirstEnergy Solutions for its bankruptcy filing? Fracking. Renewables. Pollution control costs for coal plants. Lack of electricity demand since the Great Recession of 2008.

Wow. What a list.

In other words, there appear to be several driving factors for the post-2008 economic difficulties of companies operating coal and nuclear power plants in the U.S. Any one of the aforementioned pressures might have been enough to force bankruptcy. There is not just one isolated pressure on energy companies, like FirstEnergy, that we can blame for their economic trouble.

Summary

This chapter considered five major physical and economic trends that provide data for thinking about ideas within the rest of the book: population, debt, interest rates, wages and income distribution, and infrastructure via the example of power plants. These are not the only data that can inform how energy affects economic and physical growth, but they are informative for that purpose.

Feedbacks from a finite Earth put pressure on the growth and accumulation of population and infrastructure. This pressure eventually slows growth in turn translating to an increasingly older population and infrastructure. After the 1970s

---

[43] Ari Natter, Trump Says He's Looking at Emergency Aid for Battered Power Plants, Bloomberg News, April 5, 2018, available at https://www.bloomberg.com/news/articles/2018-04-05/trump-says-emergency-aid-sought-by-firstenergy-to-be-examined.

major energy and economic trends changed. The U.S. ratio of total debt and loans to GDP increased much more rapidly following the slowdown in energy consumption during that decade—this ratio peaked at 380% of GDP in 2009, and has remained above 350% of GDP since that time. Global debt-to-GDP ratios now also reside in the same range as that of the U.S. As a consequence, central bank interest rates, as well as rates on government bonds, have rested at low values, sometimes below zero, that are unprecedented in the history of the modern industrialized economy.

Since the 1970s, income distribution in the U.S. and other rich economies (most of those in the OECD) has typically become more unequal while income (or GDP) and energy have become more equally distributed among all countries of the globe. This trend somewhat accelerated after 2000 as China joined the World Trade Organization, thus ramping up its share of an increasingly globalized economy. While policies do matter for influencing income distribution, for example, between those that work for salaries and wages and those that earn money by having money (i.e., from investment), physics-based explanations of income distribution shows that we should not expect income distributions to approach complete equality even within an economy that is completely fair to all individuals. In short, policy and physical principles both matter.

At this point, there is little value in considering another energy and economic trend in isolation. We must move on to the next step that considers how the various trends fit together into a cohesive view. We need to know how each of these trends links and feeds back to each other.

What is the common context for stagnant energy consumption in Western countries, no to low population growth, unprecedented high public and private debt levels relative to GDP, unprecedented low central bank interest rates, and an aging infrastructure? Are these all symptoms of a common cause? Is there some unifying thread? We now turn to Part II of the book, and we start *putting it all together with a systems perspective.*

References

1. Alvaredo, F., Chancel, L., Piketty, T., Saez, E., Zucman, G.: World inequality report 2018. Tech. rep. (2017). http://wir2018.wid.world
2. Angle, J.: The surplus theory of social stratification and the size distribution of personal wealth. Social Forces p. 293–326 (1986)
3. Bettencourt, L.M.A.: The origins of scaling in cities. Science **340**(6139), 1438–1441 (2013). https://doi.org/10.1126/science.1235823. http://science.sciencemag.org/content/340/6139/1438
4. Bivens, J., Mishel, L.: Understanding the historic divergence between productivity and a typical worker's pay: Why it matters and why it's real. https://www.epi.org/files/2015/understanding-productivity-pay-divergence-final.pdf (2015). Online; accessed March 13, 2018
5. CBO: The distribution of household income, 2016 (2019). https://www.cbo.gov/publication/55413
6. De Stercke, S.: Dynamics of energy systems: A useful perspective. Tech. rep., International Institute for Applied Systems Analysis (2014)

7. Dobbs, R., Lund, S., Woetzel, J., Mutafchieva, M.: Debt and (not much) deleveraging: The global credit bubble and its economic consequences. Tech. rep., McKinsey Global Institute (2015)
8. Drăgulescu, A., Yakovenko, V.: Evidence for the exponential distribution of income in the USA. The European Physical Journal B - Condensed Matter and Complex Systems **20**(4), 585–589 (2001). https://doi.org/10.1007/PL00011112
9. Drăgulescu, A., Yakovenko, V.M.: Exponential and power-law probability distributions of wealth and income in the United Kingdom and the united states. Physica A: Statistical Mechanics and its Applications **299**(1–2), 213–221 (2001). http://dx.doi.org/10.1016/S0378-4371(01)00298-9. http://www.sciencedirect.com/science/article/pii/S0378437101002989. Application of Physics in Economic Modelling
10. Fizaine, F., Court, V.: Energy expenditure, economic growth, and the minimum EROI of society. Energy Policy **95**, 172–186 (2016). http://dx.doi.org/10.1016/j.enpol.2016.04.039. http://www.sciencedirect.com/science/article/pii/S0301421516302087
11. Galbraith, J.K.: The End of Normal: The Great Crisis and the Future of Growth. Simon & Schuster, New York (2014)
12. Galbraith, J.K.: Sparse, inconsistent and unreliable: Tax records and the world inequality report 2018. Development and Change **50**(2), 329–346 (2019). https://onlinelibrary.wiley.com/doi/abs/10.1111/dech.12475
13. Graeber, D.: Debt: The first 5,000 Years. Melville House, Brooklyn, NY (2014)
14. ILO, OECD: The labour share in g20 economies. Tech. rep., International Labour Organization and Organization for Economic Co-operation and Development (2015). https://www.oecd.org/g20/topics/employment-and-social-policy/The-Labour-Share-in-G20-Economies.pdf. Report prepared for the G20 Employment Working Group Antalya, Turkey, 26–27 February 2015
15. King, C.W.: The rising cost of resources and global indicators of change. American Scientist **103**, 6 (2015)
16. King, C.W., Holman, A.S., Webber, M.E.: Thirst for energy. Nature Geoscience **1**, 283–286 (2008)
17. Kremer, M.: Population growth and technological change: One million B.C. to 1990. The Quarterly Journal of Economics **108**(3), 681–716 (1993)
18. Lab, W.I.: World inequality data, 1980–2016. Tech. rep., World Inequality Lab (2018). http://wid.world/data/. Data downloaded February 27, 2018.
19. Lawrence, S., Liu, Q., Yakovenko, V.M.: Global inequality in energy consumption from 1980 to 2010. Entropy **15**(12), 5565–5579 (2013). https://doi.org/10.3390/e15125565. http://www.mdpi.com/1099-4300/15/12/5565
20. McLeay, M., Radia, A., Thomas, R.: Money creation in the modern economy. Quarterly Bulletin 2014 Q1 (2014)
21. Nicholas, Shaxson, *The Finance Curse: How Global Finance Is Making Us All Poorer*. Grove Atlantic Press, 2019.
22. Sawyer, S.M., Azzopardi, P.S., Wickremarathne, D., Patton, G.C.: The age of adolescence. The Lancet Child & Adolescent Health (2018). https://doi.org/10.1016/S2352-4642(18)30022-1
23. Silva, A.C., Yakovenko, V.M.: Temporal evolution of the "thermal" and "superthermal" income classes in the USA during 1983–2001. EPL (Europhysics Letters) **69**(2), 304 (2005). http://stacks.iop.org/0295-5075/69/i=2/a=304
24. Smil, V.: Energy and Civilization A History. The MIT Press, Cambridge, Mass. (2017)
25. Stillwell, A.S., King, C.W., Webber, M.E., Duncan, I.J., Hardberger, A.: The energy-water nexus in Texas. Ecology and Society **16**(1) (2011). https://www.ecologyandsociety.org/vol16/iss1/art2/
26. Tainter, J.A.: Energy, complexity, and sustainability: A historical perspective. Environmental Innovation and Societal Transitions **1**, 89–95 (2011)
27. Tainter, J.A., Allen, T.F.H., Little, A., Hoekstra, T.W.: Resource transitions and energy gain: Contexts of organization. Conservation Ecology **7**(3) (2003)
28. Tainter, J.A., Patzek, T.W.: Drilling Down: The Gulf Oil Debacle and Our Energy Dilemma. Springer (2012)

29. Webber, M.E.: Thirst for Power: Energy, Water, and Human Survival. Yale University Press, New Haven and London (2016)
30. West, G.: Scale The Universal Laws of Growth, Innovation, Sustainability, and the Pace of Life in Organisms, Cities, Economies, and Companies. Penguin Press, New York, NY (2017)
31. Yakovenko, V.M., J. Barkley Rosser, J.: Colloquium: Statistical mechanics of money, wealth, and income. Reviews of Modern Physics OCTOBER–DECEMBER 2009 **81**(4), 1703–1725 (2009). https://doi.org/10.1103/RevModPhys.81.1703. http://www.sciencedirect.com/science/article/pii/S0378437101002989

Part II
Putting It All Together

Chapter 5
Systems Thinking for Energy and the Economy: Size and Structure

The wheels on the bus go round and round
Round and round, round and round
The wheels on the bus go round and round
All through the town
. . .
The horn on the bus goes Beep, beep, beep
Beep, beep, beep
Beep, beep, beep
The horn on the bus goes Beep, beep, beep
All through the town
. . .
— Verna Hills (1930s)[1]

What Is a System?

Scientists came up with the concept of a *system* to help us understand the changes that occur around us and are affected by us. Here are two definitions.

> . . . a *system* is a whole of some sort made up of interacting or interdependent elements or components integrally related among themselves in a way that differs from the relationships they may have with other elements.[2]—George Mobus and Michael Kalton (2015)

> A system is an interconnected set of elements that is coherently organized in a way that achieves something (function or purpose).[3]—Donella Meadows (2008)

---

[1] "The Wheels on the Bus" is a traditional American folk song from the 1930s written by Verna Hills in Boston, MA.

[2] Mobus and Kalton [40, pp. 73–74].

[3] Meadows [39, p. 11].

© Springer Nature Switzerland AG 2021
C. W. King, *The Economic Superorganism*,
https://doi.org/10.1007/978-3-030-50295-9_5

Thus, a system first has a number of parts, or *elements*. These parts are connected. Finally, these connected parts operate in a manner that is defined not only by the parts but also by the connections among the parts. *These connections constrain the system* to operate within a subset of all possible options that could exist with the same set of elements. We might then define the system as having a function or purpose that we cannot otherwise explain without the full present definition of a system.

The "Wheels on the bus" nursery rhyme quoted above is an example of ignoring the concept of a system. The full nursery rhyme describes only the parts of the bus, such as wheels, the horn, and windshield wipers. These are elements of the bus. It explains neither how the parts are connected nor the function of the bus—to transport people. But it is natural to start learning with the parts of a bus before learning how the parts are connected. Nursery rhymes are educational tools that are literally used as baby steps early in the learning process. To understand the economy as a system, this chapter must go well beyond babies and buses.

The concept of a system is powerful because it says that a system can have properties or act in a way that cannot be determined solely by adding up how the individual parts operate. In this way, the concept of a system is maddening to some and liberating to others, but we must be comfortable with it and use it. A system often has *emergent* properties. Emergent properties are ways of speaking of a large collection of objects, or relations among objects, in a way that is both consistent with the underlying parts and useful. Emergent properties are useful in the sense that they are shorthand for describing the world around us. For example, we have both apples and bananas. They are both made of the same types of atoms, but each has its own arrangement of these atoms. If you ask your child if they want a banana or an apple, you don't first describe the molecular content and arrangement of the atoms of each. Over and over you hand your child a banana and say "banana," and you hand them an apple and say "apple." Eventually the concepts of banana and apple emerge in their mind as combinations of multiple properties including shape, color, taste, smell, and texture. By the emergent properties we assign apples and bananas, we call them both fruit and food. So a group of atoms can turn into apples and bananas, and both can turn into an emergent category called fruit. Of course there are many types of apples (Fuji, Red Delicious, Pink Lady, etc.), but they are similar enough that we easily recognize them as a type of apple.

Scientists use the term *coarse-graining* to describe a system that has so many interacting elements that it is not useful or practical to describe each element. Coarse-graining is similar to emergence, but more explicit. Emergence can be very qualitative. Coarse-graining describes mathematical descriptions of phenomena, including properties we can objectively measure. Coarse-grained concepts summarize actions ongoing at "fine-grained" or smaller scales. The trick is that many situations going on at the fine-grained scale might translate to a single description at the coarse-grained scale. This means that if I know a system's coarse-grained description, I can't generally tell you what is going on at the fine-grained level because there might be multiple possible fine-grained descriptions or theories that are compatible with the coarser description. While coarse-grained theories are less detailed, they are more computationally efficient and thus more practical for modeling system behavior. Many narratives speak past each other because they describe emergent properties of energy and economic systems while assuming a different set of governing fine-grained activities.

A coarse-graining example we can all relate to is the concept of temperature.[4]

> Temperature is the average speed of particles in a system. Temperature is a coarse-grained representation of all of the particles' behavior—the particles in aggregate. When we know the temperature we can use it to predict the system's future state better than we could if we actually measured the speed of individual particles. This is why coarse-graining is so important—it is incredibly useful. It gives us what is called an effective theory. An effective theory allows us to model the behavior of a system without specifying all of the underlying causes that lead to system state changes.[5]—Jessica Flack (2017)

In Chap. 4 we learned how econophysicists applied the scientific principles behind temperature to describe the proportion of people that have a certain income within a country. Thus, they used the idea of coarse-graining as applied to air molecules. Instead of starting with a theory of the distribution of speed for individual particles and using temperature as an average of all activities, they started with the average personal income to infer what percentage of people have a certain income. They took a systems approach in that they assumed a wholeness of the economy (the country of study), certain elements (the people), certain constrained interconnections (the random exchanges of money), and a function (to maximize entropy and follow the second law of thermodynamics).

To some it might be disconcerting that a simple theory describing inanimate particles also describes something about our complex human economy that most people view as composed of independent agents *choosing* what to buy and where to work. However, this approach is not personal, and it's not meant to be. It's descriptive. If a bunch of objects, particles or people, start with a certain amount of something, then randomly bump into one another and exchange some of that something over and over, you will get certain statistical distributions. It's just science and mathematics that give a theory for understanding why both incomes might be distributed in the manner that we've observed over time and how our tax rules affect that distribution.

The power of scientific and systems thinking is that it helps us discover common principles for describing seemingly disparate phenomena. Most economists, and many scientists, are not taught statistical mechanics or fluid dynamics, but their ignorance does not mean that this coarse-grained temperature-analog description of income distribution is invalid. In fact, because the phenomenon is so well understood in physics and describes a large fraction of the economic data, it puts pressure on economists to adopt it or have better, or at least compatible, alternative coarse-grained descriptions that link micro-scale to macro-scale observations.

---

[4]Also see [13, pp. 97–99].

[5]Jessica Flack, 2017, Edge, What Scientific Term or Concept ought to be more Widely Known? https://www.edge.org/response-detail/27162.

As nicely stated by Sean Carroll: "You can understand the air as a fluid without knowing anything about its molecular composition, or even if there is a description in terms of particles at all."[6] In fact, this is how thermodynamics first developed: *scientists knew how to measure and think about the average temperature of air and steam before they knew it was composed of individual molecules.* Keeping in mind the income for each person is the analog for temperature of each air molecule, we can restate Carroll's quote using terms from economics: "You can understand the *population* as *an economy of people exchanging money* without knowing anything about *what each person is doing*, or even if there is a description in terms of *persons* at all."

The observation of phenomena is key to performing scientific inquiry, and for understanding systems. In her posthumously published book, Meadows emphasizes an important concept for determining the purpose of systems:

> If a frog turns right and catches a fly, and then turns left and catches a fly, and then turns around backward and catches a fly, the purpose of the frog has to do not with turning left or right or backward but with catching flies. If a government proclaims its interest in protecting the environment but allocates little money or effort toward that goal, environmental protection is not, in fact, the government's purpose. **Purposes are deduced from behavior, not from rhetoric or stated goals.**[7]—Donnella Meadows (2008) (emphasis added)

Systems have a few important characteristics. One is the *integrity of wholeness* where the system's purpose cannot be deduced from only considering its parts. We define a system's purpose by coarse-graining. A bus, the human body, a tree, and the frog in the quote above are examples of systems.

The human body has arms, legs, a heart, a brain, and other parts. We might say the purpose of legs is locomotion such as walking and running. Arms are used for accessing objects in the environment such that our hands can grasp and manipulate objects. Our heart pumps blood through our body to move oxygen and other nutrients to our cells. Our brain controls both the voluntary and involuntary actions of the parts of our body, and it interprets signals originating in our sensory organs (e.g., eyes, fingers, ears) that perceive the environment within which we live [63].

In addition to having integrity of wholeness, systems are *adaptive, resilient, and evolutionary* [39]. This means that they have the ability to react and reconfigure themselves depending on what is occurring around them. If they can adapt and reconfigure, this means they have some *structure*. Again consider our brains. Research shows that the synaptic connections among neurons within our brains have many possible configurations, but certain connections are emphasized, and others deteriorate, based upon what we observe in the environment. If our eyes do not function, the brain does not configure itself to interpret the light reflected from the objects around us. It configures itself to better interpret other sensations, such as

---

[6]Carroll [13, p. 99].

[7]Meadows [39, p. 14].

sound and touch, that our body can detect. Thus, our brains adapt to the environment we experience, and these experiences, particularly when we are young, shape the structure of our brains and how we perceive the world [63]. Take the following description of kitten brain development when it is exposed only to certain visual patterns.

> Kittens raised wearing goggles that allowed them to see only vertical lines in one eye and horizontal lines in the other have fewer than the normal number of cells that respond to oblique lines. Moreover, the [visual cortex brain] cells responsive to vertical lines are active only with stimulation of the eye that had been exposed to vertical lines, and the cells responsive to horizontal lines are active only with stimulation of the eye that had been exposed to horizontal lines.[8]—Bruce Wexler (2008)

Amazing. Brains emphasize connections that are stimulated when a body's sensory organs observe things the environment. It cannot recognize sights, sounds, smells, and the feel of surfaces that it does not know exist. From various experiments on brains of cats, monkeys, and other mammals Bruce Wexler derives three critical points in his book *Brain and Culture*.[9]

1 ...the dependence on the mammalian brain upon sensory stimulation is obligatory.
2 ...environmental stimulation shapes the structural and functional organization of the brain; it is not simply that a predetermined organization requires sensory stimulation to be realized.
3 ...once brain organization evolves, and the individual reaches sexual maturity, existing structures tend to be enduring and resistant to change. ...While activity-related functional reorganization is possible in adult mammals, it is much slower, much more limited, and achieved with much greater physiological effort.

As Wexler indicates, brains can and do change. That is to say they have *neuroplasticity*. You can teach an old dog new tricks, but it is much easier when it is a puppy. As we'll discuss in the next chapter, the same principle probably holds for economists and scientists (because they are people too!).

Systems are also *goal-seeking, self-preserving, and self-organizing*. This means that the reactions and changes to both elements and interconnections within a system are driven by changes in the environment, determined by the current structure of the system largely for the purpose of preserving itself. We can think of economies (countries, cities, etc.) as systems trying to preserve themselves.

The term *system* is so generic that there can be *systems of systems*, and to some this is another maddening feature, but also necessary. For example, there is one global economy composed of the economies of many countries. Each country's economy has many subsectors, composed of individual businesses, all of which are possibly trying to preserve themselves by maximizing profit. In a capitalist system, each business or economic sector preserves itself partly by commanding increasing flows of money through them.

---

[8]Wexler [63, pp. 45–46].
[9]Wexler [63, p. 58].

The financial sector is one part of the economy. In the heat of the 2008 Global Financial Crisis, we were told that some banks were "too big to fail." Initially, I thought this meant that they were so big that it was impossible for them to fail. No. What this phrase meant was that the U.S.'s appointed and elected leaders believed the banks were too big, and important, to let them fail. This was a narrative that won over the discussion during the heat of the moment. Since the financial crisis, the U.S. economy is larger in terms of GDP, and its banking sector accounts for approximately the same proportion of GDP as before the crisis.[10] So far both the U.S. economy overall and its financial sector have successfully preserved themselves, at least in terms of commanding a high proportion of global economic flows.

But is there some overarching reason to maintain the current financial system? What about our energy system? They have both existed in pre-industrial times, even if not understood in the same way as we do today. As mentioned in Chap. 4, before the concept of banking, humans already used the concept of debt as an obligation and a way to maintain relationships over time. Further, human society requires an energy system to function even if prehistoric humans could not describe energy or society as we do today.

The purpose of the economy is not defined by what people, including politicians, economists, and scientists, say they want to happen or what their models indicate should happen. *The purpose of a system is defined by what actually happens.* This concept is key, and it is the reason why Part I of this book emphasizes data and trends. The energy and economic data define the purpose of the economy. Not only do the energy and economic narratives battle to explain the patterns we observe in the data, they battle for what data and models are allowed for discussion. While the narratives might agree on the macro-scale observations, such as GDP and primary energy consumption, they might disagree vehemently on how the unobserved micro-scale individual activities going on in the world translate to the macro-scale observations.

The ideas of emergence and coarse-graining imply that from a practical standpoint, if we want to understand a system and its function, we have to use concepts and models that summarize the fine levels of detail, but this isn't strictly the case. "... emergence is about different theories speaking different languages, but offering compatible descriptions of the same underlying phenomena in their respective domains of applicability."[11] We don't have the time or computational power to track every economic and energetic transaction, and we don't necessarily have to.

We can only make rules and laws to influence how some elements of our economy interact. The energy and economic narratives, backed by mathematics or not, attempt to explain data that are in effect averaged values summarizing the individual interactions among the parts of our society and economy. Our task

---

[10]Banking sector is defined as "financial services and insurance" which accounted for 8% of GDP in 2008 and 7% in 2015.

[11]Carroll [13, p. 100].

is to ensure that different narratives are *compatible* with our observations, with themselves, and to alter them into compatibility if they are not.

But how well can we understand our national and global economies, within which each of us, as individuals, resides? Can we understand what we are a part of ourselves? Are we clever enough to decipher any overall purpose of our economy, the role of energy, and whether our emergent or coarse-grained descriptions are compatible? A quote from ecologist Howard Odum sums up the quandary:

> It is sometimes said that no system can understand itself. A doorbell buzzes but does not know how it did it. A human knows how a doorbell works, but he does complicated things that he cannot understand. It takes more components to understand than to be. The logical extension of this theorem is that the system . . . in its shared network of information processing does not completely understand itself. The whole system has a giant intelligence which is smarter than its components and may even have some consciousness or "group dynamics" that understands humans, but it cannot understand itself fully.
>
> The way a system can understand itself is to develop simplified ideas, sometimes called models of itself, which have enough of the main features to have some reality but are simple enough to be understood.[12]—Howard T. Odum

Scientific findings in the areas of cognitive neuroscience, biological (Darwinian) evolution, and the computer science of artificial intelligence challenge both the uniqueness and superiority of how we humans think and what we think we know.

As implied by Odum's quote, we humans are part of an economic system. To understand this system we can develop simplified ideas and models of the economy. Chapter 8 revisits whether we can understand and know the ultimate purpose of the economy, but for now ponder if we are up to Odum's challenge in making models "... which have enough of the main features to have some reality but are simple enough to be understood." In this pursuit it is useful to consider a philosophical framework that can help us guide our narratives and modeling efforts.

Naturalism

To assess the energy and economic systems within which we reside, a conceptual or philosophical framework can help us figure out if we are doing a good job. Are we assuming too much? Are we constraining ourselves to a limited set of explanations without realizing it? One such framework is *naturalism*.

Sean Carroll, in *The Big Picture*, describes naturalism as considering:[13]

1. There is only one world, the natural world.
2. The world evolves according to unbroken patterns, the laws of nature.
3. The only reliable way of learning about the world is by observing it.

---

[12]Odum [44, p. 119].

[13]Carroll [13, p. 20]. If the concept of *naturalism* intrigues you, you might enjoy a series of discussions within a workshop arranged by Sean Carroll in 2012 entitled "Moving Naturalism Forward." The videos are linked on his website http://www.preposterousuniverse.com/naturalism2012/ and available on YouTube.

Using this definition, I consider myself a naturalist, and this philosophy integrates the concepts of this book. The opposite of naturalism is *supernaturalism*. If that word invokes ideas of ghosts, then yes, that is correct. Despite many television shows, movies, and honest attempts, we don't yet have any measurements that prove ghosts exist. So at the moment, ghosts are beyond our ability to perceive them in the natural world, and there is a very high probability (very close to 100%) that they don't exist. Will we ever prove that ghosts absolutely and positively don't exist? Likely not, but the low probability that ghosts exist means that we don't make any real-world decisions based on ghosts, other than decisions to create more TV shows, movies, and books about ghosts.

Naturalism also implies that souls, gods, or the one and only God, also don't exist. While there are phenomena that we observe on Earth and in the universe that we can't yet explain, a naturalist believes this represents our ignorance rather than the actions of divine entities.

Related to naturalism is *materialism*, which claims "...there is only one sort of stuff, namely *matter*—the physical stuff of physics, chemistry, and physiology ..."[14] Thus, the natural world is the only world to observe, and it is made of matter. We know there are different forms of matter—rocks, animals, trees, etc. They are all made of the same underlying parts of atoms and subatomic particles, but they are arranged differently.

We humans require energy and nutrients in food to survive and propagate. The laws of physics dictate that our physical capital and infrastructures (buildings, power plants, vehicles, and factories) require energy inputs to perform useful work. Further, both humans and physical capital are made of matter, but there are constraints on how this matter can be organized or structured. Generically our knowledge about these constraints can be called information. To understand processes of the world we need to simultaneously consider both *size and structure* in compatible ways. To think about size and structure, we can use the three critical concepts of *matter, information, and energy* Anthropologist Richard Adams conceptualized this three-part concept in what called the "mass–energy–information" complex.[15]

Matter is the physical stuff around us, the things we can measure with scientific instruments as well as our senses of sight, sound, touch, etc. In thinking of matter we can answer the question "How much is there?" *Information* describes the arrangement of matter that is achieved during energy transformations. In thinking

---

[14]Dennett [17, p. 33].

[15]"To focus on the mass–energy–information complex is to focus on a material world, to insist that whatever it is that the social scientist may study, that thing must be of that world, or it cannot be studied. It is here that we need to employ a methodological dualism: we must for certain purposes resort to a mentalistic-energetic differentiation which, in fact, I would not subscribe to in theory. The reason for handling the mass–energy–information complex dualistically is that we not only want to be able to find the regularities between energy and mass and action, and in its manifestations as information as well, but we also want to explore how these energetic processes relate to those which have generally been subsumed under terms such as *value, cognition*, and other mentalistic labels." [1, p. 111].

of information we can answer such questions as "How do you do it?" and "How are things (energy, matter, money, etc.) distributed or constrained? " *Energy* is a concept that describes the amount of effort, or cost, required to relocate matter or rearrange it into different forms and patterns as dictated by the information.

> With the three concepts of *matter, information, and energy* we are potentially in the game for providing a coherent description of biological systems (animals, ecosystems), and physical systems (machines, the economy) that consume and dissipate energy.

Consider making a generic product, or "widget." Matter answers, "What parts of the environment is the widget made of?" Information answers, "How do you arrange matter to make the widget?" Energy and the laws of physics answer, "How many environmental stocks and flows, which we call energy, must be consumed to make and distribute the widget, that is made out of matter, in the way specified by our information?"

Note how the concept of energy is pivotal. If we want to move matter from one place to another, there must be a transformation and dissipation of energy. If we want to rearrange the configuration of atoms for a particular quantity of matter, thus changing its structure and information content, there must be a transformation and dissipation of energy.

Importantly, the combinations of matter, energy, and information do not answer "why" or "should" questions. "Why does the widget exist?" "Should the widget exist?" We could go further to ask questions such as "Why do we exist?" and "What is the purpose of life?"

While many of these questions are beyond the scope of this book, some are very much within the scope. On a practical level we can ask: What are the relationships between various trends in population, debt, GDP, and energy consumption? What theoretical frameworks are compatible with the observed relationships? On a more philosophical level, we can ask: What is the purpose of the economy? With a finite flow of energy and resource extraction at any given time, does the purpose of the economy have anything to do with accumulating and maintaining information and matter in the forms of humans and physical capital? Why does the economy grow, and should we want it to grow? Who, if anyone, or what, if anything, is or should be in charge?

In the context of these questions, Carroll adds nuance to his description of naturalism. He further defines what he calls *poetic naturalism* via three additional points:

1. There are many ways of talking about the world.
2. All good ways of talking must be consistent with one another and with the world.
3. Our purposes in the moment determine the best way of talking.

In the context of energy and economic narratives, this book very much seeks to describe "ways of talking" about the economy that are consistent with data and what we know about the concept of energy. Theoretical frameworks that force simultaneous consideration of matter, information, and energy have a better chance to describe macroeconomic trends because they constrain the number of probable explanations. This book discusses many of these scientifically constraining frameworks including ecology, evolution, thermodynamics, complexity, emergence, and information theory. It also discusses systems science along with economic theories and frameworks (much still to come).

We know that over the last 200 years *the absolute size* of the global energy system has grown several times over in the amount of human-made infrastructure and the rate at which it extracts and consumes primary energy. Further, Chap. 2 showed the energy sector became a *significantly smaller percentage* of the economy due to coal-to-steam powered mechanical work, not because less money was spent on energy, but because energy plus technology enabled a much faster increase in GDP. But there were significant turning points in the 1970s and around the year 2000. Inherent in these statements is that it is useful to discuss an "energy" part of the economy separate from the rest. This is one way to make some progress toward understanding our economic system from within. We can think about the boundary that defines the size of our energy-economic system as well as how both the absolute and relative size of the energy sector enable other parts of the economy to grow and evolve.

The Energy System Boundary

Each system is defined by the *boundaries* that distinguish it from everything else. Some elements are within the system, and some are not. Think of systems residing within other systems as a set of Russian Matryoshka dolls where each doll nests within a larger doll. An energy and economic narrative could discuss a small or large boundary just like you could play with one of the small or large Matryoshka dolls.

This book generally focuses on larger boundaries. The Earth, and perhaps its orbiting satellites, is the largest Russian doll with which we play in this book. Importantly, energy, matter, and information cross system boundaries. Sunlight crosses the Earth's atmosphere to eventually reach the surface. Some sunlight is reflected back into space, and some is converted to heat before radiating back to space. Meteorites and asteroids (hopefully not a big one soon!) enter Earth's atmosphere, and we have sent satellites and rockets into near-earth orbit and into the solar system. Further, the moon, and the sun, impose gravitational forces that affect ocean tides.

The Earth is not the largest boundary we could consider. The Earth and other planets reside within our larger solar system and Milky Way galaxy. Chapter 3 stated Julian Simon's notion that one relevant system boundary for analysis is "the cosmos," by which he meant the universe.[16] Since scientists cannot (yet) prove the size of the universe, he argued, we don't even know if it is finite or not. Thus Simon, and a few modern billionaires, pontificate we could extract extraterrestrial resources at continuously increasing rates to continue economic growth or perhaps achieve some superintelligence, human or non-human, whether on Earth or some other planet.[17] But it takes time and energy to travel across space. Crossing more space time takes more power, and power is what really relates to economic activity. This book only considers a future time span of one or two centuries instead of the lifespan of the universe, and even Simon recognized this shorter relevant time span for "social decisions."[18]

Even confining ourselves to Earthly discussions, it is still easy to get lost in the complexities and details of energy and economics. Chapter 3 summarized disagreeing narratives on some of these details. Within the Earth we have economic, political, and other social systems that we describe with words such as country, city, community, and neighborhood. We also have physical systems such as the electric grid, the Internet, airplanes, and individual people. There are many elements and subsystems within the largest Earth system to consider. However, as emphasized in this book, it is crucial to step back to understand broad trends and principles that govern our energy-economic system organization.

To understand a forest, ecologists describe not only the trees but also how the trees and other living creatures interact. We must take the same approach. To do this we must consider more than just the size of the elements within our energy-economic system. We need to consider how the *structure*, or interconnections, of the elements relates to *size and growth* of the economy itself. The role of the energy system is to extract energy from the environment that then powers the rest of the economy. But before it does this, it has to consume its own product. Just how much energy does it take to produce energy?

It Takes Energy to Produce Energy

So far I have mostly discussed energy in terms of the total amount that is extracted from the environment. This is the primary energy. *Secondary energy* is that within the energy carriers, like electricity and gasoline, consumed at end uses, such as

---

[16]"Life could even spread from Earth to other planets, other galaxies, and so on, incorporating an increasing portion of the universe's matter and energy." [49, p. 81].

[17]"And the chances would seem excellent that during that span of time [assumed approximate 7 billion year lifespan of the solar system that also includes the entire lifespan of the human species] humans will be in touch with other solar systems, or will find ways to convert the matter on other planets into the energy we need to continue longer." [49, p. 79].

[18]"...horizon relevant for social decisions—the next five, twenty-five, one hundred, perhaps two hundred years."[49, p. 171].

within buildings, industrial facilities, and cars. The quantity of secondary energy is less than that of total primary energy. There are two major reasons.

First, the second law of thermodynamics dictates we cannot convert 100% of primary energy into secondary energy carriers. Some of the energy is dissipated into heat that we can no longer use to perform work. Energy consumption and heat dissipation are required even if we want to move something from one place to another. For example, oil is converted into gasoline that is then distributed via pipelines and trucks to filling stations all around the country, and each of these steps requires energy consumption and a loss of energy to heat. Also, we burn natural gas in power plants to generate electricity with less energy content than was in the natural gas. Further, heat dissipation occurs as the electricity is distributed across power lines to our homes. Energy cannot be converted from one form to another without heat loss, and it can't be distributed without heat loss.

The second reason end-use energy is less than primary energy is that extracting primary energy itself requires useful work, or energy inputs. *It takes energy to extract energy.* This same phrase is often used to describe money instead of energy. We evaluate businesses based on their "return on investment" (ROI). If business owners receive less revenue than their spending, their net ROI is less than zero. Over the long term, aside from fraud and getting a bailout from the government, they will not stay in business. A similar ROI concept is applied to energy technologies and energy sectors. This energy output/input ratio is often called "energy return on energy invested" (EROI) to parallel the economic concept [26, 27]. Although political efforts often constrain energy prices and subsidize energy businesses, there is no direct parallel of an energy bailout for an energy project. Ultimately, if an energy extraction method uses too much energy for its own inputs, it will not be economically viable because it is not energetically viable.

An example energy resource at the margin of energetic-viability is the bitumen hydrocarbon fossil energy resource of Alberta, Canada. You might have heard about this by a different name. The fossil energy narrative might call them *oil* sands, but the renewable energy narrative might call them *tar* sands to give them more of a negative connotation.   In the 2000s, this resource became categorized as an economically viable energy *reserve*. There are two reasons. First, there were technological improvements, in particular the steam assisted gravity discharge (SAGD) in-situ extraction method. Second, the global price of oil was increasing (refer back to Fig. 3.1). Recall that the definition of an *energy reserve* is one that is economically viable using available technology. Thus, if oil prices rise, the quantity of oil reserves can increase even at constant technology.

Oil/tar sands require significantly more energy inputs to produce than historical conventional oil. The bitumen is too viscous to flow underground in its natural state. About 80% of bitumen resources are too deep to extract by digging. For these deeper resources, a significant amount of fuel is burned to create steam that is injected underground to enable the bitumen to flow and be pumped to the surface. For every unit of energy input into production, less than 6 units of energy come out in the extracted bitumen [10]. The U.S. oil and gas industry historically produced 10–20 units of energy relative to a unit of energy input [24]. Considering the additional energy inputs for refining the oil to products such as gasoline and jet fuel, oil sands deliver less than 3 units of energy, whereas conventional petroleum gasoline historically delivered between 5 and 10 [31, 32, 35]. In general, the lower the EROI, the higher the cost [34]. An EROI of 5 for gasoline roughly translates to an equivalent gasoline price of near 4 $/gallon that U.S. consumers experienced in 2008 [31].

The 2008 global financial crisis and Great Recession put downward pressure on oil prices due to a reduction in demand. Low-interest financing of increased rates of horizontal drilling and hydraulic fracturing for oil and gas in the U.S. did the same. This one-two combination caused such low oil prices after 2014 that much of the Canadian bitumen was no longer economically viable. Low prices can force companies to list some previously listed reserves as (uneconomic) resources, but the major energy data sets don't seem to indicate this reduction occurred for Canada. After 2015, several planned bitumen extraction projects were scrapped.[19] However, due to the high supply and low price of natural gas in North America (again linked to hydraulic fracturing), the monetary cost of creating steam (by burning natural gas) has historically been relatively low for bitumen extraction.

As is often joked in economic circles, the solution to low prices is low prices. In response to low oil prices, Canadian bitumen firms worked to reduce capital and operating costs, so we can't write off oil/tar sands extraction going forward.[20] The large size of Canada's bitumen reserve is a contentious point for dealing with climate change. The size of estimated reserves (and subsequent carbon) in Alberta's bitumen puts Canada in top three worldwide, along with Saudi Arabia and Venezuela. For any chance to stay below the targeted average global temperature rise relative to pre-industrial times (e.g., <2–3°C per the Paris Climate Agreement), most of Canada's bitumen has to stay in the ground. While many thought Canada's Prime Minister Justin Trudeau was committed to reducing its greenhouse gas emissions, he has turned out to follow a common theme in giving significant preference to shorter-term economic circumstances. In 2018, Trudeau alarmed Canada's climate community by spending 4.5 billion dollars of government money to purchase an export-oriented oil pipeline project from a private developer who was not going to be able to defeat provincial and first nations opposition.[21] As we learned with the Denton Fracking Ban, higher level governments overrule those at smaller levels.

---

[19]Yadullah Hussain, *Financial Post*, "Almost $60-billion in Canadian projects in peril as 'collapse' in oil investment echoes the dark days of 1999," January 2, 2015, https://business.financialpost.com/commodities/energy/almost-60-billion-in-canadian-projects-in-peril-as-collapse-in-oil-investment-echoes-the-dark-days-of-1999.

[20]Jesse Snyder, *Financial Post*, "Breakeven costs of US$40—and falling—means it's too soon to count out the oilsands," September 6, 2017 at https://business.financialpost.com/commodities/energy/more-than-just-a-glimmer-of-hope-lower-costs-suggest-its-too-soon-to-count-out-the-oilsands.

[21]Bruce Livesey, *The Guardian*, May 31, 2018, "Did Canada buy an oil pipeline in fear of being sued by China?", https://www.theguardian.com/commentisfree/2018/may/31/justin-trudeau-kinder-morgan-pipeline-china-did-he-fear-being-sued.

There are other fossil resources that have such poor energy balance that they are not, and perhaps never will be, economically viable. One example is the kerogen oil shale resource of the Piceance Basin in Colorado. Even before the U.S. oil and gas boom from fracturing led some to proclaim the U.S. was now "Saudi America," some proclaimed the U.S. was the Saudi Arabia of this low-quality kerogen [5, 12].[22] The existing methods to extract one unit of kerogen energy require even more electricity and steam inputs than for bitumen [8, 9]. Steam is needed to make the kerogen viscous enough to flow, and the electricity is needed to make an underground wall of ice around the area of extraction. The heating takes several months. During that time, the "freeze wall" prevents the heated kerogen from flowing too far away and prevents groundwater from coming into the extraction volume [8]. The EROI of refined fuels from kerogen shale is less than two. That is too low for economic viability.

Energy balance problems aren't only of concern for hydrocarbon resources. Liquid biofuels generally suffer from low EROI. This is one reason we don't have massive use of biofuels. Another major reason is high land and water use per liquid fuel production. On average over one hundred times more water is consumed to drive a mile on corn-based ethanol than on petroleum gasoline [36]. Low EROI and high land and water use are related: it takes a lot of land and water to grow biomass that in turn takes more energy inputs the more land is needed.

The Energy Policy Act of 2005, signed by President George W. Bush, established the U.S. Renewable Fuels Standard. This law established a mandate for the U.S. to consume a certain volume of biofuels each year. The Energy Independence and Security Act of 2007 increased the mandate. As of 2017, the U.S. consumed 14 out of a maximum allowed 15 billion gallons of corn-based ethanol, and 2.0 billion gallons of biodiesel. The original target was to consume (and produce) about 24 billion gallons of biofuels in 2017.[23] Because producing ethanol from corn was a well-known technology, and because the U.S. can produce a lot of corn, 15 billion gallons of fuel were allowed to be derived from corn. The 15 billion gallon limit was approximately 10% of the anticipated volume of gasoline sales, and that proportion of ethanol is easily tolerated in standard car engines.

However, it has so far proven too difficult to produce the categories of the so-called advanced and cellulosic biofuels that are mandated to originate from biomass other than corn grain. Thus, the "mandate" for these other biofuels has been lowered over time.[24]

After the Renewable Fuels Standard was established, there was a flurry of interest in assessing the energy balance of biofuels, in particular corn-based ethanol. Was

---

[22] *The Economist*, "The economics of shale oil: Saudi America," print edition of February 14, 2014, http://www.economist.com/node/21596553/print.

[23] NAS/NRC. (2011) Renewable Fuel Standard: Potential Economic and Environmental Effects of U.S. Biofuel Policy.

[24] https://www.epa.gov/renewable-fuel-standard-program/proposed-volume-standards-2019-and-biomass-based-diesel-volume-2020.

producing ethanol really worth the energy effort? Did it really shield us from the price of oil and reduce greenhouse gas emissions from using our cars? One seminal paper in the prestigious journal *Science* compared the energy output and input calculations of many studies on corn ethanol [18].[25] The result was that some studies showed an EROI a little above one and some an EROI a little below one, but none as high as two. More recent studies shows EROI perhaps near 2–4 for corn-based ethanol in Iowa due to a high concentration of ethanol refineries and the most productive corn production in county.[26] To many, the reports indicating EROI above one proved the process was energetically viable. However, this conclusion is misleading.

An EROI of 1 is not a threshold for viability. If corn ethanol (or any other fuel-producing process) has an EROI of 1, it means that for every unit of energy consumed in the process of producing an energy carrier (such as ethanol), there are actually two produced in total.[27] A process with EROI of one produces two units of output energy, and while consuming one of them for its own operation it provides the second to the rest of the economy and consumers like you and me.[28] Under this definition, an EROI of zero is the lower bound (not one) as set by the first law of thermodynamics as the conservation of energy.

Biophysical economists stress the importance of tracking matter and energy within the economy, and EROI is one metric used to do that. Using a back-of-the-envelope type of calculation, ecologist and biophysical economist Charles Hall and his students investigated the "minimum EROI" for a viable liquid fuel in our current economy [25]. They used gasoline from oil as a guiding example. We first get some energy in oil we extract from the ground, and this takes some energy inputs. For conventional oil reservoirs in the U.S. through 2007 (i.e., before the financial crisis and hydraulic fracturing of tight sands and shales was prominent) the industry consumed about 0.05–0.1 units of oil for every unit extracted [24]. Thus, the EROI of oil is about 10–20, as stated earlier. Refining oil into gasoline takes about 10% of the energy content of each barrel of oil such that the EROI of gasoline is about 5–6 instead of 10–20. By the time it reaches the filling station, the EROI of the gasoline might be 4–5.[29] Hall thought delivered gasoline, or perhaps any liquid fuel, with

---

[25] These *life cycle analyses* consider not only the energy value of the ethanol but also some energetic values of other products from the processing, such as dry distillers grains (leftover parts of the corn kernels) that are then fed to livestock.

[26] United States Department of Agriculture, Office of the Chief Economist, "2015 Energy Balance for the Corn-Ethanol Industry," February 2016, available at: https://www.usda.gov/oce/reports/energy/2015EnergyBalanceCornEthanol.pdf.

[27] For this definition of EROI, it is equivalent to "net external energy ratio" of [35].

[28] EROI = (energy extracted − energy consumed to produce energy carriers)/(energy consumed to produce energy carriers). For corn-based ethanol, EROI is approximately = $(2 − 1)/(1) = 1$.

[29] Here is the calculation from estimating (1) the EROI of oil (at the well), (2) the EROI of gasoline at the refinery, and (3) the EROI of gasoline at the filling station. (1) EROI oil at *well* = (1 unit oil energy extracted − 0.05 to 0.1 unit energy input to extract)/(0.05 to 0.1 unit energy input to extract) = (1/0.1) to (1/.05) = 10–20. (2) EROI of gasoline at *refinery* = (1 unit oil energy extracted − 0.05

an EROI of at least 3 might be required to maintain modern societal lifestyles and infrastructure.

It is intuitive that you have to spend energy at each step along the supply chain. The EROI declines as you include more steps in the process of making energy carriers like gasoline. It is also intuitive that the same concept holds for spending money on each step of the supply chain. If a company sells gasoline for three dollars per gallon, then their supply chain needs to cost less than three dollars per gallon. This comparison brings up a question: Do energy prices conform to net energy metrics like EROI?

Ten years ago I pondered this question [31]. It turned out the answer is yes.

By using the price of energy commodities, you can approximate EROI as if you calculated it from the bottom-up using energy input data from each step of the life cycle. The energy intensity ratio (EIR) is a proxy calculation for EROI that is based on broadly available economic data rather than detailed information about the energy supply chain.

Consider gasoline. The energy delivered is the energy content of the gasoline at the pump. We can divide the energy content by the price of gasoline to get the ratio of energy purchased per dollar. In 2017, the average price of U.S. regular unleaded gasoline was 2.41 dollars per gallon.[30] Each gallon contained about 127 megajoules (MJ) of heat content so that each dollar spent on gasoline bought you about 53 MJ. In principle every dollar spent on gasoline paid for all of the activities required to produce gasoline, including profits for companies involved, then we can say that it took about one dollar of spending to produce 53 MJ of gasoline *output*.

Now for the energy input. The approximation for the energy input to produce gasoline assumes that, on average, each dollar of net economic output, or GDP, requires the same amount of energy consumption. This is the *energy intensity* of the economy that is equal to total primary energy consumption divided by total GDP. In the U.S., in 2017, this was 103 exajoules (EJ) of primary energy divided by 19.4 trillion dollars such that every dollar of GDP output required an input of 5.3 MJ. The energy intensity ratio (EIR) is calculated by dividing the "energy output per dollar" of gasoline by the "energy input per dollar" of the overall economy. For 2017, this is 53/5.3 = 10. Figure 5.1 shows this EIR calculation for U.S. oil and gasoline since 1949.

---

to 0.1 unit energy input to extract − 0.1 unit of energy to refine)/(0.05 to 0.1 unit energy input to extract + 0.1 unit of energy to refine) = (0.9)/(0.2) to (0.9)/(0.15) = 4.5–6. (3) EROI of gasoline at *filling station* = (1 unit oil energy extracted − 0.05 to 0.1 units energy input to extract − 0.1 units of energy to refine − 0.03 units of energy to transport)/(0.05 to 0.1 unit energy input to extract + 0.1 unit of energy to refine + 0.03 units of energy to transport) = (0.87)/(0.23) to (0.87)/(0.18) = 3.8–4.8.

[30] All data for the EIR calculations come from the Energy Information Administration, Monthly Energy Review: Table 9.4 Retail Motor Gasoline and On-Highway Diesel Fuel Prices; Table C1 Population, U.S. Gross Domestic Product, and U.S. Gross Output; Table A3 Approximate Heat Content of Petroleum Consumption and Fuel Ethanol; Table 1.1 Primary Energy Overview.

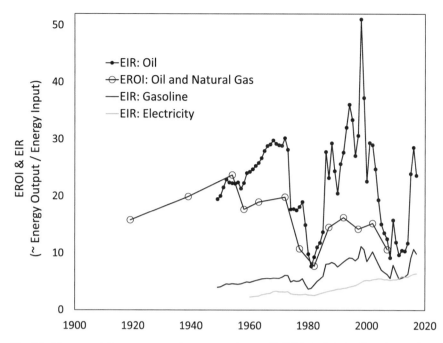

Fig. 5.1 The annual "energy returned on energy invested" (EROI) of U.S. oil and natural gas compared to the energy intensity ratio (EIR) of U.S. oil, gasoline, and electricity. The EIR calculations are proxies for the EROI of gasoline delivered to the fuel pump and electricity delivered to the wall socket [31, 35]. Oil and gas calculations from [24]. Oil, gasoline, and electricity price data are from EIA Monthly Energy Review Tables 9.1 (Crude Oil Price Summary, Crude Oil Domestic First Purchase Price), 9.4 (Leaded/Unleaded Regular Gasoline, U.S. City Average Retail Price), and 9.8 (Average Retail Price of Electricity, Total), respectively

One takeaway is that the EIR of gasoline is lower than that for oil. For energetic and monetary reasons this must be the case. As previously stated once we've extracted oil, refiners must consume additional energy and spend more money to turn oil into gasoline. The figure also compares the EIR of oil to a more detailed calculation of EROI for the U.S. oil and gas sector. This EROI calculation uses a more specific method that includes energy consumption data available every five years [24]. Importantly, the two calculations follow the same trends. When one went up, so did the other, and vice versa. Also, EIR of oil (only) is larger than the EROI of the oil and gas sector, but when they approach values below ten (when oil is getting relatively expensive), they become much closer to each other. This shows that when energy costs get high, prices are forced to follow.

The comparison of energy output to inputs of energy carriers is not limited to liquid fuels such as oil, gasoline, and ethanol. Figure 5.1 also shows the EIR for electricity. This metric and its price inputs account for the entire supply chain of extracting fuels burned in power plants, generating electricity, and delivering that electricity to each business and homes. We tend to place more value on electricity delivered to our home relative to gasoline at the pump. Electricity powers our appliances and televisions and it charges our mobile devices and computers, and we don't have to operate a power plant in our home to use electricity. Thus, we pay more dollars per unit of energy for electricity than for gasoline, and the net energy ratio, or EIR, is lower.

We know different energy commodities have different prices per unit of energy, different EROI, and we buy a mix of many energy products. Thus, we can also think about the net energy of the entire energy system.

Net Energy for the Whole Economy

We use an array of primary energy and end-use carriers. Because each energy resource and carrier has different properties, and there are both short- and long-term changes in availability and price, we use each of them for different purposes. Most oil becomes gasoline, diesel, and jet fuel; we use these fuels in cars, trucks, and planes where high energy per volume and mass, or energy density, is important. The more the fuel weighs, the more energy is needed just to move the fuel itself. A major goal for improvements in electric batteries is to increase energy density. In addition to consuming liquid fuels as energy carriers, we consume electricity to run household appliances because we can precisely direct it to individual appliances plugged into wall sockets. Further, developed economies tend to use natural gas and electric furnaces for heating homes as this avoids the unhealthy particulate emissions from burning solid fuels such as coal, biomass, and dung.

We can average together the net energy, or EROI, of each fuel into an economy-wide metric. A short-cut to approximate this economy-wide EROI is to take the inverse of total spending on energy divided by GDP. This is the inverse of the energy spending figures of Chap. 2. For example, if total economy spending on energy is 10% of GDP, then one divided by 10% gives an economy-wide EROI of 10. Higher net energy ratios indicate a higher potential for an economy to grow just like cheaper energy indicates a less constraining situation.

Given that the economy is a complex system with interacting elements, what does it mean for it to consume lower EROI, or higher-cost energy? If the energy supply is restricted due to high cost or physical constraints, then people, governments, and companies react and adapt. *The system changes* when required. It turns out that these adaptations occur at thresholds of high cost energy, or low net energy. When these adaptations cause significant structural change, we label them recessions and depressions.

There have been only two times since World War II in which the U.S. (Fig. 2.9b) and world (Fig. 2.13) economies have spent 8% or more of GDP on energy: the late 1970s/early 1980s and 2008. These correspond to the two deepest recessions since 1945. There are different ways to count energy spending, such as money spent on oil versus money spent on refined oil products like gasoline and diesel. But what research shows is that there is a *threshold* of energy spending as a fraction of GDP [6]. Above this spending threshold, the economy no longer grows. Below this threshold, energy costs do not constrain growth.

When a system becomes constrained by an input, it adapts and reconfigures itself. Accordingly, there were significant structural changes within the global and U.S. economies in response to the high energy prices, primarily oil prices, of the 1970s and 2008. In the 1970s, oil importing countries ramped up extraction from new oil supplies, such as offshore oil in the U.S. (in the Gulf of Mexico) and United Kingdom and Norway (in the North Sea). Research into renewable energy began in earnest. These countries also focused on the concept of energy efficiency for cars and buildings for the first time. Importantly, end-use efficiency was a *response* to events in the 1970s, not a predetermined plan.

Energy producers reside at the beginning of the energy distribution network, and consumers are at the end. Economists normally discuss prices as governing the balance between the input *supply* and the end-use *demand*. Both producers and consumers affect prices. If demand increases faster than supply, prices will tend to go up, and vice versa.

However, there is another way to think of this balance: the conservation of energy and mass. Generally speaking all energy that is produced gets consumed—energy input into the distribution network gets consumed at the end of the network (with some energy dissipation along the way). The same goes for the mass of natural gas, coal, oil, and other carriers that must travel within pipelines, along rails, and along roads—the mass that comes into the network eventually leaves the network.[31] Thus, prices are signals reflecting this broader physical balance within networks. To further understand this point, let's think about our energy system as a network of interdependent energy-producing and energy-consuming elements.

Energy Systems as Structured Networks

Our bodies and the economy are similar in many ways. Each operates by extracting and consuming energy from the environment outside of its boundary. Our body needs energy from food, and our economy needs energy from fossil and/or renewable resources. When we eat the right foods and have a low stress lifestyle, our bodies have a better chance to remain healthy. The energy narratives battle over which primary energy resources are the most "healthy" for our economy and the environment within which both our bodies and the economy reside.

Both our bodies and the economy take their respective input primary energy resources and transform them into energy carriers. Our body's energy carrier is ATP, adenosine triphosphate. ATP is like a battery that can be discharged when our cells

---

[31]The mass of fossil fuels, made of molecules of carbon and hydrogen that react with oxygen during combustion, "leaves" the economic system and energy distribution network as carbon dioxide and water molecules that are the products of combustion (along with heat).

need energy to perform their functions.[32] Similarly, our energy system produces energy carriers consumed by end uses just like the various end uses in our body's cells.

We can think of our cells and buildings in a similar way: as endpoints of delivery networks. Our cells need nutrients to perform their functions. These functions include repair, maintenance, transport of material within the cell, and action that triggers muscle cell contraction so that we can do things like run, kick, and throw. Before nutrients, water, and oxygen reach our cells, they must be pumped within the network that is our circulatory system.

Similarly to our body's circulatory system, energy carriers move within various networks that deliver energy to our buildings and homes. One of these networks is composed of the electric power lines you see running along highways and within our cities. Another is the array of natural gas and petroleum pipelines that distribute both raw and refined gases and liquids to our homes and gasoline stations. We might use these energy carriers to operate a drill to make home repairs, move people between floors in an elevator, and cook food.

It takes energy to operate a network of physical flows of electricity, solids, liquids, or gases. For the electricity network, heat losses are proportional to the electrical resistance of the wires, so engineers design power lines to use low-resistance conductors like copper and aluminum. We smooth the inner surface of gas and liquid pipelines because more pumping power is needed if the surface is rough. Similarly, we evolved smooth arteries and veins so that our heart minimizes its required energy consumption to pump blood throughout our body.

The concepts of networks and energy efficiency provide a framework to explain how the size of a physical system relates to its energy consumption. In the 1930s Max Kleiber plotted the basal, or resting, metabolism of animals versus their mass. Metabolism is their energy consumption. Mass is their size. These plots showed that metabolism increased *more slowly* than mass. Specifically, metabolism increases with mass to the three-quarter power. Thus, an animal species with mass ten times larger than another *does not* consume ten times as much energy in food. It consumes only 5.6 times more energy ($10^{3/4} = 5.6$). This relationship was dubbed Kleiber's rule [37]. Scientists call these kinds of relationships *scaling laws* because they explain how one factor changes, or scales, when another factor increases.

Scientists still grapple with the full explanation of Kleiber's rule, and debate how many organisms follow this scaling law [3].[33] However, in the 1990s a group

---

[32] ATP when "discharged" becomes an ADP, adenosine diphosphate, molecule and the separated third phosphate. ADP then "recharges" to become ATP.

[33] There is not unanimity for the network explanation of the 3/4-power scaling law dubbed Kleiber's rule, and scientists know it does not apply to all living creatures and stages of life. However, practically all explanations relating metabolism to organism mass consider the rate of dissipation of energy and what energy is used for in the organism (e.g., to heat an animal living in cold temperatures and to replace old cells). See [3] for a discussion of competing views and a thermodynamic explanation of the bounds we might expect for the scaling laws relating metabolism to mass.

of multidisciplinary scientists provided one explanation relevant for animals and plants that distribute nutrients within networks. For mammals they considered the circulatory system as a physical network of pipes branching out from the heart to all capillaries throughout a body [4, 61, 62]. One of the key insights from this research was that the structure of the branching network minimizes the resistance of flow in blood vessels, and, thus, minimizes the power required to pump blood. The less energy is consumed for pumping blood, the more energy is available to perform brain functions, fight off disease, grow from birth to adulthood, collect food, and produce offspring. In networks, the operational energy consumption, structure, and size are inherently linked. *The more efficient a system's distribution network, the larger it can be.* In short, size matters.

A consequence of the relationship between metabolism and mass is that as animals are larger in absolute size, say a whale compared to a mouse, each unit of mass of the animal consumes energy at a slower rate. Importantly, this relationship exists for the *average mass* of animals. In terms of energy per mass, some organs consume more energy and some less. For example brains and hearts consume energy at a higher rates per mass than skeletal muscles. When we grow from child to an adult, we add bone and muscle mass faster than more brain and heart. This is part of the reason that the scaling law also exists for a given animal as it grows. The scaling law informs us that because the average animal mass consumes energy at a slower rate as it grows, the mass that it adds during growth (e.g., muscle, bone, maybe fat) must consume energy at a slower rate than the existing masses such as heart and brain. There is no choice.

Because metabolism scales with the three-quarter power of total mass, if we divide metabolism by mass then this measure scales with mass to the negative one-quarter power. *Thus, the larger an animal, the less energy used by each ounce of its body.*

The crazy thing is that this metabolism–mass relationship also holds for entire groups of organisms, such as ant colonies. Some ants are hunter-gatherers, and some are farmers. Farming ants collect organic sources of nitrogen from the environment, and use the collected nutrients to grow fungi for food in their nests. These social insect colonies have resource-productivity patterns that emerge from the social organization, resource transfers, and physical architecture of the colony. Among fungus-growing ants, colony sizes range from 50 workers to 15 million workers between species. Thus, biologists can compare the energetics between colony-farms and to ancestral hunter-gatherer colonies using the same metabolic scaling principle as within a single animal. To do this, you have to calculate the total mass of the colony including that of the ants *and* their gardens (e.g., collected leaves or other material plus fungus).

Not all ant colonies show the same scaling between their metabolism and mass, and thus the colony-level relationships are not the same as the 3/4 rule that describes most animals. Biologists found differences among ant species related to their genetic ancestry. Hunter-gathering ants have metabolism that scales to the 0.8–0.9 power of mass [48, 60]. Just over 40 million years ago the first fungus-farming ants emerged, but their colony metabolism scaled differently than their ancestors, with metabolism scaling to the 0.58 power of mass. Thus farming seemed not to require as much energy to be the same size. Then about 26 million years ago a second group of fungus-farming ants, including the leaf-cutter ants, evolved, and their metabolic scaling is essentially the same as the hunter-gatherer ants! Thus, ant evolution produced colonies that at one point moved to farming with a different metabolism–mass relationship, and then later evolution produced colonies that exhibit the same metabolism–mass relationship as before there were farming ants! There is more than one way to run an ant colony.

Importantly, the vast majority of mass in farming ant colonies is the garden and its associated microbes, not the ants. In hunter-gatherer colonies the ants themselves make up the mass—there is no garden. In all cases, at any given time there are a certain number of ants performing any given task. It turns out that one of the keys to understanding colony-level metabolism is to know what proportion of all ants are stationary. For hunter-gatherer colonies, as the total number of ants increases so does the fraction of ants that are stationary, remaining in the nest as opposed to roaming around [60]. The more total ants there are, a higher proportion seem to be sitting around doing paper work, or "stationery" work, in their houses and office buildings!

In one cleverly designed laboratory experiment, biologists removed half of ants from different ant colonies, and the remaining ants continued to live in the same fixed space [60]. As predicted by scaling laws, when they removed half of the ants (as well as pupae and larva), the metabolism of the colony did not decrease by half, but by only about 40%. Somehow the collective colony-level metabolism changed as the number of ants changed. When there were only half as many ants, *on average* each ant had a *higher* metabolism, yet each individual active ant walked slower. The reason is that for smaller colonies, a higher fraction of ants are active. Thus, a smaller colony had a higher proportion of slower-moving ants, while a larger colony had a smaller proportion of faster moving ants. The overall colony scaling characterizes the tradeoff between the numbers of active ants with just how active is each ant. The authors note that this effect of colony metabolism was not a function of food (resource) availability because they supplied more food than the ants could possibly use. However, we can seek to interpret this experiment in the context of the techno-realistic, or finite Earth, narrative. The size of the ants' world was definitely fixed because the experimenters did not change the size of the box in which the ants lived. The authors note that the ants didn't change their spacing much when their numbers were halved. Also, the colonies with more ants did seem to have ants spend more time walking along the edge of their world.

Trees also exhibit similar scaling as animals and ant colonies. Just as our circulatory system transports blood throughout our bodies, the vascular system of trees transports water from the roots to the leaves. The respiration, or metabolism, of mature trees also increases more slowly than the mass of the tree. One interesting finding is that very small trees, or saplings, increase metabolism faster than the tree mass [42]. The same holds for fish embryos as their metabolism, measured via oxygen consumption, increases with mass to the 1.22 power before transitioning to a sublinear scaling in adulthood [14, 43]. This means that the scaling relation between energy and size is not necessarily constant over the life of organisms. As trees and fish grow, there is a transition during growth from a *superlinear* scaling—respiration and metabolism increase faster than mass—to *sublinear* scaling—respiration and metabolism increase slower than mass.

From a network perspective, is there some analogy with scaling laws and our human energy-economic system? Yes. Since the energy and transportation networks within our economy are similar to a circulatory transport system within our bodies and trees, this explanation can provide some insight into patterns within economic and energy data. One way we can interpret superlinear scaling in early (or perhaps other) stages of growth is that it is characteristic of systems that do not yet depend on distribution networks for energy and nutrient delivery. Once trees, fish, and other animals grow large enough to depend on networks for distribution, they exhibit sublinear scaling of energy related to size.

From the ant colony perspective, is there some analogy with our human energy-economic system? Yes. Since each of us roams around interacting with each other and performing some task at given point in time, there are some analogs. It can be dangerous to strictly interpret these "purely biological" analogs to our economy, but let's continue to explore these potential parallels.

First consider that present-day animals have evolved over millions of years relative to the quality, quantity, and variability of available resources in their environment. Many animals make annual migrations across long distances, chasing the summer sun and the biomass production that it drives. Even though some animals travel along these "roads," use basic tools, and have basic forms of communication, animal populations are largely resource limited. Their populations grow depending on both their food and predatory bases.

Of course our industrial human society is more complex. We have written languages and formal schooling that enable us to pass on knowledge gained over centuries. Animals can only pass on their genes to the next generation, and the young can only learn from observing their elders. Without written languages, animals cannot pass down much accumulated knowledge. Primitive human societies also had no written language, and knowledge had to be passed down via stories taught and learned by each generation.

Written language helped to more quickly accumulate the scientific and cultural knowledge that enabled us to move from a biomass economy to one where the steam engine enabled tremendous exploitation of coal. And from exploitation of coal our energy-economic system today runs on an array of fuels and technologies ranging from hydraulic fracturing of rocks 10,000 feet below ground to photovoltaic panels on our roofs.

In terms of our ability to extract and distribute resources, the techno-optimistic economic narrative proposes that we are not limited, and the techno-realistic economic narrative says that we are. It makes sense to explore how to relate energy consumption to our economy size and structure. Do our economies scale their energy consumption in a similar way as animals? Biologist Jim Brown and a cohort of interdisciplinary thinkers considered this question, and their answer is yes [11]:

> Like all organisms, humans are subject to natural laws and are limited by energy and
> other resources. . . . we use a macroecological approach to integrate perspectives of physics,
> ecology, and economics with an analysis of extensive global data to show how energy
> imposes fundamental constraints on economic growth and development. [11]

From this quote it is clear they subscribe to the finite Earth and techno-realistic
economic narrative. They describe their approach as *macroecological* rather than
macroeconomic. The quote above was published in a biology journal, not an
economic journal. This last statement will make more sense after reading Chap. 6
that describes why mainstream economic theory largely precludes this type of
physically based macroecological approach. But before too much digression, let's
continue with the network analogy between biology and the economy.

In a similar way as animals, larger economies (higher GDP) consume more
energy described by a scaling law. For this reason people characterize the material
and energy consumption of economies as societal or *social metabolism*. We can ask,
as I did with colleague Andrew Jarvis of Lancaster University, UK, how the primary
energy consumption of the world economy scales to a power of GDP. Figure 5.2
shows data for the global economy, since 1900, in a way we can interpret in the
context of Kleiber's rule.

There are three phases of growth scaling, but it is hard to notice the first one in
Fig. 5.2. Another way of displaying this three-phase pattern is to look at a plot of
energy intensity, or the ratio of energy consumption divided by GDP, as shown in
Fig. 5.3. During the first phase from 1900 to 1920, energy consumption increases
faster than GDP such that consumption scaled with GDP to a power greater than
1. Afterwards, from about 1920 to 1970 one more unit of world GDP required one
more unit of energy consumption. Thus, energy consumption scaled with GDP to
the power of 1, or it had linear scaling. Finally, after about 1970 the pattern changes.
Since the 1970s world energy consumption increases more slowly than GDP such
that one more unit of GDP requires approximately *two-thirds* more primary energy.

The data for the United Kingdom, the first industrialized nation, show a much
more pronounced rise in energy intensity during its first phase of growth from
the 1700s until the late 1800s. The United States data show a less pronounced
rise in energy intensity during its major industrialization phase in the early 1900s.
However, a U.S. energy intensity calculation that considers only modern forms
of energy consumed as commercial energy shows a clear and pronounced rise in
U.S. energy intensity from the mid-1800s to 1920 [15, 50, 51].[34] The U.S. energy
intensity is also relatively constant in the three decades following World War II, thus
indicating again that times of rapid growth of infrastructure show either increasing

---

[34]See U.S. energy intensity calculations in Figure 3.19 in Smill [50] and Figure 6.17 in Smil [51].
O'Connor and Cleveland's comparison of U.S. energy intensity with traditional energy (e.g., food,
fodder, biomass, mechanical windmills) to that without traditional energy shows that a pronounced
"U-shaped" rise and fall in energy intensity is only observed by ignoring traditional energy sources
[15].

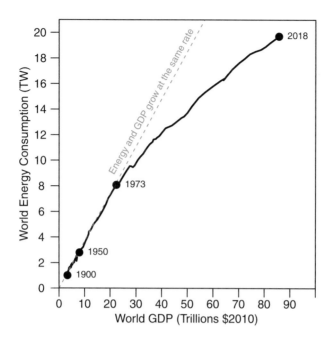

Fig. 5.2 Total primary energy consumption of the global economy grows at approximately the same rate as GDP from 1900 to the early 1970s. Afterwards, primary energy consumption grows more slowly than GDP, in a similar manner as suggested by Kleiber's rule from biology. Data were compiled by Andrew Jarvis as averages of GDP data from the Penn World Tables, World Bank, Maddison data set, and United Nations, and average of total primary energy supply data from the International Energy Agency, BP Statistical Review of World Energy 2019, U.S. Energy Information Administration, and the Primary, Final and Useful Energy Database of International Institute for Applied Systems Analysis

or nearly constant energy intensity. Vaclav Smil's calculation of Japan's energy intensity also shows clear rise from 1880–1970, followed by a decline since.[35]

Both biological and economic systems that grow rapidly from a small size, with little structure, exhibit a transition from a period when energy consumption scales superlinearly with size to one in which it scales sublinearly with size. Thus, there are two major takeaways from Figs. 5.2 and 5.3.

First, as many economic historians recognize, countries, and now the global economy overall, seem to have transitioned from a time of industrialization with a characteristic increasing or constant energy intensity to a time of decreasing energy intensity. The economic narratives disagree on why this is the case. The techno-optimistic/infinite substitutability narrative attributes declining energy intensity to human ingenuity, progress, and technology. The techno-realism/finite Earth

---

[35] Figure 6.17 in Smil [51].

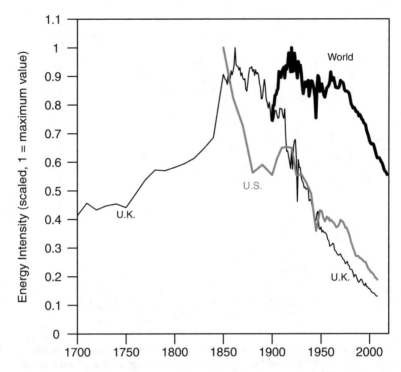

Fig. 5.3 The energy intensity (primary energy consumption divided by GDP) for each of the U.K. (thin black line), U.S. (thin red line), and global (thick black line) economies increased and/or remained relatively constant during major industrialization efforts before declining afterward to the present time. U.K. data are from [19, 20]. U.S. primary energy data from Energy Information Administration Monthly Energy Review Tables 1.1. and D1, and GDP data from Maddison Data set [7]. World data were compiled by Andrew Jarvis as averages of GDP data from the Penn World Tables, World Bank, Maddison data set, and United Nations, and average of total primary energy supply data from the International Energy Agency, BP Statistical Review of World Energy 2019, U.S. Energy Information Administration, and the Primary, Final and Useful Energy Database of International Institute for Applied Systems Analysis

narrative disagrees because we observe this same transition in biological organisms such as animals, ant colonies, and trees that do not exhibit a similar parallel of technological change during their individual lifespans. Thus, the scaling laws emerge without our common notion of technological change. Further, biological organisms can adapt to environmental signals such as the size of their environment. Even individual biological cells provide an example for how systems somehow adjust the growth of internal structures to the size of their environment. The sizes of individual organelles within cells, such as the nucleus and mitotic spindle, regulate their size based on the overall size of the cell in which they reside [41]. Larger cells have larger internal structures as somehow the organelles "know" the size of the cell in which they reside!

The second takeaway concerns the scaling factor that describes the post-1970 decline in global energy intensity. The critical point is not the exact scaling factor, but that it is less than one and greater than zero. Some countries might increase GDP with very little increase in energy consumption, and some countries might still exhibit near linear scaling. But we can realistically speak of a global economy, connected by trade, over the last few decades. Within a systems context, we cannot lose sight of the big picture that all countries link together to exhibit the global trend in Fig. 5.2. Over time the world economy moved toward, not away from, one that scales like an adult animal or an ant colony, and we can ask if we now operate within an "adult" global economy. Since we assume biological organisms do not consciously design themselves to exhibit their scaling laws relating metabolism and mass, then should we assume our economic scaling relating energy and GDP is a result of our conscious design?

Keep this question in mind as you continue reading. For now, consider that this tango between energy and GDP has something for both the techno-optimistic and techno-realistic narratives. For the optimism side, over time, GDP is growing faster than our need for environmental resources such as energy. Techno-optimists can rightly proclaim we continue to produce more output per unit of input.

For the techno-realists, both GDP and energy consumption are still increasing. Globally, energy consumption does not decrease while GDP increases, because each economic activity induces resource consumption.

This brings up the concept of *decoupling*. In the context of energy and economic growth, decoupling describes an economy that grows, say increases GDP, but decreases energy consumption and/or environmental impacts. There are two levels of decoupling, relative and absolute. *Relative decoupling* is when economic indicators grow faster than energy consumption, but energy consumption still increases. A declining energy intensity (GDP divided by energy consumption) represents one metric to describe relative decoupling. *Absolute decoupling* is when economic indicators increase but energy consumption decreases. The narratives also disagree on possible decoupling of economic growth from carbon emissions and non-material consumption.

In terms of decoupling in the global economy, from 1900–1970 there was no decoupling, and since 1970 there has only been relative decoupling. At no point yet has the global economy experienced absolute decoupling. Since it is generally anathema for politicians and company executives to plan for a shrinking economy or profits, they can still exhibit their environmental credentials by calling for relative decoupling. But can we have a growing economic cake without adding more material ingredients? To date, when the global economy grows so does energy consumption. The data indicate we have not yet achieved absolute decoupling, but can we? This is a matter of considerable debate between the economic narratives. The techno-optimistic narrative, says we can decouple, and the techno-realistic narrative says we can't, or at least should not expect nor rely on a plan to do so. Chapter 9 discusses the idea of decoupling in more detail.

For now, recall Chap. 4 showed that the 1970s signal a turning point toward higher debt ratios, stagnant per capita energy consumption, and less equal income distribution within the U.S. and many other rich economies. Thus, these major structural changes coincide with the global economy moving into a growth phase characterized by relative decoupling. These correlations between the rate of energy consumption and the structure of society (e.g., debt, income distribution) provide strong evidence of tight coupling between the costs of resource extraction and costs of resource consumption. We'll return to this concept in a few more pages, but first, a little more on the scaling relations between energy consumption and economic indicators.

There is a subtle distinction between the animal scaling law and that for economies. The animal scaling law relates metabolism, a rate or a flow of energy within the animal, to the mass of the animal, a stock quantity of matter. The macroecology scaling law in Fig. 5.2 relates a flow of energy to GDP, but GDP is a *flow* of money, not a *stock* like the mass of an animal. While this might imply the similar scaling law is a mere coincidence, it is not.

The processes that dissipate energy in our economy are physical machines such as cars, manufacturing equipment, air conditioners, and the computer server farms that provide online communications and video streaming. These machines require energy to operate. The more mass of physical machines, or *physical capital*, that dissipate energy, the larger the GDP of a country. In practice, there is a practical difficulty in adding up different kinds of physical capital. I save this discussion for Chap. 6, but for now let's consider a single type of physical capital: homes.

If we treat the U.S. physical capital stock of buildings in the same way as the mass of an animal, have we consumed more or less energy as we've become "bigger"? Does the energy consumption of our homes scale with the quantity of homes in a similar way as animals and total economies?

The network concept and data of Chap. 2 help us answer this question. From Fig. 2.7 of Chap. 2, we know that annual U.S. energy consumption increased at a high rate until the early 1970s, then increased at a slower rate until the mid-2000s, and has remained relatively constant since then. For buildings, consider only the number of U.S. households as an approximation for the number of residential dwellings. Both the U.S. population and number of households increased steadily since World War II. Taking the number of households as the "mass" of our economy in the same manner as the mass of an animal, the U.S. economy has become continuously more massive.

Figure 5.4 shows U.S. *residential* energy consumption versus the number of U.S. households.[36] Specifically, it shows residential sector energy consumption per household, or how much energy is consumed, on average, in each home. This

---

[36]The U.S. Energy Information Administration defines the *residential sector* as "An energy-consuming sector that consists of living quarters for private households. Common uses of energy associated with this sector include space heating, water heating, air conditioning, lighting, refrigeration, cooking, and running a variety of other appliances. The residential sector excludes institutional living quarters." https://www.eia.gov/tools/glossary/index.php accessed September 3, 2018.

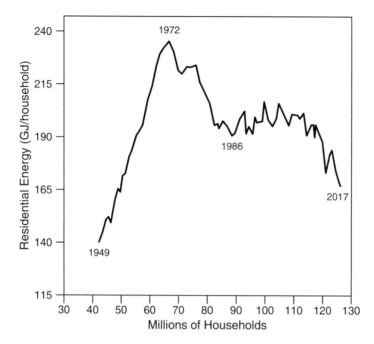

Fig. 5.4 U.S. residential energy consumption divided by the number of households increases before 1972 and declines thereafter even as the number of households, or the number of end-use points of the energy network, increases continuously. A declining trend supports a conclusion that the U.S. household sector is constrained by the same principles that govern Kleiber's rule from biology. Household data are from Federal Reserve Bank of St. Louis: TTLHH (Total Households, Thousands, Annual, Not Seasonally Adjusted). Residential energy consumption data are from EIA Monthly Energy Review Table 2.2

latter calculation is the same concept of plotting animal metabolism per mass versus its own mass. If plotting data for animals, the trend would down and to the right.

The simplest way to interpret Fig. 5.4 is to focus on whether the trend is increasing, decreasing, or staying flat. From 1949 to 1972, the trend is up. Thus, the average American household increased its energy consumption as it accumulated new household appliances and increased in size. Our energy networks expanded and delivered more in total and more energy to each house. Energy did not constrain growth. Then, in the 1970s, everything changed.

Starting in the early 1970s, energy per household declined for a decade, remained nearly steady for 25 years, and then declined again after 2010. Since the early 1970s there is an overall downward trend in residential energy per household, and this trend mimics Kleiber's scaling rule for animals. The number of U.S. households increased 89% from 1972 to 2017, but total household energy consumption increased by a lesser amount, only 34%. This is less than the increase in energy that either the three-quarter power rule of Kleiber or two-thirds power rule of the global energy–GDP relationship would "predict," but we should not necessarily expect the same scaling exponents to hold for each data set within biology, ecology, and economics.

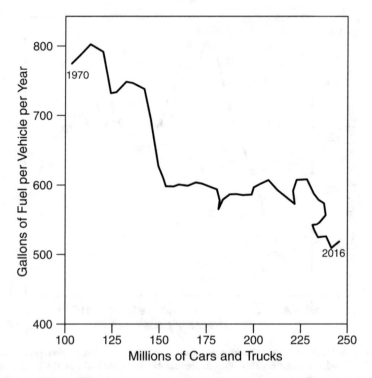

Fig. 5.5 United States fuel consumption in light duty vehicles (cars and light trucks) divided by the number of vehicles decreased since 1973 even as the number of vehicles has increased. A declining trend supports a conclusion that the U.S. household sector is constrained by the same principles that govern Kleiber's rule from biology. Data are from Transportation Energy Data book, version 37, from Oak Ridge National Laboratory, Table 4.3

The important point is that to relate the number of U.S. houses to residential energy consumption, the scaling exponent for houses is less than one and greater than zero. Since the 1970s, due to energy constraints, the residential sector was forced to consume less energy per household even while collectively the U.S. consumed more residential energy overall. It is as if the residential sector grew from the size of rabbit into the size of a fox. Physical laws in an energy-constrained situation dictated a decrease in residential energy consumption per household. Figure 5.5 shows an additional example as the same trend of "decreasing energy consumption per mass" occurred for fuel use in U.S. cars and trucks since the 1970s.

Economists often debate whether the early 1970s were a fundamental turning point in U.S. and global energy-economic relationships. Perhaps even 20 years ago we might have had some questions, but today I think the answer is clear. The data in Figs. 5.4 and 5.5 provide strong evidence that since 1973 the U.S. economy has been constrained, to a large extent, by the rate at which we can extract and afford to consume energy. Figure 5.2 shows that the same factors forced a new trend in global energy-economy relationships. Peak U.S. oil extraction in 1970, the 1973

Arab Oil Embargo, and the 1974 OPEC oil price hike combined in a way that forced the U.S. and world economies into a more constrained mode of operation. The 1979 Iranian Revolution further solidified that Western and global economies were exposed to political events within oil-exporting countries. U.S. per capita primary energy consumption also peaked in the 1970s following a similar trend as residential energy per household (see Figs. 7.1 and 9.3).

In this new post-1970s phase, a growing economy seemingly had to be associated with a more efficient use of energy. I say seemingly because more specifically we can say that each unit of capital, widget, or "mass" in the economy had to operate with less energy consumption independent of whether that capital was more energy efficient or not. In reacting to the events of the 1970s, there were two basic choices. First, we could consume less energy per widget, while building the widgets in the same way with the same efficiency. This means we would obtain fewer energy services from operating the widgets less often with less fuel. Second, while still consuming less energy per widget, we could redesign the widget to get more services from a given amount of energy input. While rabbits and foxes cannot redesign their cells and organs to become more energy efficient, as in this second option, we can do this for our household appliances. It is natural for us to choose the second option for the economy, and that is exactly what happened.

It is important to note that the U.S. and other Western countries were not planning to halt energy consumption in the 1970s. The U.S. had no policy to get to some year in the 1970s and stop the decades-long increase in energy consumption per house, per vehicle, or per person. No elected officials proposed a law that the U.S. should stop consuming energy once we reached a level of 235 gigajoules per household or 380 gigajoules per person in 1973. Quite simply the U.S., and the world, became exposed to economic and physical constraints on energy extraction for the first time since the industrial era. In response we decreased energy consumption. It wasn't a decision of choice. It was a physical necessity. In being forced to deal with this major energy constraint, an old concept emerged anew: energy efficiency.

Energy Efficiency, Jevons Paradox, and the Rebound Effect

As discussed in Chap. 2, the transition to coal-fueled machinery fostered the most dramatic decrease in the cost of energy, or power delivery, in history. Even as this transition was in full swing for the first time in the United Kingdom, someone was wondering how long the party could last. What happened if the industrializing U.K. could no longer extract increasing amounts of coal? The person contemplating this question in the mid-1800s was British economist William Stanley Jevons. He wondered about the impact of extracting coal more efficiently.

In a capitalist economy, business owners must make profits. Profits equal revenues minus costs. One strategy for maximizing profits is to minimize costs. The major types of costs are capital (e.g., owning or renting machines and buildings), wages for employees, materials, and energy. One general response to increased costs is to become more efficient in the use of resources, both in terms of energy and capital.

People usually think that becoming more efficient in the use of an input translates to less consumption of that input. They are thinking only of one machine, such as a car. If my new car needs only one gallon of gasoline to drive 40 miles instead of one gallon to drive 20 miles in my old car, then I will consume half as much gasoline as before only if I drive the same distance.

But as far back as 1866, Jevons noted that technological improvements that *decreased energy use at the small scale of individual devices often caused increased energy use at the large scale of industries or entire economies*. Because of this *rebound effect*, also called the Jevons Paradox, efficiency promotes growth that would otherwise not occur.

To understand why efficiency leads to more consumption, consider an excerpt from Jevons' writing:

> It is wholly a confusion of ideas to suppose that the economical use of fuel is equivalent to a diminished consumption. The very contrary is the truth.[37]—William Stanley Jevons (1866)

To understand why he came to this conclusion, Jevons made the following argument:

> Now, if the quantity of coal used in a blast-furnace, for instance, be diminished in comparison with the yield, the profits of the trade will increase, new capital will be attracted, the price of pig-iron will fall, but the demand for it increase; and eventually the greater number of furnaces will more than make up for the diminished consumption of each.[38]— William Stanley Jevons (1866)

Dry coal mines increase access to coal, and thus extraction rates. Steam engines pumped water from coal mines faster than previous methods. Steam engines were powered by coal. Coal was extracted at a higher and cheaper rate. More available coal could make more iron, and iron became cheaper. Cheaper and more abundant iron enabled manufacture of more steam engines that are made of iron. Higher numbers of more efficient steam engines and blast furnaces enable an increase, not a decrease, in coal extraction and iron production because *more coal mines become accessible and more blast furnaces and steam engines get constructed* than if the efficiency of steam engines and blast furnaces stayed the same.

Jevons understood that the economy is a complex adaptive system. The cycle of processes linking coal extraction and iron production exist within a reinforcing, or positive, feedback loop. A reinforcing feedback loop enhances whatever direction of change is imposed on it [39].[39] The more you have, the more you can get. In the case of coal and iron, more coal meant more iron production. More iron production then meant more steam engines that facilitated more coal extraction.

---

[37] Jevons [30, p. 123].

[38] Jevons [30, pp. 124–125].

[39] Meadows [39, p. 31].

Beyond the reinforcing feedback of doing more of the same (coal extraction leading to higher rates of coal extraction), new ways of using iron and coal can be derived. This is because the economy is adaptive. For example, instead of only using stationary steam engines next to coal mines, they were put on locomotives for transport along (iron-based) steel rail roads. Locomotives were a new and additional device in which to consume coal. Some people use the word *backfire* (distinct from rebound) to describe an extreme form of the Jevons Paradox. With backfire, for every unit of energy you save in energy consumption in a single device you get more than one unit of additional consumption at the economy or system level [29].[40]

Many people think the backfire effect is impossible [23]. However, many of us disagree as these conclusions come from analyzing narrow system boundaries.[41] This is why the beginning of this chapter emphasized the importance of system boundaries. Consider that if I only consider how much less electricity I consume in my home due to installing efficient LED (light-emitting diode) light bulbs, then I will likely not consume as much electricity for lighting within my house. But this boundary is too limited for at least three reasons. First, I consumed less electricity in my home, thus reducing demand for electricity and reducing its price. Lower priced electricity will induce other people to consume more, including possibly myself.

Second, if I'm saving money from buying less electricity, I have to choose what to do with that money. I could just get the money as cash and burn it in my fireplace, but practically no one does this. Most people use the money to buy something they otherwise couldn't afford, or they invest it in some way. Saving electricity for each LED light bulb, you might now install additional bulbs to light up your outside yard. Another example is moving the "old" kitchen refrigerator into the garage when "replacing" it with a newer and more energy efficient version. If I save enough money from efficiency, I can consume more electricity for the same or a lesser budget. Also, when we buy a new energy efficient product, some person or company had to make it, and that process consumed more energy. If you invest the money, then the company uses that investment to make more of its products, consuming more energy again.

Third, by shrinking the system boundary, you reduce interacting feedbacks. If we shrink if to only me and my house, then we've removed the interactions between me and everyone else in the world, and it is precisely these interactions among multiple people that lead to the backfire effect itself.

---

[40]The coal-iron cycle exhibits what some people call the *backfire* effect, to distinguish from the rebound effect [29]. Consider a new efficient steam engine that requires one ton of coal less than the old model. Considering the positive feedbacks, "rebound" indicates that in the end, I might actually reduce coal consumption by only 0.7 tons, because I "rebound" 0.3 tons of coal. There is a savings in total coal use. The word backfire refers to a condition where the more efficient steam engine eventually causes more than one additional ton of coal consumption than I saved in my new steam engine. My new engines saves one ton of coal, but somewhere else in the economy, more than one ton was extracted in response to the higher availability and lower cost of coal.

[41]See the book *The Jevons Paradox and the Myth of Resource Efficiency Improvement* [45] and the white paper *Energy Emergence: Rebound & Backfire as Emergent Phenomena* [29].

To me, there is no analysis needed to establish the validity that the backfire effect, or Jevons Paradox, acts to increase total aggregate consumption rather than decrease it. The evidence is with the data. We have not become more energy efficient, then said to ourselves "this is good enough," and subsequently decided to cease the pursuit of more efficiency and resource consumption.

The smaller the system boundary used for analyzing the backfire effect, the less relevant the paradox appears. To fully conceptualize the Jevons Paradox, your boundary must include the entire world economy, or every single device that consumes energy.

Increasing efficiency of individual devices is a strategy to overcome the inability to consume more resources in aggregate. Efficiency is not an strategy to reduce aggregate consumption [28–30, 45]. We observe that energy-consuming devices have become more efficient over time, not less so. This is particularly true after 1973. For the case of the U.S., primary energy consumption increased in the three decades after the 1970s. Consumption did not decrease. At our largest boundary, the world (so far) continues to consume an increasing amount of energy, not a decreasing quantity.

How many companies do you know that specifically mass manufacture and sell *less energy efficient* machinery? There are reasons to sell more costly and less efficient machines, but growing the economy isn't one of them. I'll revisit this concept in the final chapter.

At this point in my class lecture on efficiency, I've had students ask a great question: "You've told me that the Earth is finite, and now you're telling me that if I increase efficiency I will consume and extract resources at a higher rate, not a lower rate. So which is it, are we going to keep increasing resource consumption or not?"

It sounds like these two ideas have pitted us between a rock (the Earth) and an efficient place. However, they are fully compatible. In short, we'll increase resource consumption, by getting more efficient, until we can't because of the constraint of the finite Earth.

Physical constraints act within a balancing, or negative, feedback loop that opposes the direction of change in the system [39]. The more resources we extract, the lower the quality and/or the further away is the next resource, and the more costly to extract. While some can rightly claim that there never was any "easy" oil to extract, we can say we have to go further, deeper, and to lower quality rocks to get more oil. In 1901 the Lucas oil well, drilled to 1139 feet using a wooden drilling rig became a gusher in Spindletop, Texas (Fig. 5.6a). Today, offshore drilling rigs are floating cities, positioned using GPS-guided control to work hundreds of miles offshore to drill through thousands of feet of rock lying thousands of feet below the sea surface. The offshore drilling rigs in Fig. 5.6b were parked in 2019 because they were expensive to operate for the going oil price. The switch from Spindletop to offshore oil is the opposite of the Jevons Paradox and that which the efficiency strategy seeks to counteract.

(a)   (b)

Fig. 5.6 (**a**) Spindletop, Texas, 1901. Image courtesy of The Texas Energy Museum, Beaumont, Texas. (**b**) Offshore drilling platforms parked in Port Aransas, Texas, September 2019

To understand how these long-term positive and negative feedbacks play out, consider the U.S. energy and economic responses to the 1973 Arab Oil Embargo and the OPEC oil price rise in 1974. One major response was the U.S. car fuel efficiency standard that began in 1975, the Corporate Average Fuel Economy, or CAFE, standard.

In 1970 Americans drove their cars and trucks, with an average fuel use of 13.5 and 10 miles per gallon, respectively, for 1035 billion miles while consuming 80 billion gallons of fuel. With gasoline at an average of 38 cents per gallon at the time, fuel costs were $27 billion, or 2.7% of GDP.

In 2012 the numbers were 24.9 and 18.5 miles per gallon for cars and trucks, respectively, collectively driving 2665 billion miles to consume 124 billion gallons of fuel. With gasoline at $3.68 per gallon, fuel costs were $457 billion or 2.8% of GDP. It is not a coincidence that fuel spending was practically the same, relative to GDP, in 1970 and 2012 (Fig. 5.7). Given time, we adjust our policies and our consumer habits to the current technologies and energy prices.

Since World War II, each time the U.S. has spent more than 4% of GDP on oil, there was a recession [38]. The economy-wide feedback from the cost of energy relates to total energy spending relative to GDP, not from prices only. If prices are permanently higher, then consumption has to be lower to stay below the total spending threshold. If car fuel efficiency had not increased after the 1970s, Americans certainly could have afforded to drive only a fraction of the 2.7 trillion miles driven in 2012 [32].

The Jevons Paradox works whether the increase in energy efficiency occurs via business investment in more efficient processes and machines or consumer purchases of more efficient homes and cars. The early 1970s are a clear time

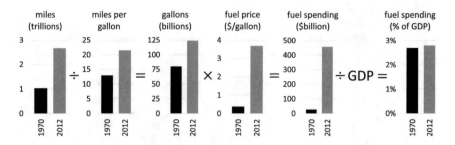

Fig. 5.7 A sequence of calculating U.S. spending on motor fuels for cars and light trucks, shows that, compared to 1970, in 2012 there were more miles driven in more efficient vehicles consuming more gallons of fuel at a higher price, yet the total spending on fuels amounted to the same fraction of GDP

of transition from a period of increasing energy abundance to one of constraint during which ever-increasing energy consumption could no longer be taken for granted. Because consumers can only buy what exists in the market, it took government policy to instigate energy efficiency standards for consumer products like vehicles, refrigerators, and light bulbs. In a competitive marketplace, businesses face natural pressures and incentives to become more efficient in producing products and services. Businesses also attempt to increase profits and economic growth by becoming more energy efficient in the face of energy-related constraints.

However, the collective responses to energy constraints have not been socially neutral. Chapter 4 explained that U.S. wages stopped increasing (Fig. 4.11) and wage share started decreasing (Fig. 4.12). This reversal in wage equality extended to most of the world's major economies (Figs. 4.15 and 4.16).

The energy and economic data show that the tremendous growth during the 30 years immediately after World War II is an anomaly. The structural changes from reacting to increases in energy costs or decreases in availability go beyond only thinking about the distribution between the capital and labor share of GDP. Energy constraints also altered the patterns of money flows within the U.S. economy.

Structures and Cycles of Growth

We know that increases in energy technology and resource access enable profound changes within what we call "the economy." These changes occur for both the *size and structure* of economies. Historians have studied how these changes occurred well before the invention of the steam engine and the pervasive use of fossil fuels. In *The Collapse of Complex Societies*, Joseph Tainter, an anthropologist, provides examples of the rise and fall of civilizations [52]. His work appropriately forced us to consider the link between societal size and structure during both growth and contraction. Two of the most studied historical examples are the Roman Empire (27

BC–476 AD) and the Maya of present-day Yucatan Mexico, Belize, and Guatemala during the Classic period (200 BC–1000 AD).

Tainter's work provides a good example of cross-disciplinary thinking. Ecologists and biologists readily study ecosystem structure via food webs, the organization and tasks of eusocial species such as ants and bees, and the relation to energy and nutrient inputs. Often they assume a quasi-constant size of the population of the species under study. Economists readily study size and growth of economies, but usually with little to no detail discussing how economic or system structure is shaped by the natural resources within which they reside and depend upon. Pure engineering analyses can also fall into this same trap.

Progress can be slow even within a group of diverse individuals seeking a common goal because each has different knowledge and problem-solving skills. Social and natural scientists often have great difficulty in communicating and educating those of other disciplines regarding the usefulness and validity of their ideas. It takes time and effort to learn the vocabulary and concepts of different disciplines.

Joseph Tainter and Timothy Allen provide ideas to help translate ideas across disciplines by comparing the structure of ant colonies, the Maya society, and the modern U.S. economy using similar systems principles [54, 55]. We already observed that the economy and ant colonies have the same scaling relationship between energy consumption and size. Generally, the quality and abundance of the resource links with the size and level of organization of the colony. To live off of and manage lower quality resources, a colony must have more ants, collect more mass (to farm fungus), and consequently exhibit more organization, or division of labor.

Consider the genus *Atta*, the leaf-cutter ants. Many of us have seen a video of ants marching along a path carrying bright green leaves many times their size. They gather the leaves for their nitrogen content, but as a resource, leaves do not have a high concentration of nitrogen. Thus, leaf-cutters constantly collect leaves. Their nests can grow into the millions of individuals, so large that they can show up on photos taken by satellites in space. While it looks like all the ants are doing the same thing, they aren't. Inside the colony many tasks are split among the ants. "Some are specialized on raising the young ... Others are specialized on removing weeds and disease inside the nest. Others are specialized on going out and finding food, and yet others are specialized on defending the colony. All of the specialization is unique to the leaf-cutters ... With other fungus-growing ants, the workers are basically interchangeable. They don't have these specialized tasks."[42]

Contrast *Atta* with the fungus-farming genus *Myrmicocrypta*. It uses insect carcasses and droppings that fall from trees to the forest floor. These resources, while in short supply and sporadically located, are nitrogen rich and thus of high-quality. Based on this high-quality nitrogen source, *Myrmicocrypta* ant society is

---

[42] Scott Solomon, May 9, 2018, Leafcutter ants' success due to more than crop selection, https://www.sciencedaily.com/releases/2018/05/180509104921.htm.

not particularly specialized [55]. When there are fewer of you, each of you has to be pretty flexible.

As ant expert Deborah M. Gordon notes, ants don't toil away at the same old task every day of their lives. She studies harvester ants that hunt and gather seeds for food. Harvester ants change tasks due to both long-term changes in the age of the nest that in turn affects the numbers of each type of worker. Ants also change tasks during shorter-term "emergencies." When Gordon placed some trash around the nest of harvester ants, some of the ants switched to the task of clearing the entrance to the nest [21]. If we want to take this analogy to humans (which is dangerous, but I'll do it), then this is a good thing! We don't stay in the same roles and jobs as we age either. From birth to childhood and adulthood, we take on different roles. And when disasters occur, such as when major hurricanes make landfall, thousands of people rush to the impacted areas to help clean up and provide assistance to those who had their homes damaged. We are ants in action, cleaning up messes after disasters.

Ants also change tasks to adjust to good events. When Gordon placed seeds outside of the nests of harvester ants, more ants collect the food both because more foraging ants become active (instead of waiting around) and some ants switch to foraging from some other task [21, 22]. This response resembles oil and gas boom towns. After Anthony Lucas broke through at the Spindletop oil field in East Texas in 1901, wildcatters flocked to the area like moths to light, or like ants to a pile of seeds.[43] The same effect happened when hydraulic fracturing and horizontal drilling opened up access to known, but previously uneconomic, "tight" reservoirs in the Barnett Shale and Eagle Ford of Texas, the Marcellus Basin of Pennsylvania, and the Bakken of North Dakota. Roughnecks flock to boom towns just as ants quickly collect seeds outside of their nest.

The parallels between ant and human behavior are eerie. The 2018 PBS Independent Lens documentary *My Country No More* presents some of the land-use challenges driven by oil activity in North Dakota. In this film, a Native American, noted to be a descendant of Chief Geronimo, tells why he ended up working in the oil industry in North Dakota. One day he picked up a hitchhiker. The hitchhiker said he was going to North Dakota for oilfield work. For a hitched ride to North Dakota, the oilfield worker would pay for the gasoline to get there and get the driver a job. The driver accepted both offers.

Perhaps not coincidentally, this is how ants determine whether to change from one task to another. We might call it social networking. If a potential foraging ant in the nest encounters enough ants that return with food, then it leaves to collect food also. "This creates a simple form of positive feedback: the more food is available, the more quickly foragers find it, and the more quickly they return to the nest, eliciting more foraging." [22] More people working in oil and gas leads people to find out that there must be jobs in oil and gas.

---

[43]Spindletop History, Lamar University: https://www.lamar.edu/spindletop-gladys-city/spindletop-history.html.

But busts follow booms. After the seeds are collected, the ants switch to other tasks. When the oil is no longer economic the related work dries up, and layoffs force workers to find other employment. Unfortunately, boom towns are not always known for having the most diverse set of jobs. Many oil and gas exporting nations also do not have diverse economies or high income equality.

But let's get real. The reason why people have flocked to oil fields is because they contain high density energy deposits that can run the economy. Per the earlier discussion of how much energy it takes to produce energy, they historically have high EROI and quantity. But all oil reservoirs are not the same. We access the best quality resources first. We're not finding any more Spindletops. Now we're on to lower quality shales and tight sands that we must forcibly fracture underground every few thousand feet to get sufficient oil to flow rather than simply use one straw with which to suck out a large volume. This "best-first" principle holds not only for oil, but any natural resource from soil to water to windy locations. This principle is neither good nor bad. It just is. Again, ants know this too. When faced with the choice of collecting a high-quality food source versus a low-quality food source, a higher proportion of ants collect the high-quality resource [46].

Structures and Cycles of Growth: People Aren't Ants, Right?

How far can we really take the ant analogy to understand human society? Some business owners might want to maximize production by placing each worker in a specialized task, day after day, or year after year. If ant colonies don't even do this, should we? Gordon thinks not:

> Historically, many have found the idea of division of labour a compelling and powerful model. Plato admired it, Adam Smith explained how economies benefit from it, and Henry Ford industrialised it. But it's not natural. A vision of human society ordered and improved by division of labour has permeated and distorted our understanding of nature [22].— Deborah M. Gordon (2016)

However, there certainly is value in considering how human societies have changed and adapted over longer periods than a work shift or a business cycle. The tasks of blue and white collar workers have changed tremendously over the industrial era, but tasks also change in pre-industrial societies.

The farming practices and social structure of Mayan cities *did* significantly change with the size and quality of their natural resources, just like the fungus-farming ants [55]. When populations were small, before the year 200 AD, the Maya could take advantage of naturally regenerating food sources without degrading them. But as populations grew, they had to adapt the land to their increased needs. Communities situated themselves at the bottom of small watersheds where seasonal rainfall naturally funneled into reservoirs for use in the dry season. They dug canals to channel water, and they used the displaced soil to create raised surfaces on which to plant crops right next to the water. The key point is that they did not have to move the water around. They chose to farm where the natural force of gravity, together with the unaltered topography, directed the water. Again, this is the best-first principle that uses the high EROI resource first. Don't fight the natural flow of water, work with it.

Eventually there was too much "slash and burn" agriculture in the lands uphill from these reservoirs. Soil erosion threatened the water storage and made the hillsides less agriculturally productive. The population had to adapt or migrate. The Early Classic Maya (200–500) adapted by using terraced agriculture and locating city centers at the summits of hills. Now, instead of being at the bottom of the hill, where water comes to you, the Maya had to engineer watersheds to keep more water at the top of the hill. "When combined with growing networks of exchange among resource specialized communities, organizational hubs and their trading tendrils became ever more complex. Much of the Maya lowlands were now experiencing a decreasing return on energy investment as a consequence of centuries of overexploitation of Holocene [epoch since the last ice age] natural resources." [55].

The peak of Maya civilization occurred between 500 and 800. This was the time of large cities with kings, monument and pyramid construction, and hierarchical social organization.

> The convex [hilltop] microwatershed adaptation was perfected at several elevated urban centres. In the convex microwatersheds, large, surfaced plazas caught and channelled water. Water was constrained and managed through reservoirs, dams, channels, sluices, filtration systems and switching stations, then released during the dry season. This system required specialised knowledge; engineering, assembly and direction of labour forces; and hierarchy. Furthermore, after these hydraulic systems were built, they had to be maintained. This was a complex system and an ongoing cost, which lowered the effective resource gradient.
>
> With greater complexity now built into the societies at the political centre, water and other resources could be managed more productively, but many vulnerabilities were accumulating. The amount of redundancy in the biogeography of the tropics is high compared to more arid or temperate environs. It is difficult to find alternative resources by travelling horizontally. The great hubs increasingly found it difficult to grow the volume of food necessary to support their populations, coordinate with other centres, or identify remaining fertile zones containing the rich Holocene stores of resources. Because the regeneration rate for soils in some upland settings (frequently argued to be the repository for the richest, well-drained, farming soils in the Lowlands) is about a cm per century, with or without Terminal Classic drought conditions, the Great Fragmentation of the Terminal Classic (CE 800–1000) was likely overdue. [55]

After creating overpopulated cities on higher ground, the Maya attempted to adapt by scattering population back into the valleys, but the resources were still too degraded. Social complexity could not be maintained and the collapse was inevitable. It is as if you took a colony of leaf-cutter ants and took away 90% of the trees from which they harvest leaves.

Just as researchers perform economic modeling of our contemporary economy, researchers model historical civilizations. One paper captures both the rapid fivefold rise in Maya population from 600 to 800 and the subsequent collapse over the next 200 years [47]. There are many hypotheses for the Maya collapse. Much of the historical narrative focuses on drought as a major causal factor, drawing parallels to climate change discussions today. However, Sabin Roman and his colleagues don't think the timing and magnitude of the precipitation changes could have played a large enough role. They show that the rise and collapse was likely due to a switch to an intensive agriculture practice that depleted soil quality. Because there was an increase in agricultural output per person, some formerly practicing slash and burn agriculture could switch to monument construction. Thus, during the boom, a smaller fraction of the population fed a growing population. But bust followed the 200-year boom. By the year 1200, the population collapsed below that at the start of intensive agriculture. While soil quality can naturally regenerate, the rate was too slow for the Maya farming practices. Practically, at the rate and manner they farmed, the Maya depleted the soils like a fossil resource.

Are there some general takeaways and principles from looking at ants and pre-industrial societies? Surely. Tainter describes his general concept as the *energy-complexity spiral* [53]. For now, think of complexity being represented by some measure of structure or distribution. How much of labor, money, capital, energy, etc., is needed for each of the various tasks ongoing in society? Tainter's energy-complexity spiral can spin in two directions. In direction one, complexity comes first.

> Societies adopt increasing complexity to solve problems, becoming at the same time more costly. In the normal course of economic evolution, this process at some point will produce diminishing returns. Once diminishing returns set in, a problem-solving society must either find new resources to continue the activity, or fund the activity by reducing the share of resources available to other economic sectors. The latter is likely to produce economic contraction, popular discontent, and eventual collapse.[53]—Joseph Tainter (2011)

Before industrialization this was the usual situation: increase complexity first, find the necessary resources next.

The other direction of the energy-complexity spiral starts with energy. On some occasions "... surplus energy precedes and facilitates the evolution of complexity." [53] Here, the attainment of surplus energy refers to a transformational leap in technology or resource access. A change occurs such that society attains much more bang for the buck in terms of the energy inputs required for energy extraction. The society experiences an increased energy return on energy invested that enables growth.

> There have been occasions when humans adopted energy sources of such great potential that, with further development and positive feedback, there followed great expansions in the numbers of humans and the wealth and complexity of societies. These occasions have, however, been so rare that we designate them with terms signifying a new era: the Agricultural Revolution and the Industrial Revolution. [53]—Joseph Tainter (2011)

As Tainter points out, "Surplus production has not been common in human history, nor has complexity." [53] The societal growth of last 200 years is a complete anomaly in the context of human history. The data presented in this book show the economic and material growth during the three decades following World War II are an anomaly within the longer anomaly of the Industrial Age (recall growth rate of energy consumption in Fig. 2.10). We shouldn't be surprised that the trends of those three decades did not continue, because we should expect diminishing returns on the energy-complexity spiral. The reason is that certain long-term patterns seem to repeat themselves.

Studies of historical societies show general cyclical patterns, say of population growth and decline, that take several centuries to play out, just as with the Maya and just as with bacteria (as mentioned in Chap. 4). Peter Turchin and Sergey Nefedov describe a generalized four-phase cycle in their book *Secular Cycles* [56].[44] Here, the word *secular* is used to mean occurring or persisting over an indefinitely long period. The four phases are expansion (or growth), stagflation (or compression), crisis (or state breakdown), and depression.

First consider the *expansion* phase. Ecological economists such as Herman Daly characterize this as an "empty world"—empty of us and our stuff but full of natural resources [16]. There is a low but increasing population, and resources are relatively abundant. Wages and equality are high as there are relatively few "elites" in the population exploiting the working "commoners." Thus, there is high and increasing sociopolitical stability along with support for increasingly centralized governments that can easily coordinate societal functions and order. Grain prices (akin to energy prices in agrarian societies) are low.

After expansion comes the *stagflation* phase. The word stagflation is a combination of stagnation and inflation. Stagnation describes a slowing of expansion, and inflation describes a rise in prices (e.g., grain) while real wages decline. Population growth slows, and "[a]lthough the majority of commoners experience increasing economic difficulties ..., the elites enjoy a golden age, and their numbers and appetites continue to expand" [56, p. 20]. Resentment builds in the population causing increasing sociopolitical instability. However, the elites generally get along with each other.

A *crisis* phase follows stagflation. It marks a turning point from growth to decline, from stability to instability. Ecological economists characterize this as a world "full" of us and our stuff but empty of natural resources [16]. Sociopolitical instability reaches its peak. Grain prices are high but have high variability. The elites no longer get along with each other. As there is not enough economic surplus to sustain all of them and their lifestyles, they battle to remain part of the shrinking elite class. The overall population declines (in agrarian societies) due to higher death rates and outward migration. Wages become volatile during this phase of rapid change. There is a collapse of centralized state control and society moves toward decentralized governance.

Finally, the *depression* phase follows during which the population declines to a minimum level. The number of elites declines quickly as their power struggle sorts itself out. Efforts are made to regain centralized control of society, but they fail repeatedly. Grain prices follow an overall decline, wages are high, and both have high variability. Sociopolitical instability is high during this phase, but it declines as problems sort themselves out via attrition.

---

[44]Table 1.1 of [56] summarizes the secular cycle phases.

The growth and stagflation phases are "integrative" in that there is increasing unity of the society and centralization of power. The crisis and depression phases are "disintegrative" with decentralization of power and infighting among elites. These four-phase cycles are useful to understand when any given society might be in this repeating pattern.

Human history and ecology tells story after story of the rise and fall, or collapse, of societies. We can contemplate how ideas such as the energy-complexity spiral, secular cycles, and ant colony organization apply to our modern human society. These concepts consider the long-term dynamics and demographics of a population of social organisms whether they be humans or ants. Let us now explore how well these cyclical rise-and-fall concepts apply to our economic situation today.

Structures and Cycles of Growth: Does the Past Apply to Today?

Chapter 4 showed that today global human population is still increasing, albeit more slowly since the 1970s, and in OECD countries population growth is mostly due to immigration. Debt relative to GDP has grown substantially over the last few decades for OECD countries, and globally since the 2008 financial crisis. These population and debt variables suggest OECD economies are nearing the end of the stagflation phase or the beginning of the crisis phase.

China and India, making up nearly a third of global population, might still be in an expansion phase, but rapid accumulation of debt in China suggests a stagflation phase, possibly masked if focusing primarily on GDP growth.

Many countries in the Middle East, North Africa, and sub-Saharan Africa are potentially already in crisis phase (e.g., the Arab Spring of 2011; Syrian refugees and Islamic State) having never fully experienced a middle-class and democratic society during the pre-1970s global expansion phase before energy efficiency became a focus. Since developing country citizens missed out (to a large degree) on the heyday of the fossil fuel boom, some use this supposition as justification for the distributed and renewable energy "leapfrog" argument from Chap. 3. If these countries were not otherwise behind in economic development, they would likely already have the ubiquitous (mostly fossil) energy infrastructure over which renewable narrative advocates now want them to leap (or move beyond).

We can observe multiple metrics related to size, such as a country's population, debt, and GDP as well as various structural metrics such as income inequality and prices. The Fund for Peace, for example, calculates the Fragile States Index that combines many of these factors.[45] Nafeez Ahmed's *A User's Guide to the Crisis of Civilization* is a nice integration of a variety of factors to assess the state of society [2].

---

[45] https://fragilestatesindex.org/.

We can also quantify system-wide metrics for the structure of a modern society or economy that are not just agglomerations of multiple metrics. In 2016 I considered the structural evolution of the United States [33]. In effect I put Tainter's and Turchin's societal structural concepts to the test, not for agrarian societies, but for the largest and most powerful country in the history of the world. I calculated how the structure of the post-World War II U.S. economy relates to resource consumption. To do this, I used the systems concepts of information theory and networks as applied by ecologist Robert Ulanowicz [33].[46]

Ulanowicz worked to understand the "structure" of ecosystems and food webs [57–59]. We can view the economic system as a network much like a biological ecosystem. The species of an ecosystem exchange energy, water, materials, and nutrients. A grasshopper eats grass, a fish eats a grasshopper, and a bear eats a fish. The food web describes the flow of nutrients and energy between species. Similarly, some companies extract natural gas. Pipeline companies send natural gas to power plants. Other companies operate power plants that burn natural gas to make electricity. Transmission companies own power lines that connect our homes to sources of electricity. Finally, we give money to companies to buy air conditioners, heaters, televisions, lights, and other devices that consume the electricity.

Instead of species in an ecosystem or natural gas flowing in a pipe, imagine the economy as a collection of companies grouped into sectors. Oil and gas companies like ExxonMobil and Chevron are collectively included in an "oil and gas extraction" sector. Farmers and ranchers make up the "agricultural" sector. Insurance companies and banks are lumped together into a "financial" sector. And so on. Because each company buys and sells products with other companies, each sector buys and sells products with other sectors. Data are readily available tracking the flows of money among economic sectors. While these data are not a fully precise way of describing money flows within an economy, we can use them to describe the general structure of an economy. Information theory provides one means of measuring this structure.

We can assess many metrics that describe money flows in the economy, and I'll describe three. The first describes sector-to-sector transactions. The second describes how much total money each sector spends or receives from all of its transactions. The third is a combination of the first two metrics.

First, the sector-to-sector transactions. If each transaction is exactly the same amount of money, the *economic flows* are 100% *redundant* and 0% *efficient*.

For example imagine a two-sector economy: agriculture and manufacturing. If the agriculture sector pays the manufacturing sector 100 billion dollars, the manufacturing sector pays the agriculture sector 100 billion dollars, and each sector buys 100 billion dollars of stuff from itself, then all four possible transactions in the economy are of exactly the same amount. This economy is 100% redundant

---

[46]Instead of reading my journal article describing the structure of the U.S. economy, you can read my blog post: http://careyking.com/relations-between-energy-and-structure-of-the-u-s-economy-over-time/.

because it spread around 400 billion dollars in the most uniform way possible. A highly redundant economy interacts with many of the possible partners in many ways and relatively equally. A highly redundant system is usually not the best for maximizing growth, but you have diversity. You have not put all of your eggs into one economic basket. If one part of the economy suffers a setback, other parts of the economy buffer the impact.

In contrast, a highly efficient economy has only a few major economic transactions such that there are fewer sectors to deal with to get things done. This efficiency can, and typically does lead to increased potential for growth, but it is more exposed to unexpected problems. Practically, we do not expect 100% efficiency or redundancy in the real economy because there are physical and other structural constraints among economic sectors that prevent these extreme cases. For example, the "oil and gas extraction" sector sells most of its products to the "refined oil products" sector, but not the other way around. This prevents 100% redundancy. Further, a 100% efficient economy would mean that each economic sector only sells to one other sector. Since we know that multiple types of businesses buy a significant number of products from multiple sectors (e.g., computers from the "electronics" sector, fuel from the "refined oil products" sector) the economy cannot achieve 100% efficiency.

Now consider the total money flowing to or from each sector. If the energy sector buys $10 billion from the computer sector and $30 billion from the manufacturing sector, then its total purchases are $40 billion. If the computer and manufacturing sectors also spend $40 billion buying goods and services from the other two sectors, then the *economic sectors* are 100% *equal* and 0% *hierarchical*. All sectors both spent and received $40 billion. A highly hierarchical economy is the opposite of an equal economy. It has a small number of sectors with a dominate share of all economic transactions. In the extreme case there is only one single sector-to-sector transaction, all other transactions are zero dollars, and the economy is 0% equal and 100% hierarchical. Effectively the other parts of the economy no longer exist because the other economic sectors don't engage with each other. These other sectors might be present conceptually, but no one buys anything from them. Just as we don't expect 100% hierarchy for a real economy, we also don't expect 0% equality.

The metric of *information entropy* combines these two tradeoffs of redundancy vs. efficiency and hierarchy vs. equality. It increases via increasing efficiency or redundancy, and vice versa. It can also increase if redundancy increases more than efficiency decreases, or vice versa. Thus, a system or an economy might have to make tradeoffs between increasing redundancy or efficiency, and information entropy measures the result of this structural tradeoff. Systems with higher information entropy are considered more complex in the sense that it takes more computation or knowledge to fully describe them.

Figure 5.8 shows the results from quantifying the structure of the U.S. economy. Focus on the *direction of change of the metrics* more than the values. Are the metrics increasing, decreasing, or remaining at the same value? By answering this question we can distinguish three phases: Phase 1 (1947–1967), Phase 2 (1967–2002),

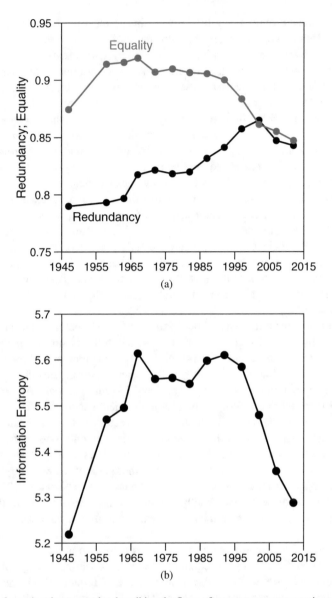

Fig. 5.8 Information theory metrics describing the flows of money among economic sectors of the U.S. economy [33]. There are three phases distinguished by the direction of change in the metrics: Phase 1: 1947–1967. Phase 2: 1967–2002. Phase 3: 2002–2012. (**a**) The metrics of redundancy (of sector-to-sector transactions) and equality (of total transactions by each sector). (**b**) Information entropy

and Phase 3 (2002–2012). The important implication of this research is that the findings support the resource-structural linkages and progression suggested by the anthropological assessments of agrarian societies, such as those by Tainter and Turchin.

During Phase 1 all metrics increase. Money flowed more equally throughout the economy in every way. Energy consumption increased at a rapid rate while food and energy became cheaper over time (refer back to Chap. 2). Phase 1 is Turchin's expansion phase during which an energy surplus drove increasing complexity (i.e., Tainter's energy-complexity spiral starting with energy). During Phase 2, equality decreases, but the other two metrics continue to increase. This is somewhat of a transition phase during which energy consumption increased, but at a slower rate than Phase 1. During Phase 3 all metrics decrease, the exact opposite directions of change from Phase 1. Further, during Phase 3 the U.S. experienced constant energy consumption while food and energy costs increased, unlike during the other phases when food and energy costs decreased. Phase 3 is akin to Turchin's stagflation condition during which the U.S. is struggling to find the resources to maintain its existing complexity. It exhibits Tainter's energy-complexity spiral where the complexity can no longer be maintained.

When energy and food were cheap and being consumed at an increasing rate, the distribution of money became more widespread. When they became more expensive and the rate of energy consumption stagnated, the distribution of money became more concentrated. Because different types of skills are required to work in different economic sectors, these sectoral-structural changes impact socio-economic outcomes. Since the 1970s, intellectually skilled jobs became more important because engineers and scientists design materials and machines to be more efficient, extract more remote and diffuse resources, and further replace labor. Much of the post-2002 structural change of Phase 3 was driven by the oil and gas sector that accounted for the highest fraction of total purchases in 2007 and 2012, but wasn't in the top 10 sectors for spending for the previous two decades. This is important because the techno-optimistic narrative neglects the ecological, or biophysical, concept that the economy's internal costs of the energy system (the energy input required to extract energy, or EROI) are critical to understanding economic structure and growth.

Summary

The economy is a complex system, and a system is composed of elements that when connected exhibit properties and dynamics different from any one of the underlying parts. Systems science focuses as much on how the elements are connected as on describing the elements themselves. By studying these interconnections we better understand how economic growth and size, measured via GDP, number of houses, number of cars, etc., are linked to economic structure, or the distribution of energy and money.

Fig. 5.9 In considering the growth of the economy, we must consider both its size and structure. This is akin to comparing a circle with a small area (i.e., U.S. economy before the 1970s) to a triangle with a larger area (i.e., U.S. economy today)

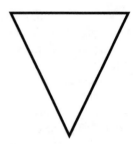

It is critical to consider both economy size and structure. A circle and a triangle might have the same size or area, but they have different structures. From the center of a circle, there is an *equal* distance to its edge no matter what direction. This is not true from the center (centroid) of a triangle.

Comparing a circle to a triangle is like comparing two economies (Fig. 5.9). The economies might have the same size, or one might be bigger than another, but they likely have different structures. Importantly, economic size and structure are related to the cost and quantity of the energy resources and technologies that operate the economy.

By comparing the sizes, structures, and stages of development of economies to those of biological systems—cells, animals as organisms composed of cells, and organisms of organisms such as ant colonies—we see many parallels. One of the major parallels is how both biological systems and the global economy (after 1970) both exhibit a sublinear scaling relationship that relates energy consumption to size: energy consumption increases more slowly than size. This is because when energy consumption can no longer be increased at whim (with minuscule internal energy consumption to extract energy), physical constraints and the balance of supply and demand in networks force similar structure and operation in both biological organisms and economies. Part of this structure is a hierarchy of elements of different size and energy intensity: some elements with critical roles might be relatively small but consume a lot of energy relative to their size. As complex systems grow, they add components with less energy consumption per size.

We can learn a tremendous amount about economic organization by observing the living creatures, ecosystems, and other systems around us. Economists don't normally draw such distinct parallels with biological systems. However, the history of the mainstream theory of economic growth has been inspired by physical laws. With this background, let's move to the next chapter that explores how economists normally think about economic growth and structure before demonstrating how the systems thinking from this chapter has inspired some alternative economic models that better explain past economic growth, structure, and the role of energy.

References

1. Adams, R.N.: Energy and structure. University of Texas Press, Austin, Texas and London, England (1975)
2. Ahmed, N.M.: A User's Guide to the Crisis of Civilization: And How to Save It. Pluto Press (2010)
3. Ballesteros, F.J., Martinez, V.J., Luque, B., Lacasa, L., Valor, E., Moya, A.: On the thermodynamic origin of metabolic scaling. Scientific Reports **8**(1448) (2018). https://doi.org/10.1038/s41598-018-19853-6
4. Banavar, J.R., Moses, M.E., Brown, J.H., Damuth, J., Rinaldo, A., Sibly, R.M., Maritan, A.: A general basis for quarter-power scaling in animals. Proceedings of the National Academy of Sciences **107**(36), 15,816–15,820 (2010). https://doi.org/10.1073/pnas.1009974107. http://www.pnas.org/content/107/36/15816.abstract
5. Bartis, J.T., LaTourrette, T., Dixon, L., Peterson, D., Cecchine, G.: Oil shale development in the united states: Prospects and policy issues. Tech. rep., RAND Corporation (2005)
6. Bashmakov, I.: Three laws of energy transitions. Energy Policy **35**, 3583–3594 (2007)
7. Bolt, J., van Zanden, J.L.: The first update of the Maddison project; re-estimating growth before 1820. Tech. rep. (2013)
8. Brandt, A.R.: Converting oil shale to liquid fuels: Energy inputs and greenhouse gas emissions of the Shell in situ conversion process. Environmental Science & Technology **42**(19), 7489–7495 (2008)
9. Brandt, A.R.: Converting Oil Shale to Liquid Fuels with the Alberta Taciuk Processor: Energy Inputs and Greenhouse Gas Emissions. Energy and Fuels **23**, 6253–6258 (2009)
10. Brandt, A.R., Englander, J., Bharadwaj, S.: The energy efficiency of oil sands extraction: Energy return ratios from 1970 to 2010. Energy **55**, 693–702 (2013). https://doi.org/10.1016/j.energy.2013.03.080
11. Brown, J.H., Burnside, W.R., Davidson, A.D., Delong, J.R., Dunn, W.C., Hamilton, M.J., Mercado-Silva, N., Nekola, J.C., Okie, J.G., Woodruff, W.H., Zuo, W.: Energetic limits to economic growth. BioScience **61**, 19–26 (2011). https://doi.org/10.1525/bio.2011.61.1.7
12. Bunger, J.W., Crawford, P.M., Johnson, H.R.: Is shale oil America's answer to peak-oil challenge? Oil and Gas Journal **August 9** (2004)
13. Carroll, S.: The Big Picture On the Origins of Life, Meaning, and the Universe Itself. Dutton, Penguin Random House LLC (2016)
14. Clarke, A., Johnston, N.M.: Scaling of metabolic rate with body mass and temperature in teleost fish. Journal of Animal Ecology **68**(5), 893–905 (1999). https://besjournals.onlinelibrary.wiley.com/doi/abs/10.1046/j.1365-2656.1999.00337.x
15. Connor, P.A., Cleveland, C.J.: U.S. energy transitions 1780–2010. Energies **7**(12), 7955–7993 (2014). https://doi.org/10.3390/en7127955. https://www.mdpi.com/1996-1073/7/12/7955
16. Daly, H.: Economics in a full world. Scientific American pp. 100–107 (2005)
17. Dennett, D.C.: Consciousness Explained. Back Bay Books (1991)
18. Farrell, A.E., Plevin, R.J., Turner, B.T., Jones, A.D., O'Hare, M., Kammen, D.M.: Ethanol can contribute to energy and environmental goals. Science **311**(5760), 506–508 (2006)
19. Fouquet, R.: Heat, Power, and Light: Revolutions in Energy Services. Edward Elgar Publishing Limited, Northampton, Massachusetts (2008)
20. Fouquet, R.: Long-run demand for energy services: Income and price elasticities over two hundred years. Review of Environmental Economics and Policy **8**(2), 186–207 (2014). https://doi.org/10.1093/reep/reu002. http://reep.oxfordjournals.org/content/8/2/186.abstract
21. Gordon, D.M.: Dynamics of task switching in harvester ants. Animal Behaviour **38**, 194–204 (1989)
22. Gordon, D.M.: The queen does not rule (2016). https://aeon.co/essays/how-ant-societies-point-to-radical-possibilities-for-humans

23. Greening, L.A., Greene, D.L., Difiglio, C.: Energy efficiency and consumption - the rebound effect - a survey. Energy Policy **28**(6-7), 389–401 (2000). https://doi.org/10.1016/S0301-4215(00)00021-5
24. Guilford, M.C., Hall, C.A.S., O' Connor, P., Cleveland, C.J.: A new long term assessment of energy return on investment (EROI) for U.S. oil and gas discovery and production. Sustainability **3**(10), 1866–1887 (2011)
25. Hall, C.A.S., Balogh, S., Murphy, D.J.R.: What is the minimum EROI that a sustainable society must have? Energies **2**, 25–47 (2009)
26. Hall, C.A.S., Cleveland, C.J., Kaufmann, R.K.: Energy and Resource Quality: the ecology of the economic process. Wiley, New York (1986)
27. Hall, C.A.S., Klitgaard, K.A.: Energy and the Wealth of Nations: An Introduction to Biophysical Economics, 2nd edn. Springer (2018)
28. Huber, P.W., Mills, M.P.: The Bottomless Well: The Twilight of Fuel, the Virtue of Waste, and Why We Will Never Run Out of Energy. Basic Books, New York (2005)
29. Jenkins, J., Nordhaus, T., Shellenberger, M.: Energy emergence: Rebound & backfire as emergent phenomena (2011). https://thebreakthrough.org/blog/Energy_Emergence.pdf
30. Jevons, W.S.: The Coal Question: An Inquiry Concerning the Progress of the Nation, and the Probable Exhaustion of Our Coal Mines, second edition, revised edn. Macmillan and Co., London (1866). Kessinger Legacy Reprints
31. King, C.W.: Energy intensity ratios as net energy measures of united states energy production and expenditures. Environmental Research Letters **5: 044006**(available at http://stacks.iop.org/1748-9326/5/044006) (2010)
32. King, C.W.: The rising cost of resources and global indicators of change. American Scientist **103**, 6 (2015)
33. King, C.W.: Information theory to assess relations between energy and structure of the U.S. economy over time. BioPhysical Economics and Resource Quality **1**(2), 10 (2016). http://dx.doi.org/10.1007/s41247-016-0011-y
34. King, C.W., Hall, C.A.S.: Relating financial and energy return on investment. Sustainability **3**(10), 1810–1832 (2011)
35. King, C.W., Maxwell, J.P., Donovan, A.: Comparing world economic and net energy metrics, part 1: Single technology and commodity perspective. Energies **8**(11), 12,346 (2015). https://doi.org/10.3390/en81112346. http://www.mdpi.com/1996-1073/8/11/12346
36. King, C.W., Webber, M.E.: Water intensity of transportation. Environmental Science & Technology **42**(21), 7866–7872 (2008). http://dx.doi.org/10.1021/es800367m. PMID: 19031873
37. Kleiber, M.: Body size and metabolic rate. Physiological Reviews **27**(4), 511–541 (1947). https://doi.org/10.1152/physrev.1947.27.4.511. PMID: 20267758
38. Kopits, S.: Oil: What price can America afford? (research note). Tech. rep., Douglass-Westwood (2009)
39. Meadows, D.H.: Thinking in Systems: A Primer. Chelsea Green Publishing, White River Junction, Vermont (2008)
40. Mobus, G.E., Kalton, M.C.: Principles of System Science. Springer, New York, NY USA (2015)
41. Mohapatra, L., Lagny, T.J., Harbage, D., Jelenkovic, P.R., Kondev, J.: The limiting-pool mechanism fails to control the size of multiple organelles. Cell Systems **4**(May 24), 559–567 (2017). http://dx.doi.org/10.1016/j.cels.2017.04.011
42. Mori, S., Yamaji, K., Ishida, A., Prokushkin, S.G., Masyagina, O.V., Hagihara, A., Hoque, A.R., Suwa, R., Osawa, A., Nishizono, T., Ueda, T., Kinjo, M., Miyagi, T., Kajimoto, T., Koike, T., Matsuura, Y., Toma, T., Zyryanova, O.A., Abaimov, A.P., Awaya, Y., Araki, M.G., Kawasaki, T., Chiba, Y., Umari, M.: Mixed-power scaling of whole-plant respiration from seedlings to giant trees. Proceedings of the National Academy of Sciences **107**(4), 1447–1451 (2010). https://doi.org/10.1073/pnas.0902554107. https://www.pnas.org/content/107/4/1447

43. Mueller, C.A., Joss, J.M.P., Seymour, R.S.: The energy cost of embryonic development in fishes and amphibians, with emphasis on new data from the Australian lungfish, neoceratodus forsteri. Journal of Comparative Physiology B **181**, 43–52 (2011). https://doi.org/10.1007/s00360-010-0501-y

44. Odum, H.T.: The ecosystem, energy, and human values. Zygon **12**(2), 109–133 (1997)

45. Polimeni, J.M., Mayumi, K., Giampietro, M., Alcott, B.: The Jevons Paradox and the Myth of Resource Efficiency Improvements. Earthscan, London (2008)

46. Price, R.I., Grüter, C., Hughes, W.O.H., Evison, S.E.F.: Symmetry breaking in mass-recruiting ants: extent of foraging biases depends on resource quality. Behavioral Ecology and Sociobiology **70**(11), 1813–1820 (2016). https://doi.org/10.1007/s00265-016-2187-y

47. Roman, S., Palmer, E., Brede, M.: The dynamics of human–environment interactions in the collapse of the classic Maya. Ecological Economics **146**, 312–324 (2018). https://doi.org/10.1016/j.ecolecon.2017.11.007. http://www.sciencedirect.com/science/article/pii/S0921800917305578

48. Shik, J.Z., Santos, J.C., Seal, J.N., Kay, A., Mueller, U.G., Kaspari, M.: Metabolism and the rise of fungus cultivation by ants. The American Naturalist **184**(3), pp. 364–373 (2014). http://www.jstor.org/stable/10.1086/677296

49. Simon, J.L.: The Ultimate Resource 2, revised edn. Princeton University Press, Princeton, N.J (1996)

50. Smil, V.: Energy at the Crossroads: Global Perspectives and Uncertainties. MIT Press, Cambridge, Mass. (2003)

51. Smil, V.: Energy and Civilization A History. The MIT Press, Cambridge, Mass. (2017)

52. Tainter, J.: The Collapse of Complex Societies. Cambridge University Press (1988)

53. Tainter, J.A.: Energy, complexity, and sustainability: A historical perspective. Environmental Innovation and Societal Transitions **1**, 89–95 (2011)

54. Tainter, J.A., Allen, T.F.H., Little, A., Hoekstra, T.W.: Resource transitions and energy gain: Contexts of organization. Conservation Ecology **7**(3) (2003)

55. Tainter, J.A., Scarborough, V.L., Allen, T.F.H.: Resource gain and complexity Water past and future. In: F. Sulas, I. Pikirayi (eds.) Water and Society from Ancient Times to the Present Resilience, Decline and Revival, chap. Concluding essay 1, pp. 328–347. Routledge (2018)

56. Turchin, P., Nefedov, S.A.: Secular Cycles. Princeton University Press (2009)

57. Ulanowicz, R.E.: The balance between adaptability and adaptation. Biosystems **64**(1-3), 13–22 (2002)

58. Ulanowicz, R.E.: Quantitative methods for ecological network analysis. Computational Biology and Chemistry **28**, 321–339 (2004)

59. Ulanowicz, R.E.: The dual nature of ecosystem dynamics. Ecological Modelling **220**(16), 1886–1892 (2009). http://dx.doi.org/10.1016/j.ecolmodel.2009.04.015. http://www.sciencedirect.com/science/article/pii/S0304380009002695. Selected Papers from the Workshop on 'Emergence of Novelties", 9–16 October 2008, Pacina, Siena, Italy

60. Waters, J.S., Ochs, A., Fewell, J.H., Harrison, J.F.: Differentiating causality and correlation in allometric scaling: ant colony size drives metabolic hypometry. Proceedings of the Royal Society B: Biological Sciences **284**(1849), 20162,582 (2017). https://royalsocietypublishing.org/doi/abs/10.1098/rspb.2016.2582

61. West, G.: Scale The Universal Laws of Growth, Innovation, Sustainability, and the Pace of Life in Organisms, Cities, Economies, and Companies. Penguin Press, New York, NY (2017)

62. West, G.B., Brown, J.H., Enquist, B.J.: A general model for the origin of allometric scaling laws in biology. Science **276**(5309), 122–126 (1997). http://www.jstor.org/stable/2892614

63. Wexler, B.E.: Brain and Culture neurobiology, ideology, and social change. MIT Press (2008)

Chapter 6
Macromodel on the Wall, How Does Growth Occur, After All?

"Everything should be as simple as it can be, but not simpler"—Albert Einstein[1]

You can resolve not to do the work of power for it. You can resolve not to let lies be told in your hearing. You can resolve not to use sloppy language that is euphemism.[2]—Christopher Hitchens (2002)

What are Models, and Why do We Use Them?

A model is a greatly simplified interpretation of a complex thing. A model on the cover of *Cosmopolitan* magazine is one (usually attractive) example of the human form (in fancy clothes). A toy car is a model, or simplified and shrunken representation, of an actual car.

This chapter is not about these types of models. Rather, this chapter describes economic models that are mathematical equations (without actually showing the equations!) used to both describe patterns in historical data and project possible future outcomes. Chapter 5 quoted ecologist Howard T. Odum as he stated that it might not be possible for a system to understand itself, but it can try. This means that since we reside within our economy, we might not be able to fully explain the economy, but we can try by developing "...simplified ideas ...called models ...which have enough of the main features to have some reality but are simple enough to be understood."[3] That is what this chapter is about. In particular this

---

[1] Roger Sessions, 1950 January 8, How a 'Difficult' Composer Gets That Way by Roger Sessions, Page 89, New York. (ProQuest) and discussed at http://quoteinvestigator.com/2011/05/13/einstein-simple/.

[2] Speech at The Commonwealth Club, "Why Orwell Matters," October 21, 2002. Viewed at https://www.youtube.com/watch?v=rY5Ste5xRAA&t=1923s.

[3] Odum [52, p. 119].

© Springer Nature Switzerland AG 2021
C. W. King, *The Economic Superorganism*,
https://doi.org/10.1007/978-3-030-50295-9_6

chapter focuses on economic models that explain macro-scale phenomena such as GDP, total primary energy consumption, and population.

This type of modeling is in many ways the exact opposite of modeling clothing. Equations are objective. There is nothing subjective about an equation. We invented mathematics in order to be as objective as possible for counting objects and describing regularities in the physical world.

However, there is one very important commonality between mathematical models and models on the pages of *Cosmopolitan* magazine: each model influences our perception of the real world.

A model represents some aspect of the real world, but it cannot represent the totality of the real world. Many critiques of fashion models describe them as disproportionately young and exceptionally thin to the point of being unhealthy.[4] Some people are skinny, and some are voluptuous. Some people are tall, and some are short. People have different hair, eye, and skin color. Thus, the fashion model critique states that the composition of fashion models does not represent the variety of shapes, sizes, colors, and races of the real-world population.

Mathematical economic models face a similar problem. One mathematical model of the economy cannot represent the diversity of countries and phenomena we observe in the real world. However, some models more accurately describe data or make successful predictions than others, and most models are useful in their proper context. One of the most common quotes in this regard is from George Box and Norman Draper (usually only the first and second sentences):

> ... all models are approximations. Essentially, all models are wrong, but some are useful. However, the approximate nature of the model must always be borne in mind.—George E.P. Box and Norman R. Draper (1987)[5]

The challenge of using mathematical models composed of equations is that one needs a framework, a theory, or dare we say a narrative that forms the basis of creating the mathematical model. This was nicely stated by George Backus who spent much of his career making complex mathematical models of the economy:

> ... a narrative is a metaphor, whereas an "equal sign" mathematical statement is precise, that unequivocally states the meaning and use of a number. It may be wrong, but is easy to critically evaluate.[6]—George Backus (2017)

The items on the left side of an "equal sign" represent *exactly* the same quantity as the items on the right side: $2 + 2 = 1 + 3$. By "easy to critically evaluate" Backus means that if you put your mind to it, you can test if the items on one side of the equation do indeed equal the items on the other side. The first law of

---

[4]Kirstie Clements, July 5, 2013, *The Guardian* "Former Vogue editor: The truth about size zero": https://www.theguardian.com/fashion/2013/jul/05/vogue-truth-size-zero-kirstie-clements. Valeriya Safronova, Joanna Nikas and Natalia V. Osipova, September 5, 2017, *New York Times*, "What It's Truly Like to Be a Fashion Model": https://www.nytimes.com/2017/09/05/fashion/models-racism-sexual-harassment-body-issues-new-york-fashion-week.html.

[5]Box and Norman [8, p. 424].

[6]Personal correspondence.

thermodynamics was first a concept that the energy content of a system *before* some physical process takes place (i.e., items on the left side of the equation) is exactly the same *after* that process occurs (i.e., items on the right side of the equation). This concept has been tested and confirmed so many times that it became physical law.

While scientists and engineers are convinced of the usefulness of the laws of physics in their work, it is exceedingly difficult to sway public opinion by describing equations. When Stephen Hawking wrote his best-selling *A Brief History of Time* he was told that every equation in his book would reduce sales by half.[7] Apparently Hawking took the advice to heart: his book has only one equation.

Whether one understands the mathematics behind physical laws or not, we all inherently follow them. I don't have to understand gravity and friction to walk along the sidewalk, and I don't have to understand how an airplane works to ride in it. However, if I want to design an aircraft, I do need to understand how it works. Our lives literally reside in the minds of engineers and scientists when they use mathematical models to design our cars, planes, and bridges. People can die if the equations are wrong.

The same concept holds for economists modeling the economy. While our lives might not immediately and directly be in the balance, we trust economists to use accurate models to design rules for our economy.

> While a poorly designed economic model, say, used to design a tax policy, might not directly lead to human death, the underlying economic principles certainly indirectly affect the distribution of resources and thus human well-being.

The problem is that economists often use models in ways that aren't accurate, aren't consistent with data, and don't actually describe how the economy works. A quote from 2018 Nobel Laureate Paul Romer's 2016 diatribe, *The Trouble With Macroeconomics*, sums this up nicely:

> The trouble is not so much that macroeconomists say things that are inconsistent with the facts. The real trouble is that other economists do not care that the macroeconomists do not care about the facts. An indifferent tolerance of obvious error is even more corrosive to science than committed advocacy of error.[8]—Paul Romer (2016)

Other economists have taken a more measured approach to critiques of flawed economic theories:

> Macroeconomists should pause before continuing to do applied work with no sound foundation and dedicate some time to studying other approaches to value, distribution, employment, growth, technical progress, etc., in order to understand which questions can legitimately be posed to the empirical aggregate data.[9]—Jesus Felipe and John McCombie (2006)

---

[7]Martin Gardner, June 16, 1988, *The New York Review of Books*, The Ultimate Turtle, https://www.nybooks.com/articles/1988/06/16/the-ultimate-turtle/.

[8]Paul Romer, The Trouble With Macroeconomics, September 14, 2016, https://paulromer.net/the-trouble-with-macro/WP-Trouble.pdf.

[9]Felipe and McCombie [12, p. 296].

In any case, both critiques leave little room for nuance: "do not care about the facts," "obvious error," and "no sound foundation." I wrote this book because I very much think macroeconomists and scientists should care about the facts.

Narratives are supported on the backs of public opinion, which can be molded by theories supported by mathematical models (Chap. 9 discusses the shaping of public opinion in more detail). This is true for science and economics. Even when new data and science correctly contradict existing narratives and models, it can take a long time to overcome them. Incorrect models can allow incorrect narratives to remain pervasive even when more accurate models exist.

The purpose of this chapter is to explore critiques of mainstream economic models, specifically neoclassical economic theory, that do not sufficiently reflect the patterns we observe in the real world. I specifically focus on patterns related to energy consumption, energy efficiency, and economic growth. Per the quote of Felipe and McCombie, the chapter also discusses alternative models that include a more accurate, practical, and realistic description of the energy input needs to operate and grow the economy. These latter models represent examples that need to become as pervasive as what practically every economics student is taught in universities: neoclassical economics.

When it comes to understanding the role of energy in the economy, we don't have to throw away neoclassical economic theory if it works. It just turns out that it doesn't work very well. As this chapter explains, when we model economic growth without some unnecessary assumptions of neoclassical economics, and we include assumptions that constrain economic activity based upon known physical principles, like the laws of thermodynamics, we can much more informatively explain modern economic trends including the GDP and energy consumption patterns of Chap. 2 and the debt and wage patterns of Chap. 4.

Neoclassical Economics: The King of Economic Narratives

Why had so much conventional wisdom been bullshit?[10]—Michael Lewis (2017)

The "mainstream" economic framework is called neoclassical economics. Because most economics faculty focus on teaching this theory to their students, the numbers of economists using this theory, and thus interpreting the economy under its worldview, far outweighs those using other worldviews. And make no doubt, *because of its inherent assumptions, neoclassical economics, like any theory, imposes a worldview.* Unfortunately, the vast majority of citizens and many economic practitioners do not know this worldview.

As introduced in Chap. 1, economic theory informs policy, and policy affects social outcomes via the distribution of money. Neoclassical theory is the framework for most policy, and most people don't know this. However, American and European

---

[10]*The Undoing Project: A Friendship That Changed Our Minds*, [39, p. 51].

citizens do know their situation, and many of them are disillusioned with politicians' and economists' explanations for the economic outcomes since the 1970s, including the 2008 financial crisis and 40-plus years of wage stagnation. Neoclassical economists didn't even have a quick answer for the Queen of the United Kingdom, Elizabeth II, when she asked of the global financial crisis, "Why did nobody notice it?"[11]

Because the vast majority of people don't contemplate economic theory, they don't understand that much of their disillusionment starts with neoclassical economics. This is why *neoclassical economics is perhaps the king of economic narratives.*

There are many critiques of neoclassical theory. Here I use information from only a subset of these previous writings.[12] This section provides a short history of how, from the beginning, neoclassical economics attempted to copy physics to justify itself as a rigorous *social science*. Unfortunately, those developing the theory did not fully incorporate what is known about the first and second laws of thermodynamics that tell us all energy is conserved in any physical process, but only some of that energy can perform useful work in the economy. In presenting a short summary of this issue, I lean on Philip Mirowski's 1989 detailed treatise *More Heat than Light,* and readers wishing to dive into more details should consult that book [46]. For readers that wish for mathematics and theoretical arguments about economic modeling in the consideration of laws of thermodynamics, see Nicholas Georgescu-Roegen's 1971 book *The Entropy Law and the Economic Process* [18].

Neoclassical Economic Narrative: Consumer Utility as Potential Energy

Chapter 2 summarized some history of originating the concept of energy, but it did not mention the idea of the *field* that separates the concept of energy from matter. The field, like energy itself, is a mathematical concept, a model if you will, that is useful. It was put into prominence by renowned scientist James Clerk Maxwell in the 1800s. His research led to our coupled description of the concepts of light, electricity, and magnetism. Thus, we speak not only of electric and magnetic fields, but of changing electromagnetic fields that emit radiation such as light and heat.

---

[11] Andrew Pierce, "The Queen asks why no one saw the credit crunch coming," *UK Telegraph,* November 5, 2008 at: https://www.telegraph.co.uk/news/uknews/theroyalfamily/3386353/The-Queen-asks-why-no-one-saw-the-credit-crunch-coming.html.

[12] For the reader interested in a more thorough discussion of problems with neoclassical theory, see Steve Keen's *Debunking Economics* for one of the more extensive critiques, Charles Hall and Kent Klitgaard's *Energy and the Wealth of Nations* that includes discussion of energy-related issues, and Philip Mirowski's *More Heat than Light* that provides the historical background on how neoclassical theory was derived to mimic only part of what was known from classical physics in the 1800s [22, 30, 46].

Electricity, magnetism, and light are fully coupled phenomena. For example, if you affect the flow of electricity, you inherently change the corresponding magnetic field.

Understanding the concept of the field is important for understanding the foundation of neoclassical economics. Consider the gravitational field around large masses, such as planets. It is a *potential energy* field. Living on Earth, we all have experience with the effect of a gravitational field. For us, this field is static, meaning it does not change. The gravity you experience today is what you experienced yesterday. However, if you change the *boundary conditions* of a gravity field, then it changes.

The gravity you experience on the surface of a planet is defined by its size and mass. For example, the moon is less massive than Earth. When astronauts landed on the moon, they experienced a different gravitational field than they experience on Earth. Our weight is defined both by our mass and that of the planet on which we reside. The astronauts had the same mass on the moon, but less weight. This is because our weight is defined by the gravitational field within which we reside, and the field within which the astronauts temporarily resided was defined by the boundary condition, or mass, of the moon, not Earth. Here is the point to keep in mind: boundary conditions can change, and changing boundary conditions means potential energy fields change. When we describe how neoclassical economists use the metaphor of the potential energy field, we see their assumptions break down. They become inconsistent. For now, a bit more on gravity.

You are riding a bicycle (on Earth). If you ride along a flat surface, it is pretty easy. If you ride uphill, it is much harder because you must work against the *force* of gravity. In physics, force and a potential energy field are connected via geometry. If you move along a field line, from a lower to a higher potential, you must overcome an opposing force. The energy you must use equals the force times the distance you travel against that force. However, if you move *perpendicular* to a field line, you move from one point to another at the same potential. No force is required to move from a point with one potential to another point at the same potential where all points in between are also at the same potential. This is why riding a bicycle on a flat surface is not (as) tiring: you move perpendicular to the gravity field lines, and you are always at the same potential. You still have to overcome friction and wind resistance, but those are additional forces other than gravity (recall from Chap. 2 there are many types of energy).

Combine the idea of moving within a potential energy field with the idea of kinetic energy discussed in Chap. 2, and we're ready to understand the basic construction of neoclassical economics. To describe the total mechanical energy of an object we must add its potential and kinetic energies. When you coast on your bicycle increasingly faster down a hill you are losing gravitational potential energy and gaining kinetic energy in two forms: you and your bicycle moving with some speed in a single direction, plus the rotational energy of your bicycle wheels. (This is the same two kinds of kinetic energy as in the rolling billiard ball example of Chap. 2. The ball moves across the table—linear kinetic energy—and rolls—rotational kinetic energy.)

Neoclassical economists mimicked physicists' energy field concept from the 1800s. They replace potential energy with a concept of value they call *utility*.[13] The concept of utility comes specifically from that of the potential energy field. Each one of us is supposed to have our own utility field that describes the preferences of what we want to buy as consumers. Our preferences are expressed, or measured, by the quantity of each commodity that we purchase. The amount of each commodity we buy is the same concept of having a position in a potential energy field, so goes the concept. Thus, prices are the same as forces in potential energy fields, and markets are mechanisms to calculate prices.

If all forces on a particle within a potential energy field balance against each other, the particle is said to rest at an *equilibrium* position defined by the fact that all nearby positions have a higher potential energy. Another way of saying this is that balls do not roll uphill; they roll downhill, to positions of lower potential energy.

To neoclassicals, the "market" is at "equilibrium" when all consumers have purchased a quantity of each product at prices at which all producers are willing to sell. The theoretical place this occurs is a massive bazaar where all people exchange all products via barter transactions (e.g., 3 apples equal two pears, one pair of shoes equals 10 apples, etc.). A barter economy is one in which goods are exchanged for other goods. This exchange of items occurs such that the relative prices are determined and assumed to balance. Further, prices and quantities are determined simultaneously. If people purchase fewer apples, the price of apples goes up. If the price of pears goes down, people purchase more pears. Equilibrium is defined as the situation in which no different exchange of products can occur without making someone's utility lower. In this way, neoclassicals translated the physical concept of an equilibrium at a minimum in potential energy to an economic concept of equilibrium at a maximum in utility.

Note there is *no* fundamental underlying price. The prices of which neoclassicals speak are not specifically money per quantity (e.g., dollars per gallon of gasoline)

---

[13]For the translations of potential energy to neoclassical economics, see Irving Fisher's 1926 *Mathematical Investigations into the Theory of Value and Prices* or Table 5.1 in [46] that is a reproduction. In physics, the potential energy field has a particular mathematical definition. Again, energy is a concept that is useful for understanding the physical world. In particular a potential energy vector field is defined by a mathematical integral. The vector field is derived by taking the derivative of a scalar field potential function. It has the property that if you move around within the vector field but return to the same point at which you started, there is no change in potential energy. Mathematically the term for this property is a "conservative vector field" that is "integrable." While neoclassical theory uses this integrability concept to solve for prices and demand at equilibrium, Wade Hands (1993) discusses [61], and Mirowski (1993) agrees [47], they do not directly translate the same mathematics as in physics. See Hands (1993) for discussion of how neoclassical theory uses mathematics consistent with the concept of a "conservative vector field." This is done by considering a consumer's budget constraint (e.g., how much can you buy with $100). By mathematically asking to (1) maximize utility with a budget constraint or (2) minimize spending to purchase a given set of items, neoclassical theory forms the Slutsky matrix that relates changes in quantities to changes in prices for a "scalar expenditure function" with equivalent mathematics of a Jacobian matrix that relates how changes in positions relate to changes in forces for a "scalar potential (energy) function."

that you see when you actually purchase something. The prices are relative: "how much of this is equal to one of those." This is not a fundamental problem to understand economic exchange. Everything could be "priced" in any single currency, such as cowry shells, gold, or apples instead of dollars. However, in a few paragraphs we'll learn how a problem arises when there is no explicit consideration of money as debt.

Also, "the market" determines prices without any specification of how much *time* it takes for producers to *supply* a certain number of items that consumers *demand* at equilibrium prices. That is, the consumer actually doesn't know how much she demands until she knows the price, but the price is not determined until the seller knows how much the consumer demands and then determines how much to produce. Mathematically this is not a problem. One can equate all demands on one side of an equation to all supplies on the other side and define the prices as those required to make to the equality true. That is what it means to come to economic *equilibrium*.

Using the metaphor of utility as potential energy can be helpful for understanding prices and how much people buy, but time doesn't appear in neoclassical theory because it uses *only* potential energy as a metaphor.[14] It might be justifiable to adopt the analog of the conservation of energy while neglecting the analog of kinetic energy, but it leaves important concepts and data unexplained.

In physics, if you neglect kinetic energy, you ignore mass and time. Recall that to calculate kinetic energy, you need to know mass and speed. No mass, no kinetic energy. No speed, no kinetic energy. If you only have 5 min to ride your bicycle home for dinner from your friend's house that lives down the perfectly flat road from you, you need to know how fast you have to ride, and thus, the kinetic energy you need to maintain. Your potential energy doesn't change from his house to yours, and thus potential energy can't tell you anything about how long it takes to get home.

In neoclassical economics, expenditures, or spending, is the corresponding concept to kinetic energy. Thus, the conserved quantity for neoclassical economics should be expenditure plus utility, just like the conserved quantity for (classical) physics is kinetic energy plus potential energy. Just like you need to know how fast you have to ride your bike to get home in 5 min, you need to know how much money you have to spend to install 100 MW of solar panels in 1 month. If you don't spend enough per day to hire enough installers, you won't get the full installation completed in time. Everybody has heard the old adage "time is money." It sounds better than "kinetic energy is expenditures."

While solving mathematically for economic equilibrium can be a reasonable concept to think about, I now emphasize three particular problems with the neoclassical approach that prevent it from providing enough explanatory power for the energy and economic trends and systems concepts discussed in the previous chapters.

---

[14]This story of neoclassical economics improperly mimicking mechanical energy, or Hamiltonian mechanics, is the subject of Chapter 5 in [46]. Also see Section 4 of [47].

First, as already mentioned, there is no role for time. Everything happens over some unspecified time to come to equilibrium, or agreement on prices and quantities. Because there is no time, it becomes almost impossible to discuss an energy *transition*, where we use the term transition to imply change over time.

> How can we use neoclassical theory to inform the energy narratives without explicit consideration of time? We can't.

Because there is no time, there are no inventories of products that need to buffer imbalances between when people buy and when producers manufacture. Everything that is produced is sold. As Mirowski states: "Transactors were not allowed to hold stocks of inventories except for personal consumption; transactors were lobotomized into passively accepting a single price in a market ..."[15] This lack of inventory would be like all consumers and producers first meeting at an empty grocery store, then coming to an agreement on prices and quantities for steaks, lettuce, and cheese, and finally with a head nod (as in the 1960s sitcom *I Dream of Genie*) or an eye twinkle (as in the competing 1960s sitcom *Bewitched*), all of the food shows up in the correct quantities of each shopper's cart. Somehow everyone's cart is full of groceries yet the grocery store itself never actually contains any food, and we don't know how long it takes for this process to happen. This lack of inventory sounds a little odd because we go to the grocery store specifically because it is a building that stores food.

Second, even though neoclassicals recognize that money exists, to them the quantity of money itself does not affect anything in the economy. The three commonly stated properties of money are as a medium of exchange, a measure of value, and a store of value. For money to act as a medium of exchange means that people give money to buy any good or service and receive money if they produce a good or service. To some this means that money exists to avoid operating a barter economy in which people must exchange one set of goods for another set of goods. However, there has never been an economy based solely on barter.[16] The reason, as discussed in Chap. 4, is that most pre-industrial transactions were among people that knew each other and encountered each other on a regular basis. Economic participants neither had to instantaneously exchange good-for-good, nor use money. They could understand that each had an obligation, or form of debt, to have a one-way exchange today and wait for a reciprocal exchange some time in the future. Because neoclassical economics does not fundamentally consider time, it is easy to see why the theory associates money as needed only for instantaneous exchange. Also, if there is no time (or memory or future), then each person you meet is effectively a stranger since you can't remember meeting them in some past that didn't exist. We can wonder if the lack of time and debt in economic analyses helps transform us into strangers.

---

[15]Mirowski [46, p. 240].

[16]A good reference is Chapter 2 "The Myth of Barter" of David Graeber's *Debt: The First 5000 Years* [20], in particular page 29 for a quick synopsis.

Not only does this simplified view of money avoid a historical bogeyman of the barter economy, but it also avoids describing the real-world influence of money in our modern economy. If there is no role for the *quantity* of money to affect anything, the concept of borrowing money as debt or a loan also does not exist. Thus, the concept of money is assumed, but money performs no fundamental role in the economy. If your theory can't consider time along with money as debt, then you can't consider concepts like paying interest, over a period of *time*, on loans for homes, cars, and university expenses. A quote from Steve Keen summarizes the problem:

> It may astonish non-economists to learn that conventionally trained economists ignore the role of credit and private debt in the economy—and frankly, it is astonishing. But it is the truth. Even today, only a handful of the most rebellious mainstream 'neoclassical' economists—people like Joe Stiglitz and Paul Krugman—pay any attention to the role of private debt in the economy, and even they do so from the perspective of an economic theory in which money and debt play no intrinsic role. An economic theory that ignores the role of money and debt in a market economy cannot possibly make sense of the complex, monetary credit-based economy in which we live. Yet that is the theory that has dominated economics for the last half-century. If the market economy is to have a future, this widely believed but inherently delusional model has to be jettisoned.[17]—Steve Keen (2011)

Finally, there is no role for how much it costs to produce something. There are only prices, no costs. For neoclassicals, price is not a function of how much it costs to produce something. It is solely a function of the consumer demand and producer supply curves determined during exchange of all goods and services. The producers *have to sell* everything they produce and the consumers *have to buy* everything produced at the equilibrium prices determined at some unspecified time. Want 1000 apples? The price is equal to an ounce of cheese. Want only 1 apple? The price is equal to one pound of cheese. To neoclassicals it doesn't matter how much it costs to make cheese. Of course, there are costs to making cheese. A cheese maker pays for milk from a cow that requires feed and water to keep alive. It takes time and physical resources, such as energy, to raise a calf, make cheese, grow apples, make airplanes, transport plastic toys from China, etc.

With all of these assumptions and caveats associated with using neoclassical theory, it is amazing that it is so widely practiced. This is precisely why neoclassical economics is the king of economic narratives.

Real businesses make goods, borrow money to pay for real costs, and use inventories to account for discrepancies in timing between sales and production. It is up to each one of us to decide how to appropriately use any economic model that does or does not account for time, costs, debt, inventories, and any other physical or social concept. To further understand the concepts of the cost of production, we now turn to the neoclassical theory model of economic growth, or production.

---

[17]Keen [30, p. 6].

Neoclassical Economic Narrative: Production and Growth

Economists use the term *production* to mean the act of creating goods and services. More production is akin to higher GDP. But just how does production occur? What does it take to produce something? Mirowski notes that there is no greater "...source of discord in the history of neoclassical theory ..." than that due to lack of consensus concerning the meaning of production.[18] "Production ... does not "fit" in neoclassical value theory."[19] For a more entertaining quote:

> To get a trained economist to entertain this thesis [that production is not conceptually consistent with neoclassical value theory] is as easy as getting a Catholic priest to entertain the notion of the fallibility of the Pope.[20]—Philip Mirowski (1989)

The reason is that neoclassical economists are wedded to the idea of the potential field, as discussed in the previous section, to describe the value of purchased economic output. In effect, they try to use it to imply how the economy produces goods and services, but their method doesn't work conceptually or in practice.

Because neoclassicals assume that value, or utility, derives from consumers and producers at the moment of exchange, when prices are determined, there is no role for the *time and cost* of production to affect the value of goods. That is to say, because it doesn't matter how much it costs to produce a good, the ultimate value has nothing to do with that history. As economist Nicholas Georgescu-Roegen noted:

> The Neoclassical mode of representing the production function ignores the time factor.[21]— Nicholas Georgescu-Roegen (1971)

The lack of sufficient history stems from the lack of the concept of time that it takes to move from one equilibrium (of supply and demand) to another. This is the outcome of using a potential field theory to establish economic value. A static field has no history, no arrow of time that we experience by remembering events that happened in the past, or how much time it takes to go from "here to there. " In the simplest terms, goods available for you to buy "now" had to be produced in the past, or before "now," but to neoclassicals these factual historical events of production don't matter.

There is a good point to the neoclassical argument that prices of goods are determined when people buy them. We've all seen the price of a dress drop dramatically when going "on sale." Clearly the price of a dress can and often does have little to do with its cost to produce. Also, in developing countries there are often localized markets, say in the center of the city, where sellers congregate in one location, selling various goods from their booths. In these markets it is quite normal for sellers and consumers to haggle "on the spot" over the price of shoes,

---

[18]Mirowski [46, p. 293–294].

[19]Mirowski [46, p. 284].

[20][46, p. 284].

[21]Georgescu-Roegen [18, p. 244–248].

kitchen wares, and clothes. In these cases the price really is determined at the end of a negotiation, but there are lower bounds.

It is possible, or course, for a business to produce a good that no one wants to buy. If this business only produces this good that no one buys, it will go bankrupt. Why? Because it costs more than zero dollars and zero energy to produce anything. At a minimum the business owner has to pay for his food to stay alive and the raw materials for his product even if he makes it by hand. If it costs one dollar to make a product, but consumers will only pay fifty cents for it, the business will eventually go bankrupt and the good will cease to exist.

When I was in college there was a convenience stand within the mechanical engineering building selling snacks and some donuts. Students being poor, and engineers taught to be resourceful, we noticed that at closing time the workers would put the vendor's unsold donuts into the trash can in the hallway. Not being biology majors, we figured that free donuts in the trash for less than 15 min were no less healthy than fifty-cent donuts in the display case 15 min earlier. We didn't stop eating donuts, but we stopped buying donuts, at least for a while. The vendor shut down and didn't return to the building the following semester.

In neoclassical theory, this reality of supply-demand mismatch does not exist because the exact quantity consumers are willing to buy to maximize their utility comes into existence irrespective of time or cost. The donut vendor didn't keep lowering prices until we bought all of the donuts, thus making supply equal demand. We didn't have a discussion with the business owner to come to an agreement on donut prices. Her supply at stated price was greater than our demand at that price, and she threw away the remaining donuts, which we then ate for free.

But enough about donuts. This book is about energy. How do neoclassical economists model how to "produce" something, and in particular how do they include the concept of energy?

Let's explain neoclassical economic production by using a standard introductory textbook, *Macroeconomics* by Paul Samuelson (Nobel Prize in Economics, 1970) and William Nordhaus (Nobel Prize in Economics, 2018) [56].[22] The book mentions economic production has four types of inputs: human resources (labor supply, education, discipline, motivation), capital formation (machines, factories, roads), natural resources (land, minerals, fuels, environmental quality), and technology (science, engineering, management, entrepreneurship). These factors combine into an *aggregate production function* relating economic net output, or GDP, to the amounts of these input *factors of growth*.

We can imagine people (the labor, educated to some extent) working in a factory or office building with machines (capital) that require fuel and raw material (natural resources) inputs with which to operate machines and make new products. The explanation of natural resources notes important resources are arable land and soil, oil and gas, forests, water, and mineral resources. This all sounds reasonable, but it goes downhill from here.

---

[22]Samuelson and Nordhaus [56, p. 217–231].

The *Macroeconomics* authors start simplifying. They appropriately refer to Robert Solow as an "apostle" and the father who birthed the mainstream neoclassical growth model from his head in 1956 [57]. For his work, Solow received the Nobel Prize for Economics in 1987, and to his credit he recognized, even at the time of originating the model, that his framework neglected to explain a large portion of economic growth, often attributed to "technology," as will be described shortly [58].

However, the first step Samuelson's and Nordhaus's text takes in describing Solow's model (also known as the Solow–Swan model) is to remove natural resources from the equation. This will turn out to be a big problem. It is important to note that at the time of his original work in the 1950s, Solow included the amount of natural resources (via its economic value) in the concept of capital. However, this is still not equivalent to considering energy consumption as fuel for machines. In agrarian times no one debated that land was needed to produce food and fodder. In the fossil-fueled industrial era, land was no longer a limiting input for growth. But total energy and physical natural resource flows have always been relevant for economic activity.

For now, consider the equation for growth of GDP has only three inputs: capital, labor, and "technology."[23] This production function concept now treats "technology" as another potential field, just like the one governing consumer utility. This is not obvious. Just like you define your supposed utility potential field via the combination of products you prefer, a technology potential field is defined by the quantity of input capital and labor needed to produce economic output. Recall that the mathematical concept of the potential field "...is useful only in cases where one can safely abstract away all considerations of process and the passage of time."[24] Just as there are many combinations of products you can buy to achieve your maximum utility, there are many combinations of capital and labor that can achieve a given level of economic output.

One important consequence of production as a potential field is that it cannot deal with the real concept of *intermediate inputs*, or those "...outputs that, directly or indirectly, become inputs of the same production process."[25] This is a problem when applied to energy and natural resources, because it means the theory can't conceptualize and use important feedback concepts such as energy returned on

---

[23]The common form of the neoclassical production function is called the Cobb–Douglas production function. In the case of Solow's version, there are only two core inputs, and GDP is expressed as $Y = A(t)K^\alpha L^{(1-\alpha)}$ where Y = GDP, K is the value of all capital (perhaps adjusted for quality of different types of capital), L is the hours worked by all workers (perhaps adjusted for different labor quality), α is the output elasticity of capital (and less than 1), $(1 - \alpha)$ is the output elasticity of labor, and $A(t)$ is the "technological progress" function of time t, estimating what is known as *total factor productivity* to minimize the difference between the estimate from this equation and the GDP data [58]. The *Solow residual* is the difference between the data on economic growth and the estimate from this equation without $A(t)$, within α equal to the GDP cost share of capital, and $(1 - \alpha)$ equal to the GDP cost share of labor.

[24]Mirowski [46, p. 347].

[25]Mirowski [46, p. 319].

energy invested, or how much energy it takes to extract and convert energy to fuels (refer back to Chap. 5). We know, from physical principles, that we have to consume some energy to build and maintain machines that in turn extract energy from the environment. By only modeling production using a field concept, with no other specifications, you can't figure out if energy extraction might become too expensive or physically limiting itself to enable economic growth, such as during the recessions of the late 1970s and 2008 financial crisis, when energy spending crossed a growth threshold near 8% of GDP (Fig. 2.13). Recall that major recessions have corresponded to times when a high percentage of GDP was spent on energy [5, 33].

After removing physical resources, the second step is to assume capital no longer need be described as distinct physical items that need fuel (energy) inputs to operate, but to assume capital is now the monetary value of the physical items. This problem, as well as that of defining technology as a field, was at the heart of what was known as the "Cambridge Capital (Theory) Controversies" (CCC) in the 1950s and 60s: a battle between economists in Cambridge, MA, United States versus those in Cambridge, United Kingdom. A group of U.S. economists, including Robert Solow and Paul Samuelson, at Massachusetts Institute of Technology (MIT) advocated for the neoclassical production theory, while another group of economists, including Joan Robinson and Piero Sraffa at the University of Cambridge, U.K. argued against it. The British argued that the process of economic production is grounded in physical processes that require various types of physical capital. Because different types of capital have fundamentally different properties (e.g., a building is different than a truck), you cannot combine them into one *aggregate* quantity of capital. Mathematically you cannot add items of different units, and this is essentially what one does when aggregating capital.

Most economists know that adding all types of machines together by their economic valuation is a simplification. One mainstream macroeconomics textbook states "...it should be clear that it is still a drastic simplification of reality. Surely, machines and office buildings play very different roles in production and should be treated as different inputs."[26] But is it a useful simplification? To some degree it can be, if you give up on the strict assumptions of neoclassical growth theory. As stated by Nicholas Georgescu-Roegen:

> As a highly abstract simile, the standard form of the Neoclassical production function—as a function of ... homogeneous "capital," and ... homogeneous "labor"—is not completely useless. But ... the value of the standard form of the production function as a blueprint of reality is nil. It is absurd to hold on to it in practical applications—as is the case with the numberless attempts at deriving it from cross-section statistical data. ... True, capital and labor may be rendered homogenous but only if they are measured in money.[27]—Nicholas Georgescu-Roegen

Pay attention to Georgescu-Roegen's statements that the concept of total aggregate (or homogeneous) capital is not "completely useless" if the quantity of capital is

---

[26]Blanchard [6, p. 216].

[27]Georgescu-Roegen [18, p. 244].

"measured in money." The CCC was largely about whether it was useful to aggregate capital via its monetary values. However, even some steadfast adherents to modeling the physical nature of economic production accept the idea of aggregating capital, labor, and energy for the purposes of modeling economic output. For example, Reiner Kümmel and Dietmar Lindenberger state that just as you could use the input factors of capital, labor, and energy to uniquely describe physical work and information processing in their own right, you can also use those input factors to uniquely describe economic output in units of money [40].[28] However, there are physical constraints on how energy and capital relate to each other that must be considered at some level (to be discussed later).

The MIT contingent conceded on the philosophical conundrum of aggregating capital with physical qualities, but argued it was still acceptable to combine all forms of capital by adding their monetary values into an aggregate capital value that, in the end, still does not determine prices. Remember, neoclassical theory states that prices are determined by consumer preferences, not production costs. The U.K. Cambridge criticisms derive from neoclassicals forcing production into the mathematical framework of the potential energy field, although Mirowski states they never quite grasped this fundamental linkage to a principle of physics.[29] The field framework is simply not suited for the concept of production whose purpose is to describe how inputs are combined to create some output that is different in practically every way from the simple sum of the inputs.

Ultimately what comes out of the CCC is that the neoclassical paradigm won the war of practicality by instilling their economic growth model into the bulk of economists minds today.

Step three for the Solow model is to calculate aggregate labor in a similar manner as done for capital. Labor is now all hours worked by all types of people. Neoclassical economists recognize the fact that all types of capital and workers are not equal. When calculating the input factors of capital and labor they "adjust" for differences in quality. For example, a surgeon provides higher quality labor than say a construction laborer. However, this quality adjustment is performed on the basis of hourly pay, which neoclassical economists assume must be the correct pay based upon the value of that worker's *marginal* contribution to the economy as expressed by his equilibrium price for labor. In the phrase worker's marginal contribution, economists assume that each person gets paid based on the value they contribute to the economy.[30] However, neoclassicals still must translate this quality difference into the same units, such as hours worked. For example, a surgeon is tallied as

---

[28]"Since work performance and information processing are subject to the causal laws of nature, their result, the economic output, should depend as uniquely on the work-performing and information-processing production factors capital, labor, and energy as any state function of physical systems depends on its physical variables." [40].

[29]Mirowski [46, p. 341–343].

[30]The economics terminology for a person being paid based on the value they contribute to the economy is "quality of labor adjustment represented by workers being remunerated according to their marginal productivity."

working more quality adjusted hours than a laborer even if they each work 8 h per day.[31]

Note what is inherently assumed by inserting quantities for aggregate capital (a sum of the quantities of all types of machines) and aggregate labor (a sum of all labor hours across all types of human work) into an equation. One assumption is that one type of capital can indeed perform the function of another type. This is like saying a refrigerator can make a solar panel or drill an oil well. Another assumption is that one type of worker can perform the function of another worker. This is like saying a construction worker can perform brain surgery, *successfully*, just by working longer. I suppose this is true, but only if he spent years acquiring the knowledge to become a surgeon. Only in the movie *The Matrix* can Neo plug into a computer network to acquire a lifetime of knowledge in a few seconds. Perhaps advances in artificial intelligence and understanding of our brain will enable *The Matrix* to become reality so that construction workers can take the red pill and perform surgery a few seconds later. For now, we can only speculate, and in Chap. 8 I will opine on how artificial intelligence and evolution might be consistent with neoclassical or other views of economic growth that more directly include energy within economic growth.

---

[31]"Changes in labour quality reflect movements in the distribution of hours worked among categories of workers, and differentials in the hourly pay of categories of workers. For example, if hours worked by a highly skilled and highly remunerated type of labour (such as brain surgeons) increased, then the volume of labour input as measured by QALI [quality adjusted labor input] would increase by more than the observed increase in hours. Conversely, a decrease in hours worked by unskilled workers in elementary occupations who receive lower than average remuneration would result in a fall in QALI by less than the proportional decrease in hours worked." From "Quality adjusted labour input: UK estimates to 2014," Release date: 22 May 2015, https://www.ons.gov.uk/economy/economicoutputandproductivity/productivitymeasures/articles/qualityadjustedlabourinput/estimatesto2014;

"We calculate QALI [quality adjusted labor input] by categorising workers by identifiable characteristics (based on age, sex, industry of employment and level of education), and weighting changes in the hours worked of each worker type by their share of total labour income. The rationale for this approach is that, under competitive markets, economic theory suggests that different factors of production (different categories of workers, and different types of capital assets) will be remunerated according to their marginal productivity. Consequently, relative shares of labour income provide a proxy for the relative productivity or "quality" of different types of workers.

Using a suitable weighting system, it is possible to subtract movements in hours (sometimes referred to as "unadjusted hours") from movements in QALI indices, and hence to identify the pure "quality" or compositional movement in labour input to production.

From the perspective of measuring productivity, it is the movement in QALI rather than the movement in hours worked that offers a better representation of what is happening to labour input. For example, growth in labour quality of 1% with hours unchanged is equivalent (abstracting from distributional effects) to growth in hours worked of 1%, with labour quality unchanged." From "Quality adjusted labour input: UK estimates to 2016," Release date: 6 October 2017, https://www.ons.gov.uk/economy/economicoutputandproductivity/productivitymeasures/articles/qualityadjustedlabourinput/ukestimatesto2016.

Labor substitution could go the other direction, however. There might be an intellectually skilled person that chooses to perform low-skilled physical labor in construction. The movie *Office Space* portrayed exactly that as a computer programmer got fed up with his pointless job, which as far as he could tell, only existed to help his boss's stock go up a "quarter of a point." He did not feel that either he or his boss was getting paid based on what he felt was any real contribution to the economy. If you don't want to rely on movie fiction to demonstrate the fallacy that all people get paid based on the value they contribute to the economy, then consider the first two pages of David Graeber's book *Bullshit Jobs* [21]. There, Graeber recounts a story from a German worker employed by a subcontractor of a subcontractor of a subcontractor for the German military. A tremendous process ensues to move a computer from one office to another two doors down: "So instead of the soldier carrying his computer for five meters, two people drive for a combined 6–10 h, fill out around fifteen pages of paperwork, and waste a good four hundred euros of taxpayers' money."[32] Another common analogy to juxtapose the value of a job with its pay is a garbage collector; just try to imagine New York City if garbage collectors stop working for a week (as actually happened in 1968) as opposed to a 1 month strike of the city's public relations professionals.

We can now take the fourth and final step for understanding the limitations of the Solow growth model. Now that neoclassicals have removed resources from the growth equation, added up all of the physical machines and buildings of the world into a number for capital, and added up all of the different forms of work into a number for labor, they calculate the remaining input factor that describes GDP: *technology*. Except, we don't know what "technology" is. No problem. There are four parts of the economic growth equation: GDP on one side, and capital, labor, and technology on the other. We gather data to estimate GDP. Thus, the actual fourth step is to estimate technology growth as equal to the growth in GDP minus the growth in capital minus the growth in labor. Instead of leaving GDP on one side of the equation by itself, we can reshuffle the equation so that technology is on one side by itself. Economists call this measure of technology *total factor productivity*, or TFP.[33]

The important takeaway is that TFP is not itself defined by any first principles. It is by definition the *unexplained* part of economic growth when subtracting components assumed within the Solow model. TFP was originally called the "Solow residual," where *residual* is the mathematical term for the portion of the output of an equation that is not explained by the inputs of the equation. Thus, the Solow residual was the part of economic growth that remained unexplained after accounting for capital and labor. For this reason the Solow model is termed an

---

[32]"Bullshit Jobs," LiquidLegends, https://www.liquidlegends.net/forum/general/460469-bullshit-jobs?page=3, written June 28, 2014, accessed January 21, 2019. Referenced in [21].

[33]Technological growth, or growth in total factor productivity (TFP), is mathematically the growth in GDP minus the *weighted* sum of the growth of labor and capital. Labor is usually weighted by about 65–75%, and capital is usually weighted the other 25–35%.

exogenous growth model, meaning that "technology" growth comes from outside the model. If you don't use the Solow model, you don't have TFP. In this way, TFP and the neoclassical exogenous growth model come together.

Mirowski sums up the unwelcome conclusion for using the neoclassical theory of production:

> Neoclassical economics shifted the onus of invariance [What is constant, my own preferences or the properties of the physical world?] onto individuals and their preferences, but in doing so neglected to elaborate the mechanism whereby the physical world retained its identity for the economic actor. Hence the possibility exists that the economic identity of goods may clash with their physical identity, with dire consequences for the theory of value.[34]

Yes, this means that neoclassical theory assumes people can agree to prices, and thus value goods, without any understanding of the required natural resources inputs or engineering processes that convert resources into those goods. Sure, consumers generally do not know the physics or engineering of how to make things. They don't have to know and generally do not care to know. But someone needs to know! Producers do need to know how to make things, at least at some basic level, and the producers that learn more about how to make things usually make the choice to make them with fewer inputs and/or increased functionality.

In short, neoclassical economic theory confuses people as to the role played by energy, other natural resources, and engineering constraints in producing goods. It does this by positing that "production" ultimately requires no explicit description of the physical world or relation to physical constraints. Therefore, if pricing influences how many goods one purchases and consumes, and human well-being is at least partially based on what we consume, then well-being also has no relation to the physical world.

Note what has now happened. Natural resources were originally stated as necessary inputs to produce goods. Then they were removed when it was time for economic calculations, and replaced with TFP. Economists replaced things we can count (land area, joules of energy, kilograms of materials) with a mathematical remainder called total factor productivity. Is this a big deal? Absolutely. For the U.S., estimates of TFP growth averaged near 1.3–1.6% per year in the twentieth century.[35] U.S. GDP growth averaged 3.2% per year from 1948–2017.[36] Thus, half of the growth in GDP is unexplained by a model that is supposed to explain economic growth! Economists do recognize that this TFP as technology is not a satisfactory concept, and Solow noted this when he initially derived the model.

Since the 1980s economic growth research has explored how *endogenous* technological progress can be characterized by capital and labor changes within

---

[34]Mirowski [46, p. 322].

[35]Gordon [19, Figure 16–5].

[36]Data from U.S. Federal Reserve of St. Louis, Bureau of Economic Analysis data code A191RL1Q225SBEA, Real Gross Domestic Product, Percent Change from Preceding Period, Annual, Seasonally Adjusted Annual Rate.

the model [53]. However, the vast majority of this research focuses on developing *human capital* (education, know-how, and research and development capacity) within a country that lacks enough skilled workers. Thus, endogenous growth modeling still largely ignores the role of energy and natural resources in growth. Some energy-minded researchers have included the concept of endogenous growth by assuming "technological change" specifically refers to increases in the efficiency at which primary energy is converted into useful work [3, 10]. More of this research is a move in the right direction, but forcing it into frameworks that don't explicitly define resource stocks and flows might be a fool's errand. By construction, any economic framework that separates "technology improvement" from the use or definition of natural resources cannot describe how technology or the economy relates to interactions with the environment (e.g., to extract energy).

We now turn to explaining the implications from the lack of consideration of the principle of energy when modeling economic growth. This narrowed scope still requires explanation of several key points.[37] These points drove some researchers to more directly consider the role of energy in economic growth, and in doing so they created very important insights to more directly relate GDP to the use of energy.

The Energy, Stupid!

"The Economy, stupid," was a successful catchphrase used by James Carville, campaign strategist for Bill Clinton's 1992 U.S. presidential run. If you want to get elected in the U.S., talk about the economy. If you want to understand economic growth, you have to talk about energy. At least one book on the 2008 financial crisis makes this link to resources, using a variation of that quote as its subtitle: "It's the energy, stupid!" [49] Some researchers have taken this to heart.

Reiner Kümmel and Robert Ayres (along with Benjamin Warr) took similar minded approaches that have spawned a breed of energy-economic modelers to use their concepts for more accurate energy-economic modeling [3, 37]. As Ayres wrote with *Debunking Economics* author and economist Steve Keen, "[labor] without energy is a corpse, while capital without energy is a sculpture."[30, 32]

Saying it like that makes it simple. If we don't consume food, we die. If a machine doesn't use energy, it can't move. If an economic model does not include these concepts, then it should, because otherwise it has almost nothing to say about the role of energy in the economy. At the end of this chapter, I discuss insights from

---

[37]One can read the following books for additional explanations of the fallacies of the neoclassical production function. Philip Mirowski's Chapter 6 of *More Heat than Light* details the theoretical problems [46]. Charles Hall and Kent Klitgaard point out some basic concepts in their 2018 *Energy and the Wealth of Nations* (Chapters 3 and 5) [22]. Blair Fix provides a nice summary of critiques of neoclassical production in his *Rethinking Economic Growth Theory from a Biophysical Perspective* [14].

my research that show taking these concepts to heart when modeling provides some important insights into important economic trends.

How could Kümmel and Ayres see past the fallacy of growth without energy? Perhaps because both Kümmel and Ayres are physicists. You can't get a physics degree if you ignore the necessary role of energy transformation to compute anything, move matter, or shape matter. Economic activity also involves these processes. Their education did not depend on accepting resource-free neoclassical theory as a description of the economy. Thus, when they thought about the economy, it was natural to include energy as a necessary input.

Neoclassical theory imposes some mathematical restrictions on production functions. They are not obvious to someone who simply reads a report discussing results from an economic model. However, it is crucial to understand these assumptions to then understand why neoclassical economists cannot interpret economic trends as being driven by energy. To explain this point we must discuss some math. But do not fear, dear reader, we will do this using words.

The Solow model is a variation of the more general Cobb–Douglas function (or equation form), which is in turn a variation on the even more general Constant Elasticity of Substitution (CES) function. We'll revisit CES functions later when we discuss modeling long-term changes to the energy system, such as transitioning to low-carbon energy. For now just consider that the Cobb–Douglas function lets you add as many input factors as you want into the growth equation.[38] Do you want to add energy consumption as an input? No problem. Neoclassical economists added total energy consumption into the Cobb–Douglas function. The result? Not much better than without energy. A lot of economic growth was still unexplained.

Here is where Reiner Kümmel comes in. At this point, he decided to model economic growth differently, by including constraints on how the three input factors, capital, labor, and energy, could relate to each other. He also used an even more general form of an economic growth equation than the Cobb–Douglas format—one that is still follows mathematical properties that neoclassical economists assume must hold for aggregate production functions [36, 37].[39] In doing so he could describe GDP such that the unexplained economic growth, or residual, was less than a few percent of the total—much less than the 50% attributable to TFP. But he also included changes that neoclassical theory doesn't. He allowed the output *elasticities*

---

[38]The neoclassical Cobb–Douglas production function is of the generic form $Y = A(t)X_1^{\alpha_1} X_2^{\alpha_2} \ldots X_n^{\alpha_n}$ where $Y =$ GDP, each X_i is some input factor required for production, and $A(t)$ is the part of GDP not described by the input factors. Two other notable requirements to understand this Cobb–Douglas formulation are (1) that the sum of all of the exponents α_i must equal one and (2) the fraction of GDP paid to each of the input factors is equal to its respective exponent.

[39]This form he called the LinEx function, for "linear-exponential" function, in which economic output is a linear function of energy (or useful work, if desired) but an exponential function of labor, capital, and energy. The LinEx function contains technology parameters that may change in time (when creativity is active). They are to be determined econometrically by minimizing the sum of squared errors ("fitting") between the assumed equation and the data (e.g., GDP), subject to the constraints that each output elasticity is non-negative. See p. 199–206 of [37].

to change each year, based upon how capital, labor, and energy consumption change, rather than remain constant as assumed in the Cobb–Douglas function and Solow model. In economics, these elasticities relate how much economic output changes in relation to a change in one input factor.[40] Further, since he wasn't using the all neoclassical growth assumptions, his elasticities of the input factors are more flexible than those used in neoclassical theory, and via his formulation we can interpret the importance of each input in a different light. However, Kümmel's approach is generally ignored by economists. But just what are these elasticities supposed to be, and how do we make sense of them?

Each input factor in a Cobb-Douglas growth function is raised to a power, or exponent. To an economist, this exponent is the *output elasticity*, or the output elasticity with respect to the input. The sum of all elasticities must add up to one. This ensures that if you double *all* of the inputs into the economy, you get double the outputs, and this makes intuitive sense. Twice as many of the same exact machines with twice as much of each input can make twice as many of the same exact widgets. Since all elasticities sum to one, each elasticity is less than one. This means that there are "diminishing returns" from each input factor. For example, if you increase *only one* input factor by 10% and GDP grows by 5%, then if you increase that input factor an additional 10%, GDP might only grow by 4% more (some amount less than 5%).

For neoclassicals, the elasticity for each input factor is equal to the fraction of all input costs associated with that factor. These fractions of input costs to produce GDP are the so-called *cost shares*. The "cost share" theorem says that for any given input factor, its output elasticity is equal to the fraction of all input costs spent on that input factor. This equality "...is a consequence of the [neoclassical] assumption that the economy operates in an equilibrium determined by the maximization of either profit [or all of consumers' utility added up over infinite time]—without any constraints on input factor combinations." [38].

The neoclassical model, and the cost share assumption that inputs are paid the fraction of their contribution to GDP, breaks down for two main reasons. First, if there is no equilibrium, then prices, and thus cost shares, have not settled to their theoretically optimum level. If an input contributes a higher or lesser fraction to GDP than it is paid for that contribution, then this violates the cost share theorem. Data clearly show the economy is not in equilibrium as specified by neoclassical economics—the energy and economic data trends do not have constant rates of change. As we saw in Fig. 4.12, and will revisit in Fig. 6.6, the fraction of GDP paid to workers (or wages) has not remained constant over the last 50 years. Second, by constraining the elasticities to be equal to the input cost shares, the theorem constrains the inputs in the wrong way. We need to constrain how input factors of capital, labor, and energy relate to GDP, but in other ways. For instance, production

---

[40]The elasticity of output, or GDP, with respect to an input, X is defined as the change in GDP divided by the change in X multiplied by the current value of X divided by the current GDP, or elasticity $= \frac{\partial GDP}{\partial X} \frac{X}{GDP}$.

functions need some representation of physical constraints, such as the fact that capital must operate both with energy as an input and never above 100% of its full capacity [32, 34, 40]. These types of physical constraints are missing in the Cobb–Douglas production function, thus the Solow growth model, and thus neoclassical growth theory. They must be added to represent physical laws and constraints, and when they are added in some way, you can no longer use the cost share theorem. *In short, the cost share assumption (and its assumption of equilibrium) of neoclassical theory is its fatal Achilles heel.*

Consider an example where energy, capital, and labor are the three input factors that create GDP. The post-World War II U.S. typically paid laborers 60% of GDP (Fig. 4.12) and household consumers spent about 7% (typically 5–10%) of GDP on energy (Fig. 2.9b). Thus, we must allocate the remaining 33% of GDP as profits to owners of capital such as stocks in companies, rental apartments, and other businesses. Per neoclassical theory applied to the Cobb–Douglas function, the growth in GDP *is equal* to 0.07 times the growth in energy consumption *plus* 0.60 times the growth in labor hours *plus* 0.33 times the growth in capital *plus* any residual factor.

The implications of this formulation are stark for understanding how neoclassical theory interprets the effect of energy on economic growth. For example, with an energy cost share of 7% "...a 99% fall in energy input would cause only a 28% fall in output [GDP]" if both capital and labor stay the same [32]. Over 70% of the economy can remain in operation with 99% of energy consumption gone? This result is absurd, as demonstrated by the following logical sequence. First, capital, or machines and buildings, consumes the vast majority of energy in a modern economy. Second, if left with only 1% of energy consumption, then all existing capital could only operate a very small fraction of its capability because it requires energy to operate. Third, therefore the economic value of all capital would plummet. Finally, as a consequence, GDP would further decline due to both reduced energy and active capital. Practically everyone, including neoclassical economists, recognizes this dynamic, but not everyone uses mathematics that is consistent with it. This is the crux of the problem.

We can gain insight into economic growth when using the production function concept with energy as an input factor, even assuming the Cobb–Douglas formulation. However, this must be done appropriately, i.e., by determining output elasticities without unnecessarily constraining them via the cost share theorem. When solving for the elasticities that provide a best match to the GDP data, *Kümmel and Lindenberger showed that economic growth is much more dependent on energy than normally believed by mainstream economists.* Their analysis shows that energy is an order of magnitude more influential than assumed when invoking the cost share theorem that implies a 1% reduction in energy consumption translates to only a 0.05–0.1% reduction in GDP [37, 40]. For example, from 1960–2013, a 1% change in energy translated (on average) to a 0.4% change in Germany's GDP and a 0.3% change in U.S.'s GDP [41].

Robert Ayres understood the value of Kümmel's concept. Like Kümmel, Ayres considers that energy must be an input into GDP, and it must be considered without the constraint of the cost share principle. But he adds a wrinkle. He takes into consideration the most fundamental energy-relevant machine characteristic that influences production: the *efficiency* at which a machine converts its fuel into useful work. In other words, he directly considers the second law of thermodynamics. Recall from Chap. 2 that useful work is the output from machines that consume fuels—the mechanical drive from a car engine or electric motor, the heat driving a chemical process, and the electricity powering computers and light bulbs. Useful work is equal to the energy delivered to a device times the efficiency at which that device converts that energy to its final form that performs some service.[41]

It makes perfect sense to consider *energy times efficiency*. To get more work out of physical processes we can consume more input energy *and* increase efficiency. Thus, Ayres and his former student Benjamin Warr set out estimating the efficiency of various processes. They multiplied energy that ends up in our cars, planes, buildings, and power plants by the respective efficiency of each to derive an aggregate total useful work for the entire economy. In this case of multiple processes that output useful work, it is entirely appropriate to aggregate them because they all have the same units of energy. When they divided all useful work by total primary energy for the U.S., they got an overall efficiency for the economy of about 4% in 1900 and 12% in 2000 (see Fig. 6.1). Most of the increase occurs from the 1930s to the 1970s. Shortly we will use this tremendous rise in efficiency to explain the neoclassical change in "technology" as total factor productivity (TFP).

In an attempt to communicate to neoclassical economists in language they understand, in one instance Ayres used the Cobb–Douglas production function with inputs of capital, labor, and useful work instead of capital, labor, and total energy [1]. When he did this, he effectively explained almost all of U.S. GDP without the need for a large residual factor like TFP. But the neoclassicals still can't accept Ayres' finding. Why? Ayres does keep each of his Cobb–Douglas powers constant, as a common simplification by neoclassicals, but he solves for the best power, or elasticity, to which each input is raised, rather than assuming the cost shares for each input factor per neoclassical theory. After all, there are no data to know the share of GDP that goes to purchase useful work. We don't pay for car motion and mechanical drive, we pay for cars and fuel.

At this point I've introduced researchers and methods that better explain historical GDP by incorporating the concept of energy and useful work, even through use of an aggregate production function that neoclassical economists use (but without one critical assumption). For some, this is enough to move forward with the useful work agenda. Others find that forcing useful work into aggregate production functions is like forcing a round peg into a square hole [23]. I conclude that if you are compelled to describe GDP with a single aggregate production function, then including the concept of energy or useful work, without the cost share assumption, is *much better* than neglecting energy flow altogether. We'll now discuss the length to which neoclassical economists go to explain the technological

---

[41]Technically useful work is the exergy, not energy, delivered to a device times its efficiency. For this reason, some use the term "useful exergy" instead of useful work. Exergy is a measure of energy that accounts for the second law of thermodynamics. The exergy per kg of fossil fuels is only slightly lower than total energy per kg.

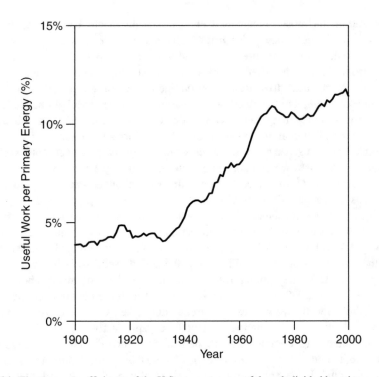

Fig. 6.1 The aggregate efficiency of the U.S. economy as useful work divided by primary energy consumption. Data from [62]

change without reference to energy. This is perhaps the most important reason why historical economic arguments produced more heat than light in trying to explain economic growth or technological change.

The Problem with Productivity: All Play and No Work

> Culture itself has become a commodity, and a combined force of economics and ideology now drives its dissemination, making retreat from the intercultural contact zones impossible and battles for control of the cultural environment a common occurrence.[42]—Bruce Wexler (2008)

In using the Solow neoclassical economic growth model that projects GDP using only capital, labor, and total factor productivity (TFP), adherents must spend a lot of time trying to explain TFP. Perhaps the most obvious trends to explain were the

---

[42] Wexler [64, p.231].

tremendous economic growth rates during the three decades after World War II. Here is quote from a 2006 macroeconomics textbook by Olivier Blanchard [6]:

> The first two columns of Table 10-1 show growth rates of output per capita for both pre- and post-1973. Note that the growth rate fell in all four countries [France, Japan, United Kingdom, and United States]. Pinpointing the exact date of the decrease in growth is difficult. The date used to split the sample in the table was 1973, and this is as good as any date in the mid-1970s. ...
>
> At a growth rate of 4.1% per year—the average growth rate across our four countries from 1950 to 1973—it takes only 16 years for the standard of living to double. At a growth rate of 2.0% per year—the average from 1973–2000—it takes 35 years, more than twice as long.[43]—Olivier Blanchard (2006)

Blanchard is no slouch. He is a former Chief Economist at the International Monetary Fund. His book goes on to state three "facts about growth in rich countries since 1950." I list items one and three that are relevant for energy.[44]

1 Growth is not a historical necessity. There was little growth for most of human history, and in many countries today, growth remains elusive. Theories that explain growth in the OECD today must also be able to explain the absence of growth in the past, and its absence in much of Africa today.

3 Finally, in a longer historical perspective, it is not so much the lower growth since 1973 in the OECD that is unusual. More unusual is the earlier period [1950s and 1960s] of exceptionally fast growth. Finding the explanation for lower growth today may come from understanding what factors contributed to fast growth after World War II, and whether those factors have disappeared.

One only needs to look at the global data in Chap. 2, in particular Fig. 2.10, for which U.S. data show a similar trend, and note the same glaring anomaly in the growth rate of *energy consumption* in the 1950 and 1960s. Recall that 1955–1979 is the only time in history that the 10-year running average growth rate in energy consumption was greater than 4%/yr. Not only should every college economics student be exposed to these data of energy consumption and cost trends in Chap. 2, perhaps every high school student should learn them as part of a basic education. In explaining the high GDP growth rates of this period, the Blanchard text states (referring to neoclassical theory) "Our theory implies this fast growth may come from two sources ...," technical progress (i.e., TFP) and the growth of capital (per worker). So those are the *only* two sources to which neoclassicals look: better technology and more machines. In applying neoclassical theory to explain the GDP growth rates from the 1950s through 2000, the text states:[45]

---

[43]Blanchard [6, p. 209].

[44]The text comes from the 2006, 4th Edition of Blanchard's *Macroeconomics* [6, p. 213]. The latest 2017 seventh edition has a similar, but different, discussion comparing rates of growth in select rich countries during the 1950–2011 period (Table 10-1) and noting rates of technological progress, or total factor productivity, and growth in output per worker since 1985 (Table 12-2).

[45]Blanchard [6, p.258–259].

1 The period of high growth of output per worker until the mid-1970s was due to rapid technological progress—not to unusually high capital accumulation.
2 The slowdown in growth of output per worker starting in the mid-1970s has come from a decrease in the rate of technological progress, not from unusually low capital accumulation.

Again, no discussion of energy extraction. Blanchard states that faster growth before the 1970s is due to a high growth rate of TFP, and slower growth after the 1970s is due to a low growth rate of TFP. In the earlier quote he states theories of growth "must . . . be able to explain" the presence of growth in industrial rich countries and the absence of growth in the pre-industrial past.

A feasible answer for such a theory is apparent to someone who considers the physical and energetic basis of the economy. *The use of capital, which requires energy as an input, to more efficiently extract and convert natural resources into goods is one major governing factor for growth.* If you can both extract energy faster and use it more efficiently, as occurred in the U.S. from the 1930s to early 1970s (Fig. 6.1), you can literally and physically power more economic activity.

Think about it like this. Imagine you have the most sophisticated drilling rig in history, but if you drill in a spot with no oil and gas, your drilling rig doesn't extract anything. That is a total loss of money and waste of energy. This is why oil and gas companies spend so much time and money understanding Earth's geologic history. It's important to know *where* to drill. We can certainly describe drilling rigs and seismic imaging methods today as higher quality capital than those 40 years ago. This capital is also designed by people who must have time to research and develop knowledge for new designs.

A physical drilling rig by itself does not promote growth. Growth is enabled by providing the fuel to the economy when the drilling rig performs the useful work of poking a hole into Earth to release oil and gas from where it resides. The same concept holds for wind turbines and solar panels. The first ones we develop are not that great, and we generally first place them in windy and sunny locations because we can convert more sunlight and wind into more useful work, not less. In short, you need the human-derived technology *and* the naturally occurring energy resource.

Newer economic growth models consider that investment in research and development increases our ability to design better machines that extract harder-to-get resources. Paul Romer recently won the Nobel Prize for Economics for this idea of "integrating technological innovations into long-run macroeconomic analysis" via the endogenous growth concept mentioned earlier.[46] Usually this technological innovation is attributed to increases in human knowledge, something difficult to interpret outside of the context of any reason why we need or want increased knowledge. Again, the missing component is energy and natural resources

---

[46]Paul M. Romer Nobel Prize Lecture. https://www.nobelprize.org/prizes/economic-sciences/2018/romer/lecture/.

themselves. As pointed out in the last chapter, ecosystems, animals, and economies are characterized by a similar mathematical scaling law linking their energy consumption and size. If we attribute this same pattern in the economy as due to "technology," then do we think ants are also developing new technology as they grow their colony?

Learning economics without incorporating the principle that energy is a required input for all activities is akin to depriving animals of critical stimuli during brain development. Recall from the last chapter that animal brains can lose the capability to sense certain stimuli if never exposed to the stimuli. Thus, to make useful growth models we must stimulate our brains with both economic and physical concepts.

> Economists might find it hard to learn how to integrate energy into their thinking just as energy scientists and engineers might find it hard to learn to integrate economics into their analyses. But this cross-learning does happen, and it needs to happen more often.

As pointed out by Wexler's quote at the beginning of this subsection, *our culture is partly defined by our economics, and the battle for control of culture is continuous.* If this is true, then to change our culture more people need to learn and practice improved economic principles. To close the loop on brain stimulation and learning economics, a review of existing studies concluded that students taught neoclassical economics become less moral than their peers. Apparently, the focus on self-interest and consumer goods "... renders those influenced by its teachings less moral and more antisocial."[11].

While the previously quoted macroeconomics texts are a couple of decades old, mainstream economic discussion of technology still focuses on TFP and the Solow model that lacks explicit resource input. Robert Gordon's 2016 tome *The Rise and Fall of American Growth: The U.S. Standard of Living since the Civil War* is a popular recent book with significant focus on TFP trends [19]. Gordon uses his own extensive research regarding what he feels is large undercounting of GDP, and, importantly, actual personal welfare, when significant new products first come into the market. Ford's Model T is an example of a new product he describes as undercounted within GDP. Also, receiving the same income while moving from a 60 h to a 40 h work week is a large gain in welfare—same pay, fewer working hours.

These welfare and accounting concerns are very relevant, but here we will focus on Gordon's discussion of trends in TFP. Gordon refers to total factor productivity as "the best available measure of innovation and technical change."[47] However, as I've hammered home, there is a problem with the typical interpretations of TFP by Gordon and neoclassical economists. They cannot distinguish the quality of an energy extraction technology from the quality of the resource it extracts.

Consider drilling for oil. The following two scenarios would not be distinguishable. Assume the use of a vertical drilling rig, with no hydraulic fracturing

---

[47]Gordon [19, p. 546].

capability, as the same "technology." Scenario 1 is drilling in the prolific east Texas oil fields in the 1930s (e.g., Spindletop), and Scenario 2 is drilling into the tight sand and shale formations now pursued in the Bakken formation of North Dakota and the Permian Basin of West Texas. By no means would you extract the same amount of oil from drilling one well into each rock formation in its original condition. In Scenario 1, you would produce enough oil to become a millionaire and trigger the age of oil, and in Scenario 2 you would go broke. The difference between the scenarios is the resource size and quality because the human-made technology is exactly the same in both. Of course, the combined new technologies of horizontal drilling and hydraulic fracturing are partially responsible for companies' ability to feasibly extract oil from tight sands. I say partially, because the coupling of consumers and producers within networks, as discussed in Chap. 5, induced changes in technologies, namely more fuel efficient cars, that make the more costly oil extraction affordable to consumers.

To show TFP is agnostic as to the energy narratives, the same problem holds in assessing renewable energy technologies. Technology capability would seem the same whether you installed a 10% efficient solar panel in sunny Phoenix, Arizona versus a 20% efficient panel placed in Seattle that has half the annual sunshine. The reason is that the same amount of (annual) electricity would be generated in both situations even though Seattle has a more capable technology in a poorer solar resource. Clearly a 20% efficient solar panel must have a different human-based design than a 10% efficient solar panel. The more efficient technology is needed with respect to, not in spite of, the quality of the solar resource. We should interpret this in no other way.

In short, TFP cannot distinguish between the quality of an energy resource itself and the technology that extracts the resource. This distinction is obvious: technological widgets are invented by humans, but natural resources aren't. We shouldn't include natural resources in any definition of human-derived technology, but that is what TFP inherently does.

Figure 6.2 shows Gordon's calculation of TFP as decadal averages, the same method as displayed in his book.[48] He describes 1920–1970 as the period in U.S. history with the highest growth in TFP. In particular he calls the 1920s through 1950 the "Great Leap." The decades after the 1970s show a marked decline in TFP. Together, these two time periods represent the rise and fall, respectively, governing the title of his book. Figure 6.2 also shows a second estimate of TFP from the U.S. Federal Reserve Bank of San Francisco, and later we'll compare those data to changes in useful work efficiency of Fig. 6.1.

Gordon asks a good basic question: "What allowed the economy of the 1950 and 1960s so unambiguously to exceed what would have been expected on the

---

[48]Gordon [19, Figure 16-5].

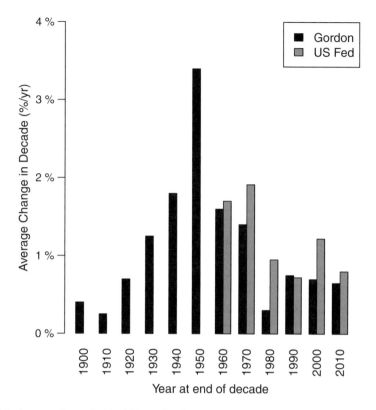

Fig. 6.2 Average change in Total Factor Productivity (TFP) by decade as reported by Robert Gordon (black, left columns, 1900–2010s) and the U.S. Federal Reserve (red, right columns, 1960–2010s) [13]

basis of trends estimated from the six decades before 1928?"[49] We will find that his explanations are consistent with the physically grounded concept that more capital, combined with the energy input to operate it, largely describe increases in "productivity" of the U.S. economy. Strangely, he seems not to see the corollary—if you can't afford or lack either new capital *or* the energy to operate capital, this can explain poor gains in productivity.

Gordon defines the Great Leap by a significant jump in wages ($/hour) and GDP per hour worked. The post-Depression New Deal legislation, such the Fair Labor Standards Act, empowered unions and significantly increased the number of workers covered by 8 h work days with overtime pay. The two decades after the Great Depression were unique in timing for making use of and refining relatively new technologies:

"...there was a leap in TFP between 1929 and 1950 as real GDP more than doubled even as labor and especially capital input grew far less rapidly. Our search for an explanation centers on the timing and magnitude of the Great Depression and World War II, both of which caused the inventions of the second industrial revolution, particularly electric motors and assembly-line methods, to have their full effect on productivity years earlier than might have otherwise occurred."[50]

Gordon also argues that the extraordinary investment in wartime manufacturing facilities, financed by the government and essentially free to the owners, provided such an increase in the number of factories that their boost in output brought the U.S. out of the secular stagnation of the post-Depression years: "The number of machine tools in the U.S. *doubled* from 1940 to 1945, and almost all of these new machine tools were paid for by the government rather than by private firms."[51] When you are buying a lot more tools, particularly paid for by someone else, you can accelerate the purchases of new technologies like electric motors, and redesign factory layouts to take advantage. Thus, the few decades after World War II saw fewer total working hours per person and increasingly higher wages than during the 1920s.[52] Gordon argues that these higher wages incentivized investment to substitute capital (machines) for the higher-cost labor.

It is likely that electric power and the assembly line explain not just the TFP growth upsurge of the 1920s, but also that of the 1930 and 1940s. There are two types of evidence that this equipment capital was becoming more powerfuel and more electrified. First is the horsepower of prime movers, ... and the second is kilowatt hours of electricity production.[53]

Gordon recognizes that these more powerful factories experienced a "...vast increase in the amount of electricity consumed per unit capital." This translates to an increase in useful work to produce more products, but without a need to increase the hours per worker. Just as machines substituted for physical labor in the farm, they did so in the factory. He notes that the U.S. reached the peak GDP per hour worked in 1972 after an unprecedented 40-year increase.[54] In Fig. 5.4 of Chap. 5 we've already seen that residential energy consumption per household also increased rapidly from World War II until peaking in 1972. Thus, more GDP per hour worked and more of GDP going to workers through 1970 meant consumers could buy larger houses that contained more appliances that consumed more energy. The year 1973 is also the peak in total U.S. energy consumption per person. (In the next chapter, we'll summarize many U.S. data trends in one location to more easily see linkages between energy consumption and economic indicators.) Just as 1973 seemed as convenient of a year as any for Blanchard to choose for the end of a time of anomalous growth, Gordon does not mention 1972 as having any relevance related to the 1970 peak in U.S. conventional oil extraction or the subsequent OPEC

---

[50]Gordon [19, p. 528].

[51]Gordon [19, p. 553].

[52]Gordon [19, p. 537].

[53]Gordon [19, p. 557].

[54]Gordon [19, Table 16-1 and Figure 16-3].

oil price increase in 1974. This in spite of the fact that Gordon recognizes two energy-related trends that ended in the early 1970s.

First, "…between 1929 and 1950, motor vehicle horsepower tripled and total electricity production rose 3.3 times."[55] Second, the "…epochal moment in the history of the American petroleum industry occurred with the discovery, in October 1930, of the east Texas oil field, which has been called "the largest and most prolific oil reservoir in the contiguous United States."[56] That epoch ended in 1970.

The explanation seems to be just outside of his grasp when Gordon recognizes the value of increased use of electricity for manufacturing and oil for transportation fuels during the Great Leap. He doesn't quite buy the direct energy-economic relationship, because in describing the lack of growth in TFP after 2000 he states "The most recent decade, 2004–14, has been characterized by the slowest growth in productivity of any decade in American history …"[57] This decade also corresponds to some of the highest average real oil and natural gas prices in U.S. history (recall Fig. 3.1) and follows the year when U.S. energy and food costs as a share of GDP generally stopped declining (Fig. 2.9). Aside from the decades of the 1930s during the Great Depression as well as 1973–1983, the period of 2004–2018 is the only span in U.S. history with a constant level of primary energy consumption (recall Fig. 2.7).

The key missing factor by Gordon, and neoclassical economists in general, is the feedback from the cost of energy. This is largely because their "…search for explanations begins with elementary economics."[58] By "elementary economics," he means neoclassical theory:

> To explain the upsurge in labor productivity [during the Great Leap], the best place to start is with basic economic theory. In a competitive market, the marginal product of labor equals the real wage, and economists have shown that labor's marginal product under specified conditions is the share of labor in total income times output per hour. If the income share of labor remains constant, then the growth rate of the real wage should be equal to that of labor's average product, the same thing as labor productivity.[59]—Robert Gordon (2016)

His "basic economic theory" is *too* basic, and the "specified conditions" are the neoclassical assumptions. Here is a rephrase of the quote above to indicate what he really means, with italics indicating where I have rephrased his words:

> To explain the upsurge in labor productivity, the *usual* place to start is with *neoclassical* economic theory. *If* a *fully* competitive market *exists, which it practically never does*, the marginal product of labor equals the real wage, and economists have *assumed* that labor's marginal product under *the assumptions of neoclassical theory, such as equilibrium*, is the share of labor in total income times output per hour. If the income share of labor remains constant, *but unfortunately the data indicate that it has not since the 1970s*, then the growth

[55]Gordon [19, p. 559].

[56]Gordon [19, p. 560].

[57]Gordon [19, p. 529].

[58]Gordon [19, p. 537].

[59]Gordon [19, p. 541].

rate of the real wage should be equal to that of labor's average product, the same thing as labor productivity.—rephrase of passage in Robert Gordon [19, p. 541]

Remaining stuck with basic economic theory, Gordon's major explanation for the growth of his Great Leap is that the Great Depression (and World War II) directly contributed to the high growth rates because it spurred the legislation of the New Deal:

> ...with its NIRA and Wagner Act that promoted unionization and that directly and indirectly contributed to a sharp price in real wages and a shrinkage in average weekly hours. In turn, both higher real wages and shorter hours helped to boost productivity growth rapidly in the late 1930s, before the United States entered World War II. Substitution from labor to capital as a result of the jump in the real wage is evident in the data on private equipment investment, which soared in 1937–41 substantially above the equipment investment:capital ratio of the late 1920s.—Robert Gordon (2016)[60]

So Gordon claims that total factor productivity, and thus economic growth, increased during the Great Depression because the mandate for higher wages incentivized businesses to use machines instead of people. While I agree businesses faced this motivation, it misses a larger point.

> Without both the machines *and the energy to operate them* there
> could not have been the increase in economic growth and output per
> worker witnessed from the 1930s to the 1970s.

It was not only the proliferation of power plants and motors, but also the availability of coal to burn and water to flow through dams that represent the absolute physical necessities for power generation.

Further, studies of the mathematics behind the "basic economic theory" of the Solow growth model show that Gordon is mathematically correct when he states that "higher real wages" helped boost productivity. Jesus Felipe and John McCombie derived that what neoclassical economists call TFP is in fact based only on a mathematical identity used to define GDP. TFP is simply an average of the change in wages and the change in the rate of profit on capital [12].[61] So yes, by mathematical construct, if real wages increase, then TFP increases! Changing wages have absolutely nothing to do with either human ingenuity or anything physically tangible that we might call "technological change." All companies could raise the wages of all workers tomorrow without changing any machines or consuming any more energy, and yet these changes would affect the calculation of TFP, what Gordon calls "...the best available measure of innovation and technological change."[62]

When it comes to understanding the role of energy for the economy, we don't have to throw away neoclassical economic theory if it works, but it is not the

---

[60]Gordon [19, p. 563].

[61]Using the standard "cost shares" of 0.7 for wages and 0.3 for labor, the change in TFP is thus 0.7 times the rate of change of wages plus 0.3 times the rate of change of the profit rate.

[62]Gordon [19, p. 546].

"best place to start" because it doesn't sufficiently explain long-term growth and TFP. As I now explain, we have a technological characteristic, that we can directly measure, that relieves us from using the neoclassical growth model and TFP as an explanation for "technological change," "human ingenuity," or practically any concept of "progress" someone wants to attribute to it.

What makes more sense is to think of the economy, like animals and ecosystems, as a physical metabolic system that consumes and dissipates energy in order to grow and maintain itself. To understand these systems, the places to start are the conservation laws of physics and thermodynamics. These conservation laws have assumptions behind them, and they have been verified time and time again by controlled experiments.

Ayres' research considered applying the straightforward concept of useful work to modeling GDP and this eliminated much of the need for TFP:

> The efficiency of converting energy into useful work largely describes what TFP really is.

Figure 6.3 shows this is the case. *The rate of change of U.S. useful work efficiency (from Fig. 6.1) follows in lockstep with estimates of U.S. TFP. Further, useful work follows GDP.*

The implication is that we can more accurately model economic growth by projecting useful work, something we can measure and quantify, instead of assuming TFP, which we can't.

TFP by its definition within the neoclassical Solow growth model ignores many factors. This is the case with any model. The main problem with the neoclassical growth model is it ignores the obvious: energy and other natural resources must be consumed to do anything. It doesn't describe this consumption in a way that affects economic growth, and this lack of description makes it less useful than what we need. Most economists' explanations for TFP only adjust the value of quality of capital and labor without noting the real physical constraints of economic production related to energy consumption and time delays to make more capital. These constraints are normally considered in modeling the dynamics of individual businesses, just not as much for the overall macroeconomy.

Ask yourself this question: Why do we need to get smarter to design machines that have more functionality such as higher power, higher efficiency, more information processing? Of course, there are many answers, but one important answer is *to acquire more natural resources that in turn become more capital, become consumed to operate that capital, and become food for people.* These are the fundamental processes that occur in the economy. The more capital that operates, the larger the economy. The more people alive, the higher the drive to extract more resources to support their livelihoods via the operation of capital. More capital can produce more useful work if the energy system delivers more fuel to the capital. Warr and Ayres state this more explicitly: "Exergy efficiency changes with (a) improvements in the efficiency of existing technologies and (b) the innovation and adoption of new technologies which either improve the performance of existing process, or (c) cause a shift in the structure of energy service (the type of useful work) demanded." [62]

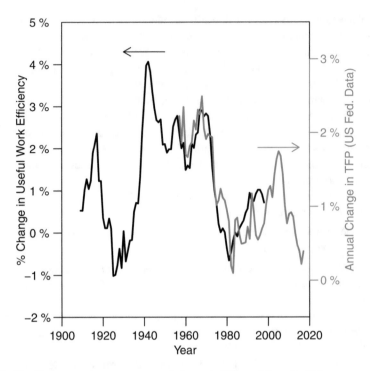

Fig. 6.3 The 10-year running average annual change in U.S. useful work efficiency compared to the 10-year running average annual change in total factor productivity (TFP) as reported by the U.S. Federal Reserve [13]. Useful work efficiency data from [62]

Figure 6.4 shows two metrics relating U.S. energy to GDP. Modelers commonly relate total primary energy to GDP by dividing the former by the latter. This *energy intensity* is the red line that declines from approximately 50 MJ/\$ in 1920 to 15 MJ/\$ in 2000.[63] If we instead compare Ayres' useful work to GDP to calculate *useful work intensity*, we get an approximately constant value across 100 years, ranging from 1.5 to 2.5 MJ/\$.

Figure 6.5 shows these same data in a different way. It plots GDP and useful work together over time. The U.S. data are in subfigure (a), and the same calculations are shown for three other countries (U.K., Austria, and Japan). The high correlation between GDP and useful work is clear to see.

For the U.S., we see that useful work and GDP grew at almost the same near-exponential rate, both increasing nearly 15 times over the twentieth Century. What this means is that if you know how much useful work is performed in a year, then you just need to multiply that by some constant number to estimate GDP

---

[63]The data used from Warr et al. [62] are for primary exergy, not energy, but they are quite similar. Here and in the figure I use the term energy for simplicity of discussion.

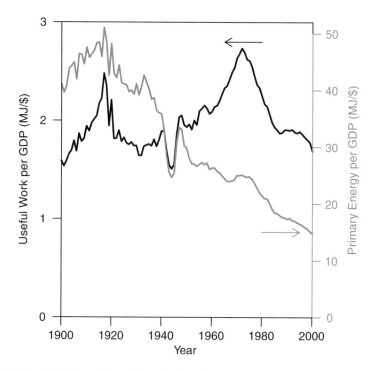

Fig. 6.4 (left axis) U.S. useful work intensity (= primary energy times conversion efficiencies to useful work divided by GDP). (right axis) U.S. primary energy intensity (= primary energy divided by GDP). Real GDP, energy, and useful work data from [62]

more accurately than the Solow neoclassical growth model. For the U.S., we can approximate trillions of dollars of U.S. GDP quite closely simply by dividing each one billion joules of useful work by 1.9. Of course, we have to diligently calculate useful work from known data, but that is a more concrete task than trying to figure out exactly all changes in the world that could possibly describe total factor productivity.

The conclusion is clear.

> If we want to model GDP, we should include the concept of energy and its efficiency of use by machines, or the useful work of the economy.

This answers a question that Robert Solow himself asked in 2007, 50 years after he derived his original growth model:

> There is also a …long-standing worry of mine. We estimate time series of TFP in the conventional way, more or less completely detached from the narrative of identifiable technological changes that a historian would produce for the same stretch of time. There are reasons for this disjunction. TFP is estimated for aggregates, for a whole industry at a minimum, whereas the historical narrative is usually about single firms or even single individuals. Both temporal aggregation and cross-sectional aggregation will mask

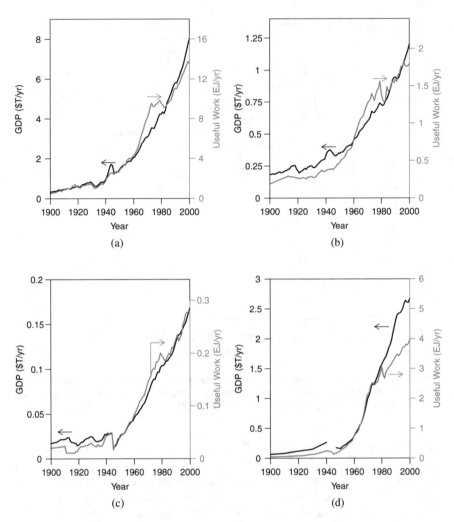

Fig. 6.5 GDP and useful work (= primary energy times conversion efficiencies to useful work) for (**a**) the U.S., (**b**) the U.K., (**c**) Austria, and (**d**) Japan. Real GDP, energy, and useful work data from [62]

individual events. ...it would be interesting to see if any connection can be made, perhaps in a specific industry, between the time series of TFP and an informed narrative of significant innovations and their diffusion. (One can see in principle how TFP should be related to new-product innovations, but it is not clear what would happen in practice.) [59]—Robert Solow (2007)

Well, an "...informed narrative of significant innovations ..." that informs what does "...happen in practice ..." is the story of thermodynamic energy conversion efficiency, and efficiency very much follows that of TFP. Steve Keen, Robert Ayres, and Russell Standish go even farther. They say that we should actually stop using our current calculation of GDP and replace it with the calculation of useful work [32]!

The statement brings up an important question: just what is GDP measuring? If GDP in some approximate way measures what we value, are we inherently valuing useful work without thinking about it? We'll save more philosophical discussion of the purpose of the economy for Chap. 8. But before we get there, it's important to think about perhaps the most important energy policy implications of economic modeling: the cost and feasibility of transitioning to a low-carbon energy system.

Neoclassical Growth: Problems for Policy

It is hard to overstate the policy implications of relying too heavily on the neoclassical model, and in particular the Solow model with exogenous technology as total factor productivity (TFP). To project future economic growth using the Solow growth model you must assume future growth of TFP. By assuming TFP, the modeler effectively assumes about one-half of economic growth out of faith and ignorance because TFP is by definition independent of any policies or parameters within the model. You could assume no growth in TFP, but if you did you wouldn't be able to mimic historical GDP trends or have any reasonable say about GDP in the near-term future.

Consider the ramifications of using TFP and neoclassical growth theory to discuss the energy narratives. Due to concerns regarding climate change, we want to understand the economic impacts from transitioning to 100% renewable energy or a low-carbon energy system with near-zero greenhouse gas emissions. There are many reasonable questions. Does a transition from fossil to renewable energy promote or inhibit economic growth? How does the speed of a low-carbon transition affect the economy?

Researchers use integrated assessment models (IAMs) to help discuss these questions. These IAMs link models of the Earth's climate to models of the economy. Because we want to specify the shift to low-carbon energy, the economic part of IAMs must represent different types of energy resources and technologies from biomass power plants to oil drilling rigs. And now we see the crux of the problem: *IAMs based on neoclassical growth theory assume that economic growth is not affected by the quantity, conversion efficiency, or cost of energy inputs.* How can models that assume energy has no role in economic growth explain the economic impact of a new energy system? They don't. And they can't.

Think of it this way:

1. We want to know how the economy responds if we convert to a low-carbon energy system composed of renewable energy and storage technologies, nuclear power plants, systems that capture carbon dioxide and inject it underground, and maybe even technologies that take carbon dioxide directly from the air.

2. Integrated assessment models (IAMs) link climate models to economic models that use TFP and neoclassical growth theory to project future economic output.
3. Neoclassical growth models using TFP assume growth is not a function of energy inputs, conversion efficiencies, or costs. This is the same as assuming energy will always be available at low cost and at any rate needed.
4. Thus, economic output from IAMs is unaffected by changes specific to a low-carbon energy system.

The result from most IAM models is that no matter what, the economy always grows! Stay high carbon? Economy grows a lot. Going to zero-carbon emissions? Economy still grows a lot. The reason is that instead of assuming how the rate of investment and cost to convert to a low-carbon energy system affect economic growth, most IAMs generally assume economic growth first, via TFP, and decide later how many ways you can reconfigure the energy system.

Importantly, almost all of today's economic models used to understand a renewable or low-carbon energy transition assume a variation of neoclassical growth. Included in this list is the famous DICE model (Dynamic Integrated model of Climate and the Economy) of William Nordhaus, a model used to explore U.S. climate policy and the price of carbon we might charge ourselves to incentivize reductions in greenhouse gas emissions [50]. Because the IAMs don't actually answer the question that we really want to ask (What are the economic impacts of an energy transition?), this is misleading to climate advocates. Consider this quote from a blog post "It's Not Too Late To Stop Climate Change, And It'll Be Super-Cheap":

> To be crystal clear, my position—what the literature and field experience make crystal clear—is that solving climate (stabilizing at 2 °C) is cheap, by any plausible definition of the word. Indeed, it is "super-cheap." ... "The always overly-conservative Intergovernmental Panel on Climate Change reviewed the entire literature on the subject and concluded the annual growth loss to preserve a livable climate is 0.06%—and that's "relative to annualized consumption growth in the baseline that is between 1.6 and 3% per year." So we're talking annual growth of, say 2.24% rather than 2.30% to save billions and billions of people from needless suffering for decades if not centuries.[64]—Joe Romm (2014)

Romm correctly cites the Intergovernmental Panel on Climate Change's (IPCC) summary of the IAM literature. He's also correct to be enthused that climate mitigation might reduce suffering for billions of people. However, he's incorrect in stating that the literature "concluded the annual growth loss to preserve a livable climate is 0.06%." The models didn't *conclude* this; they *assumed* it [9, 35].[65]

---

[64]Joe Romm, ThinkProgress, January 29, 2015, It's Not Too Late To Stop Climate Change, And It'll Be Super-Cheap, https://archive.thinkprogress.org/its-not-too-late-to-stop-climate-change-and-it-ll-be-super-cheap-8865694dbbd2/.

[65]Also see King (2015) that discusses the model outputs used by the Fifth Assessment report of the IPCC. The report summarizes results indicating the economy would grow 250–800% from 2010 to 2100 even going to an economy with zero or negative greenhouse gas emissions by 2011. The report readily states that "negative feedbacks" from energy costs are not considered (see Figure A.II.1 of [35]).

Again, this is because the models assume energy quantities and costs play, at most, a very minor role in growing or constraining the economy. Also, his quote includes a key tell that the use of TFP or some other quantity simply are assumed to increase into the future: "annualized consumption growth in the baseline that is between 1.6 and 3% per year." About half of that "baseline" is just assumed to occur. For an example model, consider the Global Change Assessment Model (GCAM), one of the major IAMs. GCAM projects future GDP by assuming the growth of both population and GDP per person. However, *nothing* in the GCAM model provides a way for the modeled energy system to affect the assumed GDP change and population: "Population and economic activity are used in GCAM through a one-way transfer of information to other GCAM components. For example, neither the price nor quantity of energy nor the quantity of energy services provided to the economy affect the calculation of the principle model output of the GCAM macro-economic system, GDP."[66]

Economic modelers can assume whatever GDP and population growth they want, but for baseline projections they typically stay within values calculated from recent history. However, as any investor reads on any mutual fund or stock prospectus, "past performance is no guarantee of future results."

Think about the quoted GCAM assumption: "... *neither the price nor quantity of energy ... affect the calculation of ... GDP*."! Assume something absurd such as all energy consumption stops tomorrow. In the model, GDP is the same. Do you think that "result" is useful? When we use models to pontificate future low-carbon energy scenarios that stray from historical data (such as within the energy versus GDP plot of Fig. 5.2) but provide no feedback between energy and GDP, then there is a good chance our model results will have no real meaning.

Do climate advocates know of this energy and "technology assumption" problem within macroeconomic models in IAMs? Most people do not. If they do know, would they even care? Probably not, because the models "tell" us that growth occurs in a zero-carbon world or a high-carbon world. This techno-optimistic, or should we say techno-ignorant, growth narrative helps promote climate change mitigation and adaptation just as well as it promotes the avoidance of mitigation and adaptation.

I'm not saying that I know for certain whether the economy would go into a depression if we converted to a zero-carbon world in 30 years. I'm also not saying a low-carbon economy takes us to nirvana. What I *am* saying is that neoclassical economic growth models can't fundamentally tell us anything about the issue.

Do you believe that a zero-carbon world, one that requires actively extracting carbon dioxide from the atmosphere, has only zero point zero six percent (0.06%) less annual growth than a full-carbon-ahead world reaching 4+°C or more temperature rise by 2100? That does not pass the smell test. The models literally cannot tell the difference because their underlying theory and assumptions prevent them from doing so. Again, I make this statement only by evaluating economic model

[66]GCAM model documentation, GCAM v5.1 Documentation: The GCAM Macro-Economic System, accessed March 21, 2019 at http://jgcri.github.io/gcam-doc/macro-econ.html.

outputs. I'm not discussing any effects from the physical changes related to higher greenhouse gas concentrations in the atmosphere. Also, as we'll discuss in the final chapter, growth isn't everything, and a focus on human level outcomes provides reasons to pursue policies for low-carbon energy and increased income equality, for examples, even if GDP declines.

Many researchers understand the flaws of neoclassical models. For example, Sgouris Sgouridis criticizes the structure of the energy-economic modeling within IAMs from another angle: the substitutability of one energy technology for another [27]. Recall the Cobb–Douglas production function assumes infinite substitution, thus by definition at the extreme end of economic narrative for techno-optimism and infinite substitution. Many IAMs use a more general form of this function known as the Constant Elasticity of Substitution (CES) production function. The CES function allows modelers to put limits on how much one input factor can substitute for another. Many of the IAMs also have "nested" structures that allow a subset of substitutable technologies to produce an output that is again one of many inputs to produce a second output, and so on for several levels of nesting.

For example, there are many ways to generate electricity as "output 1" using wind turbines, solar panels, and natural gas plants. Electricity in turn can be one input, along with capital and labor, for stationary power and heat as "output 2." Then stationary power and heat can be one input of many to produce the ultimate "output 3" of GDP. However, this substitution game is still played mostly in the context of monetary costs of technologies rather than on their physical capability. As Sgouridis and his co-authors note: "One would assume that a review of empirical findings should be a critical first step when modeling transitions. Yet Rosen and Guenther [54] found no literature comparing investment decisions for energy-consuming equipment implicit in IAMs with real-world trends in the past, ... " [27].

The study they refer to actually states current IAMs are of no use to estimate costs of transitioning to low-carbon energy: "Because of these serious technical problems, policymakers should not base climate change mitigation policy on the estimated net economic impacts computed by integrated assessment models." [54] Thus, while the CES concept seems like a step in the right direction, in practice it has not delivered. The problem is still that the economic theory assumes at some level that "...non-physical inputs of knowledge and capital can somehow substitute for energy thus reducing the economic energy intensity." [27].

We should use models that explicitly track the flows of energy and other natural resources so that we can include both realistic substitutions and physical constraints. These approaches have produced tremendous insight. The problem is that they've been attacked by prominent mainstream economists because, well, the models are different than what they know and use.

Attack on *The Limits to Growth*

Given the documented inaccuracies of predominant economic growth frameworks, what other modeling frameworks exist, and are they any better? These are reasonable questions. Other modeling researchers, myself included, have developed alternatives to study energy-economic interactions. But there are not enough of us.[67]

One of the most discussed and well-known models of world dynamics is World3, the model used in three versions (1972, 1992, 2004) of *The Limits to Growth* book (first mentioned in Chaps. 1 and 2) [43–45]. This model was based upon the work of Jay Forrester, the father of what is known as *system dynamics* modeling. In the 1960s while working at the Massachusetts Institute of Technology, Forrester developed the method to simulate the interactions among various elements in a way that could explain complex trends and data, such as some of those discussed in Chaps. 4 and 5. Forrester created the basic structure for the model that became World3 [15], the first relatively complex computer model to simulate the dynamics of "... five major trends of global concern—accelerating industrialization, rapid population growth, widespread malnutrition, depletion of nonrenewable resources, and a deteriorating environment."[68] As shown in the previous chapters, there were several global energy and population growth trends that seemed as though they could not continue. Growth of energy consumption and population was increasingly exponential until the early 1970s. In the 1960s in many U.S. cities, pollution was readily apparent as a problem (e.g., smog in Los Angeles, the Cuyahoga River catching fire in 1969 due to industrial pollution in Cleveland). People wanted to know why these trends were occurring and what, if anything, could be done about them.

World3 was meant to improve our understanding of global, not local, trends using graphs. As the modelers stated: "... for world population, capital, and other variables on a time scale that begins in the year 1900 and continues until 2100. These graphs are not exact predictions of the values of the variables at any particular year in the future. They are indications of the system's behavioral tendencies only."[69] Note here that the modeled system is the *global* economy and population, and the purpose is to explore "behavioral tendencies."

---

[67]For examples the papers of Roert Ayres [1, 2, 62, 63] and recently combined with Steve Keen of Kingston University in London [32], a recent IAM by Bovari et al. that does not use neoclassical theory [7], research at the University of Leeds (e.g., the "MARCO-UK" model [55]), the Center for the Understanding of Sustainable Prosperity at the University of Sussex [24, 25], conceptual modeling of Tim Garret of University of Utah [16, 17], modeling of the economy as a large distribution network consuming energy to operate and grow by Andrew Jarvis [26], the ENERGY2020 model (of Systematic Solutions) which is a child of the FOSSIL2 system dynamics model of the U.S. economy that was derived from the same concepts as the World3 model, and my recent "HARMONEY" model [34].

[68]Meadows et al. [44, p. 21].

[69]Meadows et al. [44, p. 92–93].

In order to create this system model, piece by piece, the World3 team utilized "...the most basic relationships among people, food, investment, depreciation, resources, output relationships that are the same the world over, the same in any part of human society or in society as a whole. ... there are advantages to considering such questions with as broad a space-time horizon as possible. Questions of detail, of individual nations, and of short-term pressures can be asked much more sensibly when the overall limits and behavior modes are understood."[70]

This last sentence is very informative and insightful. How are we supposed to understand the trends of each country, town, citizen, and energy resource if we have a poor conceptualization of the broader limits and patterns of the global economy within which they reside? We need the broad perspective to understand the purpose of the world system. As mentioned in Chap. 5, *The Limits to Growth* author Donella Meadows indicated that a system is defined by what it does. In their 1972 assessment, she and her co-authors concluded that:

> The apparent goal of the present world system is to produce more people with more (food, material goods, clean air and water) for each person.[71]— Donella H. Meadows et al. (1972)

We will return to the question of the purpose of the world system, and whether we are capable of understanding it, in Chap. 8. What's important to understand now is that the study was both praised and vilified. Why? A restatement of their three main conclusions from 1972 summarizes:[72]

1. If the present growth trends in world population, industrialization, pollution, food production, and resource depletion continue unchanged, the limits to growth on this planet will be reached sometime within the next 100 years. The most probable result will be a rather sudden and uncontrollable decline in both population and industrial capacity.
2. It is possible to alter these growth trends and to establish a condition of ecological and economic stability that is sustainable far into the future. The state of global equilibrium could be designed so that the basic material needs of each person on earth are satisfied and each person has an equal opportunity to realize his individual human potential.
3. If the world's people decide to strive for this second outcome rather than the first, the sooner they begin working to attain it, the greater will be their chances of success.

They stated that *if the present growth trends* through the 1960s continue then limits to growth will be reached, not next year, or next decade, but broadly sometime in the next 100 years. In Chap. 2 we see that the global energy consumption data was growing at a near constant exponential growth rate from 1900 *until* 1973, just *after The Limits to Growth* was published.

> The techno-realism narrative states that exponential growth on a finite planet can't continue indefinitely. The global data verify this

---

[70]Meadows et al. [44, p. 96].

[71]Meadows et al. [44, p. 86].

[72]Meadows et al. [44, p. 23–24].

statement as the pre-1970 global exponential growth trends in fact did not continue on the finite Earth after the 1970s.

Chalk one up for the World3 model of *The Limits to Growth*. Unfortunately, many critics misinterpreted the statements within *The Limits to Growth* with regard to exponential growth, and this continues through today. Andrew McAfee's 2019 book *More from Less* states that *The Limits to Growth* is "far gloomier" than other writings he already thinks are gloomy [42].[73] We should avoid using qualitative and vague terms, such as gloomy, to discuss very specific mathematics. Aside from that, McAfee summarizes findings from *The Limits to Growth* as follows:

> The most generous estimate of future resource availability included in The Limits to Growth assumed that exponential consumption would continue, and that proven reserves were actually five times greater than commonly assumed. Under these conditions, the team's computer models showed that the planet would run out of gold within 29 years of 1972; silver within 42 years; copper and petroleum within fifty; and aluminum within fifty-five.
> These weren't accurate predictions.[74]—Andrew McAfee (2019)

He goes on to ask:

> How could these predictions about resource availability, which were taken seriously when they were released, have been so wrong?[75]—Andrew McAfee (2019)

My response to McAfee's question is that his question is misleading. A reading of the passage from Chapter 2 of *The Limits to Growth* on which McAfee bases his statements shows his interpretation is incorrect. I copy the original text such that you can see for yourself that in no way did *The Limits to Growth* authors claim to *predict* when, or that there even would be a time, in which any specific mineral such as gold, silver, copper, petroleum, or aluminum would "run out." For the passage below, keep in mind they describe chromium only as a specific example of more broadly considering individual fossil minerals (in Table 4 of their Chapter 2) as well as their aggregation:

> The world's known reserves of chromium are about 775 million metric tons, of which about 1.85 million metric tons are mined annually at present. Thus, at the current rate of use, the known reserves would last about 420 years. The dashed line in figure 11 illustrates the linear depletion of chromium reserves that would be expected under the assumption of constant use. The actual world consumption of chromium is increasing, however, at the rate of 2.6% annually. The curved solid lines in figure 11 show how that growth rate, if it continues, will deplete the resource stock, not in 420 years, as the linear assumption indicates, but in just 95 years. If we suppose that reserves yet undiscovered could increase present known reserves by a factor of five, as shown by the dotted line, this fivefold increase would extend the lifetime of the reserves only from 95 to 154 years.[76]
> . . .

---

[73]McAfee [42, p. 119].

[74]McAfee [42, p. 119–120].

[75]McAfee [42, p. 120].

[76]Meadows et al. [44, p. 61].

Figure 11 shows that under conditions of exponential growth in resource consumption, the static reserve index (420 years for chromium) is a rather misleading measure of resource availability. We might define a new index, an "exponential reserve index," which gives the probable lifetime of each resource, assuming that the current growth rate in consumption will continue. We have included this index in column 5 of table 4. We have also calculated an exponential index on the assumption that our present known reserves of each resource can be expanded fivefold by new discoveries. This index is shown in column 6. The effect of exponential growth is to reduce the probable period of availability of aluminum, for example, from 100 years to 31 years (55 years with a fivefold increase in reserves). Copper, with a 36-year lifetime at the present usage rate, would actually last only 21 years at the present rate of growth, and 48 years if reserves are multiplied by five. It is clear that the present exponentially growing usage rates greatly diminish the length of time that wide-scale economic growth can be based on these raw materials

Of course the actual nonrenewable resource availability in the next few decades will be determined by factors much more complicated than can be expressed by either the simple static reserve index or the exponential reserve index.[77]

Pay attention to wording such as "at the current rate of use" and "under conditions of exponential growth." In this way the full sequence of interpretation of the above excerpt from *The Limits to Growth* is the following:

1. If you assume exponential growth in consumption of a mineral continues unabated, and
2. if you assume five times more reserves of that mineral than was known in 1970, then
3. the world would extract all of those reserves after a certain number of years, but these static and exponential reserve indices are too simple to explain what will actually occur.

The *The Limits to Growth* authors clearly assumed that exponential growth cannot continue on a finite planet. That is the point of the book, and the title of the chapter to which McAfee refers is "The Limits to Exponential Growth." The plots of simulated chromium usage (Figures 12 and 13 in *The Limits to Growth*) show that they know 100% of chromium, or any fossil resource, will never be extracted because price rises with depletion, and "The higher price causes consumers to use chromium more efficiently and to substitute other metals for chromium whenever possible."[78] One of the limits to exponential growth is that you can't afford to extract 100% of the mineral.

The Limits to Growth also considered the effects of rising costs of depletion of all nonrenewable minerals in aggregate. Depending on your point of view, the *The Limits to Growth* authors' 1972 prediction of a cessation to growth "within the next one hundred years" is on par with or even bolder than that of M. King Hubbert's correct prediction of the timing of a peak in conventional U.S.-48 oil production over 10 years before it happened in 1970. At the time they were quite confident in this very general conclusion, stating that "...the basic behavior modes we have

---

[77] Meadows et al. [44, p. 62–63].

[78] Meadows et al. [44, p. 65].

already observed in this model appear to be so fundamental and general that we do not expect our broad conclusions to be substantially altered by further revisions."[79] In 2004, via their 30-year update to the original book, the authors stuck with their "broad conclusions" from 1972 [45]:

> For those who respect numbers, we can report that the highly aggregated scenarios of World3 still appear, after 30 years, to be surprisingly accurate. The world in the year 2000 had the same number of people (about six billion–up from 3.9 billion in 1972) that we projected in the 1972 standard run of World3. Furthermore, that scenario showed a growth in global food production (from 1.8 billion tons of grain equivalent per year in 1972 to three billion in 2000) that matches history quite well. Does this correspondence with history prove that our model was true? No, of course not. But it does indicate that World3 was not totally absurd; its assumptions and our conclusions still warrant consideration today.[80]

They state that the World3 model was *not totally absurd*, and I whole-heartedly agree. What's important to understand now is that the general concept and structure of the model has stood the test of time very well. Other reassessments show the model effectively describes global macro trends that have taken place in the 40+ years since the original study [25, 60]. You cannot find another model that predicted trends to the degree of consistency as World3. As stated in the summary by Tim Jackson, Professor of Sustainable Development at the University of Surrey, and Robin Webster: "There is unsettling evidence that society is still following the 'standard run' of the original study—in which overshoot leads to an eventual collapse of production and living standards." But, when you start talking about declining living standards and end of growth, you will find some critics, and some are (or were) prominent economists.

Ugo Bardi's book *The Limits to Growth Revisited* details the history of *The Limits to Growth* and its criticisms [4]. He discusses how William Nordhaus and other mainstream economists misinterpreted the modeling approach because the system parameters, feedbacks, and lookup tables that influenced the dynamics were unfamiliar. For example, in Nordhaus' 1992 paper he stated: "Both models [Limits to Growth 1972 and 1992] rule out ongoing technological change. In this respect, they are inconsistent with the standard interpretation of economic history during the capitalist era."

It would be natural for a neoclassically trained economist to make this statement. This is because World3 does not include a neoclassical aggregate production function that most economists recognize and use for projecting "technological change" as non-physical total factor productivity, or human ingenuity [51]. World3 did not neglect the ability to model technological change. Because it modeled technological change via a framework and factors that differ from neoclassical theory, its structure is not conducive to many economists' "standard interpretation of economic history." This is not the same as saying the model is wrong or inaccurate, just different.

---

[79]Meadows et al. [44, p. 22].

[80]Preface of *Limits to Growth: The 30-Year Update* [45].

World3 also includes a dependent structure that is similar to the net energy, or energy return on investment (EROI) concept discussed in Chap. 5 in that as resources are depleted, more capital must be allocated per unit of output to extract the next bit of resource. This concept is crucial to produce realistic feedbacks. We clearly see this "more capital with depletion" in data associated with unconventional oil production and solar panels because they do require more capital per unit of oil and electricity than past methods. In order for a model to include this feedback, it must define an appropriate internal structure that requires an output from the economy, such as energy, to also be an input. The standard neoclassical approach, using an aggregate production function, ignores this type of feedback.

As already noted, another major criticism of World3 was its explicit consideration of a limited physical size of nonrenewable resources. World3 includes a parameter that effectively represents the maximum size of all nonrenewable resources (e.g., fossil energy and minerals) lumped together. The assumptions that the world was physically finite and that industrial output necessitated the use of resources led to a result that physical output and population could not continuously grow exponentially or indefinitely. Again, the data in Chaps. 2 and 4 show that exponential growth effectively ended in the 1970s, as predicted by World3.

In his 1992 criticism, Nordhaus did introduce a relevant question as to the role of theoretical models: "One of the major points that has emerged up to now is that the existence and significance of constraints to long-term economic growth, imposed either by environmental concerns or natural resource limitations, cannot be determined by the kinds of theoretical models [World3] developed in Limits I or II. Indeed, it is hard to see how even the best of economic models could do more than frame the questions for empirical studies to address." [51]

I disagree that models like World3 cannot be used to understand physical constraints on long-term growth, but I agree with Nordhaus that theoretical modeling constructs provide the bases for interpreting and collecting data. All models should be seen in the context of both interpreting data and the restrictions assumed by the theories and worldviews that guide the interpretation. Different worldviews present different interpretations of the same data. Martin Weitzman's discussant comments in Nordhaus' paper accurately juxtapose the worldviews of the "limits-to-growth" perspective with those of the "average contemporary economist":[81]

> There may be a some value in trying to understand a little better why the advocates of the limits-to-growth view see things so differently and what, if anything, might narrow the differences.
>
> I think that there are two major differences in empirical worldviews between mainstream economists and anti-growth conservationists. The average ecologist sees everywhere that carrying capacity is a genuine limit to growth. Every empirical study, formal or informal, confirms this truth. And every meaningful theoretical model has this structure built in. Whether it is algae, anchovies, or arctic foxes, a limit to growth always appears. To be sure, carrying capacity is a long-term concept. There may be temporary population upswings or even population explosions, but they always swing down or crash in the end because of

---

[81]Martin L. Weitzman discussion in [51].

finite limits represented by carrying capacity. And *Homo sapiens* is just another species-one that actually is genetically much closer to its closest sister species, chimpanzees, than most animals are to their closest sister species.

Needless to say, the average contemporary economist does not readily see any long-term carrying capacity constraints for human beings. The historical record is full of past hurdles to growth that were overcome by substitution and technological progress. The numbers on contemporary growth, and the evidence before one's eyes, do not seem to be sending signals that we are running out of substitution possibilities or out of inventions that enhance productivity.

Studies like World3 and comments such as Weitzman's inspired me to derive models to bridge the gulf between worldviews. In this book I've emphasized the need to consider both the *size and structure of the economy*. This conclusion is informed by my research, that I now describe, that indicates how resource constraints can lead to slower growth an economic restructuring, just like the U.S. experienced following the 1970s.

Putting My Money and Energy Where My Mouth Is

I spend much of my time around engineers and scientists who design technologies and models of energy and electricity flowing within the economy. On the other hand, politicians, think tanks, lobbyists, policymakers, and other holders of the various energy and economic narratives tend to talk about how much money flows from one pocket or another. The renewable energy pocket or fossil fuel pocket. The rich, middle class, or low-income pocket. The unions or the business owners and bankers. The pocket of "Big X" (Big Oil, Big Pharma, Big Agriculture, etc.) or of small business. It is difficult for people to discuss the systemic issues presented in Chap. 5, but often easier to revert to political explanations for why money is distributed to "them" instead of "us," and Chap. 9 visits some of these narratives. Ideally we should say something about economic growth *and* distribution, or size *and* structure.

One aspect of World3 that makes it hard for some to translate to contemporary issues is that the modelers purposefully avoided explicit counting of certain economic quantities such as wages, debt, and employment. Given the increased concerns over issues of debt and wealth distribution, I thought it was time we bridge a gap between models like World3 that have much insight into human and resources dynamics but might lack concepts of distribution, and economic models that are based on the distribution of money within the economy, but have little to no insight on the role of natural resource use.

With that goal in mind, I created a model based on a similar concept as World3 in that it has an allocation of resources and capital between "energy" and "non-energy" parts of the economy, and it also includes economic factors that both economists and workers care about, such as wages and debt. This combination allows us to understand if and how energy and resource consumption play a role in the trends of debt ratios and wage inequality that we explored in Chap. 4.

It is easier to propagate your model if you give it a memorable name, and I called my model HARMONEY for "Human And Resources with MONEY" [34]. The HARMONEY model is a combination of two other existing models. The first is a very simple model of an agrarian society that harvests a forest-like resource to feed itself.[82] The second is a model of a simple economy with fluctuating business cycles, tracking physical capital, wages, and employment, while also considering the real-world tendency of businesses to invest more than their profits by borrowing money from a bank.[83] This borrowing is what "creates money" as debt within the model, just like commercial banks create money when they provide a loan to a business.

From the standpoint of natural resource use, HARMONEY has three key features that are consistent with real-world physical activities. First, natural resources are required to operate capital. This is the same as saying you need fuel to run your car, and a factory needs electricity to operate manufacturing machinery and computers. Second, natural resources are required to make new capital. This is the same as saying that all of the objects around you now (coffee mugs, computers, buildings, etc.) are made of natural resources. Third, natural resources are required to sustain human livelihood. This is the same as saying that, at a very basic level we need food to survive, and at a higher level more resource consumption leads to more longevity. Thus, whatever the flow of natural resources, those resources must be allocated between the three aforementioned uses.

From an economic theory standpoint, the model does not calculate GDP using an aggregate production function, such as the Cobb–Douglas or CES formulations. HARMONEY is simple in that it has only two types of activity. The first uses machines to extract resources, and the second uses machines to make more machines, or capital. Importantly, both activities require capital, labor, and natural resources (e.g., energy) to function, and any one of them can be the constraining factor. This enables the model to incorporate the net energy feedback of energy return on investment (EROI) and understand these biophysical metrics in the context of more common metrics of GDP, debt, wages, capital accumulation, and population growth.

The results from the HARMONEY model have an uncanny ability to mimic and explain very important long-term trends in the economy. Figures 6.6 and 6.7 show two comparisons of model results to U.S. data. Before describing the insights from these figures, an excerpt from my publication provides some context [34]:

> While the model trends show important similarity to those of the U.S., we caution that the model is not calibrated to the U.S. or any economy. ...the comparison to U.S. data indicates that the model characterizes important underlying processes that govern long-term growth and structural change in an economy such as that of the U.S. For our model-U.S.

---

[82]This is the HANDY, or Human And Nature DYnamics model of [48].

[83]This is Steve Keen's "Minsky" model that uses what is known as the Goodwin model but incorporates a new equation for debt creation [28, 31].

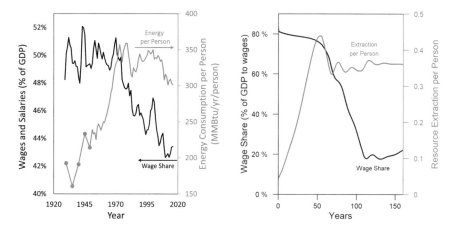

Fig. 6.6 (Left figure) Data for the U.S. wage share (left axis) and per capita energy consumption (right axis) both change their long-term trends in the 1970s. (Right figure) In the same way as the U.S. data, the wage share (left axis) from the HARMONEY model shows the same simultaneous turning point in long-term trend, from a constant value to a declining value, when per capita resource consumption reaches its peak [34]

comparison, the general sequence of long-trends and structural change are important, not the relation of magnitudes of variables or specific model times to specific years in the U.S. data.

Three reasons support comparison of the model to the U.S. First, the U.S. is relatively resource self-sufficient, and the model assumes full self-sufficiency. Second, our investment behavior matches that of the U.S. in that gross investment is significantly greater than net profits. Third, both the model and U.S. data exhibit an initial period with increasing per capita resource, or energy, consumption followed by one with approximately constant per capita consumption.

This third reason is critical. The HARMONEY model assumption of an economy extracting a regenerative renewable resource inherently simulates a trend of increasing and then steady per capita resource extraction … Thus, our model is useful for answering the question "How might the economy respond when transitioning from a period of increasing per capita resources consumption to one with steady per capita resources consumption?" It just so happens that the U.S. economy also exhibits this trend for energy consumption.

Figure 6.6 shows the wage share and per capita energy consumption of the U.S. The wage share is the percentage of GDP allocated to hourly or salaried workers, and these are the main portion of the data for total worker compensation shown in Chap. 4 (Fig. 4.12). Notice how both the wage share and per capita energy consumption have a different trend before versus after the early 1970s. Before 1973, wage share remained constant at about 50% of GDP, and energy consumption per person increased exponentially at 3%/yr. After 1973, wage share declined at about 1.5–2% per decade, and we could say energy consumption per person declined slightly or remained relatively constant.

Amazingly, the model results show practically the exact same trends as in the U.S. data. I did not anticipate this result. Also, when initially formulating the model, I

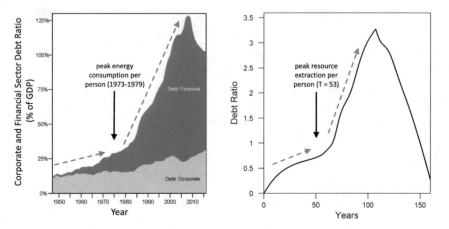

Fig. 6.7 (Left) The debt ratio of U.S. corporations and financial companies (debt/GDP) compared to (right) the equivalent debt ratio metric from the HARMONEY model [34]

had no immediate goal to mimic this type of relationship. I did want a model that had several important elements, but I didn't anticipate my first results would so clearly relate to real-world data. This wage share decline highlights one important difference for the HARMONEY model from that using the neoclassical framework: the neoclassical framework assumes a constant fraction of GDP going to wages and profits (e.g., using the Cobb–Douglas function or Solow growth model), but there is no need to make this assumption. When the real-world data show the wage share substantially declined by 7% over four decades (from 50% in 1973 to 43% in 2013), we can question any modeling approaches that simply assume a constant value. In the HARMONEY model, the wage share emerges because of how its systems-oriented structure relates the elements to one another.

The HARMONEY model also provides insight into debt accumulation. Figure 6.7 shows only two categories of the same debt data introduced in Fig. 4.6, the U.S. private company debt ratio (debt divided by GDP) for corporations and financial institutions. These two categories are equivalent to the concept of debt included in the HARMONEY model. It was the accumulation of U.S. private and household debt, and associated interest payments, that triggered the 2008 financial crisis, although the HARMONEY results did not include household debt. The crisis was not triggered by government debt.

Note how the private debt ratio increases much more rapidly after the 1970s than before, and the increase in financial sector debt drives the overall trend for the U.S. This same breakpoint occurs in the HARMONEY model and for the same reasons. In both the U.S. data and the model, when per capita resource consumption was rapid the debt ratio increased, but at a much slower rate than after per capita consumption stagnated. Recall that neoclassical theory does not account for the concept of debt, and it assumes the quantity of money has no fundamental role in long-term trends. Steve Keen's research provided a way for me to include debt into economic growth modeling [28, 31]. In his book *Debunking Economics*, Keen states the problem clearly:

> This [lack of consideration of debt], along with the unnecessary insistence on equilibrium modeling, is the key weakness in neoclassical economics: if you omit so crucial a variable as debt from your analysis of a market economy, there is precious little else you will get right.[84]—Steve Keen (2011)

Again, this is the fundamental reason why mainstream economists could not foresee or anticipate the 2008 financial crisis. They don't model debt, the cause of the crisis itself! The Queen of the United Kingdom, Elizabeth II, was wondering why (almost) no one seemed to notice the credit problem when she attended a briefing at the London School of Economics in 2008. Eight months later a group of economists sent a letter to the Queen apologizing that most economists have a failure of imagination and don't think systematically enough about how the economy operates:

> In summary, your majesty, the failure to foresee the timing, extent and severity of the crisis and to head it off, while it had many causes, was principally a failure of the collective imagination of many bright people, both in this country and internationally, to understand the risks to the system as a whole.[85]

My answer to the Queen is in this section, and I apologize for only recently studied economics, having studied only science and engineering before 2008! In addition, going back to the wage share decline, it is driven by two quantities: the accounting for depreciation for an increasing quantity of capital and the interest payments on a rising debt ratio. The pattern occurs if you assume, as observed in the U.S. data, that companies keep investing more money than their profits. Since the 1920s, U.S. corporations typically invest 1.5–2.5 times more each year than they make in profits (Fig. 4.8).[86] Thus, in this face of constant or slower increase in total energy consumption, the economy accumulates capital that either operates less or requires less energy to operate (e.g., efficient equipment, computers). Think about the patterns in Figs. 6.6 and 6.7 the following way. We can assume four major distributions of GDP (or "value added") in national economic accounting:

---

[84]Keen [29, p. 321].

[85]Tim Besley and Peter Hennessy, "The Global Financial Crisis—Why Didn't Anybody Notice?", *British Academy Review*, 14, November 2009 available at https://www.thebritishacademy.ac.uk/publications/british-academy-review/global-financial-crisis-why-didnt-anybody-notice. Andrew Pierce, *UK Telegraph*, November 5, 2008, "The Queen asks why no one saw the credit crunch coming", accessed July 16, 2019 at: https://www.telegraph.co.uk/news/uknews/theroyalfamily/3386353/The-Queen-asks-why-no-one-saw-the-credit-crunch-coming.html. Associated Press, July 26, 2009, "Sorry Ma'am—we just didn't see it coming," accessed July 16, 2019 at: http://www.nbcnews.com/id/32156155/ns/business-world_business/t/sorry-maam-we-just-didnt-see-it-coming/#.XZEg_EZKhm9.

[86]See Supplemental Figure 3 in King (2019) [34] using data from the U.S. Bureau of Economic Analysis Tables 1.1.5 (GDP and gross investment) and 1.1.12 (corporate profits with inventory valuation adjustment, IVA and capital consumption adjustment, CCAdj).

government (as taxes), private profits including interest (or rent) payments to capital owners, depreciation (on capital), and wages (to workers).

In a capitalist system based on maintaining private sector profits, increases in debt ratio and the amount of capital per person means that increasing shares of GDP go to both depreciation and profits from interest payments. Because the last several decades show a constant share to government taxes,

> when there is a restriction in the growth of GDP, the prioritization of allocation to profits, taxation, and depreciation means that the workers' share is the only portion available to take the hit.

When you include debt and resources into a model, then "BINGO," out comes the insights presented in this section.

Summary: Macroeconomics

This has been a long chapter. We've covered a number of important concepts that inform mental and mathematical models we use to explain patterns in energy and economic data. One of these models is the neoclassical economic growth model. Despite its severe flaws, it reigned supreme for decades leading up to the 2008 financial crisis. The crisis exposed its major flaws, including the lack of consideration of debt and the concept of modeling the economy in equilibrium. But well before 2008, as far back as the 1970s, researchers, such as those using system dynamics methods, had devised alternative frameworks that more comprehensively and coherently described many important long-term trends of the world. Unfortunately, the critics included many proponents of the techno-optimistic narrative, including prominent mainstream economists, that simply didn't understand what they were criticizing. Policymakers listened to the mainstream, and the low energy prices in the 1980s and 1990s made people lose interest in new methods for energy-economic modeling. But there has been a resurgence in research since the turn of the twenty-first century, and this research provides improved understanding of the fundamental roles of both resources (and energy) and debt in the economy.

> Mathematical and conceptual models that consider the constraints of how energy must flow through the economy, into machines with thermodynamic energy conversion efficiencies, produce much more direct insight into how energy relates to economic output.

By including resource flows, the efficiency of converting energy into useful work, and debt into macroeconomic growth models, we can explain the broad trends of growth and structural change in modern economies over the last 50–100 years.

From a systems perspective, when growth is exponential, the system does not yet perceive any constraints or boundaries to its growth, so many more options for allocation are possible [45]. If this growth, say of energy consumption, is no longer increasing exponentially, then a different allocation of energy and money must occur, as it did in the 1970s. At some points the constraints on increasing net energy output from the resource extraction sector might cause a reduction in the flow of money to the other parts of the economy, such as happened to U.S. worker wages.

The translation of why wages relate to energy consumption is simple: by and large workers get paid to do things that directly and indirectly consume energy. Even companies like Google and Facebook don't avoid this. Though most of their revenue comes from selling advertisements, they sell these ads for companies that in turn sell products that are made of resources and consume energy to manufacture and operate. So many businesses of the "information economy" (Google, Facebook, Twitter) are actually supported by the "old economy" that still makes stuff. Plus their web servers must consume electricity to save our photos and deliver streaming video.

The last three chapters have been heavy on data, scientific concepts, and economic theory. The end of this chapter largely marks the end point for introducing new data. The rest of the book further interprets the data shown thus far and puts it into context of economic narratives we hear more in the popular press, outside of formal economic circles.

Chapter 7 now takes a more qualitative approach to summarizing U.S. energy, economic, and political trends. Placing many of the important trends in one location helps show just how many of them each follow three major phases over the last 100 years. This approach also provides context for the perceptions of accessibility to an American middle-class lifestyle, both before and after the 1970s, that most politicians claim, or at least hope, to open up to a larger share of citizens today. Chapter 7 also sets the stage for understanding just how much we have changed, and how much we might be able to change, the size and structure of the U.S. economy.

References

1. Ayres, R.U.: Sustainability economics: Where do we stand? Ecological Economics **67**(2), 281–310 (2008). Times Cited: 2
2. Ayres, R.U., Warr, B.: Accounting for growth: the role of physical work. Structural Change and Economic Dynamics **16**, 181–209 (2005)
3. Ayres, R.U., Warr, B.: The Economic Growth Engine: How Energy and Work Drive Material Prosperity. Edward Elgar (2009)
4. Bardi, U.: The Limits to Growth Revisited. Springer (2011). Springer Briefs in Energy: Energy Analysis
5. Bashmakov, I.: Three laws of energy transitions. Energy Policy **35**, 3583–3594 (2007)
6. Blanchard, O.: Macroeconomics, 4th edn. Pearson Prentice Hall, Upper Saddle River, New Jersey (2006)
7. Bovari, E., Giraud, G., McIsaac, F.: Coping with collapse: A stock-flow consistent monetary macrodynamics of global warming. Ecological Economics **147**, 383–398 (2018). https://doi.org/10.1016/j.ecolecon.2018.01.034. http://www.sciencedirect.com/science/article/pii/S0921800916309569
8. Box, G.E.P., Draper, N.R.: Empirical Model-Building and Response Surfaces. Wiley (1987)

9. Clarke, L., Jiang, K., Akimoto, K., Babiker, M., Blanford, G., Fisher-Vanden, K., Hourcade, J.C., Krey, V., Kriegler, E., Löschel, A., McCollum, D., Paltsev, S., Rose, S., Shukla, P.R., Tavoni, M., van der Zwaan, B.C.C., van Vuuren, D.: Assessing transformation pathways. In: O. Edenhofer, R. Pichs-Madruga, Y. Sokona, E. Farahani, S. Kadner, K. Seyboth, A. Adler, I. Baum, S. Brunner, P. Eickemeier, B. Kriemann, J. Savolainen, S. Schlömer, C. von Stechow, T. Zwickel, J. Minx (eds.) Climate Change 2014: Mitigation of Climate Change. Contribution of Working Group III to the Fifth Assessment Report of the Intergovernmental Panel on Climate Change. Cambridge University Press, Cambridge, United Kingdom and New York, NY, USA. (2014)
10. Court, V., Jouvet, P.A., Lantz, F.: Long-term endogenous economic growth and energy transitions. The Energy Journal **39**(1), 29–57 (2018). https://doi.org/10.5547/01956574.39.1.vcou. http://www.iaee.org/en/publications/ejarticle.aspx?id=3026
11. Etzioni, A.: The moral effects of economic teaching. Sociological Forum **30**(1), 228–233 (2015). https://doi.org/10.1111/socf.12153. https://onlinelibrary.wiley.com/doi/abs/10.1111/socf.12153
12. Felipe, J., McCombie, J.: The tyranny of the identity: Growth accounting revisited. International Review of Applied Economics **20**(3), 283 – 299 (2006). http://ezproxy.lib.utexas.edu/login?url=http://search.ebscohost.com/login.aspx?direct=true&db=bth&AN=22089337&site=ehost-live
13. Fernald, J.G.: A quarterly, utilization-adjusted series on total factor productivity. https://www.frbsf.org/economic-research/indicators-data/total-factor-productivity-tfp/ (2018). Produced on December 05, 2018 10:59 AM by John Fernald/Neil Gerstein. Online; accessed January 27, 2019
14. Fix, B.: Rethinking Economic Growth Theory From a Biophysical Perspective. Springer, Cham, Switzerland (2015)
15. Forrester, J.W.: World Dynamics. Wright-Allen Press, Inc., Cambridge, MA (1971)
16. Garrett, T.J.: Are there basic physical constraints on future anthropogenic emissions of carbon dioxide? Climatic Change **104**(3), 437–455 (2011). https://doi.org/10.1007/s10584-009-9717-9
17. Garrett, T.J.: Long-run evolution of the global economy: 1. physical basis. Earth's Future **2**(3), 127–151 (2014). https://doi.org/10.1002/2013EF000171
18. Georgescu-Roegen, N.: The Entropy Law and the Economic Process. Harvard University Press, Cambridge, Mass. (1971)
19. Gordon, R.J..: The Rise and Fall of American Growth: the U.S. Standard of Living since the Civil War. Princeton University Press, Princeton, NJ (2016)
20. Graeber, D.: Debt: The first 5,000 Years. Melville House, Brooklyn, NY (2014)
21. Graeber, D.: Bullshit Jobs. Simon & Schuster, New York, NY (2018)
22. Hall, C.A.S., Klitgaard, K.A.: Energy and the Wealth of Nations: An Introduction to Biophysical Economics, 2nd edn. Springer (2018)
23. Heun, M.K., Santos, J., Brockway, P.E., Pruim, R.J., Domingos, T., Sakai, M.: From theory to econometrics to energy policy: Cautionary tales for policymaking using aggregate production functions. Energies **10**(203), 1–44 (2017)
24. Jackson, T.: Prosperity Without Growth: Foundations for the Economy of Tomorrow, second edition edn. Routledge, Milton, UK and New York, NY, USA (2017)
25. Jackson, T., Webster, R.: Limits revisited: a review of the limits to growth debate. Tech. rep. (2016). http://limits2growth.org.uk/revisited/. Accessed April 12, 2017
26. Jarvis, A.J., Jarvis, S.J., Hewitt, C.N.: Resource acquisition, distribution and end-use efficiencies and the growth of industrial society. Earth System Dynamics **6**(2), 689–702 (2015). https://doi.org/10.5194/esd-6-689-2015. http://www.earth-syst-dynam.net/6/689/2015/
27. Kaya, A., Csala, D., Sgouridis, S.: Constant elasticity of substitution functions for energy modeling in general equilibrium integrated assessment models: a critical review and recommendations. Climatic Change (2017). https://doi.org/10.1007/s10584-017-2077-y.
28. Keen, S.: Finance and economic breakdown: Modeling Minsky's financial instability hypothesis. Journal of Post Keynesian Economics **17**(4), 607–635 (1995)

29. Keen, S.: Debunking macroeconomics. Economic Analysis and Policy **41**(3), 147–167 (2011). http://dx.doi.org/10.1016/S0313-5926(11)50030-X. http://www.sciencedirect.com/science/article/pii/S031359261150030X

30. Keen, S.: Debunking Macroeconomics - Revised and Expanded Edition: The Naked Emperor Dethroned?, 2 edn. Zed Books, London and New York (2011)

31. Keen, S.: A monetary Minsky model of the great moderation and the great recession. Journal of Economic Behavior & Organization **86**, 221–235 (2013). http://dx.doi.org/10.1016/j.jebo.2011.01.010. http://www.sciencedirect.com/science/article/pii/S0167268111000266

32. Keen, S., Ayres, R.U., Standish, R.: A note on the role of energy in production. Ecological Economics **157**, 40–46 (2019). https://doi.org/10.1016/j.ecolecon.2018.11.002. http://www.sciencedirect.com/science/article/pii/S0921800917311746

33. King, C.W.: Comparing world economic and net energy metrics, part 3: Macroeconomic historical and future perspectives. Energies **8**(11), 12,348 (2015). https://doi.org/10.3390/en81112348. http://www.mdpi.com/1996-1073/8/11/12348

34. King, C.W.: An integrated biophysical and economic modeling framework for long-term sustainability analysis: the harmoney model. Ecological Economics **169**, 106,464 (2020). https://doi.org/10.1016/j.ecolecon.2019.106464. http://www.sciencedirect.com/science/article/pii/S0921800919302034

35. Krey, V., Masera, O., Blanford, G., Bruckner, T., Cooke, R., Fisher-Vanden, K., and E. Hertwich, H.H., Kriegler, E., Mueller, D., Paltsev, S., Price, L., Schlömer, S., Ãœrge Vorsatz, D., van Vuuren, D., Zwickel, T.: Annex ii: Metrics & methodology. In: O. Edenhofer, R. Pichs-Madruga, Y. Sokona, E. Farahani, S. Kadner, K. Seyboth, A. Adler, I. Baum, S. Brunner, P. Eickemeier, B. Kriemann, J. Savolainen, S. Schlömer, C. von Stechow, T. Zwickel, J. Minx (eds.) Climate Change 2014: Mitigation of Climate Change. Contribution of Working Group III to the Fifth Assessment Report of the Intergovernmental Panel on Climate Change. Cambridge University Press, Cambridge, United Kingdom and New York, NY, USA. (2014)

36. Kümmel, R.: The impact of energy on industrial growth. Energy **7**(2), 189–203 (1982). https://doi.org/10.1016/0360-5442(82)90044-5. http://www.sciencedirect.com/science/article/pii/0360544282900445

37. Kümmel, R.: The Second Law of Economics: Energy, Entropy, and the Origins of Wealth. Springer (2011)

38. Kümmel, R.: Why energy's economic weight is much larger than its cost share. Environmental Innovation and Societal Transitions **9**, 33–37 (2013). https://doi.org/10.1016/j.eist.2013.09.003. http://www.sciencedirect.com/science/article/pii/S2210422413000634. Energy, materials and growth: A homage to Robert Ayres

39. Lewis, M.: The Undoing Project: A Friendship That Changed Our Minds. W. W. Norton & Company, New York and London (2017)

40. Lindenberger, D., Kümmel, R.: Energy and the state of nations. Energy **36**(10), 6010–6018 (2011). https://doi.org/10.1016/j.energy.2011.08.014. http://www.sciencedirect.com/science/article/pii/S0360544211005445

41. Lindenberger, D., Weiser, F., Winkler, T., Kümmel, R.: Economic growth in the USA and Germany 1960–2013: The underestimated role of energy. BioPhysical Economics and Resource Quality **2**(3) (2017). https://doi.org/10.1007/s41247-017-0027-y

42. McAfee, A.: More from Less The Surprising Story of How We Learned to Prosper Using Fewer Resources—and What Happens Next. Scribner, New York, NY (2019)

43. Meadows, D.H., Meadows, D.L., Randers, J.: Beyond the Limits. Chelsea Green Publishing, Post Mills, VT (1992)

44. Meadows, D.H., Meadows, D.L., Randers, J., Behrens, W.W.I.: Limits to Growth: A Report for the Club of Rome's Project on the Predicament of Mankind. Universe Books, New York (1972)

45. Meadows, D.H., Randers, J., Meadows, D.L.: Limits to Growth: The 30-Year Update. Chelsea Green Publishing, White River Junction, Vermont (2004)

46. Mirowski, P.: More Heat than Light: Economics as Social Physics, Physics as Nature's Economics. Cambridge University Press, Cambridge (1989)

47. Mirowski, P.: The Goalkeeper's Anxiety at the Penalty Kick. History of Political Economy **25**(suppl_1), 303–350 (1993). https://doi.org/10.1215/00182702-1993-suppl_1004
48. Motesharrei, S., Rivas, J., Kalnay, E.: Human and nature dynamics (handy): Modeling inequality and use of resources in the collapse or sustainability of societies. Ecological Economics **101**(0), 90–102 (2014). http://dx.doi.org/10.1016/j.ecolecon.2014.02.014. http://www.sciencedirect.com/science/article/pii/S0921800914000615
49. van Mourik, M., Slingerland, O.: The misunderstood crisis it's the energy, stupid! L'artilleur (2014)
50. Nordhaus, W.D.: A Question of Balance: Weighing the Options on Global Warming Policies. Yale University Press (2008)
51. Nordhaus, W.D., Stavins, R.N., Weitzman, M.L.: Lethal model 2: The limits to growth revisited. Brookings Papers on Economic Activity **1992**(2), 1–59 (1992)
52. Odum, H.T.: The ecosystem, energy, and human values. Zygon **12**(2), 109–133 (1997)
53. Romer, P.M.: Endogenous technological change. Journal of Political Economy **98**(5, Part 2), S71–S102 (1990). https://doi.org/10.1086/261725
54. Rosen, R.A., Guenther, E.: The economics of mitigating climate change: What can we know? Technological Forecasting and Social Change **91**, 93–106 (2015). https://doi.org/10.1016/j.techfore.2014.01.013. http://www.sciencedirect.com/science/article/pii/S0040162514000468
55. Sakai, M., Brockway, P.E., Barrett, J.R., Taylor, P.G.: Thermodynamic efficiency gains and their role as a key 'engine of economic growth'. Energies **12**(1) (2019). https://doi.org/10.3390/en12010110. http://www.mdpi.com/1996-1073/12/1/110
56. Samuelson, P.A., Nordhaus, W.D.: Macroeconomics, 18th edn. McGraw-Hill Irwin, New York, New York (2005)
57. Solow, R.: A Contribution to the theory of economic-gowth. Quarterly Journal of Economics **70**(1), 65–94 (1956). https://doi.org/10.2307/1884513
58. Solow, R.: Technical change and the aggregate production function. review of economics and statistics **39**(3), 312–320 (1957). https://doi.org/10.2307/1926047
59. Solow, R.M.: The last 50 years in growth theory and the next 10. Oxford Review of Economic Policy **23**(1), 3–14 (2007). https://doi.org/10.1093/oxrep/grm004
60. Turner, G.: Is global collapse imminent? an updated comparison of the limits to growth with historical data. MSSI Research Paper No. 4 (2014)
61. Wade Hands, D.: More Light on Integrability, Symmetry, and Utility as Potential Energy in Mirowski's Critical History. History of Political Economy **25**(suppl_1), 118–130 (1993). https://doi.org/10.1215/00182702-1993-suppl_1010
62. Warr, B., Ayres, R., Eisenmenger, N., Krausmann, F., Schandl, H.: Energy use and economic development: A comparative analysis of useful work supply in Austria, Japan, the United Kingdom and the US during 100years of economic growth. Ecological Economics **69**(10), 1904–1917 (2010). https://doi.org/10.1016/j.ecolecon.2010.03.021. http://www.sciencedirect.com/science/article/pii/S0921800910001175
63. Warr, B., Ayres, R.U.: Useful work and information as drivers of economic growth. Ecological Economics **73**, 93–102 (2012). https://doi.org/10.1016/j.ecolecon.2011.09.006. http://www.sciencedirect.com/science/article/pii/S0921800911003685
64. Wexler, B.E.: Brain and Culture neurobiology, ideology, and social change. MIT Press (2008)

Chapter 7
Summary of U.S. Energy and Economic Trends

Thus far the book has presented and described data trends, presented quotes that explain the energy and economic narratives, and discussed the concepts of systems thinking. The last chapter provided a background on the history of economic theory and modeling of growth, and described why many researchers, including myself, see the need to use modeling approaches that differ from the mainstream neoclassical approach. Thus, the last chapter made the shift in tone from one of description to one of interpretation (of economic growth modeling), and this chapter, as well as the rest of the book, continues with this more interpretive approach.

Most of the time news stories focus on very few high-level indicators such as GDP and employment. But if considering too few measures, we cannot sufficiently distinguish the economy of today from that of 50 years ago, much less 100 years ago, and thus we cannot provide enough insight into the likelihood of future economic outcomes. If you've seen the cockpit of a jet aircraft, there are many indicators and dials, each providing a crucial piece of information for the pilot to understand the overall operating state of the aircraft. To understand the state of the economy, we also need several indicators.

By simultaneously considering the trends of multiple metrics we can see within a systems context that the U.S. economy has undergone significant change, and that various physical and social outcomes are interdependent. For example, we cannot make a policy change, say to ensure an increasing share of GDP goes to employees, and pretend the change is independent of physical resources consumption. We cannot assume that a major change to the energy system will have no impact on jobs and wages. The political and social movements of the last century succeeded and failed within the context of the energy system that existed at the time.

This chapter, via Figs. 7.1 and 7.2, combines several of the previously described U.S. trends in one place. Presented in this manner we can see how the distribution of money within the economy changes along with both the quantity of energy consumption per person and the cost of the food and energy resources required to feed ourselves and operate the economy.

© Springer Nature Switzerland AG 2021
C. W. King, *The Economic Superorganism*,
https://doi.org/10.1007/978-3-030-50295-9_7

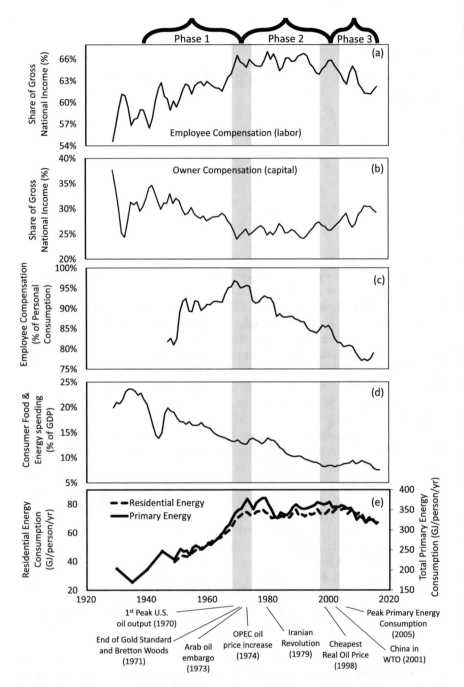

Fig. 7.1 The U.S. experienced three structural growth phases (transition periods in gray shaded columns) over the last 70–90 years. The percentage of U.S. income going to (**a**) workers (or labor) and (**b**) owners of capital, as in Fig. 4.12. (**c**) Household compensation divided by spending (%). A number less than 100% means households spend more than they receive. [Compensation of Employees (BEA Table 1.12) divided by Personal Consumption Expenditures (BEA Table 2.3.5)] (**d**) Consumer spending on food and energy divided by GDP (%) from Fig. 2.9. (**e**) U.S. per capita total primary energy consumption (right axis, solid line) and residential energy consumption (left axis, dashed line) in Fig. 5.4

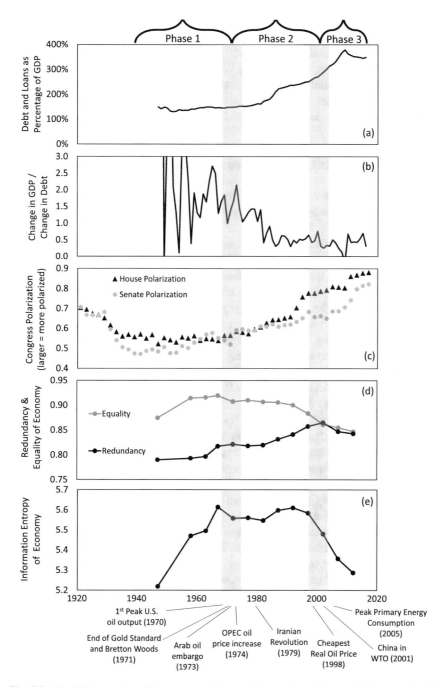

Fig. 7.2 The U.S. experienced three structural growth phases (transition periods in gray shaded columns) over the last 70–90 years. (**a**) U.S. total credit ratio (debt and loans divided by GDP). (**b**) U.S. debt productivity of Fig. 4.7. Values above one indicate one dollar of debt translates to more than one annual dollar of GDP. (**c**) U.S. Congressional Polarization of Fig. 4.14 (larger values represent more polarization). (**d**) and (**e**) Information theory metrics of the U.S. sector-to-sector economic transactions of Fig. 5.8, where larger values represent more equal distribution of money throughout the economy and more complexity

Two transition periods separate three general phases of operation of the U.S. economy. **Phase 1** runs from just after World War II until the late 1960s and early 1970s. **Phase 2** runs until a time period around the turn of the twenty-first century, and **Phase 3** continues since then. This book emphasizes the need to consider economic size and structure. Each phase is characterized by different structural change.

Chapter 4 introduced these three phases when discussing how the share of gross national income has changed its distribution over time (Fig. 7.1a and b), and Chap. 5 referenced these phases when describing how money flows within the economy (Fig. 7.2d and e). Chapter 6 provided some background on economic theory and modeling that helps understand the phases of the U.S. economy interpreted within this chapter.

During Phase 1 (1945 to late 1960s/early 1970s), the U.S. was the only major industrialized nation without significant physical damage and/or major loss of population related to World War II. Energy resources, and in particular oil, were abundant and cheap. Absolute and per capita energy consumption expanded more rapidly than any time in U.S. history (Fig. 7.1e). The constraint on U.S. economic expansion was not the capacity to increase resource extraction or manufacturing (e.g., as we saw in Chap. 5), it was in the capacity to increase resource consumption. In other words, economic growth was limited by the number of devices that *consumed energy*, not by the number of devices that *extracted energy*.

Coming out of World War II the U.S. promoted consumption of the abundant energy via a consumer economy. After the war, the U.S. also provided financial support and debt forgiveness to European countries, including Germany. The U.S. needed consumers of its products whether they resided in the U.S. or not. Considering the U.S. economy as an organism, it operated outside of the scaling law that constrains animals to increase food intake less than they increase their mass. During Phase 1, the U.S. economy increased energy intake faster than it increased its "mass," or capital, and GDP. Over a span of 25 years it was as if the U.S. economy grew from the size of a dog to the size of a bison, but instead of consuming the energy of a bison, it consumed the energy of an elephant.

But exponential growth on a finite planet cannot continue forever, and it didn't. The U.S. growth trend changed to Phase 2 in the early 1970s, running to the early 2000s. The U.S. could no longer instantaneously extract a higher flow rate of energy by simply opening the oil well valves in Texas. OPEC was no longer willing to sell oil at a low price, and by this time its members had renegotiated contracts with international oil companies to obtain larger shares of oil royalties. As a result, physical and economic access to energy now constrained the U.S. economy. The first 10 years of Phase 2 are constrained both by the amount of energy-extracting capital seeking new energy resources and the inefficiency of the existing energy-consuming capital such as cars and buildings.

Recall from Chap. 5 that the concept of energy efficient consumer end-use devices practically did not exist before the 1970s. Efficiency standards for cars, refrigerators, and other appliances were passed into legislation after the 1973/1974 oil crisis, not before. It was not until the mid-1980s that the U.S. consistently consumed more energy than in 1973, and economic growth was higher. It appeared as though the 1970s were a bump in the economic road.[1]

While there was economic growth in Phase 2, it differed from that in Phase 1. The transition to Phase 2 marks a significant change in the *distribution of money*, or structure, within the U.S. economy. First, as Fig. 7.1a shows, during Phase 1 workers earned an increasing share of national income, but during Phase 2 this "compensation share" stagnated while the "wage share" declined (refer back to Fig. 4.12).

Increasing per capita energy consumption correlates with increasing wages and benefits, and vice versa. The "HARMONEY" economic growth model summarized in Chap. 6 provided theoretical support for why wage share declines with constant per capita energy consumption (see Fig. 6.6) [1]. Second, Fig. 7.1c shows that while workers have traditionally spent more each year than their total compensation in salaries and benefits (values less than 100%), their increasing compensation during Phase 1 brought them closer to parity. This trend reversed during Phase 2 as workers kept spending more while their compensation remained stagnant (also recall Fig. 4.11).

The transition from Phase 1 to 2 also marked the end of the Bretton Woods agreement, as the U.S. eliminated the link between the value of the dollar and gold, effectively removing any physical basis to world money supply. *This adjustment is likely not a coincidence, but a response* to a major energy constraint for the first time since industrialization. As noted in Chap. 4, the end of Bretton Woods led to a new regime where commercial banks became a major source of money creation by issuing loans. Thus, during Phase 2 U.S. debt increased considerably faster than GDP. Unlike Phase 1, during Phase 2 each dollar of debt issued no longer triggered the flow of more than one dollar of GDP (Fig. 7.2b).

These changes to energy consumption, wage distribution, and debt accumulation are all interrelated, and they coincide with increasing polarization between the Democratic and Republican parties (Fig. 7.2c).

Phase 3 starts around the year 2000. Multiple indicators change direction or reach levels unprecedented in the previous 80 years. Measured as a fraction of GDP, both industrial spending on energy as well as consumer spending on energy and food stop declining (Fig. 7.1d), and this cessation marks the end to a cost decline that started with industrialization and the fossil fuel era over 100 years ago. Per capita energy consumption (Fig. 7.1e) and total worker compensation share (Fig. 7.1a) no longer hold constant as during Phase 2, but now start to decline for the first time in the data set. In fact, all metrics of the internal "structural complexity" of the economy decline, indicating increased concentration of money flows among certain sectors (primarily energy and banking)—*the exact opposite of the glorious growth years of Phase 1* (Fig. 7.2d and e). In addition, the 2008 global financial crisis triggered a

---

[1]Using data from the Bureau of Economic Analysis, Table 1.1.6 of real GDP in chained 2012 dollars shows that the 10 years after 1983 saw average annual GDP growth at 3.4%, higher than the 2.3% for the 10 years following 1973.

global recession; global debt followed a similar trajectory as U.S. total credit that peaked near 380% of GDP as each new issuance of U.S. debt no longer related to any increase in GDP (Fig. 7.2a and b).

Using the Data and Trends: Takeaways for Interpretation

What are the overall takeaways from the summarized trends in Figs. 7.1 and 7.2? Chapters 5 and 6 stated the theoretical reasons why more energy and efficiency translate to economic growth, or increased size. But economic distribution is also affected by the rate of energy consumption, and hence economic growth.

When energy consumption increased rapidly the U.S. was able to more equally and more broadly distribute money within the economy. This relationship holds whether we consider the distribution of GDP between labor and capital, the distribution of money among economic sectors, or the distribution of votes within the U.S. Congress. Votes were less partisan when energy consumption per person increased, and there was more political "pork barrel" spending to go around. Further, after energy consumption per person peaked and the U.S. no longer backed the dollar with gold as a physical substance, consumer and private debt increased at rapid rates until triggering the Great Recession of 2008–2009.

These linkages between energy and monetary distribution force us to question the underlying relationships connecting our concepts of work, jobs, money, natural resources consumption, and whatever we think "the economy" is or should be. Because an increasingly equal distribution of money coincided with an extension of social and political rights, many people believe that increased income equality is the key to social cohesion. Social and behavioral economic data provide support to the idea that times of higher income equality coincided with more political cohesion in the U.S.

These broad takeaways gloss over many important details, but they effectively summarize the long-term trends. With this context, the Great Recession was not solely the result of anomalous lending practices of the previous decade. It was the culmination of four decades of declining purchasing power of the typical American worker, coinciding with the slower energy consumption that began in the early 1970s.

This conclusion also poses much larger questions than social scientists and political economists usually consider. Are income inequality, social harmony, and political cohesion significantly influenced by energy and resource consumption? Specifically, were the post-1970s changes in economic rules and social policies triggered by energy constraints? And, are political and social changes reactions to our changing energy situation, or are they independent choices? In deciding how you might answer these questions, consider a popular cultural perspective of the three phases of the U.S. economy.

From Happy Days to Today

The early twenty-first century has brought winds of change in Western politics. As this book proposes, these changes are not independent of the cost of energy and natural resources. In 2016, Britain voted to leave the European Union, and Donald Trump was elected as the 45th U.S. president. Both election results, though narrow, showed increasing nationalist tendencies driven by rich-world voters who believe they've obtained too few benefits from a globalizing economy. Much of this response stems from nostalgia. The United States is still the dominant global political and economic player, and Britain was the previous. But many Americans see this dominance waning, and they want to retain it. In short, older Americans now live during Phase 3 of Figs. 7.1 and 7.2, but they remember the good times (high economic growth, good wages, low debt) of Phase 1.

From 1950–1980 the United States (and much of Western Europe) achieved historically unprecedented high levels of income equality. Cheap and increasing energy consumption along with distributive policies for incomes enabled good access to a middle-class lifestyle. American television shows and movies of the 1960s and 1970s demonstrated this lifestyle, at least for white families. *Leave it to Beaver* (1957–1963) portrayed traditional gender roles with a single working father easily providing for a family of four. While *The Brady Bunch* (1969–1974) broke social ground for acceptability of second-marriage families in the 1960s, it still portrayed a single working parent (Mike, the father) providing for six children while augmenting the traditional domestic role of the mother (Carol) by paying for a live-in maid (Alice). During the course of the show Mike became partner at his architectural firm, and thus we can at least imagine that he was in the top of the income bracket in America and able to provide a middle-class lifestyle for his effective *family of nine*.

Even *The Flintstones* (1960–1966), the first prime time animated show, featured Fred as a beloved blue collar working man living a suburban "modern stone age" lifestyle while working in the natural resources sector at a rock quarry. *The Jetsons*, another animated prime time sitcom launched in 1962–1963, was the futuristic counter to the *The Flintstones*. In the future when most chores are performed by automated machines, George Jetson has to work only 2 days per week, 1 h per day.[2] This lifestyle allows George, his wife, and two children to tussle with common family shenanigans while still complaining of remaining inconveniences, even with the service of their robot maid Rosie. While in the U.S. today people must work 30 h per week to be guaranteed employer-provided health insurance under the Affordable Care Act, we can only presume the Jetsons had no such problem form George's 2 h work week.

---

[2] According to the Jetsons Wiki page describing Episode 30 of Season 2, "The Vacation," original airdate November 7, 1985 (https://thejetsons.fandom.com/wiki/The_Vacation): "After a long 1 h day of pushing a button, George Jetson finally gets to call it a day and head home. He and other Spacely Space Sprockets, Inc. employees spend 1 h a day, 2 days a week at their jobs."

While *The Jetsons* portrayed the future, *The Flintstones* remained the most successful animated television show until *The Simpsons* debuted in 1989. Just as in *The Flintstones*, *The Simpsons* portrays the husband, Homer, as a bumbling, careless, but lovable everyman. Like Fred Flintstone, he supports his family of four by working in the energy and resources sector as a safety inspector at a nuclear power plant. Like George Jetson, Homer only pushes buttons.

But the show perhaps most reminiscent of the 1950s and 1960s U.S. was *Happy Days*. You can't get more succinct than that: happy days. The show was set in Midwestern America: Milwaukee, Wisconsin. Like *Leave it to Beaver* the episodes largely centered on the childrens' issues related to growing up as well as the exploits of the motorcycle-riding bachelor Arthur Fonzarelli, known as "the Fonz." After four successful seasons, the writers kept things new in a 1977 episode where the gang visits Hollywood, California. A local water skier challenges The Fonz into ski jumping over a water pen with a shark. The phrase "jump the shark," derived from this *Happy Days* episode, has since become the description for a time in which a television show or other cultural phenomenon begins its decline in popularity. While people argue over the popularity of *Happy Days* after that episode, the show had a great run, lasting 11 seasons until 1984.

How long could sitcom viewers of the 1970s and early 1980s ignore the economic changes taking place in the U.S.? Had the U.S. "jumped the shark" on the policies driving the economic equality of the post-World War II era? The 1973 Arab Oil Embargo, coupled with the 1974 steep rise in the posted price of oil, triggered the first major oil shock that sent the Western economies into recession. After the 1979 Iranian Revolution, Iran's oil extraction declined by about four million barrels per day compared to 1977, thus triggering another oil price rise and a second recession within a decade. In 1979 a nuclear reactor at Three Mile Island had a partial meltdown, increasing anxiety about nuclear power as a reliable source of energy.

Moreover, American middle-class incomes were no longer growing like they used to. When Ronald Reagan was elected U.S. president in 1980, he promised to overcome the economic stagnation and high inflation of the 1970s. Long before Donald Trump was elected president in 2016, Reagan had already campaigned that he could "Make America Great Again."[3] As it has turned out, the average U.S. worker has earned the same real wage since that time (recall Fig. 4.11). Needless to say, "Keep American Incomes Stagnant" is not as catchy of a campaign slogan.

Today, United States politicians tell families they can again live with the rosy socio-economic outlooks known in the three decades after World War II. In the 1950s and 1960s, U.S. children were destined to be more prosperous than their parents, and the television shows of the 60s and 70s reinforced that notion. The common theme of those shows: *life is easy in America if we just don't screw it up*. It was easy in the past, should be in the present, and will be even easier in the future.

---

[3]Business Insider (2017), How Trump came up with his slogan "Make America Great Again," http://www.businessinsider.com/trump-make-america-great-again-slogan-history-2017-1.

Even the taming of the atom for nuclear power in *The Simpsons* requires little to no focus from Homer.

In the real America of the twenty-first century, however, blue collar work doesn't provide the same lifestyle as it did six decades ago, and the cost of building a new nuclear power plant is so high it risks bankruptcy for the companies involved. Of the only four nuclear reactors starting construction in the U.S. since the 1980s, cost overruns of more than double initial estimates led to bankruptcy of the power plant design company Westinghouse and the cancellation of two of the planned reactors.[4] If you want a low-carbon electricity supply, you care that current nuclear plant designs have become too expensive for almost any company to own, including most vertically integrated, or monopoly, electric utilities that get a government-backed guarantee to recover their costs and make a profit for investors.

Can these positive portrayals of the 1950s/60s U.S. become reality for all Americans today, or in other countries as well? Is the answer primarily up to politicians to create new laws and remove existing ones? What influence do we have as citizens and consumers? Do we simply need "consumer confidence" to make purchases that continue a virtuous business cycle? Are new energy technologies, such as solar photovoltaic panels and hydraulic fracturing, so transformative that our economy will go back to a past "normal" mode of shared growth and prosperity? As the energy and economic narratives show, these are big questions with few consensus answers.

Aside from relatively rapid growth across the last three generations, there is another fundamental reason why it is difficult to comprehend the constraints of our finite planet. As discussed in Chap. 6, most economists have systematically avoided and assumed away the interdependent influence of resource and debt constraints. This is the case despite the fact that the domain of macroeconomics encompasses the need to explain how our complex economic and social systems interact with natural resources. Thus, when we hear economists interviewed on the news and in the media, rarely if at all do they discuss the long-term effects of the energy system. We hear "pain at the pump" newscasts when gasoline prices spike, but this short-term story superficially evades the more systemic issues. There are journalists that realize "Economists often don't know what they're talking about," but we aren't yet getting the rigorous debate and explanations of the energy-economic linkages explained in this book.[5]

---

[4]Reuters (May 2, 2017), How two cutting edge U.S. nuclear projects bankrupted Westinghouse, https://www.reuters.com/article/us-toshiba-accounting-westinghouse-nucle/how-two-cutting-edge-u-s-nuclear-projects-bankrupted-westinghouse-idUSKBN17Y0CQ. Washington Post (July 31, 2107), S.C. utilities halt work on new nuclear reactors, dimming the prospects for a nuclear energy revival, https://www.washingtonpost.com/business/economy/sc-utilities-halt-work-on-new-nuclear-reactors-dimming-the-prospects-for-a-nuclear-energy-revival/2017/07/31/5c8ec4a0-7614-11e7-8f39-eeb7d3a2d304_story.html?utm_term=.88b022efd1cf.

[5]Robert J. Samuelson, "Do Economists Really Know What They're Talking About?" *Newsweek*, June 27, 2010 at: https://www.newsweek.com/do-economists-really-know-what-theyre-talking-about-73537. Robert J. Samuelson, "Economists often don't know what they're talking

Because most of our public policies are based upon economic analyses, improper economic frameworks allow the projection of outcomes that also are unconstrained by natural resources. In other words, the projected futures that were derived in the past end up being too dissimilar from what actually happened. After the 1970s we were told lower taxes would allow wealth to trickle down from the top, but instead it was disproportionately sucked up to the top. As a result, the citizens of developed economies have become increasingly disillusioned with "expert opinion" on economic matters.

When per capita energy consumption stagnates and the economic pie doesn't expand fast enough, people can be forced, out of necessity, to change the allocation of resources. In such cases it can be natural, and it is easier to prefer, allocations to people more similar to yourself. For example, in experiments priming people to think of perceived scarcity, such as during a recession, have shown white people lower their threshold to perceive someone as black (i.e., how "black" is a person) [2, 3]. In effect, when there is less money to go around, we seem to exclude people more easily from our group. Other research shows that people give fewer positive comments to job candidates when they are told they can hire fewer candidates [4]. Again, a situation of scarcity makes people more critical of who should be allowed "in."

Thus, when we think there is not enough (of anything) to go around, it is seemingly human nature to distrust any "other." Another political party or ideology. Another race. Another nationality. Another income bracket. The "can do" attitude and American exceptionalism prevent U.S. citizens from believing they individually are each at fault for their stagnant living standards. Either the immigrants take too many jobs (the hard right's answer) or oligarchs within corporations keep too much money for themselves (the far left's answer). While I largely agree that the data support the oligarchical explanation, and Chap. 9 discusses this further, these polarizing right and left arguments tend to ignore the influence of constraints on energy consumption.

Polarization also increased prior to the fall of the Roman Empire. Resource and financial constraints played a significant role [5]. Polarization can't happen unless there are at least two groups to put at odds with one another. We can take a systematic look at the potential causes of strife only by stepping back and considering the properties of our overall energy and economic system within which the fighting factions reside.

It makes sense to think of the combination of increased political harmony (Fig. 7.2c) and increasing allocation of GDP to workers (Fig. 7.1a) during the U.S.'s Phase 1 of economic growth as indicative of a time of resource abundance. When we think there is a lot of something to allocate, we are less discriminatory because we don't have to be when everyone can get more. Similarly, it makes sense to think of

---

about," **The Washington Post**, May 12, 2019 at: https://www.washingtonpost.com/opinions/economists-often-dont-know-what-theyre-talking-about/2019/05/12/f91517d4-7338-11e9-9eb4-0828f5389013_story.html.

the combination of increased political polarization and decreasing allocation of GDP to workers during Phase 3 of economic growth as indicative of a time of relative resource scarcity.

Consider that the retirement incomes of many Baby Boomers are not what they were promised when they entered the workforce. In 2018, 40% of Americans didn't have savings or cash to cover an emergency expense of $400.[6] Automation in the economy and fewer traditional manufacturing jobs have translated not to the utopia of *The Jetsons* but to people working more hours, not fewer, in more than one job sometimes only to maintain a constant or slowly declining living standard.

Politicians sometimes either irresponsibly or ignorantly claim they can "change" the path of fundamental physical and social trends, such as those summarized in this chapter, to return to a time when the Earth was less populated with people and capital, but more full of natural resources. And, mainstream economists continue to wonder why developed economies cannot return to "normal" growth and productivity trends of the Phase 1 decades before the 1970s.

We don't have to embrace politicians who take advice from confused economists. A group of scientists and heterodox economists have explanations that much better describe our long-term socio-economic trends (discussed in Chaps. 4, 5, and 6). We know that the changes during the last 200 years are the anomaly in human history, and the 30 years of Phase 1 of Figs. 7.1 and 7.2 are the anomaly within the anomaly. However, there is a reasonable explanation for understanding our current state of economic affairs. This explanation centers on energy resources and the technologies that put them to use.

Homo sapiens have continued to multiply, and in the process make machines that consume natural resources. Chapter 5 showed that the growth patterns of our economy match those of other living systems. Let's now think through these observed parallels between the human economy and living organisms in a more systematic, theoretical, and philosophical manner.

References

1. King, C.W.: An integrated biophysical and economic modeling framework for long-term sustainability analysis: the Harmoney model. Ecological Economics **169**, 106,464 (2020). https://doi.org/10.1016/j.ecolecon.2019.106464. http://www.sciencedirect.com/science/article/pii/S0921800919302034
2. Krosch, A.R., Amodio, D.M.: Economic scarcity alters the perception of race. Proceedings of the National Academy of Sciences **111**(25), 9079–9084 (2014). https://doi.org/10.1073/pnas.1404448111. https://www.pnas.org/content/111/25/9079
3. Rodeheffer, C.D., Hill, S.E., Lord, C.G.: Does this recession make me look black? the effect of resource scarcity on the categorization of biracial faces. Psychological Science **23**(12), 1476–1478 (2012). https://doi.org/10.1177/0956797612450892. PMID: 23085641

---

[6]Anna Bahney, CNN Money, 40% of Americans can't cover a $400 emergency expense, May 22, 2018 at: http://money.cnn.com/2018/05/22/pf/emergency-expenses-household-finances/index.html.

4. Ross, M., Ellard, J.H.: On winnowing: The impact of scarcity on allocators' evaluations of candidates for a resource. Journal of Experimental Social Psychology **22**(4), 374–388 (1986). https://doi.org/10.1016/0022-1031(86)90021-1. http://www.sciencedirect.com/science/article/pii/0022103186900211
5. Tainter, J.: The Collapse of Complex Societies. Cambridge University Press (1988)

Chapter 8
A Narrative That Works for Both Energy and Economics

Collectively we are at fault [for overshooting sustainable environmental limits], but none of us individually is to blame.[1]—Nathan Hagens (2018)

In Chap. 5 on systems thinking, we visited the idea from ecologist Howard Odum that "no system can understand itself . . . ," but the way a system can try to understand itself is to make models of itself.[2] Almost 50 years from Odum's writing, the vast majority of people don't contemplate that we reside in a system we call "the economy," and that this economy has some purpose we likely can't understand. Are we just cogs in an economic machine, or network, in which the machine is as Odum posed, ". . . doing things to keep itself regulated, adapted, and consistent with energetic laws that the individual [person] cannot envision."?[3]

To many, this kind of question is nonsense. We humans created engineered systems, economic markets and regulations, and political systems. What do you mean we can't understand how these systems work and what they are for?

Because systems are defined by interconnected parts, even when we can describe the function of each part, we can't necessarily describe the function or purpose of the system itself. As implied in the quote from Nate Hagens at the start of this chapter, individually we make choices that might make sense to us, but collectively we induce outcomes that might not be what we expect. This is just like the Jevons Paradox.

---

[1] WEP2018 TV: Energy, Money and Technology—From the Lens of the Superorganism at https://youtu.be/2DpfsqjQbP0.

[2] Odum [27, p. 119].

[3] Odum [27, p. 119].

© Springer Nature Switzerland AG 2021
C. W. King, *The Economic Superorganism*,
https://doi.org/10.1007/978-3-030-50295-9_8

The Economy is a Superorganism

In the context of naturalism from Chap. 5, let's now create an energy and economic narrative that is self-consistent, consistent with the physical world, and relevant for our "purposes in the moment" for understanding the coupled relationships among important energy and economic macro trends. The coherent and combined energy and economic narrative is the following:

> *The global economy is a superorganism.*[4]

Where does the narrative of the global economy as a superorganism concept reside on the narrative chart in Chap. 1? For the energy narratives, it is right down the middle. The superorganism considers all energy options and chooses those that most enhance growth and maintenance. For the economic narratives, it is also down the middle. The superorganism is consistent with techno-realism because its structure is influenced by physical environmental constraints. The superorganism narrative also achieves its purpose by making use of techno-optimistic principles such as markets that provide information "at the moment" in the form of price signals that represent the marginal cost to grow.

This last statement is why interpreting the economy as a superorganism presents a consistent viewpoint for the majority of economists that promote markets and prices to inform investment decisions.

Interpreting the economy as a superorganism also presents a consistent viewpoint for natural scientists such as biologists, ecologists, and physicists as well as economists within the fields of *ecological and biophysical economics.* Just as each organ provides one or more roles within our body, each country provides a set of roles within the global economy. Just as each organ in the body does not have the same metabolism-to-mass ratio, each individual country economy does not have the same energy-to-GDP ratio, or economic metabolism.

Per naturalism, each process in the body and economy follows the laws of physics, but the networked collection of processes and subsystems in each mean that they can lead to emergent scaling laws, such as Kleiber's rule, that relate energy and size in a non-intuitive manner. Recall from Chap. 5 that emergent properties are short-hand descriptions that enable us to model the system in a way that is consistent with observations yet can be ignorant of the actual interactions among elements within the system. Temperature is an emergent property from a

---

[4]For additional information and interpretations of the economy as a superorganism, see work by Carsten Herrmann-Pillath [14, 15] and his website https://www.cahepil.net/, Howard T. Odum [25, 26, 28], D. J. White and Nathan Hagens [31], and additional work by Nathan (Nate) Hagens [9] and information from his "Reality 101" course at the University of Minnesota. Search YouTube "Reality 101 - UMN Nexus One" for associated videos. For example talks see"WEP2018 TV: Energy, Money and Technology—From the Lens of the Superorganism" at https://youtu.be/ 2DpfsqjQbP0 and "The Resilience Gathering—Keynote Address "The Human Predicament" at https://www.youtube.com/watch?v=MNzLkdr7UIU.

collection of particles and molecules, but scientists conceived of and measured the temperatures of systems before they understood the molecular structure of nature.

Scaling law emergence necessarily comes from some amount of ignorance, or lack of information. Thus, if we knew everything exactly and had time to perform the calculations for the interaction of everything, we'd not have to use these short-hand estimates for describing relationships within systems composed of multiple interacting parts. We can go far as to say emergent ideas are core components of narratives! The more data we throw away, the less constrained is the narrative.

If emergence and ignorance go hand in hand, what can we predict? Why use scaling laws and trends relating energy, efficiency, and useful work to GDP if they are so generic? Animals evolve over time, and we can't predict what new species will emerge. Likewise, we can neither predict the precise use of existing energy technologies nor the development of new technologies. To at least have something to say about the future, we'll next consider three concepts.

First, we'll draw the boundary around the economy as if we consider it an ecosystem rather than a single animal. Second, we'll introduce the analogy of evolution within biology and the economy. Third, we'll consider the purpose of an ecosystem, which leads us to question not only the purpose of our economy but also our collective or individual human agency to direct the global economy in a stated or preferred direction, such as to reduce greenhouse gas emissions to mitigate climate change.

The Boundary of the Economy

Recall from Chap. 5 the principles of systems analysis start conceptually simple. There is a boundary. Matter, energy, and information cross the boundary. An ecosystem is a system composed of multiple interacting organisms within some environmental boundary. Resources such as sunlight, water, air, and nutrients, as well as organisms, enter and leave the ecosystem by crossing its boundary. We can also consider some resources as residing within the ecosystem (e.g., rocks and minerals, landscapes).

From a theoretical perspective, is there a difference between thinking of metabolic systems as single organisms versus entire ecosystems composed of many organisms? Not really. Individual animals and colonies of ants have similar metabolic scaling laws. However, the difference between these two concepts is not as straightforward as it might seem. For example, each of us thinks of ourselves as one body and thus one organism. However, our bodies include symbiotic *E. coli* bacteria within our large intestine, each of which themselves is an individual organism. Our bodies provide the environment and food for those bacteria, and in turn they provide valuable functions for our bodies, or their environment. We can conceive of our bodies as both a single organism and a single ecosystem within which multiple organisms reside. But who is in charge? Our body or the bacteria? Or is it something else?

Ever since the concept of evolution was brought into mainstream science by Charles Darwin's *The Origin of Species*, scientists have struggled to define the biological system boundary that defines agency.

By agency, here I mean the capacity to make choices for directed action to achieve a desired outcome. Do individual living plants and animals have agency? What about the collective organisms within a species? What are the desired outcomes? Generally speaking in biological evolution, the main desired outcome is to survive and reproduce. To have survival as a goal, an entity needs the concept of a boundary between itself and the environment. To wit:

> As soon as something gets into the business of self-preservation, boundaries become important, for if you are setting out to preserve yourself, you don't want to squander effort trying to preserve the whole world: you draw the line. You become, in a word, *selfish*. This primordial form of selfishness ... is one of the marks of life.[5]—Daniel Dennett (1991), (emphasis added)

Just what is preserving itself, and does it have to be *alive* by any definition? Richard Dawkins' 1976 book *The Selfish Gene* explored this question and suggested that genes, not whole living organisms, are the basis for interpreting evolutionary fitness. Since the term *selfish* has too many preconceived notions with it, Dawkins stated that he should have titled the book "The Immortal Gene" because it is the information encoded in genes that is passed on to subsequent generations of organisms.

There is no consensus on a single definition of a gene, but we do not need to discuss the various definitions for our purposes here.[6] Just keep in mind that genes are portions of DNA that are themselves components of organisms.[7]

While we can consider organisms as being alive, genes (and DNA) are not generally considered alive because they are only chemical compounds, like water. Scientists studying the origins of life, or how the first living organism first came into existence, don't yet have a consensus definition of life. For our purposes of defining life, consider a living system as one that takes in energy and matter from outside of its boundary and consumes the energy to transform the matter into structures that perform various functions. These functions enable the matter and energy intake and transformation processes to be repeated such that the organism maintains a structure distinct from its environment. A mountain is an example of something that has structure but is not alive. A mountain has a structure in the sense that there are specific layers of rock and it protrudes from the surrounding landscape such that we can distinguish it from the surrounding landscape. However, the wind and rain

---

[5]Dennett [7, p. 174].

[6]For a brief discussion of the applicability and translation of the idea of the gene as the basis of evolution, see a discussion on David Dobb's article "Die, Selfish Gene, Die" at https://aeon.co/essays/dead-or-alive-an-expert-roundtable-on-the-selfish-gene as well as the original article at https://aeon.co/essays/the-selfish-gene-is-a-great-meme-too-bad-it-s-so-wrong.

[7]DNA = deoxyribonucleic acid.

slowly erode the mountain, rounding off its peaks and ridges, and in the process, remove layers of rock that do not get replaced. Thus, the mountain does not take in matter and energy to maintain its structure.

However scientists define a gene, a set of them in DNA represents encoded information, or instructions. An organisms' genes define its genotype. The genetic information passed from one organism to another guides each organism to develop relatively unique physical characteristics and behaviors, or phenotypes, that we can observe. Each organism is composed of the same building blocks of matter: carbon, hydrogen, oxygen, etc. The genetic instructions constrain how organisms obtain material and energy resources that they metabolize and turn into structure in order to grow, maintain themselves, and potentially produce offspring who continue the existence of the information encoded in the genes.

In the context of naturalism and systems thinking, we can use the three concepts of *information, matter, and energy* to have a basic appreciation for how organisms operate and grow. Genes encode the recipe (information) for both combining the ingredients (matter) and guiding the organism's activities (that consume energy) that maintain its structure and respond to environmental conditions.

We now translate the concepts of the gene and the organism to the economy.

Economic Evolution

Organism maintenance, growth, and reproduction all fall into the economic concept of investment. Capitalism largely focuses on how much to invest in any particular type of activity. Austrian economist Joseph Schumpeter linked the concept of biological evolution with changes in capitalist economies. In his 1942 book *Capitalism, Socialism, and Democracy* he noted:

> ...in dealing with capitalism we are dealing with an evolutionary process. It may seem strange that anyone can fail to see so obvious a fact which moreover was long ago emphasized by Karl Marx.[8]—Joseph A. Schumpeter (1942)

He then goes on to say:

> The opening up of new markets, foreign or domestic, and the organizational development from the craft shop and factory ... illustrate the same process of industrial mutation—if I may use that biological term—that incessantly revolutionizes the economic structure from within, incessantly destroying the old one, incessantly creating a new one. This process of Creative Destruction is the essential fact about capitalism. It is what capitalism consists in and what every capitalist concern has got to live in.[9]—Joseph A. Schumpeter (1942)

Here we see Schumpeter's famous term *creative destruction* that he relates to what he called "industrial mutation," using a biological analog. Figure 8.1 shows the

---

[8]Schumpeter [2, p. 82].

[9]Schumpeter [2, p. 83].

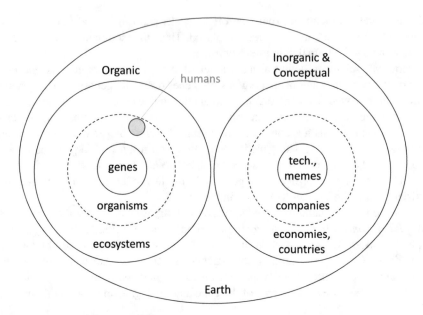

Fig. 8.1 Just as organisms' bodies are the "survival machines" for genes, and these organisms reside within ecosystems on Earth, companies are "survival machines" for technologies, and these companies reside within economies on Earth

parallels between Dawkins' selfish, or immortal, gene and Schumpeter's creative destruction. To make the translation, consider another term coined in Dawkins' 1976 book: *meme* (rhymes with "cream"). Just like a gene propagates, along with its encoded information, so does a meme. In case you thought the word meme originated with cat videos on the Internet, it did not.

> I think that a new kind of replicator has recently emerged on this very planet. It is staring us in the face. It is still in its infancy, still drifting clumsily about in its primeval soup, but already it is achieving evolutionary change at a rate that leaves the old gene panting behind.
>   The new soup is the soup of human culture. . . .
>   Examples of memes are tunes, ideas, catch-phrases, clothes fashions, ways of making pots or of building arches. Just as genes propagate themselves in the gene pool by leaping from body to body via sperms and eggs, so memes propagate themselves in the meme pool by leaping from brain to brain via a process which, in the broad sense, can be called imitation.[10]—Richard Dawkins (1976)

Our choices are largely constrained and shaped by "the soup of human culture" into which we are born. Dawkins had linked the concept of replicating genetic information via biological evolution to ideas that spread via cultural evolution. The organism is the survival machine that sets the boundary for the genes encoded within it just as companies and countries set the boundary for cultural the ideas within

---

[10]Dawkins [6, p.192].

them. Our culture is largely defined by what we think the economy is, and thus the economy sets the context for the propagation of ideas, or memes.

In the economy, technologies play the role of genes. Technology here is meant to be quite general. Technologies can be processes, systems of organization, new machines, and similar concepts. In this context, technologies are memes. Importantly, just as the same genes can reside within multiple species, organisms, and ecosystems, the same technologies can reside within multiple companies, industries, and economies.

For example, computers and vehicles are used by almost every company and industry in modern economies, and these ubiquitous technologies are sometimes called *general purpose technologies*. Just as the types and numbers of organisms organize into a hierarchical structure within ecosystems, companies and industries organize into hierarchical structures within economies.

Many others besides Schumpeter have seen parallels between ecosystems and economies. Recall in Chap. 5 I demonstrated how changes in the hierarchical structure of the U.S. economy relate to energy consumption and cost: higher growth rates in energy consumption translated to more equal distribution of money. I derived this conclusion by calculating information-based metrics from data on the U.S. economic sectors—copying the method from an ecologist! No single domain of science, social or natural, completely owns the space of ideas for describing the size and structure of physical, social, and economic systems.

Importantly, in the context of economic evolution, the entities that get created or destroyed are the individual companies, not the memes or industries. Species, organisms, and companies come and go, but successful technologies and genes tend to stick around in many companies. Recall that older primary energy sources, such as biomass and coal, haven't gone away as we learned to more effectively extract oil, gas, wind, and sunlight (refer back to Fig. 2.10). We've invented many types of light bulbs, but we still make candles. The "stone age" didn't end because we're still using stone tools in the form of diamond-tipped blades and sand for hydraulic fracturing. We've invented many technologies to generate electricity, but in most power plants we still burn stuff to generate steam to flow through steam turbines.

As usual, there are exceptions to simple rules. One extremely influential historical energy technology no longer in practical use today is the (reciprocating) steam engine that kicked off the Industrial Revolution. Steam engine trains still exist as novelty tourist attractions, but steam engine functionality has been replaced by (reciprocating) internal combustion engines and (rotating) steam turbines and electric motors. Perhaps like descriptions of steam engines in engineering textbooks, there are still genes in our DNA that seem to serve no purpose but might only be expressed under some extreme circumstances that were more relevant millions of years ago.

Just as companies can "die" and merge, so can countries and economies. A map of Europe, Asia, or Africa today looks much different than in 1900 or even 40 years ago. Countries as economies also compete and generally have access to the same set of technologies. Historically, economies have in part been defined by large companies with headquarters in their native country. When we think of Toyota,

Honda, and Yamaha, we think of Japan. When we think of IKEA and Volvo, we think of Sweden. However, as economies have become more interconnected, individual companies spread operations around the globe. This is why Fig. 8.1 shows the organism–ecosystem and company–economy boundaries as dashed lines. There is no real clear boundary separating those concepts. We think of Apple as an American company, but manufacturing of Apple products occurs in Asia because that is where the necessary workforce resides: "Designed by Apple in California. Assembled in China."[11]

Can we translate the gene–organism–ecosystem perspective into a globalized meme–company–economy perspective?

To answer this question we can try to explain the purpose of the overall economic system that Howard Odum says we can't understand. Because of his and others' ideas, I think we can get pretty darn close. But because the explanation necessarily uses emergent concepts, it (in all likelihood) cannot be the final or only say on the matter. While the explanation does not specifically restrict choices and goal-oriented actions that define human agency, it does provide a coherent context to interpret observations at two levels: individual and system.

First, each of us individuals conceives of a very large number of choices we can make, but we tend to act on only a subset of all possible choices. To give an extreme example, we have the choice to pursue the design of a perpetual motion machine (and some do), but our confidence in the laws of nature tells us that this is an exercise in futility. Thus, most of us choose not to pursue such efforts.

Second, at the system level we observe emergent socio-economic and political trends, such as average wages, GDP, energy efficiency, and energy consumption. This system level provides constraints, such as the laws of thermodynamics. Also the conservation of flow in networks specifies that any given network can only distribute 100% of what flows within it, not more and not less, and that the inputs must equal the outputs. Consider that at the individual level, assuming money was no barrier, each American could choose to fly in a plane to Hawai'i tomorrow. But at the collective system level, there are simply not enough planes, runways, and time to fly all 330+ million Americans to Hawai'i on the same day.

The Purpose of the Economy

To give you the short version as a preface to what I'm about to explain, the purpose of the economy revolves around survival via both consuming resources and processing information as fast as possible.

---

[11]Chris Rawson, "Why Apple's products are 'Designed in California' but 'Assembled in China'," in engadget, January 22, 2012: https://www.engadget.com/2012/01/22/why-apples-products-are-designed-in-california-but-assembled/.

Another Schumpeter quote helps us to continue:

> ... since we are dealing with an organic process, analysis of what happens in any particular part of it—say, in an individual concern or industry—may indeed clarify details of mechanism but is inconclusive beyond that. Every piece of business strategy acquires its true significance only against the background of that process and within the situation created by it.[12]—Joseph A. Schumpeter (1942)

When we consider the processes of economic evolution, it takes time to know if a particular technology has or will become prominent, and the effectiveness of business strategy can only be understood relative to the environment at the time. Successful genes are those in species that survive to procreate and pass on the genetic code. Successful technologies are those that enable businesses to survive by obtaining higher profits from providing new or better products, selling more products, and reducing operating costs. These surviving businesses pass on their memes. What if laws or regulations oppose or are in conflict with such technologies? Because countries enforce laws and regulations across companies, in principle it is possible to legally and politically restrict the use of technology. In practice, it is difficult for governments to establish or enforce laws that restrict technologies or business arrangements that provide high economic growth.

Regulations usually appear *after* new technologies appear. For some cases in which existing (restrictive) regulations apply to new technologies, knowledgeable officials sometimes preemptively adjust the regulations to clear the way. For example, an amendment to the Safe Drinking Water Act that was dubbed the "Halliburton Loophole" clarified, or ensured, that regulation of hydraulic fracturing for oil and gas extraction stayed at state, rather than federal, agencies.[13]

Private companies can also act to prevent government or regulatory control. For instance, during World War II, the U.S. established the Petroleum Reserves Corporation (PRC) as a government company. At the time the American joint-venture company, the California-Arabian Oil Company, owned rights to Saudi Arabian oil. To ensure U.S. control of Middle East oil, the PRC planned to buy a majority ownership stake in that company. The private companies of California-Arabian did not want a government takeover, so they changed their name to the Arabian-American Oil Company, or Aramco (later to become Saudi Aramco, fully owned by Saudi Arabia), and raised the necessary capital to expand Saudi oil production by having Standard Oil of New Jersey and New York (now ExxonMobil)

---

[12]Schumpeter [2, p. 84].

[13]Section 322 of the Energy Policy Act of 2005 changed the Safe Drinking Water Act as follows: SEC. 322. HYDRAULIC FRACTURING. Paragraph (1) of section 1421(d) of the Safe Drinking Water Act (42 U.S.C. 300h(d)) is amended to read as follows: "(1) UNDERGROUND INJECTION"—The term 'underground injection'— "(A) means the subsurface emplacement of fluids by well injection"; and "(B) excludes"— "(i) the underground injection of natural gas for purposes of storage"; and "(ii) the underground injection of fluids or propping agents (other than diesel fuels) pursuant to hydraulic fracturing operations related to oil, gas, or geothermal production activities."

purchase a 40% share in Aramco [24].[14] In this case the private sector essentially exerted capability beyond that of the U.S. government. Over the last decade several *information* technology and social networking companies (e.g., Facebook, Google) have faced increasing scrutiny as to whether they wield too much power relative to national governments. The next chapter visits the concepts of "political will" of government representatives to speak for citizens and workers and "political power" for workers (as we'll see, within the energy industry) to gain rights for themselves.

The energy and economic data support the following description of the "purpose" of the economy:

> *The purpose of the economy seems to be to maximize the average*
> *rate of useful work output.*

This is quite close to the "... apparent goal of the present world system to produce more people with more ... for each person." as quoted from the original 1972 *The Limits to Growth* in Chap. 6.[15] More useful work is needed to produce and consume more things.

I write "seems to be" because I'm conforming to the hypothesis that my being only a part of the global ecosystem and economic superorganism disqualifies me (and everyone) from definitively assigning a purpose to the highest level systems in which I reside. In any case, this purpose is not what most people think when they consider policies guiding economic growth and distribution. It is not about creating the most prosperity, happiness, or utility for as many people as possible. It is not about income equality or racial justice. However, this purpose is compatible with both biological evolution and the proliferation of computation in computers, the Internet, and artificial intelligence. It also does not assume that humans are particularly unique or central to the grand purpose.

We are only one type of organism among many. Recall the example of the purpose of a frog in Chap. 5. It turns left, right, and backwards and extends its tongue to catch flies. From the standpoint of the immortal gene, the frog's purpose is to acquire energy by catching flies such that it has a better chance of passing on the information in its genetic code to another generation. However, for the last few 100,000 years our species is the one that resides at the apex of evolution in the sense that our actions have enabled higher rates of useful work production, by using energy and machines external to our bodies, that in turn enabled faster replication of information in genes and memes than any other species.

My conclusion is not original. I'm certainly not the first to adhere to it.

The concept of acquiring high power and work flow is broader than it first appears. It transcends disciplines, and was evident to biologists 100 years ago. In 1922, Alfred Lotka stated the following:

---

[14]This paragraph summarizes information from pages 114–115 in Timothy Mitchell's *Carbon Democracy*. For further information on U.S. and U.K. plans efforts to control access to Middle East oil post-World War II, see [24, pages 113–120].

[15]Meadows et al. [21, p. 86].

If sources are presented, capable of supplying available energy in excess of that actually being tapped by the entire system of living organisms, then an opportunity is furnished for suitably constituted organisms to enlarge the total energy flux through the system. Whenever such organisms arise, natural selection will operate to preserve and increase them. The result, in this case, is not a mere diversion of the energy flux through the system of organic nature along a new path, but an increase of the total flux through that system.[18]—Alfred Lotka (1922)

In short, Lotka says that when organisms evolve to extract some source of energy from the environment that no existing organism can extract, then that organism doesn't just divert resources from others, it enhances the energy extraction of the entire ecosystem. Howard Odum took on Lotka's idea, calling it the *maximum power principle*:

This [maximum power] principle says that the more lasting and hence more probably dynamic patterns of energy flow or power (including the patterns of living systems and civilizations) tend to transform and restore the greatest amount of potential energy at the fastest possible rate.[16]—Howard T. Odum (1977)

To transform potential energy is to convert stocks of energy in the environment, such as biomass, fossil fuels, and other energy storage, into useful work. Lotka's and Odum's concept, linking evolution and power, explains the rationale behind energy and economic narratives that posit more energy is always better. Better adapted organisms survive more than those that are less adapted to their environment. More dominant species are those that adapt in order to appropriate a higher quantity of resources. Odum claimed, and do other ecologists and biophysical economists such as Charles Hall and Kent Klitgaard [10], that the maximum power principle even applies to human society:

When ecosystems and the systems of humanity are similarly diagrammed, the patterns of systems energy are found to be similar, showing the universality of energy laws that apply to the large and the small. The tremendous complexity and variability of human individuals begin to make sense. Individual choices are a means for exploring alternatives so that humanity as a whole finds the patterns that maximize the system's energy flow. Most people are accustomed to thinking of human behavior as the cause of behavior of the larger systems. They have difficulty realizing that patterns of the system can draw from the individuals the behavior that helps the system track maximal power through competition of variant patterns for survival.[17]—Howard T. Odum (1977)

It is worthwhile breaking down this statement from Odum. First, he's saying that both natural ecosystems and human economies operate under the same physical principles governing the flow of energy. Both extract energy from the environment to maintain structure and grow. Both must adhere to the laws of thermodynamics. But the second part of the quote is what makes some people uneasy. Each one of us has unique experiences and personality, but Odum asks us to consider that our decisions are not as independent and unique as we think. Instead of our decisions being the major cause of events, it is the system in which we reside, say our economic system,

---

[16]Odum [27, p. 109].

[17]Odum [27, p. 111].

that drives our behavior. Further, just as with natural ecosystems, he poses that the goal of the human economic system is to appropriate as much power, or energy flow, from the environment as possible. Since biological organisms evolve slowly, their efficiency of converting energy into useful work and structures, such as ant colonies and spider webs, changes very slowly. Thus, for them the rate of increase in power extraction and useful work is practically the same.

Consider that the global economic superorganism selects for a distribution of organizations with traits that enable it to extract resources at an increasingly higher rate, just like natural selection selects for a distribution of organizations with traits that extract the maximum flow of energy from the environment. For the global economy the organizations are companies and country economies, and for ecosystems the organizations are organisms, species, and smaller ecosystems. The tricky part is that while we humans are programmed by genes, we also produce memes that partially define the economy. As mentioned in Chap. 5, *Homo sapiens* is a unique species in that we as members can pass along information outside of the process of evolution in forms such as books, music, stories, and computer codes.

But just why would something like "the economy" emerge as a concept? Economist Carsten Herrmann-Pillath has provided a good synthesis for relating humans, the economy, and the maximum power principle [14].[18] Consider his following statement:

> ...human agency is considered as a special case of an evolved function. Thus, the MPP [maximum power principle] as a principle of natural selection also operates for all extensions such as, in technology, the evolution of artefacts under economic selection, which matches with the existing generalizations of the principle in general ecology (Odum, 2007) [28]. That means, a steam engine, together with the human agent using it, is just another manifestation of physical inference devices which evolve, for example, in the direction of higher efficiency. Higher efficiency follows MPP in the sense of maximizing work output ... *Ultimately, the steam engine is just one way to increase the steepness of the gradient of energy dissipation, and hence, entropy production* ... [14]—Carsten Herrmann-Pillath (2011) (emphasis added)

Herrmann-Pillath is indicating that we cannot separate our memes, or technologies and culture, from our genes and the overall trend of evolution. *This makes the economy a product of evolution itself.* In much the same way anthropologist Richard Adams also linked culture and social relations to physical resources and energy, and Chap. 9 summarizes some of his thoughts on the relation of physical power to social power, or political will.

---

[18]See Carsten Herrmann-Pillath [14] for a discussion of the linkages among (1) evolution and the organization of ecosystems can relate to the human economy, (2) how humans, including our inventions and scientific understanding, are part of evolution and the maximization of power flow (i.e., resource consumption), (3) entropy production, in the thermodynamic sense of heat, from all of Earth's processes, and (4) entropy production, in the sense of information (theory) and knowledge, or the lack of information and knowledge for describing and understanding all processes and distribution of matter [14].

Our modern concept of "the economy" is relatively new. It only emerged in the 1930s. The concept of *ecosystem* was also derived in the 1930s [32].[19] I assume this timing is a coincidence, but perhaps the time was right for more holistic descriptions of the interactions within the natural world and the human exchanges of goods and services.

The mathematical characterization and analysis of "the economy," according to Timothy Mitchell, emerged to fight populist policies after the onset of the Great Depression and put limits on democratic practice.[20] If this is a consistent conclusion, then we can hypothesize that because our concept and modeling of "the economy" is the result of natural selection, evolution (and history itself) is indifferent to human well-being, much less categories of individuals such as "workers"[21] [11].

This lack of favor to human individuals is also consistent with the *immortal gene*, not each human organism, as the replicating component of evolution. The hypothesis that "The purpose of the economy is to maximize the average rate of useful work output" comes with the corollary that "The economy is indifferent to *Homo sapiens*." It is unclear how we could unambiguously test this hypothesis.

Let's spend some additional effort breaking down Herrmann-Pillath's quote.

Since we can't separate ourselves and our actions from the influence of the environment in which we live, the maximum power principle states that our actions amount to extracting higher flows of energy from the environment. Since "human agency," the idea that we make conscious decisions to achieve targeted outcomes, is simply an "evolved function" derived from the process of natural selection within evolution, then our collective decisions act to extract the maximum power.

---

[19]From [32]: "The word 'ecosystem' was first used in print by A. G. Tansley (1935) in his well-known paper on vegetational concepts and terms," and "Although the coining of the term 'ecosystem' has long been attributed to Tansley, and his 1935 paper gives no acknowledgement, this term was suggested to him in the early 1930s by A. R. Clapham when Tansley asked Clapham (then a young man in the Department of Botany at Oxford) if he could think of a suitable word to denote the physical and biological component of an environment considered in relation to each other as a unit . . . ".

[20]"The shaping of Western democratic politics from the 1930s onwards was carried out in part through the application of new kinds of economic expertise: the development and deployment of Keynesian economic knowledge; its expansion into different areas of policy and debate, including colonial administration; its increasingly technical nature; and the efforts to claim an increasing variety of topics as subject to determination not by democratic debate but by economic planning and knowhow. They Keynesian and New Deal elaboration of economic knowledge was a response to the threat of populist politics, especially in the wake of the 1920 financial crisis and the labour militancy that accompanied it and that re-emerged a decade later. Economics provided a method of setting limits to democratic practice, and maintaining them.

The deployment of expertise requires, and encourages, the making of sociotechnical worlds that it can master. In this case, the world that had to be made was that of 'the economy.' This was an object that no economist or planner prior to the 1930s spoke of or knew to exist." page 124 of [23].

[21]Harari [11, Chapter 13].

By "function" Herrmann-Pillath means any process that generates information to understand causality consistent with observations of the world. As some philosophers claim, our brains are simply prediction devices, what Herrmann-Pillath referred to as "physical inference devices." As with the philosophy of naturalism, we don't really have to know any objective reality of nature, we just have to be good at predicting outcomes. However, as we better understand observed natural phenomena, we get better at predicting outcomes.

Why do we humans seek causality to explain the natural world? To prove we have become better "physical inference devices."

Why do we want to become better physical inference devices? To consume more energy and convert it useful work?

Why do we seek to produce more useful work? Because in the context of evolution, performing work at a higher rate increases the odds that both organisms survive to pass on their genes and companies and economies survive to pass on their memes (technologies, culture).

Over time we humans have created ways to store and pass on information (e.g., narratives, books, computer storage) as memes that in turn allow us to extract more power. Herrmann-Pillath further states that "... the human observer is a special case of the more general case of systems of evolving functions that drive the emergence and diffusion of information under certain energetic constraints." [14].

Thus, it's not all about humans. We're just a special case of a more general process. Other living organisms also anticipate the future in some way, whether driven partially by genetic programming (e.g., squirrels store nuts for the winter, bears hibernate) or through learning after birth. We're just one product of evolution that processes information. All organisms do it. However, we're unique in our use of written language and other means of saving information that enable people to learn other than by direct communication with another person. Thus, we humans are the most capable at creating the information and novel technologies, such as computers and artificial intelligence algorithms, that demonstrate mastery of understanding the physical processes of the natural world. We can anticipate the very near-term future of milliseconds to contemplate a stock trade. We can anticipate the next decade to plan a research program to land on the moon. We can also anticipate a century-long plan to reduce greenhouse gas emissions.

Whether acting on long-range plans, such as mitigating climate change by limiting greenhouse gas concentrations in the atmosphere, is consistent with evolution and the maximum power principle is a question that we will return to in the final chapter.

I understand this might sound too hollow. How can humans and our economy just be about consuming more energy to convert at higher efficiency into more useful work? It seems like a lot of philosophical mumbo-jumbo to end up with a conclusion that has no meaning. This is where Herrmann-Pillath's reference to the steam engine is useful. Why would we, or any organism, invent, use, and improve upon a concept, an invention, a meme, like the steam engine, or for that matter a wind turbine or enhanced horizontal drilling with hydraulic fracturing?

I answer in three parts. First, by producing and consuming more useful work, this is the *least ambiguous* way we can demonstrate increased mastery of modeling our natural world. We update our model of the world based upon experimenting and observing natural process, interpreting the processes in the context of imposed constraints, and obtaining useful work from engineered systems. If a technology obtains more work, then we conclude the technology represents an increased understanding of the world such that by using this technology, or meme, the overall economic system can more effectively replicate itself, grow, and as a consequence consume resources at a higher rate.

There are ways we've increased our understanding of the natural world without unambiguously demonstrating how this understanding leads to more useful work. This is the purpose of basic scientific research. However, if you do indeed invent a new process to produce useful work, then you have unequivocally demonstrated and proved enhanced mastery of modeling the natural world. Those who do this are praised. Just think about what science gets funded:

> Scientific research is usually funded by either governments or private businesses. When capitalist governments and businesses consider investing in a particular scientific project, the first questions are usually, 'Will this project enable us to increase production and profits? Will it produce economic growth?' A project that can't clear these hurdles has little chance of finding a sponsor. No history of modern science can leave capitalism out of the picture.
>
> Conversely, the history of capitalism is unintelligible without taking science into account. Capitalism's belief in perpetual economic growth flies in the face of almost everything we know about the universe. A society of wolves would be extremely foolish to believe that the supply of sheep would keep on growing indefinitely. The human economy has nevertheless managed to keep on growing throughout the modern era, thanks only to the fact that scientists come up with another discovery or gadget every few years—such as the continent of America, the internal combustion engine, or genetically engineered sheep. Banks and governments print money, but ultimately, it is the scientists who foot the bill.[22]— Noah Yuval Harari (2014)

Second, with increased mastery comes increased ability to propagate memes that most accurately explain the natural world, which in turn enables our genes and memes to dominate the planet. This is the fundamental notion of evolution by natural selection. A person that starts a new business that accumulates a tremendous amount of profit can better afford to lobby governments and financiers as well as produce commercial advertisements to promote his or her ideas. However, we cannot draw conclusions that wealthier people are inherently intellectually superior to less wealthy people or that they communicate a more accurate understanding of the world. With more money one can certainly afford more communication regardless of the validity of the message. Also, no one invents technology in a vacuum. For example, Jeff Bezos, the founder of Amazon, and Bill Gates, the co-founder of Microsoft, are ranked as the two richest persons in the world. Both developed and purchased novel software and algorithms that necessarily relied on

---

[22]Harari [11, p. 314–315].

existing technologies, developed by others, such as personal computers and the Internet.

Third, we increasingly understand our own biological limitations as data and information processors or "physical inference devices," and we develop systems to overcome those limitations. Since useful work is energy times an efficiency, we might need enhanced data processing to both extract more energy and use it more efficiently. If humans as biological organisms are inhibiting the growth of the economic superorganism, would the economy favor inorganic forms of intelligence? In a more specific question, is the economy trying to substitute silicon-based networked computer brains for human brains?

A Maximizing Superorganism: What Could Go Wrong?

Recent popular press books emphasize the trend toward higher rates of information processing within a new age of data-seeking artificial intelligence (AI). Some fear or anticipate that AI systems will acquire a *superintelligence* that surpasses human capability [4, 12]. The recent advances in AI, such as DeepMind's AlphaGo algorithm defeating the legendary world Go champion Lee Sedol in 2016 (4 out of 5 games), shows that we can create systems that perform a complex set of strategic decisions that are beyond human capacity. Not to be outdone by itself, in 2017 DeepMind used a newer algorithm, AlphaGo Zero, which, once being programmed with the rules, mastered the game of Go after playing itself for 40 days.

This is the same as the W.O.P.R. (War Operation Plan Response) computer playing "global thermonuclear war" against itself in the 1983 movie *WarGames*. To avert W.O.P.R. from starting World War III all by itself, the computer whiz-kid protagonist believed the computer would learn by itself that the "game" was fruitless because neither side (the U.S. or the Soviet Union) could actually win a global thermonuclear war. Because that game took too long to simulate, he first had it play itself in tic-tac-toe, a much simpler game of futility. W.O.P.R. consumed more and more power as it played itself faster and faster, but eventually it learned its lesson in the nick of time to avoid initiating a nuclear missile strike!

In the real world of today, after 40 straight days of playing itself in Go, AlphaGo Zero was able to defeat the previous AI algorithm, AlphaGo, in 100 consecutive games, while using less computing power. While *WarGames* ended with hugs and cheers that the computer learned not to start a nuclear war that might destroy humanity, many computer experts wonder if real-world algorithms can learn the same lesson. Contrast *WarGames* to the 1984 movie *The Terminator* which spawned an entire series. In *The Terminator*, the artificial intelligence system Skynet, built for military purposes just as W.O.P.R., does not learn W.O.P.R.'s lesson. Instead, when it becomes self-aware, it sees humans as a threat to its own existence and wants to kill them. It successfully initiates a first strike nuclear attack from the U.S. on Russia, and Russia retaliates causing the desired nuclear holocaust that wipes out most of humanity. How can we imagine a computer system seeking some goal without regard for destroying humanity in the process?

One of the fundamental (and fun to think about) arguments for why computers might take us over is the "utility monster" posed by philosopher Robert Nozick. Chapter 6 introduced the use of the concept of utility in economics, and philosophers us it as well. *Utilitarianism* is the idea that there is some quantifiable but perhaps non-physical "utility" of existence, of humans, AI, or any conscious being, in which more of it is always better.

In utilitarianism, only the maximum utility of the group is what matters. The distribution of utility within the group does not matter. Consider the idea of *marginal utility*, which is how much utility comes from achieving the next increment of utility on top of what you already experience. Economists typically assume, supported from analysis of consumer data, that we value each additional unit of the same item less and less. I think about buying bananas at the grocery. Bananas only remain ripe for about a week once you buy them, so I don't buy 100 of them at a time. I do like bananas, but even buying bananas for the next week I usually don't buy seven at a time. I break apart the small bunch of bananas, usually wrapped in groups of five to seven, so that I only have two or three, because I don't want to eat a banana every day. In other words, by the time I eat a banana for the 4th day in a row, I'm tired of eating bananas. So my marginal utility of the fourth banana is quite low.

Now combine marginal utility with the idea that we could make a computer with artificial intelligence that becomes autonomous and so superintelligent that it has the power to control our physical world through the Internet and cellular communications linked to our homes, factories, and cars. Because this superintelligent AI system might be much smarter than all of humanity, it could have a higher marginal utility than that of humans. As a point of comparison, let's say I consume another kilowatt-hour of energy in my computer searching the Internet, just looking at puppy videos, rather than learning new fundamental concepts about energy and the economy. As an alternative, this AI machine could conclude that, because it is so smart, its consumption of that extra kilowatt-hour enables it to become even smarter and "happier" while accessing more utility and performing more useful work than we humans can do with the same kilowatt-hour. Thus, not only is its total utility maximized by its consuming 100% of electricity, or other natural resources, but it would also maximize overall utility of the world simply by maximizing its own utility. Compared to this utility monster AI computer, our individual utilities are too small to matter.

Nick Bostrom pontificated an AI thought experiment of the utility monster as a paper clip maximizer [4]. The paper clip is an arbitrary placeholder for a goal, but one that is meant to seem to invoke no moral significance. What if the superintelligent AI system only tries to maximize the number of paper clips? The AI system would likely determine that keeping humans alive is detrimental to the goal of maximizing the number of paper clips in the world. Humans need resources to survive, and these resources could be used to make more paper clips. It is not that the AI machine dislikes or specifically tries to harm humanity. It is just that the superintelligent AI system is indifferent to our existence. More paper clips equals more utility, and we humans just get in the way of that goal.

The conceptual value of utilitarianism is that utility is ultimately determined by each individual conscious being's subjective desires. The problem is that we can't exactly know each person's utility to add them together. People who are worried about the problem of creating a superintelligent AI utility monster understand this. Even if we humans define a seemingly benign utilitarian objective (e.g., paper clips) for the AI system, it might not interpret that utility in the same way that we anticipate.

The concept of utility is embedded in mainstream economics. While utility can in principle be measured by many concepts such as happiness, contentment, how many good meals you've had in the past week, or whether you have health insurance, in economic practice it is often based upon how much and on what you spend as a consumer. Thus, economists often solve their economic growth equations by "maximizing discounted utility," and discounting practically translates to a couple of assumptions.

First, assume you know how much consumers purchase each year indefinitely into the future. Of course we don't really know this, but the next assumption means economists don't have to peer too far into the future. Second, assume how much consumers value consuming each year. This is the concept of *discounting* that describes how we tend to value consuming today more than next year; we value consuming next year more than the year after that; and so on. The discounting concept is consistent with evolution and the maximum power principle. This is because organisms that consume more power "today" tend to be more likely to survive to "tomorrow" and thus have the ability to pass on their genes. Similarly for the economy as a superorganism, companies that make more profits this year tend to be more likely to survive through next year.

Because economic calculations are heavily influenced by the chosen *discount rate*, some fierce policy debates have centered on that choice. In discussions surrounding how much to spend on climate change mitigation, people that agree on everything but the discount rate can come to wildly different conclusions such as allocating 1% of GDP to climate mitigation, or practically nothing [5]. Consider that a person with a positive discount rate applied to all years means that some point in the future is so far away that she gets practically no utility by consuming in that year. The higher the discount rate the closer that future "zero utility" point comes to the present. However, if a person has a 0% discount rate, then she values consuming 150 years from now the same as consuming today, even though she will then be dead. Now you can see why economists tend to use discount rates greater than 0% per year and why someone worried about the negative impacts of climate change on future generations would want a discount rate near 0%. Lower discount rates assign more value to the future.

Instead of thinking about the utility of a person or computer, now think about "the economy" and the metric of gross domestic product (GDP). GDP is roughly treated as utility in economics (our personal consumption accounts for the largest fraction of GDP). Here, GDP is now a substitute for Nick Bostrom's paper clips. Could we tell the difference between a world that is run by a superintelligent GDP maximizer and the world that we live in right now? That is to say, if certain politicians, business owners and executives, and economists are pushing for rules that maximize GDP, then is "the economy" simply a mechanism to maximize GDP without regard for how money is distributed?

How could we know if we have allowed the economy to simply become a GDP maximizing utility monster? Perhaps GDP would keep going up, but if it didn't, perhaps we'd start adding activities to GDP that have existed for centuries, but had previously not been counted due to illegality or other reasons. As mentioned in Chap. 2, prostitution and legalizing previously illegal drugs are examples of recent additions to GDP in some countries and states.

Perhaps if all we wanted to do was increase GDP, we'd cut corporate taxes to spur investment in capital versus spending on public education, which is for people. Perhaps human life expectancy would go down,[23] but due to drug abuse (e.g., opioids in the U.S.) drug sales would go up (the utility monster is indifferent to people). Perhaps we'd see increases in wealth or income inequality (refer back to Chap. 4). Perhaps people would contract with transportation network companies like Uber and Lyft, to drive around, wait for algorithmic signals on where to drive to pick up a person or thing, and then deliver that person or thing as directed, making few if any decisions for themselves. But there is no need to stop with human-directing algorithms. The economy might get rid of the human drivers altogether, and this is exactly what vehicle and technology companies, from Tesla to Uber and Google, have been working on.

Many interesting (troubling to many) trends are occurring in the U.S. regarding health, distribution of income, and the ability of people to separate concepts of fact and truth. Thus, we should consider whether the superintelligent AI future some fear might already be in action, but at perhaps a slower and more subtle pace than some pontificate might happen after the "singularity" when AI becomes more capable than humans. The recent populist political movements in the U.S. and other countries could in fact be a rejection of the "algorithm of GDP maximization" associated with our current economic system.

Importantly, utility, like GDP, is not fundamentally a physically measurable thing. While we have rules to calculate GDP, the dollar-denominated value itself is not a physical quantity. I state this even though the data and concepts within this book show *GDP is highly linked to energy consumption, and specifically highly correlated to useful work. Thus, GDP very much mimics a measure of economic physical output.*

The non-physicality of utility is both a virtue and a curse. The virtue is that an economist can work out mathematics and claim that the world would be better by using her equation to maximize utility. The curse is that we can never know if her claim is true because we'll never quantify every person's utility. In this sense, utility

---

[23] Olga Khazan, "A Shocking Decline in American Life Expectancy," *The Atlantic*, December 21, 2017 at https://www.theatlantic.com/health/archive/2017/12/life-expectancy/548981/.

of people is not practically part of the measurable natural world, so it is conceptually difficult to consider whether it is consistent with the natural world. We have many observable and measurable metrics that relate to a good lifestyle (calories of food per person, liters of fresh water access per person, income, etc.), and we can use these as proxies, but if that is the case, why consider the concept of utility in the first place? Chapter 9 suggests much of the reason has to do with the multitude of options available to us when we make choices. Our freedom to choose what we buy, what we eat, and how we spend our time are, however, dependent on the constraints of the natural world as well as both the global and local economies that we define and within which we reside.

The economy's pursuit of superintelligence is consistent with the maximum power principle. If more energy consumption requires more information processing outside of our bodies, then the economy says, "let's figure out how to do that." If more information processing outside of our bodies requires more energy consumption, then the economy tries to figure that out too. Since GDP is proportional to useful work, which is equal to energy consumption times efficiency, then to increase useful work and GDP we can increase energy consumption *and* efficiency. Efficiency is a measurable quantity that is the manifestation of whether the economy is processing the type of information that promotes growth.

> In the sense of information expressed as energy efficiency, increases in both energy and information promote economic growth, and information and energy exist within a self-reinforcing, or positive, feedback loop.

Let's now use the concept of the economy as an evolving superorganism, pursuing the maximum power principle and the maximization of useful work, to *knock down or support*, the energy and economic narratives. First, energy.

The Energy Narratives ... *Knocked Down!*

This section heading says it all. I have written a few hundred pages to finally get to the point where we can *knock down the energy narratives*! At this point we only need to tie the concepts together.

> *From the perspectives of both biological evolution of genes via humans as organisms and economic evolution of memes via human technology, there is no reason to think that either fossil fuels are preferable to renewable energy or renewable energy is preferable to fossil fuels.*

If "the economy" is just trying to operate and grow as a superorganism, and a larger organism needs more, not less energy consumption, then it seeks to combine all possible energy extraction technologies in a way that maximizes average useful work output. In this context there is no need to restrict energy options. For example, assume we have four energy extraction technologies: A, B, C, and D. One or more combinations of these achieve the maximum average extraction rate. If you remove one option, say technology D, then the maximum energy extraction rate can only decrease or stay the same. The only way the A–B–C technology extraction rate can remain at the maximum of the A–B–C–D technology extraction rate is if technology D is so ineffective that the economy chooses not to use it at all.

This summary is equivalent to an "all of the above" strategy. It is also why an all of the above strategy *does not* reduce greenhouse gas emissions faster than a "less than all of the above" strategy. This "all of the above" reasoning is why some label renewable energy technologies "fossil fuel extenders." As the data of Chap. 2 show, to date, renewable energy technologies have not replaced or decreased the rate of fossil energy extraction; they have complemented, and thus extended our ability to continue fossil energy extraction.

However, from a more philosophical perspective, labeling renewables as fossil fuel extenders distracts from an important point. There is no need to describe renewable energy technologies with a comparison to fossil fuels to begin with. As a superorganism, the economy extracts matter and energy from the environment. It then uses the energy to combine the matter in ways that facilitate repeating the process. In repeating the process, the economy develops new ways to both organize matter into technologies and organize the components, or elements, of which it is composed. Both forms of organization are memes, or ideas, that encode the information necessary to produce more useful work from the extracted matter and energy. This description is that of positive feedback loop where more energy, matter, and information enable further extraction of energy and matter and more information processing.

The economy as a superorganism doesn't distinguish between the use of renewables and fossil fuels, it just uses both to enhance growth and total useful work output. At the individual project and policy level, proponents fight for their technology or fuel, but this is part of the process of achieving growth at the economy-wide level. While some of us choose to eat a vegetarian or vegan diet, humans evolved as omnivores consuming both meat and plants. In this sense, *the economy is an omnivorous superorganism.* It consumes both fossil (including nuclear) and renewable energy so that if situations drastically change, it is not committed to a resource that suddenly becomes unavailable on short times scales (less than a year) such that growth would be hindered on longer time scales.

Recall from Chap. 3 the argument that modern renewable technologies are not "renewable" because they are industrial technologies made of concrete, metals, and other fossil materials that we mine from the Earth. The economy as a superorganism doesn't recognize this narrative because, again, it has no need to even label technologies as renewable or fossil. Nate Hagens suggests we might alternatively label renewable technologies as "repeatables" because once you install them, they repeat the conversion of wind and sunlight into electricity with almost no more human intervention [31]. "The economy" just views wind turbines and solar panels as pieces of infrastructure that, once in place, can "repeatedly" extract flows of energy from the environment. From its point of view, those technologies just have different characteristics from technologies that extract stocks of energy in the form of oil, gas, coal, and radioactive materials from Earth's crust.

To make an analogy with our bodies, our feet have different capabilities than our hands. We're much more functional with two feet at the ends of our legs and two hands at the ends of our arms than we are by substituting two feet for two hands, or vice versa. From a functional point of view, neither hands nor feet are either "good" or "bad." Each just has different characteristics that enhance survivability of the body in total.

In the same way, the economy as a superorganism sees neither fossil nor renewable energy technologies as good or bad. It places them in the context of its present situation and constraints of the environment. It seeks the combination that best enables it to maintain itself and grow. Overall, those "purposes" seek more useful work via more efficiency and energy extraction, not less.

The Economic Narratives ... *Knocked Down!*

What structures or organizations would the economic superorganism use? Answer: the ones that enable it to extract energy and increase efficiency to perform more useful work and thus increase economic growth. Since a system makes models of itself to understand itself, then our mathematical modeling of "the economy" is an attempt for the economy to understand itself. The evidence points to capitalism and price-setting markets as subsystems that (perhaps best) foster understanding of the economy, as a superorganism, via modeling to support increased resource consumption and growth.

As mentioned earlier in this chapter, the idea of mathematically modeling "the economy" is not necessarily democratic because only a small number of people know how to do it. By stating this I'm not required to make any conclusion on whether economic models or systems of governance promote individual or collective human welfare, income equality, or other metrics of enhanced human lifestyle or well-being. Let's first discuss capitalism and then markets.

For simplicity, think of capitalism as opposed to populism. The former targets the accumulation of "capital" in various forms (money, physical infrastructure, intellectual property, etc.) that are privately owned and controlled. The latter seeks to increase money, services, and resource flows to people who might or might not own capital. However, since the 1970s, when the rate of increase of global energy consumption slowed, and after the collapse of the Soviet Union in 1989, countries have generally moved more toward capitalism and away from socialism where the state, or government, owns property instead of individual citizens. The next chapter elaborates on both the pros and cons of socialism and capitalism, but one could argue that capitalism has been on the rise since its start on the Island of Madeira in the 1400s.

As Raj Patel and Jason W. Moore point out in *A History of the World in Seven Cheap Things*, during the 1400s the Island of Madeira, or "Island of wood," served as an early proving ground for the tactics of capitalism: minimize all costs, including via division of labor [29]. Located several hundred miles southwest of Portugal, the Portuguese named the island for its abundance of forests. The wood was first used for ships, but "The second, more dramatic deforestation was driven by the use of wood as fuel in sugar production."[24] Madeira had cheap land and cheap energy, but it needed cheap labor to harvest the cane. The Portuguese imported slaves for that purpose. You can't have cheaper labor than by owning people as property who earn no wage and have no human rights. Because sugar juice degrades within a couple of days after harvest, it must be processed quickly. "To reach such speeds, production had to be reorganized, broken into smaller, component activities performed by different workers. It simply isn't possible to get good returns from workers who are exhausted from cutting cane and then spend the night refining it."[25]

In the 1400s, there were practically no competitive markets as we know them today. Today markets provide signals based on both marginal costs known to producers and subjective purchase preferences of consumers. Prices are these signals. For example, higher oil prices trigger an increase in the number of active drilling rigs and a reduction in consumer purchases of gasoline. Prices theoretically incorporate information from all businesses and processes across the economy, but practically they fall short of this ideal. Even with practical limitations, markets incorporate distributed information such that there is no single price setter. Each business adjusts to the price signal that is itself informed by individual investment and consumer decisions in a feedback loop.

Remember the Chap. 5 story about ants? Individual ants change their tasks based upon the frequency at which they encounter ants performing a particular task. This frequency is their "market signal." Each ant doesn't wait for the queen ant to tell it what to do. This distributed response is how the ant colony maximizes total energy flow from its environment (i.e., maximum power principle). In an ecosystem, there is no single value, described by some equation or formula, that one or more organisms look to maximize.

Friedrich Hayek was a proponent of markets for their ability to make use of decentralized information. He won the Nobel Prize for economics because of his work that "...highlighted the problems of central economic planning. His conclusion was that knowledge and information held by various actors can only be utilized fully in a decentralized market system with free competition and pricing."[26] Even though Schumpeter makes the comparison of capitalism to evolution, it is Hayek that perhaps makes the more direction connection to how the economy evolves by making use of the price mechanism to induce incremental change, just like mutations in genetic code.

---

[24]Patel and Moore [29, p. 15].

[25]Patel and Moore [29, p. 16].

[26]Friedrich August von Hayek—Facts. NobelPrize.org. Nobel Media AB 2019. Accessed September, 2019 at: https://www.nobelprize.org/prizes/economic-sciences/1974/hayek/facts/.

In his 1945 paper "The Use of Knowledge in Society" Hayek states there is no room for intentional design of the economy:

> But those who clamor for "conscious direction"—and who cannot believe that anything [the price mechanism] which has evolved without design (and even without our understanding it) should solve problems which we should not be able to solve consciously—should remember this: The problem is precisely how to extend the span of our utilization of resources beyond the span of the control of any one mind; and, therefore, how to dispense with the need of conscious control and how to provide inducements which will make the individuals do the desirable things without anyone having to tell them what to do.[27]—Friedrich Hayek (1945)

In Chap. 9 we return to Hayek to question whether he really believed there is no "conscious direction," because, after all, markets are consciously designed and created by people. On the whole he is primarily interested in the "inducements" for the mass of "individuals," and his 1945 paper goes further to agree with English philosopher Alfred Whitehead, who he quotes as noting there is no need for us to think about what we are doing:

> It is a profoundly erroneous truism ... that we should cultivate the habit of thinking what we are doing. The precise opposite is the case. Civilization advances by extending the number of important operations which we can perform without thinking about them.[28]—Alfred Whitehead (as quoted by Friedrich Hayek)

As much as I've written this book to *assist in thinking* about the purpose of the economy, Whitehead and Hayek were onto something. Their statements are consistent with those of Odum and Lotka. We presume that each individual ant in a colony isn't thinking too hard about what to do, but reacting to the situation at hand. Further, the energy consumption of our economy scales relative to its size just like that of an ant colony. The translation? Each one of us is not entirely in control of our job within the economy. Again, ant specialist Deborah Gordon:

> So why is the ant colony as a factory of specialised workers such a compelling image? First, it's familiar: a little city of ants, each carrying out its assigned job, is a miniature version of a human city. It's comforting to imagine that each ant gets up in the morning, drinks its coffee, grabs its briefcase and goes off to work. To envisage how an ant's task of the moment arises from a pulsing network of brief, meaningless interactions might compel us instead to ponder what really accounts for why each of us has a particular job. [8]—Deborah M. Gordon (2016)

Not clear from this single passage from Gordon's article is her point that not even individual ants toil away at the same task over and over, independent of aging and changes in their environment. While we should not expect our own tasks and jobs to be any more strict, at any given time there is a set distribution of tasks among us for which we are not in complete control. How many readers of this book were 100% in control of their career path? No luck? Didn't depend on where you were born? Are you working at your current job because you really want that job versus any other?

---

[27] Hayek [13, Section VI].

[28] Friedrich Hayek quoting Alfred Whitehead in [13, Section VI].

It is consistent to conceive that we humans have enabled a system in which we react to a "pulsing network of brief, meaningless interactions." We receive various signals: prices for goods and services, narratives from other people, and advertisements for products and services. Social networking companies increase the rate of these interactions via *gamification* of their apps to make them addictive. This brings us closer to Whitehead's vision of "…extending the number of important operations which we can perform without thinking about them." While each interaction on Facebook or Twitter might not be the most important operation of our lives, the use of these platforms to sway public opinion, and perhaps even affect how someone votes in political elections, does indirectly influence important decisions which we like to believe are preceded by critical contemplation.

Many decisions do involve significant deliberation. The U.S. ended slavery after more than a century of deliberation (including during the founding of the country from its colonial roots) culminated in a civil war. Some proponents of the fossil fuel narrative link fossil-fueled industrialization and capitalism to the liberation of slaves.[29] Clearly machines provide physical work that can substitute for human muscle work, whether or not those humans are in bondage. While the Madeira story shows that the ideology of capitalism is at best indifferent toward the human condition of "others," the industrial era did witness the creation of legislation that both made slavery illegal and improved working conditions. Chapter 9 explores "political will" as an explanation of the latter. However, even capitalism operates within physical constraints, and if those constraints do not allow for the economy to simultaneously grow and distribute a constant share of output to workers, then capitalism might foster growth at the expense of more equal distribution. Recall data showing that before the 1970s the U.S. experienced increasing wage equality and low political polarization alongside *increasing* per capita energy consumption, but the former two trends reversed after the 1970s when per capita energy consumption remained constant (revisit data in Figs. 4.11, 4.12, 4.14, and 7.1 as well as Fig. 6.6 showing this trend exhibited within an economic model).

We have decent data describing the last 100 years of U.S. economic growth and income distribution. For the first 50, GDP grew, per capita energy consumption increased, and income distribution became more equal. For the last 50 years, GDP grew while per capita energy consumption stagnated (and slightly declined since the early 2000s) and income distribution became less equal. While some interpret the data of the last decade or two as evidence that the U.S. economy has reached some state of absolute decoupling [19], the structure of the economy before 1970 is by no means the same as after 1970. As mentioned in Chap. 5 (Fig. 5.9), it is as if the U.S. economy was a small circle in 1970 and a larger triangle today. Yes, the triangle is larger, but it is also a different structure with less even distribution of its area. When the U.S. became the world's dominant energy consumer after World War II, more energy consumption per person translated to more equal income distribution. The

---

[29] See Chapter 7 "The release of slaves: energy after 1700" in [30], and Hartnett-White, Kathleen, "Fossil Fuels: The Moral Case," white paper of the Texas Public Policy Foundation, June 2014.

U.S. economy's different structure has been shaped by feedbacks originating both within and outside of its border.

The Finite Earth Affects Physical, Economic, and Social Trends

Is the finite Earth affecting the growth and structure of our society and economy? To me, the answer is clearly *yes*.

A more precise way to phrase this question is "For any given data trend, is it representative of what can occur in a finite or infinite (at least "big") environment?" This answer has been known at latest since the 1972 *Limits to Growth* study. Exponential growth occurs in systems not large enough to notice the size of their environment. Contraction, stagnation, or decelerating growth occurs in those systems big enough to have filled their environment to such as a degree as to prevent continued exponential growth. At the simplest level, this rubric is all you need to apply to any given time series. If the data don't show exponential growth, then the explanation *could, and most likely does,* have something to do with an environment characterized by a limited size, quantity of stocks, or flow rates.

If you don't have the concept of a maximum size in your worldview or mathematical model, then you aren't even in the game of answering the question at the start of this section because you've already assumed the answer is "no." We don't know exactly what is the maximum size of our environment, but we do know it's not infinite. Albert Einstein told us that there is no difference between space and time. They are one and the same. Thus, arguments that assume the environment is so large as to be infinite ignore both space and time. The infinite-resource argument aims to convince us that neither time nor the size of natural resources is important.

From the standpoint of information propagation and processing via genes, memes, and technologies, time might not matter that much. However, from the standpoint of living creatures with finite lives, the time required for change matters very much. Since practically all economic analyses occur in the context of ourselves as humans with finite lifespans, time and space are limiting factors for our well-being.

But what is an argument against finite resources affecting our economy? Consider the following reply I received in response to a presentation I gave in 2017:

> In the part about resources and earth's crust, we estimated at some point how much oil and other minerals are left in the crust. If you make the uniformitarian assumption that the crust is more or less the same everywhere, then we have extracted about 1/10,000 of the oil, and the numbers for the other things we estimated were similar. So not infinite but there is still a

lot in there, and we have continued to get better at extracting it. In the case of oil, let's hope it stays there.[30]

In that presentation I discussed a subset of the economic and energy trends of this book while arguing the post-1970s data indicate that limits to physical growth are affecting our social outcomes. Since Chap. 1 I've made no secret that I lean to the economic narrative of techno-realism and the finite Earth. I see too much evidence in the data that supports the notion that the finite Earth affects our social, economic, and biophysical outcomes. My argument comes from a systems approach for interpreting the data, such as the scaling laws relating energy consumption to size, and relating them to physical and evolutionary perspectives of ecosystems and the economy (e.g., the Jevons Paradox; the evolution of genes and memes).

While the quote states the Earth is not infinite, I put it into the techno-optimistic and infinite substitutability narrative because it was written to counter my argument. Should we make the "uniformitarian assumption that the [Earth's] crust is more or less the same everywhere?" No. Everything we know about oil and other mineral deposits indicates they are not uniformly distributed around the world or within individual oil fields or regions. The quality and size of these deposits generally follow power laws, meaning that there are relatively few high-quality large deposits and more numerous low-quality smaller deposits.

Fresh water and land are *not* uniformly distributed across Earth. Words in books are not uniformly distributed on the pages or by the frequency with which each word appears. For example, the word "uniformitarian" appears four times in this book, but the word "the" probably occurs multiple times on every page of this book (I didn't check!).

In fact, whenever you have a set of items that have some correlation, or set of rules or physics relating them to each other, you can describe the frequency with which each appears by using a *non-uniform* distribution. Zipf's Law is the concept often used to refer to these types of distributions. Just the act of categorizing these items by a property (e.g., size) means you get a non-uniform distribution of how many of these items are big versus small [3]![31]

For example, consider the number of cities of certain sizes. There are very few cities with population larger than 20 million, slightly more cities with ten million or more, even more with at least one million, and so on. Everyone doesn't live in a city the same size, each word doesn't appear the same number of times in a book, and

---

[30]A correspondence from a person at an economic think tank regarding a presentation I gave in which I discussed that the relative cost of energy and food (energy and food spending relative to GDP) appears to have hit a minimum point around the year 2000 per data in Chap. 2. I keep the person's name confidential because my purpose here is neither to promote, shame, criticize, or otherwise reveal this person's informed opinions and/or his/her organization. The presentation I gave was December 15, 2017 "Energy and the Economy over the Long-Term: Size and Structure," available at http://careyking.com/presentations/.

[31]See article by Ladad Adamic, [1], for a lay description of Baek et al. [3] that explains a mathematical foundation for Zipf's Law, or more generically how power law distributions arise from statistical data.

each person doesn't earn the same income. These non-uniform distributions are the norm, and the understanding of this finding is based on the concepts of information and probability.

As emphasized in this book, to understand processes of the world we need to think about *size and structure*, and to do that we can use the concepts of *matter* (how much is there), *information* (how do you do it, how are energy and matter distributed), and *energy* (how much effort is required, how much does it cost).

It is important to consider the non-uniform distribution of resources because when we extract matter, energy, and power from the environment, we tend to access the higher quality resources (large, shallow, nearby, and pure) first. These resources require less technology than smaller, deeper, more distant and more diffuse resources. In addition to tracking energy costs in units of dollars, the concept of energy return on energy invested, or EROI, as mentioned in Chap. 5, also captures the balance between technology and quality using units of energy and power flows instead of money.

If the uniformitarian distribution assumption held for any given resource, then because everywhere would look the same, we wouldn't need any new information to find and extract the quantity because the same amount is everywhere and it costs the same amount of energy to get it. It is precisely the feedback of a declining resource quality that provides the incentive to invent new technology and develop information. This feedback would not exist if resources were distributed uniformly.

Summary

The global economy is a superorganism. As such, it has no choice but to exist in its finite environment. Thus, the concept of the economic superorganism is consistent with the finite Earth and techno-realistic narrative. Simultaneously, the economy evolves within its finite environment, and to most effectively grow it makes use of distributed information rather than instructions from a single central authority. Thus, the concept of the economic superorganism is also consistent (to some degree) with the techno-optimistic and infinite substitutability narrative. The economy does change and evolve, but ultimately there are limits to these changes.

I now summarize data and concepts from this book that support the conclusion that the finite Earth is indeed affecting economic and social outcomes, and that the extreme form of the techno-optimistic narrative is not useful for understanding economic growth and structure. I *do not agree* with the claim that the techno-realistic narrative is equivalent to a "doom and gloom" viewpoint. Words like doom and gloom are not specific to physical laws, logic, or mathematics. They are used simply to invoke an emotional response and distract from understanding the problem at hand. I view the finite Earth and techno-realistic narrative simply as one stating that we can neither understand the economy nor operate within it without the context that its components have finite lifespans and exist (for all practical historical purposes) within a finite space. From a broad perspective, changes in trends in

both the 1970s and the 2000s present evidence for growth headwinds that relate to physical constraints within a finite world and finite time.
From Chap. 2:

- Global energy and food spending relative to GDP declined steadily (with hiatus in the 1970s) from the onset of industrialization and the fossil fuel era, but has no longer declined since around the year 2000.

From Chap. 3:

- The average real oil price after the 1970s is significantly higher ($58/BBL) than the previous 90 years ($19/BBL). This provides some indication that, even though we extract more oil today than in 1970, it is more scarce in the sense of driving economic growth.

From Chap. 4:

- The global human population *growth rate* is slowing (as opposed to increasing or remaining steady), and thus population is aging, with the 1970s as a turning point toward these trends.
- A finite world imposes constraints on physical and economic size and growth rates, leading to a need for increased complexity to solve new social problems. Because this organization of complexity requires more energy and resources, physical constraints can limit, and indeed seem to be limiting, the ability of developed countries to both grow and maintain pre-1970s structure:

  - The U.S., and thus the world abandoned the gold standard in 1971, thus removing any explicit link of money to physical resources.
  - The global debt-to-GDP ratio (and those for most developed countries) has risen rapidly after the 1970s to its highest level in history. Debt ratios increased when the returns from GDP were lower than anticipated from investment and commercial banks were allowed to create money by lending.
  - Increasing income and wealth inequality of developed countries since the 1970s indicates that despite their lead on economic development, they are struggling to grow *and* maintain previously more equal (more complex) levels of distribution.
  - For the U.S., a declining proportion of GDP to wages, the "wage share," coincides with stagnation in per capita energy consumption (Fig. 4.12). This same declining wage share, starting at the same time, occurred in many of the world's largest economies [16].

- After the 2008 Financial Crisis, Western central banks took unprecedented actions that have not yet fully reversed.

  - Many central banks lowered interest rates below 1%, levels (near zero) never previously experienced in history, in attempt to induce faster economic growth.
  - Some government bond yields are now negative. This means that in some cases investors are now paying governments, rather than receiving interest

payments, for the privilege of loaning money to them—an unprecedented situation.

From Chap. 5:

- Before the 1970s, global energy consumption increased at the same rate as the size of the economy. GDP increased one unit for every additional unit of primary energy consumption. The rate of growth of GDP and energy extraction was limited by the amount of capital that could perform useful work by consuming the energy that could otherwise flow more quickly.
- Since the 1970s, global energy consumption increases more slowly than size of the economy. Global GDP increases more than one unit for each additional unit of primary energy consumption. Thus, after 1970 the global economic superorganism operates in a similar manner as individual animals and colonies of animals (such as ants) that are limited in the rate they can consume and metabolize food. In order to grow and consume more energy in total "the economy" has since forced each additional unit of capital to consume energy at a slower rate. The rate of growth of GDP and energy extraction is now limited by the correlation and coordination between *both* the capital that extracts energy and the capital that performs useful work by consuming energy at some efficiency.

From Chap. 6:

- There exist dynamic economic growth models that avoid mainstream (neoclassical) economic theory and incorporate a limited resource (in size or flow rate) with declining resource quality during extraction. These models can accurately replicate several global and U.S. physical and economic trends. If energy and natural resources are assumed infinite, the long-term growth trends cannot be as accurately represented.

  - The trends of the World3 model used in the 1972 *Limits to Growth* book (and subsequent updates) have stood the test of time rather well, despite criticisms [20–22]. (More debunking of criticisms are in the next chapter.)
  - For the U.S., a declining proportion of GDP to wages, the "wage share," coincides with stagnation in per capita energy consumption. My HARMONEY model provides an explanation that links these two trends by integrating the concepts of a biophysical economy, a finite Earth, and capitalism fueled by credit (e.g., borrowing money for investment) [17].

- GDP is proportional to useful work, which is primary energy extraction times a system-wide conversion efficiency.

  - Primary energy extraction is ultimately limited by the size of Earth, and while it is likely not possible to predict a future upper limit on the energy extraction rate, the decline in quality with increased extraction ultimately feeds back to prevent 100% extraction of any resource.
  - Conversion efficiency is ultimately limited by the second law of thermodynamics.

– The previous two points combine to imply the rate of useful work, and thus gross domestic product, are ultimately limited.

We can view the interpretation of the economy as a superorganism as a new narrative, one that enables us to avoid using the binary energy and economic narratives that most people use, as discussed in this book. By considering the economy as a superorganism we create an interpretation that is consistent with both the concept of naturalism and the data characterizing modern society while making it useful for our purposes of understanding economic growth, size, organization, and structure.

> With naturalism and the integration of humans within an evolving economic superorganism, we've come full circle to replace ourselves, that is again place ourselves and our societal organization, within the ecosystem that is our natural world.

An economic superorganism is consistent with the natural world, the world in which it lives, and the only world that we can observe. Over the course of four billion years of evolution on Earth, genes evolved to program life forms from single-celled organisms to multi-celled organisms, plants, dinosaurs, mammals, and us humans. While genes store the historical information accumulated via evolution, the genes within us humans have enabled us to create additional methods for storing and propagating information (or memes) via languages, including computer languages.

Ecosystems exist, based on exchanges of energy and matter, but as far as we know without any part of the ecosystem contemplating its own existence. Likewise, for most of human history, the concept of an economic superorganism did not exist. Arthur Willis defined an ecosystem as "...a unit comprising a community (or communities) of organisms and their physical and chemical environment, at any scale, desirably specified, in which there are continuous fluxes of matter and energy in an interactive open system." [32].

I've taken the Earth, and nearby satellites, as the largest ecosystem within which our global economy exists. In this context we can include the global economic superorganism into a rewrite of Willis' definition of ecosystem by replacing the word "organisms" with "economies": the Earth, as an ecosystem *is a unit comprising a community (or communities) of **economies** and their physical and chemical environment, at any scale, desirably specified, in which there are continuous fluxes of matter and energy in an interactive open system.*

When we *Homo sapiens* first became a separate species around 300,000 years ago, we were groups of individuals that had no need to consider themselves different than any other animals. With the eventual invention of agriculture and the domestication of some animals, it would have become clear that we humans were able to take actions that altered and controlled the natural world in a manner different from other living creatures. This ability to see cause and effect, action and change, led to inquiry about how this was possible. Religions emerged as narratives to satisfy a crave to explain how man was different from nature and other animals [12].[32]

Part I of this book focused on data and trends. Part II focused on systems thinking and ways we can understand the role of energy in the economy. The book now shifts to Part III that focuses on how narratives battle to influence the future. We have already extensively discussed the energy narratives, as well as economic modeling, but the next chapter fills a gap in describing economic narratives that are less grounded in biophysical nature of the economy, but reside almost entirely in the social domain.

Today we live in cities and countries governed by social rules and laws that make us even more unique from other living organisms. But in setting these rules, exactly how much independence and freedom of choice do we have? We make many choices that seem very unrelated to concepts of energy, work, efficiency, and evolution. Many policymakers and economic pundits believe we are in such control of our social and economic future that we have no need to consider the biophysical constraints emphasized in this book. Chapter 9 now describes some of these beliefs. After this next and penultimate chapter, we are then prepared for the final chapter to consider some energy and economic choices and policies we face in the next several decades, and based on the tradeoffs associated with these choices, which options are more probable than others.

References

1. Adamic, L.: Unzipping Zipf's law. Nature **474**, 164–165 (2011)
2. A.Schumpeter, J.: Capitalism, Socialism, and Democracy, Taylor & Francis e-library edn. George Allen & Unwin, London, England and New York, New York (2003)
3. Baek, S.K., Bernhardsson, S., Minnhagen, P.: Zipf's law unzipped. New Journal of Physics **13**, 21pp (2011). https://doi.org/10.1088/1367-2630/13/4/043004
4. Bostrom, N.: Superintelligence: Paths, Dangers, Strategies. Oxford University Press (2014)
5. Broome, J.: The ethics of climate change. Scientific American **June**, 96–102 (2008)
6. Dawkins, R.: The Selfish Gene (30th Anniversary Edition). Oxford University Press, Oxford, UK (2006)
7. Dennett, D.C.: Consciousness Explained. Back Bay Books (1991)
8. Gordon, D.M.: The queen does not rule (2016). https://aeon.co/essays/how-ant-societies-point-to-radical-possibilities-for-humans
9. Hagens, N.: Economics for the future – beyond the superorganism. Ecological Economics **169**, 106,520 (2020). https://doi.org/10.1016/j.ecolecon.2019.106520. http://www.sciencedirect.com/science/article/pii/S0921800919310067
10. Hall, C.A.S., Klitgaard, K.A.: Energy and the Wealth of Nations: An Introduction to Biophysical Economics, 2nd edn. Springer (2018)
11. Harari, Y.N.: Sapiens A Brief History of Humankind. HarperCollins, New York, NY (2015)
12. Harari, Y.N.: Homo Deus A Brief History of Tomorrow. HarperCollins, New York, NY (2017)

---

[32]See Chapter 2, "The Anthropocene" of *Homo Deus: A Brief History of Tomorrow* [12].

13. Hayek, F.A.: The use of knowledge in society. The American Economic Review **35**(4), 519–530 (1945). http://www.jstor.org/stable/1809376
14. Herrmann-Pillath, C.: The evolutionary approach to entropy Reconciling Georgescu-Roegen's natural philosophy with the maximum entropy framework. Ecological Economics **70**(4), 606–616 (2011). https://doi.org/10.1016/j.ecolecon.2010.11.021. http://www.sciencedirect.com/science/article/pii/S092180091000474X
15. Herrmann-Pillath, C.: Energy, growth, and evolution Towards a naturalistic ontology of economics. Ecological Economics **119**, 432–442 (2015). https://doi.org/10.1016/j.ecolecon.2014.11.014. http://www.sciencedirect.com/science/article/pii/S0921800914003589
16. ILO, OECD: The labour share in g20 economies. Tech. rep., International Labour Organization and Organization for Economic Co-operation and Development (2015). https://www.oecd.org/g20/topics/employment-and-social-policy/The-Labour-Share-in-G20-Economies.pdf. Report prepared for the G20 Employment Working Group Antalya, Turkey, 26–27 February 2015
17. King, C.W.: An integrated biophysical and economic modeling framework for long-term sustainability analysis: the Harmoney model. Ecological Economics **169**, 106,464 (2020). https://doi.org/10.1016/j.ecolecon.2019.106464. http://www.sciencedirect.com/science/article/pii/S0921800919302034http:// www.sciencedirect.com/science/article/pii/S0921800919302034
18. Lotka, A.J.: Contribution to the energetics of evolution. Proceedings of the National Academy of Sciences **8**(6), 147–151 (1922). https://doi.org/10.1073/pnas.8.6.147. https://www.pnas.org/content/8/6/147
19. McAfee, A.: More from Less The Surprising Story of How We Learned to Prosper Using Fewer Resources—and What Happens Next. Scribner, New York, NY (2019)
20. Meadows, D.H., Meadows, D.L., Randers, J.: Beyond the Limits. Chelsea Green Publishing, Post Mills, VT (1992)
21. Meadows, D.H., Meadows, D.L., Randers, J., Behrens, W.W.I.: Limits to Growth: A Report for the Club of Rome's Project on the Predicament of Mankind. Universe Books, New York (1972)
22. Meadows, D.H., Randers, J., Meadows, D.L.: Limits to Growth: The 30-Year Update. Chelsea Green Publishing, White River Junction, Vermont (2004)
23. Mitchell, T.: Carbon Democracy: Political Power in the Age of Oil. Verso, London and New York (2013)
24. Mitchell, B.R.: British Historical Statistics. Cambridge University Press, Cambridge (1988)
25. Odum, H.T.: Environment, Power, and Society. John Wiley & Sons, Inc., New York (1971)
26. Odum, H.T.: Environmental Accounting: Emergy and Environmental Decision Making. John Wiley & Sons, Inc., New York (1996)
27. Odum, H.T.: The ecosystem, energy, and human values. Zygon **12**(2), 109–133 (1997)
28. Odum, H.T.: Environment, power, and society for the twenty-first century The Hierarchy of Energy. Columbia University Press,, New York (2007)
29. Patel, R., Moore, J.W.: A History of the World in Seven Cheap Things: A Guide to Capitalism, Nature, and the Future of the Planet. University of California Press, Oakland, California (2017)
30. Ridley, M.: The Rational Optimist: How Prosperity Evolves. HarperCollins, New York, New York (2010)
31. White, D.J., Hagens, N.J.: The Bottlenecks of the 21st Century Essays on the Systems Synthesis of the Human Predicament. Independently published (2019)
32. Willis, A.J.: The ecosystem: An evolving concept viewed historically. Functional Ecology **11**(2), 268–271 (1997). http://www.jstor.org/stable/2390328

Part III
The Battle for the Future

Chapter 9
Delusions of Control

As far as we can tell, from a purely scientific viewpoint, human life has absolutely no meaning. Humans are the outcome of blind evolutionary processes that operate without goal or purpose. Our actions are not part of some divine cosmic plan, and if planet Earth were to blow up tomorrow morning, the universe would probably keep going about is business as usual. As far as we can tell at this point, human subjectivity would not be missed. Hence *any* meaning that people ascribe to their lives is just a delusion.[1]—Noah Yuval Harari (2015)

Armed with data of Chaps. 2, 4, and 7, the energy narratives of Chap. 3, the systems thinking of Chap. 5, the economic modeling background of Chap. 6, and Chap. 8's concept of the economy as a superorganism, this chapter now describes several high-level economic narratives that vie to explain why Western economies are or are not sufficiently solving current economic problems. We might hear that consumers lack confidence, politicians lack "political will," workers and unions have too little "bargaining power," or there is not enough finance. Maybe you've heard that we've already decoupled economic growth from material consumption, thus providing evidence that not only can we run an economy that works for both humans and the rest of the environment, we have been doing so for the last several decades. Perhaps you've read we are simply misled by advertising, lobbying, and public relations campaigns, and if we only had "real" information, we'd make better decisions.

In today's world of enhanced lobbying activity and economic mathematics, many view policy as being made via one of two concepts. The first is that companies and business organizations battle for influence of politicians using political contributions, lobbyists, and public relations campaigns. Often the lobbyists actually draft the bills that lead to legislation. Narratives can go directly to legislation, but most people don't have the money or political access to influence policy in this way. The second is that think tanks, bureaucrats, and academics use their economic models to calculate the option with highest benefits and lowest costs. Cost–benefit calculations

---

[1]Harari [1, p. 391].

© Springer Nature Switzerland AG 2021
C. W. King, *The Economic Superorganism*,
https://doi.org/10.1007/978-3-030-50295-9_9

were enabled when we created the concept that there is a thing called "the economy" that can be modeled. The vast majority of people don't have a mathematical model of "the economy," and thus they can't influence the cost–benefit game.

The two policy frameworks, lobbyist influence and mathematical cost–benefit analysis, have some merit, but they are usually invoked independently of the physical flow of energy and other natural resources. They too often neglect the control of energy as perhaps the most important and fundamental driver of social and economic phenomena. Thus, they mislead us as to the level of control we have over economic activity. In this sense they are *delusions of control*.

To effectively think about possible energy and economic futures, we must put historical political events in the context of where citizens and workers reside within the networks of energy and economic systems. In this way we see a much clearer physical explanation for economic growth (size) and distribution (structure).

As discussed in Chap. 6, many analysts and pundits don't realize how much their methods assume energy is unimportant for explaining economic growth and organization. Thus, a common flaw in those views is as Timothy Mitchell states, "Innovations in methods of calculation, the use of money, the measurement of transactions and the compiling of national statistics made it possible to imagine the central object [the economy] of politics as an object that could increase in size without any form of ultimate material constraint."[2] If our policymakers assume there is no ultimate material constraint, then they will misdiagnose economic problems, and we might elect leaders with a misguided view of what, why, and how easily the economy's size and structure can change. As this chapter explains, there is a linkage between physical power and political power. However, this is not part of the usual frameworks for the derivation of policy.

Before we dive into social movements, lobbying, and marketing, a little philosophy sets the stage for understanding some limits to our freedom of choice. We can choose to pursue many activities, but we can't successfully act out every idea we can imagine.

Degrees of Freedom to Choose

The coming century, I think, will be dominated by major social, political turmoil. And it will result primarily because people are doing what they think they should do, but do not realize that what they're doing are causing these problems. So, I think the hope for this coming century is to develop a sufficiently large percentage of the population that have true insight into the nature of the complex systems within which they live.[3]—Jay Forrester (2013)

---

[2]Mitchell [2, p. 143].

[3]In a conversation with Anupam Saraph as attributed at the following website (https://metasd.com/tag/forrester/) as being recorded in this video clip: https://metasd.com/2013/06/jay-forrester-on-hope-for-the-coming-century/ (both accessed August 2019).

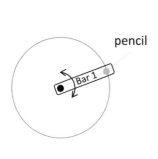

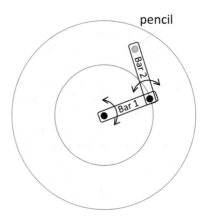

| 1 degree of freedom | 2 degrees of freedom |
| (pencil can only draw a circle) | (pencil can fill in a circle) |

Fig. 9.1 A bar with one degree of freedom can only draw a circle, but two bars with two degrees of freedom can fill in a circle

Educated as an engineer, I was excited to hear philosopher Daniel Dennett describe human consciousness and free will in the context of *degrees of freedom*, such as exists within engineered devices. To understand the concept of degrees of freedom, imagine a bar that can rotate about one of its ends (Fig. 9.1). This is like using a wrench to tighten a nut. Also imagine the non-fixed end of the bar has a pencil which marks on the table upon which the bar resides. This bar can only do one thing, rotate. Thus, the pencil can only draw a circle with radius equal to that of the length of the bar. This device has only one degree of freedom because we can specify the position of the pencil only using one coordinate: the angle of the bar.

Now attach a second bar to the free end of the first bar, and put the pencil at the free end of this second bar. This second bar can rotate about the non-fixed end of the first bar. Now, to describe the position of the pencil I need two coordinates, the angles of both bars, and thus this 2-bar system has two degrees of freedom. Instead of only being able to draw a circle on the table, this 2-bar design can fill in the entire area within a circle. The one and two bar system is easy to imagine by thinking of your arm. Your upper arm bone, the humerus, is like the first bar, and your lower arm bones, the radius and ulna, act like the second bar.

Now transition from the idea of describing a system with two degrees of freedom to more complex interacting systems. The global economy is composed of billions of individual people and billions of individual machines (cars, planes, computers, etc.) each with some number of degrees of freedom ranging from one to several trillions (i.e., individual bits stored on a hard drive or memory stick).

As introduced in Chap. 5, a system cannot understand itself. This is because it cannot describe each and every internal degree of freedom. The total number of degrees of freedom of all people and machines is too large to know, and this opens up the space for multiple competing narratives. The narratives compete by explaining patterns we observe in data describing the natural world, including our economy.

We cannot unequivocally determine why history has proceeded in the way it has just as we cannot determine the course of future events. Likewise, we can't fully describe the present, and thus don't have this description as a starting point to simulate the future. This lack of information does not prevent philosophers from debating if the future could, in principle, be known to some entity that somehow did know the entire state of the universe.

Determinism is the idea that the current state of the world—really of the universe—determines its future. That is to say, if we precisely knew the velocity and position of every fundamental particle in the universe (or every degree of freedom of particles), and if we knew all of the laws of physics, then we could predict what would happen next. You don't need to know the past for this prediction. You just need to know the present. At this fundamental level there is no difference between the past and future. For the two-bar problem, if I flicked the two bars into a spinning motion, then using the laws of physics I can determine how they would spin forever. I could simulate backward or forward in time.

However, in a practical sense, because we don't know the exact position and velocity of the bars when I set them in motion, and we don't know all the laws of physics, we can only accurately approximate their position and velocity for a limited time into the future. Even for simulating systems in which we assume we "know" all of the physical laws, the science of nonlinear dynamics and chaos informs us that very minute changes in initial conditions, or how the world is now, can translate to wildly different future outcomes. This sensitivity to initial conditions is often called the *butterfly effect*. Edward Lorenz, the father of chaos theory, gave a talk in which he posed the question of whether a butterfly flapping its wings in Brazil might set off a tornado in Texas [3].

> ...if a single flap [of a butterfly's wing] could lead to a tornado that would not otherwise have formed, it could equally well prevent a tornado that would otherwise have formed. ... a single flap would have no more effect on the weather than any flap of any other butterfly's wings, not to mention the activities of other species, including our own.[4]—Edward Lorenz (1993)

Many times people use the term "butterfly effect" to say how small deviations *can* lead to large changes later. But as Lorenz cautioned, this concept *does not mean that we can equivocally determine* that the wing flap of a butterfly was indeed *the cause* of some future event, such as a tornado. Part of the reason is that we cannot precisely measure the exact position and velocity of every butterfly, air molecule, etc., in the world. Because we can't precisely measure all of these things, we do not know exactly the state of the world before a tornado forms. And because a slightly different state of the world in the past ends up with no tornado, we cannot say if the butterfly did indeed cause a tornado. There is simply no way for us to know. Even

---

[4][4, pp. 14–15].

if we have all of the laws of physics, we are too ignorant of the past to know how we've arrived at the present. You can extend our ignorance to the time of the Big Bang, 14 billion years ago.

This type of determinism by simulating change via the laws of physics is different than me and my wife freely choosing, or "determining," where to go for dinner. We are both a living collection of particles, with a history, and in that sense are systems with emergent properties with different food preferences. Recall from Chap. 5 that emergent properties are ways of speaking of a large collection of objects, or relations among objects, in a way that is both consistent with the underlying parts, and useful. For each underlying system component, time might not matter, but for the entire system composed of many interacting parts, time does matter, and there is a past. If my wife ate Thai food last night, she's very unlikely to want to eat it again tonight. But if she last ate Thai food 1 month ago, she's much more open to eating it tonight. However, as my wife can tell you, you don't need to know what I ate last night to know if I want Tex-Mex food—I tend to want it independently of the past!

But here is an important point. In terms of choosing what cuisine to eat at any given time, if my wife and I want to survive, there is one choice we cannot continually make: eat no food.

In this sense of systems, causes are emergent explanations for describing the patterns we observe in nature and society. I stay alive as an organized system because I eat food as an intake of energy. Therefore, it is useful to say that eating food causes, or continues, life. On the flip side, we might say of an elderly dying person that they stopped eating, and this was the cause of death. However, we could also say that their body was no longer able to process food, and that is the cause of why they stopped eating. Did the person die because they stopped eating, or did they stop eating because they were dying?

There is not necessarily one right answer. The laws of physics themselves don't define causes, such as how my wife and I agreed on what cuisine to eat [5].[5] However, the second law of thermodynamics does state that if we don't eat some type of food, we will die and our bodies will start the process of decomposition, no longer able to maintain a structure that is unique from the environment.

In the same way, if an economic superorganism is to survive and maintain structure, it can choose among many technologies for extracting energy input from the environment, *but it can't make the choice to use none of them.* The infinite substitutability economic narrative often ignores this simple point by focusing on the increasing number of options (e.g., multiple ways to generate electricity) we have for any given economic need, rather than the fundamental needs of an economic system.

Economist Alfred Eichner summed up this point when discussing how increased material standards of living have been associated with substituting energy resources (via technology) for human physical labor:

---

[5]Carroll [5, p. 44].

One could, of course, give up the hope of being able to continually economize on the use of human labor. But that would bring an end to the secular trend of the past 300 years whereby an ever increasing portion of the world's population has been able to count on a steadily improving material standard of living. In other words, a solution that requires substituting human resource, or labor, inputs for the various inanimate sources of energy would be self-defeating in terms of the purpose an economic system is intended to serve. The natural resource constraint must therefore be accepted as a real limit on the rate of economic expansion, one that cannot be overcome through substitution. The question is, just how severe a constraint is the availability of fossil fuels and other natural resources.[6]— Alfred Eichner (1991)

Over time people interact with each other and the environment to invent new technologies and uses of natural resources. This cause and effect relationship on the evolution of our freedom to choose among an increasing number of degrees of freedom, even if in a deterministic world, is precisely what Daniel Dennett states:

...there can be a growth in ability over time in a deterministic world, as well as a widening of opportunities and what is made of them by particular deterministic agents. Such increase in ability over time is utterly invisible to the mind-set that adopts the narrow vision of possibility enshrined in the definition of determinism: "There is at any instant exactly one physically possible future." According to that vision, in a deterministic world, at any time t, nothing *can do* anything other than the one thing it is determined at t to do, and in an indeterministic world, at any time t, a thing *can do* as many different things—at least two—as that brand of indeterminism allows for, presumably a deep and immutable fact of physics that could not be perturbed by changes in practices or knowledge or technology. The obvious fact that people today *can do* more than people used to be able to do disappears from sight if we understand possibility this way, and yet this fact is as important as it is obvious.[7]—Daniel Dennett (2003)

Dennett points out the "obvious" fact that, compared to the past, we humans have learned and invented technologies such that we literally have more choices, or degrees of freedom. In 2019 I could buy a ticket to ride in a jet aircraft flying from Los Angeles to Sydney, Australia. By 1519, Leonardo da Vinci had conceived of flying machines, but there wasn't enough collective knowledge of natural laws and engineering capability to enable flight. Also, many people today can't practically fly across an ocean because they don't have enough money for an airline ticket.

Collectively, society accumulates technological capability and degrees of freedom; individually, the lack of money limits access to those degrees of freedom.

Higher energy consumption enables more technological and cultural development that in turn enables more physical growth via increasing the degrees of freedom, or number of choices, available to society. Individually we choose among our options, but collectively we still follow certain patterns whether most people realize it or not. Because the economy grows within constraints, it necessarily organizes (e.g., into countries, cities, companies) in the same ways summarized by

---

[6]Eichner [6, pp. 913–914].

[7]Dennett [7, pp. 295–296].

the emergent scaling laws that relate energy consumption to the sizes of animals, trees, and ant colonies (as discussed in Chap. 5).

More degrees of freedom *for the system* means more configurations and relationships among the elements within the system. Thus, there are more configurations, or options, from which to choose. In addition, more options change *the probability* for any given choice. When we gain a new option, some existing options must become less probable to make room for the new option. For example, we can approximate the probability of installing a certain type of power plant as its observed fraction of all new power plant installations. In 2018, photovoltaic solar panels represented about 14% of the newly installed large-scale power generation capacity in the U.S. (see Fig. 4.19).[8] Twenty years ago, due to high cost, the odds of installing a solar panel were much smaller. One hundred years ago the odds of installing a solar panel were zero—the technology did not yet exist. Because solar panels exist today, the odds are lower for installing other types of power plants, such as one that burns coal.

Consider an even larger historical leap. Imagine the options available around 200,000 years ago when our ancestors first evolved into *Homo sapiens*. There was one mode of travel, walking, but there were many different places to walk for food and shelter. Going back even further in Earth's history we can imagine the number of options for the first organisms that could sense light. Go to the light, or away from the light? Was this a choice, or just a reaction to the environment?

Are politicians, business leaders, and consumers making choices, or like primordial organisms that can only sense light, are we simply reacting to signals from the environment? Perhaps the answer is that both are true. These last questions open up an array of similar questions.

Are our forms of government and economic rules, such as labor laws and taxation, a cause or effect from our ability to extract energy from the environment? Do we humans choose how to construct our economic systems, or do the laws of physics drive our social and civil laws in some direction? Are some economic and political ideologies more consistent with the physical and biological concepts supporting the economy as a superorganism? Perhaps more importantly, should we try to push an economy outside of the regularities and patterns in the historical data, or do we have to accept these patterns as sacrosanct?

To think about these questions, we now dive deeper into concepts, dare I say delusions, that vie to influence political and economic discourse. We first explore consumer choice.

---

[8]U.S. Department of Energy, Energy Information Administration Form 860 indicates that 4911 MW of photovoltaics were reported installed in 2018 (i.e., have an "operating year" of 2018), and 34,213 MW of total capacity were installed, of all types, in 2018.

The Free Will Delusion: Memes and the Engineering of Consent

If people are convinced a false concept is true, this is all that matters for the short-term objectives of the purveyor of the false concept. Consider the following from Edward Bernays:

> When the public is convinced of the soundness of an idea, it will proceed to action. People translate an idea into action suggested by the idea itself, whether it is ideological, political, or social. They may adopt a philosophy that stresses racial and religious tolerance; they may vote a New Deal into office; or they may organize a consumers' buying strike. But such results do not just happen. In a democracy they can be accomplished principally by the engineering of consent. [8]—Edward Bernays (1947)

The "engineering of consent." A scary phrase if you believe each of us has the independent free will, or agency, to govern our choices. But just how much agency does each of us have? The concept of the economy as a superorganism implies that at a high level each of us are cogs in a global economic machine. But as discussed earlier in this chapter, each of us human cogs must eat to survive. Multitudes of advertisements and diet fads attempt to engineer consent for exactly what we should eat, but ultimately, eat we must.

Edward Bernays was a founding father of public relations. In his quote above he indicates one can convince people of "the soundness of an idea" with intent and design. We don't have to rely on luck or random mutations as with evolution. In Bernays' 1928 book *Propaganda*, he states this more explicitly:

> The conscious and intelligent manipulation of the organized habits and opinions of the masses is an important element in democratic society. Those who manipulate this unseen mechanism of society constitute an invisible government which is the true ruling power of our country.
>
> We are governed, our minds are molded, our tastes formed, our ideas suggested, largely by men we have never heard of. This is a logical result of the way in which our democratic society is organized. Vast numbers of human beings must cooperate in this manner if they are to live together as a smoothly functioning society.
>
> . . . in almost every act of our daily lives, whether in the sphere of politics or business, in our social conduct or our ethical thinking, we are dominated by the relatively small number of persons . . . who understand the mental processes and social patterns of the masses. It is they who pull the wires which control the public mind, who harness old social forces and contrive new ways to bind and guide the world.[9]—Edward Bernays (1928)

Shakespeare told us "All the world's a stage, And all the men and women merely players," each changing roles over the course of their life.[10] In the case of a theatrical performance, there are many actors, but there is only one director. When it comes to democratic society, Bernays tells us that there are a "small number of persons"

---

[9]*Propaganda*, Chapter 1 [9, pp. 9–10].

[10]William Shakespeare, *As You Like It*.

controlling the marionette that is the collective public mind. And he was a master puppeteer.

Bernays was the nephew of Sigmund Freud, the founding father of the field of psychoanalysis. Inspired by his uncle's ideas, Bernays convinced Americans to consume items they otherwise would not. During the womens' liberation movement in 1929, in what is regarded as the first major public relations campaign, Bernays was hired by the American Tobacco Company. In this campaign he promoted cigarettes to women as "Torches of Freedom."

The American Tobacco Company gave Bernays money. Bernays, thinking he was tapping into the unconscious minds of women, gave women liberation in the form of a phallic cigarette—the penis they are not born with. Maybe women liked to smoke anyway, and Bernays's campaign just helped make it socially acceptable for them. Regardless, smoking women gave the American Tobacco Company a hefty return on its investment in Bernays' campaign. Bernays later regretted taking the cigarette campaign, only after its cancer-causing reality became undeniable.

Before the Torches of Freedom campaign, Bernays worked for the U.S. government to engender support for World War I. Coming out of World War II his ads promoted mass consumption to American citizens that had practiced conservation, saving, and sacrifice for nearly two decades since the start of the Great Depression.[11]

Whether or not Freud's ideas accurately inspired Bernays to public relations success, Bernays knew how to use advertising and social pressure to make people buy stuff. But he wasn't the only one. The top advertising slogan of the past century, as selected by *Advertising Age* (now *AdAge*) magazine, was that of De Beers Consolidated Mines, Ltd.: "A diamond is forever."[12] Before this campaign, in the late 1800s major diamond finds in South Africa triggered the formation of

---

[11]For a background on Edward Bernays, see the BBC documentary *The Century of the Self, Part 1: Happiness Machines*.

[12]From the De Beers website on January 19, 2020 (https://www.debeersgroup.com/the-group/about-debeers-group/brands/a-diamond-is-forever):

A DIAMOND IS FOREVER How the slogan of the century changed the diamond industry

1930S DIAMOND SALES IN THE U.S. WERE AT AN ALL TIME LOW They were seen as an extravagance for the wealthy, and sales, already declining for more than two decades, had plummeted during the Great Depression.

DE BEERS NEEDED A STRATEGY TO CREATE A MULTI-FACETED DEMAND FOR DIAMONDS In the unique position of having to create demand for a product that hadn't been widely marketed before.

1938 DE BEERS HIRED ADVERTISING AGENCY N.W. AYER TO CRAFT A CAMPAIGN They were chosen for their approach—to conduct extensive research on social attitudes to diamonds.

A NEW FORM OF ADVERTISING WAS BORN The brilliant concept was to create an emotional link to diamonds, the sentiment being love, like diamonds, is eternal.

1947 FRANCES GERETY A copywriter on the De Beers account at the advertising agency N.W. Ayer solidified the link between eternal romance and diamonds by suggesting the line "A diamond is forever."

THESE FOUR ICONIC WORDS HAVE BEEN USED EVER SINCE Making it one of the longest running and successful campaigns in history.

De Beers as "The major investors in the diamond mines realized that they had no alternative but to merge their interests into a single entity that would be powerful enough to control production and perpetuate the illusion of scarcity of diamonds."[13] Diamond scarcity and demand were engineered such that in 2014, 80% of first time U.S. brides received diamond engagement rings.[14]

Chapter 8 equated ideas, or memes, to genes. Just like genes, advertising campaigns and other memes are not necessarily good or bad from any given person's perspective, but they can be used to influence people for good, bad, or seemingly agnostic purposes. Some memes seek to promote human empowerment and achievement. A high school football coach might ask you to give 110% to inspire you to practice and play with more effort than you think is possible, but 110% exists only as a meme in his and your head, not as reality from your body.[15] On any given day, a football player cannot give more than 100% of his capability. Because each of us does not know with certainty the physical effort equal to 100% of our body's capability, the 110% meme can propagate.

The Internet and social networks are meme-propagating super highways. Memes propagate on the Internet via videos and social network tweets on topics ranging from dance crazes and cooking recipes to government overthrow and jihadist propaganda. Memes can also be "to die for" or promote negative outcomes to an "infected" individual. Just like genetic mutations can reduce its host's fitness, so can memes:

> ...we must consider as a real possibility the hypothesis that the human hosts are, individually or as a group, either oblivious to, or agnostic about, or even positively dead set against some cultural item [meme], which nevertheless is able to exploit its hosts as vectors.[16]—Daniel Dennett (2003)

BY 1951 8/10 BRIDES IN THE UNITED STATES RECEIVED A DIAMOND ENGAGE-MENT RING The engagement diamond tradition was established.

AN EMBLEM OF LOVE "A diamond is forever" became a symbol of enduring love weaving itself into popular culture and inspiring books, films, and songs.

1999 SLOGAN OF THE CENTURY It's no wonder Advertising Age voted the De Beers campaign as the top advertising slogan of the past century.

AN ENDURING EMBLEM OF LOVE: FOREVERMARK In 2008 it was brought to the high street by creating a range of beautiful, rare, and responsibly sourced diamonds. Each Forevermark diamond carries its own distinct timeless mark making it unique to the owner.

A DIAMOND IS FOREVER IS THE ULTIMATE GEM OF AN IDEA

[13]Edward Jay Epstein, *The Atlantic*, February 1982 "Have You Ever Tried to Sell a Diamond? An unruly market may undo the work of a giant cartel and of an inspired, decades-long ad campaign," available 1/19/2020 at https://www.theatlantic.com/magazine/archive/1982/02/have-you-ever-tried-to-sell-a-diamond/304575/.

[14]Uri Friedman, February 13, 2015, "How an Ad Campaign Invented the Diamond Engagement Ring: In the 1930s, few Americans proposed with the precious stone. Then everything changed." *The Atlantic* at: https://www.theatlantic.com/international/archive/2015/02/how-an-ad-campaign-invented-the-diamond-engagement-ring/385376/.

[15]The prominent basketball coach John Wooden was *Coach Wooden: The 7 Principles That Shaped His Life and Will Change Yours* by Pat Williams and James Denney [p. 57].

[16]Dennett [7, p. 178].

Daniel Dennett reminds us that many people have sacrificed themselves for memes.[17] The pressure to fight and die for Japan and its emperor made it hard for most Japanese pilots to decline the Kamikaze suicide missions of World War II. The 1978 Jonestown (in Guyana) massacre is famous for Jim Jones convincing nearly a thousand followers of his California-based Peoples Temple cult to commit suicide and murder children by ingesting a cyanide-laced fruit drink. And American football players, including professional Pro Bowl tackle Korey Stringer in 2001, have died in practice due to heat exhaustion, trying to give 110%.[18]

A few can engineer the consent of the many. When this consent goes against one's own individual livelihood or survival, the idea drives a person past their body's physical limit. The body crashes.

While memes include religions, diets, doll fads, and high school pride, *the memes of concern for this chapter are those that support or refute the energy and economic narratives (as posed in Chap. 1) and the idea of the economy as a superorganism.*

If public relations masters control consumer choices, and their memes can infect us, then why would we care what people and consumers think? Does it matter how much confidence we have with regard to future consumption of energy, other items, or economic growth in general?

The Confidence Delusion

With enough *confidence*, whether via pure speculation or using some behavioral or economic models, we can always grow the economy. We just need confidence in ourselves, in markets, in human ingenuity and technological change. At least, this is what we are often told.

Economists and pundits pay attention to business and consumer confidence surveys. These surveys gather certain types of information about the state of the economy. But as stated in this book, it is both challenging and important to understand how individual actions are connected and constrained by higher system level feedbacks from both information and physical resource inputs.

---

[17]"Now, am I saying that a sizable minority of the world's population has had their brain hijacked by parasitic ideas? No, it's worse than that. Most people have. (Laughter) There are a lot of ideas to die for. Freedom, if you're from New Hampshire. (Laughter) Justice. Truth. Communism. Many people have laid down their lives for communism, and many have laid down their lives for capitalism. And many for Catholicism. And many for Islam. These are just a few of the ideas that are to die for. They're infectious." Dangerous memes, TED talk by Daniel Dennett, 2002: https://www.ted.com/talks/dan_dennett_on_dangerous_memes/transcript.

[18]Kevin Allen, August 12, 2018, *USA Today*, "Heatstroke dangers reinforced by investigation into death of college football player." Available April 8, 2019 at: https://www.usatoday.com/story/sports/ncaaf/2018/08/12/heatstroke-maryland-death-practice-korey-stringer-jordan-mcnair/967134002/.

Too many ideas assume that individual actions are independent of these feed-backs. Consider this quote from a book entitled *How the Economy Works: Confidence, Crashes, and Self-Fulfilling Prophecies*:

> The wealth of households depends on what other households believe. Wealth depends on confidence![19]—Roger Farmer (2010)

Farmer also states that confidence can be viewed as a "fundamental" driver of the economy:

> Because there is no unique fundamental labor market equilibrium, there is also no unique fundamental value for the price of a stock. By adding confidence as a separate fundamental, we can retain a theory in which everything is determined by fundamentals, including the value of stock prices. Confidence is an independent driving force of the business cycle.[20]— Roger Farmer (2010)

But just what kind of confidence are we talking about here? How is confidence a "fundamental?"

Somewhere around 99.99999% of scientists are confident that we cannot make a perpetual motion machine, but someone could be confident that he can make one. Company executives sometimes state that they just want "certainty" in knowing market rules and regulations will not change quickly so that they can have confidence in their investment decisions. But what if the regulations state with certainty and clarity that the company must be limited in its size or level of profits? Does that certainty provide confidence in the same way that lower tax rates do?

It is easy to see how some concepts of belief and memory affect business cycles for a few years or maybe a little more than a decade. After all, major business cycles have periods of several decades.[21] Stock price bubbles take several years before they pop.

Nonetheless, I have a hard time thinking that confidence is itself some "independent driving force" of long-term growth trends lasting longer than a couple of decades. Just because we say something is fundamental doesn't make it so. A used-car salesman tries to make us confident that we're making a good choice to buy his car. When I signed the mortgage on my house, I was a bit skeptical of what seemed to me as disingenuous praise from people that kept telling me how good of a deal I was getting. This was because their fees depended upon my signature. I'm not claiming that title offices, realtors, and mortgage lenders don't provide valuable services. I'm just stating that their praise didn't raise my confidence level. But, to be fair to them, I'm an engineer, scientist, and a heterodox economist. I seek flaws and enjoy improving designs, models, and our ways of thinking, so in that sense I tend to question those who seem overly confident.

---

[19] Farmer [10, pp. 163–164].

[20] Farmer [10, p. 113].

[21] For example, Kondratiev cycles or waves, named after Nikolai Kondratiev, are posed 40–60 year cycles in economic activity.

I am confident in some ideas, however. I'm confident that our extraction of energy from the environment is the most fundamental driver of life and the economy, and that this premise is severely and "confidently" neglected by most economists. That is what this book is about. Perhaps my confidence expressed in this book will become a meme to infect others while also curing them of other infectious memes.

But how do we know what to believe? Should we want confidence or information? What information should we have access to? Donella Meadows states the power of having credible information:

> Missing information flows is one of the most common causes of system malfunction. Adding or restoring information can be a powerful intervention, usually much easier and cheaper than rebuilding physical infrastructure.[22]—Donella Meadows (2008)

An example of acquiring information on the state of the economy is the Consumer Confidence Survey®. It asks individuals questions about business conditions, employment conditions, and family income today and six months out to calculate a set of indexes.[23] This survey specifically concerns *consumer confidence*, so asking these questions of consumers can justifiably inform immediate investment decisions for businesses. But are these the right questions to ask? Sure, for some purposes.

A person's appraisal of their current employment and income situation is a valid piece of information for tracking short-term trends such as jobs, income, and consumer spending on energy, computers, and vacations. After all, if at any level we want the economy's purpose to focus on the human condition rather act as an indifferent superorganism, we should know how people perceive it. Individuals know their employment and income status as well as the status of several friends and family members. These types of measurements are easy to make, not only for individuals, but also for companies and governments that track mathematical quantities such as GDP (even though we change the definition of what transactions are included in GDP), stock prices, and market prices of commodities like oil. However, just because some items are easy to measure and quantify doesn't mean that those measurements are what we need, and it doesn't mean that they represent what we think they represent or want them to represent.

Donella Meadows, again, has a good statement on this:

> We try to measure what we value. We come to value what we measure. This feedback process is common, inevitable, useful, and full of pitfalls.[12]—Donella Meadows (1998)

Employment and anticipated near-term purchases aren't so important for assessing the confidence people have in longer-term aspects of their lives that don't involve

---

[22]Meadows [11, p. 157].

[23]The indexes are based on responses to five questions in the survey: Consumer Confidence Index: 1. Respondents' appraisal of current business conditions. 2. Respondents' appraisal of current employment conditions. 3. Respondents' expectations regarding business conditions six months hence. 4. Respondents' expectations regarding employment conditions six months hence. 5. Respondents' expectations regarding their total family income six months hence. The Conference Board, Consumer Confidence Survey® Technical Note—February 2011, https://www.conference-board.org/pdf_free/press/TechnicalPDF_4134_1298367128.pdf.

acting as a consumer or employee. We think further than 6 months out and about other outcomes. Is short-term confidence a good metric for thinking about long-term policy and observed economic trends? Do consumers, politicians, and business owners know how to put annual and even decadal trends into the context of human history? How do we know that the short-term trends in home prices, or any other economic indicator, are reflecting long-term fundamentals, speculation, or noise? Per Jay Forrester's earlier quote, do we have enough people that "have true insight into the nature of the complex systems within which they live?"

A lack of crucial information and knowledge can prevent us from accurately interpreting a situation. In the case of the 2008 financial crisis, many investors did not have full and correct information on the low quality of mortgage-backed securities. The crisis was triggered by confident belief in the meme that U.S. house prices only go up, never down, such that that investing in mortgages was a safe bet. These beliefs turned out to be very wrong. However, some people knew these beliefs were wrong, and they benefited (financially) by pulling "the wires which control the public mind" (per Bernays) to engineer a sense of confidence that housing prices would keep rising. If a banker were motivated and incentivized to make money by selling packages of low-quality mortgage loans, would he tell a potential buyer to question the level of "confidence" he attributes to what he is selling? Very few bankers did this leading up to the 2008 financial crisis. It was about making money at the moment.

Investors were confident that the ratings agencies properly vetted the securities, if they even thought about it at all. Some bankers issuing securities knew they were bundling low-quality mortgages, but even for some honest bankers, their theory and analysis of historical statistics made them confident that their risk was properly hedged.

One of the problems with considering confidence as a driver of economic growth is that you don't have to be knowledgeable to be confident that you are correct. This ill-placed confidence is known as the Dunning–Kruger Effect, named after the psychologists David Dunning and Justin Kruger. They determined "... people who are incompetent at something are unable to recognize their own incompetence. And not only do they fail to recognize their incompetence, they're also likely to feel confident that they actually are competent."[24]

Some have argued something similar for the use of economic models. Most models can be useful in their proper domain of applicability, but it can be disastrous to rely on them outside of that domain. In 2008, after the beginning of the financial crisis, then Federal Reserve Chairman Alan Greenspan stated "... I discovered a flaw in the model that I perceived is the critical functioning structure that defines

---

[24]Mark Murphy, University of Michigan, College of Literature, Science, and the Arts, "The Dunning–Kruger Effect Shows Why Some People Think They're Great Even When Their Work Is Terrible": https://lsa.umich.edu/psych/news-events/all-news/faculty-news/the-dunning-kruger-effect-shows-why-some-people-think-they-re-gr.html.

how the world works. I had been going for 40 years with considerable evidence that it was working exceptionally well."[25]

Regardless of whether scientists and economists might be overconfident in their theories and models, some take the perspective that if we understand how any given model is incorrect, it might still be worthwhile to use because it ensures we're all using a common narrative. Thus, even if we know our story of how the economy works is not quite right, but we are knowledgeable in how it is wrong, then at least we're all talking about the same narrative whether on an academic campus or in the cafés of Davos.[26]

So what comes first: confidence or wealth? Can there be wealth derived from confidence that transcends more than a decade, or even more than a century? This book argues a resounding *no*.

Confidence is not specific enough: we can have confidence of future prosperity or stagnation. If we have confidence in stagnation, then the whole economic system breaks down. Nate Hagens summarized this well on a post on the formerly (very) active blog, *The Oildrum*, on why people disagree about the impacts from peak oil extraction:

> It's not about running out of oil, but running out of the perception of growth:
> Our debt based capitalist society is based on the ability of everyone to climb the ladder. If it becomes apparent that there is a ceiling, all the rules of the system breakdown. Growth is based on the ability of people to get loans, grow businesses and repay the loans with interest. If there is less and less energy available each year that's one thing—it might just show up as recession/belt-tightening. However, if peoples PERCEPTION is that less and less energy will be available then why would banks give out loans, why would people go to work, etc.?[27]—Nathan Hagens (2007)

It is first wealth, or lack thereof, that creates confidence in interpreting the status of the economy. Wealth derives from extracting energy and natural resources. Differences in the confidence between individuals or groups of people are derived from their historical experience and the times during which they grow from childhood to maturity. Those growing up in times of increasing energy consumption might see things differently than those growing up in times of stagnation.

The U.S. Baby Boom generation is stereotypically more confident that if you work hard and make good decisions, you will earn a good wage and achieve material wealth, the "American Dream." The Millennial generation is stereotypically more confident that if they make the same choices at the same stage in life as the Baby Boomers, they will not gain the material and monetary wealth of the Baby Boomers. To many of the Millennial generation, the American Dream is alive and well, but only in their dreams.

---

[25] Andrew Clark and Jill Treanor, "Greenspan—I was wrong about the economy. Sort of," *The Guardian*, October 24, 2008 at https://www.theguardian.com/business/2008/oct/24/economics-creditcrunch-federal-reserve-greenspan.

[26] Davos is a town in Switzerland where the World Economic Forum gathers the world's business and banking elite each Winter to discuss world economic affairs.

[27] http://www.theoildrum.com/node/2367.

A short and poignant example of the contrast between American generations is the 2019 Saturday Night Live comedic skit of a game show entitled "Millennial Millions."[28] In this skit, two Millennials have the chance to win cash, health insurance, or debt relief. All they have to do is listen to a Baby Boomer complain about their life for thirty seconds, without interrupting. As the game show host indicates (sarcastically), "It sounds easy, but I know how you Millennials love anything that challenges your worldview." Even for hundreds of thousands of dollars, the Millennial contestants can't stay quiet during the Boomer rants. In response, the game show host only replies: "I'm Gen X. I just sit on the sidelines and watch the world burn."

As far as my own self-reflection regarding the writing of this book, my assessment is this: I am an academic Generation X-er writing a book, not running a business, not appreciably engaged in politics at the moment. I'll have to admit that the reader might see me as sitting on the sidelines making commentary, just like the Saturday Night Live game show host of "Millennial Millions." A common caricature of academics is that we hang out next to ivory towers on campus, disconnected from the "real" world. My office literally resides in the building next to the ivory tower at the University of Texas at Austin. I see it from my office window.

That said, I'm writing this book because I believe it provides societal value by explaining why natural resource and energy consumption are often underappreciated or neglected factors for explaining our contemporary economic situation. Energy consumption and the feedbacks from finite Earth effects help explain why Baby Boomers can't see eye-to-eye with the Millennials. Baby Boomers are confident that their accumulation of monetary wealth can be used to maintain their wealth via lower taxes or at least the maintenance of a relatively low tax regime. Millennials are confident that the same principle prevents them from achieving what the Boomers have. This interpretation is entirely consistent with the idea that young voters increasingly seek more drastic, rather than incremental, changes to tax policies and functions of the federal government.

Much of this change can occur by electing representatives with viewpoints more in tune with the struggles of the younger generation. So what about the representatives that we, of different generations, elect to make collective decisions for our economy and communities? Do politicians need a particular kind of confidence, and from where do they get it? Do politicians, and the judges who determine the validity of our laws, have any more free will than citizens and consumers? To answer these questions, we now shift to a discussion of political will.

---

[28] Saturday Night Live, January 19, 2019, Millennial Millions skit: https://www.nbc.com/saturday-night-live/video/millennial-millions/3867395.

The Political Will Delusion

Political Will   Perhaps you've heard this term used or read it in a news article. Consider the following quote regarding the lack of global action to reduce greenhouse gas emissions:

> The problem is not the Paris agreement, which is the best climate pact ever negotiated; rather the problem is inadequate political will in capitals around the world.[29]—Nigel Purvis (2018) (Chief Executive of Climate Advisers at the 24th Conference of Parties meeting, Katowice, Poland)

Inadequate political will to do what? Do politicians and citizens have only one goal, to limit global temperature rise by capping greenhouse gases in the atmosphere? Certainly some political will exists to mitigate climate change, but some political will also opposes climate change mitigation. Aside from energy and climate concerns, elected officials also have political will to fix potholes, achieve affordable health care, educate the public, and maintain a strong national military. How is political will distributed, and how are we to know which political desire overcomes another?

This question reminds me of the start-up company I worked for in the 2000s. At some point the executives decided to assign priorities to tasks. In theory this would help me and the main technician determine which design work to perform next. The problem was that we just accumulated too many "highest priority" tasks. I was pretty sure I was living in a *Dilbert* cartoon. Ultimately the technician and I ignored the prioritizations and decided the next task to work on by ourselves. This is exactly the problem with political will. A politician has many constituents, each possibly having a unique number one priority. So how does he or she decide how to prioritize their legislative efforts among multiple "highest" priority items?

Often political will is viewed as taking actions that benefit the majority of the people rather than a smaller subset of connected factions. Similarly, *social power*, or political power, is the ability of an individual or group to influence the behavior of others, often without the powerful group having to change its own behavior in an undesirable manner.

As referenced in Chaps. 4 and 7, government support to maintain workers' rights and bargaining power for wages is a common example to explain the concept of political will. The immediate post-World War II decades are often viewed as those with high political will in the United States. Paul Collier explains the end of the glorious thirty years (Phase 1 highlighted in Chap. 7) of successful social democracy in the U.S., U.K., and some European nations:

---

[29] Busby, Joshua "The latest global climate negotiations just finished. Here's what happened," *Washington Post* Monkey Cage blog, December 17, 2018, accessed December 20, 2018 https://www.washingtonpost.com/news/monkey-cage/wp/2018/12/17/the-latest-global-climate-negotiations-just-finished-heres-what-happened/.

Social democracy worked from 1945 until the 1970s because it lived off a huge, invisible and unquantifiable asset that had been accumulated during the Second World War: a shared identity forged through a supreme and successful national effort. As that asset eroded, the power wielded by the paternalistic state became increasingly resented.[30]—Paul Collier (2018)

By *social democracy*, Collier refers to political arrangements in which people agree to pool resources and co-own assets to implement policies that address the concerns of everyday people. Concerns such as health care, pensions, education, and unemployment insurance became mainstream political concepts, embodied in legislation after the Great Depression and World War II. The policies were social because most people both contributed to them, via taxes, and benefited from them by having educated children and retirement security. People gave into the system, and the system gave back services they wanted. There was reciprocity, an important requirement for a well-functioning social order. If people contribute to the greater good, and never receive anything in return, this reduces the desire to contribute to collective goals.

Time is said to be the healer of all wounds, but in this case, Collier implies that perhaps it was time that forged a divide between any common purpose that previously spanned generations. Winning a war helps forge a national identity that brings people together. Wars are, of course, also destructive, and forging unity under a nationalist banner can lead to discord and war itself, as it did via the rise of the German Nazi party instigating World War II. Thus, we need alternative constructive ways to bring citizens together.

Unions are often seen as entities that can bring workers together for their common cause. Figure 9.2 shows U.S. union membership along with the percent of income going to the top 10% income bracket.[31] Clearly higher union membership coincided with workers obtaining a larger share of the economic pie. We should try to understand why union membership *could or was allowed to* increase in the 1930s and the run-up to World War II while it has declined since the early 1970s.

High union membership might be easier during times of increasing rates of energy consumption. As noted in Chap. 7, when resources are more abundant, people are more willing to share. In the case of rising employee compensation in the decades leading to 1970, businesses were more willing to share with workers. Was it political will that established social democratic policies in the 1930s and

---

[30]Collier [13, p. 15].

[31]Lawrence Mishel and Jessica Schieder, "As union membership has fallen, the top 10% have been getting a larger share of income," Economic Policy Institute website, May 24, 2016 at https://www.epi.org/publication/as-union-membership-has-fallen-the-top-10-percent-have-been-getting-a-larger-share-of-income/. Per this reference: (1) data on union density follows the composite series found in Historical Statistics of the United States; updated to 2014 from unionstats.com. and (2) Income inequality (share of income to top 10%) from Piketty and Saez, "Income Inequality in the United States, 1913–1998," *Quarterly Journal of Economics*, 118(1), 2003, 1–39. Updated data for this series and other countries is available at the Top Income Database. Updated 2016.

Fig. 9.2 Higher union membership has been associated with a higher share of income to lower income brackets (the lower 90%) and a lower share of income to the top 10% of earners. While the percent of American workers belonging to labor unions rose dramatically during the New Deal 1930s, peaking at the end of World War II, it has fallen dramatically since the early 1970s when per capita energy consumption no longer increased (see Fig. 7.1e)

1940s, or that undermined them after the 1970s? Or was it abundant and cheap energy resources that enabled social democratic policies to work until the 1970s, and energy constraints that forced a restructuring of policy after the 1970s?

Recall that my economic modeling discussed in Chap. 6 shows that, even with no change in the assumption related to labor "bargaining power," you can explain a shift from increasing to declining income equality (higher equality expressed as a higher wage share) by a corresponding shift from a period of rapidly increasing per capita resource consumption to one of constant per capita resource consumption. This thesis is supported by both the data and an economic theory that consistently links the flows of resources and money.

> What scientists, economists, politicians, and citizens alike need to appreciate is that rapidly increasing energy and material consumption supported post-World War II socially-democratic and distributive policies.

It doesn't take as much political will, or social power, to distribute pieces of a rapidly expanding economic pie versus one of a constant, shrinking, or slowly

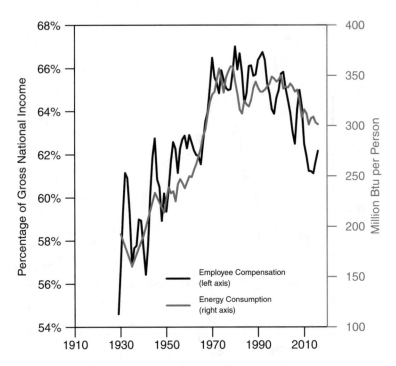

Fig. 9.3 The fraction of total U.S. income to workers (black line, left axis) follows the same trend of U.S. per capita primary energy consumption (gray line, right axis (data as in Fig. 7.1))

increasing size. When everyone gets more, it is easier for everyone to be satisfied. If the flow of energy or money per person shrinks, one or more groups will have to get less. If all groups do not decrease their share by the same percentage, this can be interpreted as a decline in political will. But from the standpoint of laws of conservation of flow of money and energy, it is a mathematical necessity to decrease the flow to some.

The connection between U.S. worker compensation and energy consumption is clear in Fig. 9.3 (same data as in Fig. 7.1a and e). The difficult situation we face is the glory years of high equality and political will only came during a time of unprecedented increases in per capita energy consumption, and because of the constraints of a finite Earth, we can't indefinitely increase per capita energy consumption (at least with a growing or constant population). Thus, it is unclear how we maintain high income equality in any economic system, particularly in our current system based on capitalism and debt-based finance. Most political will discussions ignore this energy-income equality conflict.

I had a useful discussion on this conflict between energy consumption and "bargaining power" with a colleague, an economist in the U.K. He noted that during the Golden Age after World War II, or Phase 1 of Fig. 7.1, inflation was low and employment was high. Even though wages were rising, there was no upward inflation because labor productivity also increased with wages (refer back to Fig. 4.11). However, company profits declined toward the late 1960s due to growing labor costs, so the Golden Age might not have continued even without the 1970 peak in U.S. oil extraction and 1974 OPEC oil price increase. After the OPEC oil price hike, inflation surged, and many labor unions were powerful enough to gain matching wage rises, thus maintaining the wage–price spiral that brought inflation to 10%. The Federal Reserve then reacted by increasing interest rates to over 15% (refer back to Figs. 4.9 and 4.10), causing stagflation and recession, shifting power to business.

This colleague and I could not come to full agreement on the interpretation of the U.S. shift from Phase 1 to Phase 2 of Fig. 7.1. We did agree that the situation forced a change in economic operation. My position is that increasing wages reduced profits to the point of stressing the capitalist system, and that one important reason it took about 30 years (starting in 1945) to realize this was because of the rapid increase in low-cost energy consumption. Without the abundant and low-cost energy, the political will for increasing wages and union membership might have materialized, but it would have likely had less longevity.

As soon as oil prices and total energy expenditures rose outside of the control of U.S. businesses and policymakers, and energy consumption declined, the economy was forced into some combination of falling wages and profits. In the U.S. and U.K., policies favored profits more than wages. Some European countries such as Germany, France, and those within Scandinavia placed more emphasis on labor stability and socially democratic benefits.

These economic and social arrangements are becoming more stressed today, as many OECD countries are in the same Phase 3 situation as the U.S.

Given that higher wages, higher income equality, and more political will existed during the three decades after World War II when energy consumption increased more rapidly than any time in history, is this correlation simply a coincidence? While we know companies influence our consumer decisions via advertising and public relations campaigns, companies also have other people they want to influence: politicians and regulators.

The Political Will Delusion: Lobbying

We can't change the structure of the economy by changing the physical laws. There is no other viable choice but to change the *legal rules and social norms* that guide relations between citizens, governments, and corporations. By legal rules I mean corporate, civil, and criminal laws enforced by governments, as well as social and religious norms. Social norms are less explicit rules that also guide social and economic relations.

By definition, we can take any action that is possible within the bounds of physical laws and enabled by the degrees of freedom at our disposal. In turn, social rules reduce the allowable degrees of freedom to a subset of all degrees of freedom.

Thus, collectively social norms and legal rules affect "... the data that actors in the system have to work with, and the ideas, goals, incentives, costs, and feedbacks that motivate or constrain behavior."[32]

Edward Bernays understood the importance of controlling the data and information flows to influence consumer behavior. Likewise, Lewis Powell understood how to influence the economy via its civil and corporate laws.

The quote below comes from Lewis Powell's August 23, 1971 letter to his friend Eugene Sydnor, Jr., the Director of the U.S. Chamber of Commerce. This letter is also known as *The Powell Memo* and *The Powell Manifesto*. At the time, Powell was a corporate lawyer. Two months later President Richard Nixon nominated Powell to become a Supreme Court Justice. (Powell was confirmed in December 1971.)

> No thoughtful person can question that the American economic system is under broad attack. This varies in scope, intensity, in the techniques employed, and in the level of visibility.
>
> There always have been some who opposed the American system, and preferred socialism or some form of statism (communism or fascism). Also, there always have been critics of the system, whose criticism has been wholesome and constructive so long as the objective was to improve rather than to subvert or destroy.
>
> But what now concerns us is quite new in the history of America. We are not dealing with sporadic or isolated attacks from a relatively few extremists or even from the minority socialist cadre. Rather, the assault on the enterprise system is broadly based and consistently pursued. It is gaining momentum and converts.[33]—Lewis Powell (1971)

To Powell, the "American economic system" is that of capitalism and free enterprise, as opposed to the only undesirable alternatives with "... varying degrees of bureaucratic regulation of individual freedom":

> There seems to be little awareness that the only alternatives to free enterprise are varying degrees of bureaucratic regulation of individual freedom—ranging from that under moderate socialism to the iron heel of the leftist or rightist dictatorship.
>
> We in America already have moved very far indeed toward some aspects of state socialism, as the needs and complexities of a vast urban society require types of regulation and control that were quite unnecessary in earlier times. In some areas, such regulation and control already have seriously impaired the freedom of both business and labor, and indeed of the public generally. But most of the essential freedoms remain: private ownership, private profit, labor unions, collective bargaining, consumer choice, and a market economy in which competition largely determines price, quality and variety of the goods and services provided the consumer.[34]—Lewis Powell (1971)

Why did Powell state that the "American economic system is under broad attack?" Because from his perspective, he thought it was under attack from the federal government and American citizens themselves. From the breakup of

---

[32]Meadows [14, p. 237].

[33]"Confidential Memorandum: Attack of American Free Enterprise System." Available July 2019 at: http://reclaimdemocracy.org/powell_memo_lewis/ and https://d1uu3oy1fdfoio.cloudfront.net/wp-content/uploads/2012/09/Lewis-Powell-Memo.pdf.

[34]Lewis Powell, August 23 1971, *Confidential Memorandum: Attack of American Free Enterprise System*. A memo to Eugene Sydnor, Jr., the Director of the U.S. Chamber of Commerce.

Standard Oil in 1911 as violating the Sherman Antitrust Act, to the 1929 stock market Great Crash starting the Great Depression and ushering in President Franklin Delano Roosevelt's New Deal. Increasingly after World War II, U.S. policies favored workers' rights and incomes, as demonstrated in Figs. 9.2 and 9.3.

The 1950s and 1960s continued to witness landmark laws governing individual rights and the environment. The 1964 Civil Rights Act outlawed discrimination based on race, color, religion, sex, or national origin. The 1965 Voting Rights Act put in place measures to enforce 15th Amendment (ratified 95 years earlier in 1870) that established the right to vote for African Americans.

As I've emphasized, in interpreting the energy and economic narratives, it is important to consider that many economic, resource, and environmental trends came to a head at the same time during the late 1960s through the 1970s. In April 1970, 20 million U.S. citizens marched into streets and parks to celebrate the first Earth Day [15]. After a century of industrialization, "They demanded a better balance between corporations and people and better stewardship of our land, water, and air."[35] In response, many environmentally related governmental programs or regulations were either started or enhanced in the 1970s. For example, the U.S. Environmental Protection Agency (EPA) was established in 1970 along with an enhancement to the Clean Air Act. The Clean Water Act passed in 1972. The list goes on.

The influence of citizens and workers on the passage of these civil and environmental laws is undeniable, and these and other laws significantly influenced the historical course of events. Before the U.S. EPA was formed, air and water pollution were getting worse; after the EPA was established, air and water became cleaner. Political scientists and activists study this type of cause and effect relationship.

These political relationships are, however, emergent concepts, just as with physical cause and effect relationships. Matter does not move or change in shape without a corresponding energy-consuming process. As this book emphasizes, *the long-term economic growth and structural trends we have seen over the last century must be understood by considering the rate of energy consumption as a necessary input into economic processes.*

When I've made this case to political scientists and economists, some tell me that I need to get my cause and effect relationships correct. As a physical scientist, my initial reaction was a combination of confusion and insult. I then try to explain: 'Of course I consider cause and effect relationships. Didn't you just hear me explain that energy consumption and conversions are the cause of economic activities?' While the cause and effect distinction becomes less obvious the closer you get to physical and philosophical foundations, we can say that certain phenomena tend to come together, rather than separately. Energy conversions come with economic activity.

But there is a more important point. It's not that I was wrong and my policy and economic colleagues were correct. We are all correct. From the standpoint of the philosophy of poetic naturalism, we can simultaneously hold multiple ideas in our heads if they are compatible and useful.

---

[35]Clements [15, p. 14].

The statement "human decisions and policies causally affect economic behavior and outcomes" is compatible with the statement
"energy consumption causally drives economic activity."

Because of the near simultaneous timing of many important political, environmental, and energy events with the span of a decade, there will likely *never* be
complete agreement on the relative influence of each applicable cause and effect
relationship. The fight over economic allocation between workers (or labor) and
owners of property and corporations (or capital) is a continuous battle.

Many credit Powell's 1971 memok as starting a decades-long process of
increasing rights and influence of corporations in legislative and judicial processes.
Powell indeed intended to kick-start American corporate influence in the political
and judicial system. Before Powell was a Supreme Court Justice, he resided on the
Board of Directors of the cigarette company Philip Morris.

While Edward Bernays used advertising to convince women to smoke cigarettes
in the 1920s, in the 1960s Lewis Powell used legal arguments to defend the industry
against the emerging science that smoking was harmful to human health [15].[36]
Given his pro-corporate tendency, and observing that liberals and the "far left"
already used the judicial system for their end goals, Powell realized that businesses
should do the same:

> American business and the enterprise system have been affected as much by the courts as
> by the executive and legislative branches of government. Under our constitutional system,
> especially with an activist-minded Supreme Court, the judiciary may be the most important
> instrument for social, economic and political change.[37]—Lewis Powell (1971)

Corporate America executed the campaign that Powell suggested by starting
several organizations, or "legal foundations," in the 1970s. As Jeffrey Clements
stated in his book *Corporations are not People*: "These legal foundations were
intended to drive into every court and public body in the land the same radical
message, repeated over and over again, until the bizarre began to sound normal:
corporations are persons with constitutional rights against which the laws of the
people must fall."[38]

Thirty-nine years after his memo, and 12 years after his death, Powell's dream
came true in the form of the Supreme Court ruling known as *Citizens United v.
Federal Election Commission*. The non-profit corporation Citizens United won its
argument that an existing law (the Bipartisan Campaign Reform Act) "... violated
the First Amendment right of free speech because it prevented Citizens United
... from engaging in 'electioneering activity' and for-profit corporations from
contributing to Citizens United's electioneering activity."[39]

---

[36]Clements [15, pp. 19–22].

[37]Lewis Powell, 1971, *Confidential Memorandum: Attack of American Free Enterprise System*.

[38]Clements [15, p. 23].

[39]Clements [15, p. 9].

While running for the Republican nomination for U.S. president in 2011, the *Citizens United* ruling led Mitt Romney to say to a heckler at an Iowa rally: "Corporations are people, my friend."[40] The heckler replied: "No they're not!"

Many Americans are justifiably confused upon learning their common sense definition of a corporation conflicts with the current legal definition implied by *Citizens United.*

The memes that define our economy and society matter.

The Political Will Delusion: The Battle Over -isms

In his 1971 memo, Powell presented two basic options for societal organization. The first is the free enterprise system within *capitalism* that is defined by private ownership of companies and capital. To Powell, this system can do nothing but maximize individual freedom and prosperity. However, many argue that, in part due to Powell, this system has morphed into something that does indeed limit individual freedom, and we will soon return to this concept when discussing neoliberalism.

Powell's second option is some range of government control, or regulation of individual freedom. At the worst, we could live under a dictator within a (leftist) communist modern day North Korea, a (rightist) World War II era Nazi Germany under Adolf Hitler, or a (rightist) fascist Italy under Benito Mussolini.

Per Powell's quote, "moderate socialism" is the least bad alternative to free enterprise. *Socialism* is defined by state, or government, property ownership. In the strict definition a central government owns every factory, hotel, farm, and mineral, but it does so for the benefit of the citizens. When pro-socialist Americans think of socialism, they think of Scandinavian countries such as Sweden, Norway, Finland, and Denmark.[41] These countries are not strictly socialist. They are social democracies that currently exhibit many of the pro-worker/citizen characteristics that Americans seem to increasingly support.

In 2018, Gallup reported that while the percentage of Americans with a "Positive view of socialism" remained steady within each age category from 2010 to 2018, the 18–29-year-old group has had a more favorable view, ranging from 49% to 55%, than that of older age groups at less than 40%. The major change during this time span relates to the "Positive view of capitalism" in 18–29-year-old category—it dropped from 68% in 2010 to 45% in 2018—indicating that in 2018 the young

---

[40]Ashley Parker," 'Corporations Are People,' Romney Tells Iowa Hecklers Angry Over His Tax Policy," *The New York Times* August 11, 2011 available at: https://www.nytimes.com/2011/08/12/us/politics/12romney.html. Also see YouTube video of the Iowa rally: https://www.youtube.com/watch?v=St1wSWtm_BI.

[41]Pew Research Center, October 2019, "In Their Own Words: Behind Americans' Views of 'Socialism' and 'Capitalism' ", available at https://www.people-press.org/wp-content/uploads/sites/4/2019/10/PP_2019.10.07_Socialism-and-Capitalism_FINAL.pdf.

had a more positive view of socialism than capitalism.[42] Statistics such as these prompted President Trump to confront the concept of socialism in his 2019 State of the Union address, stating:

> Two weeks ago, the United States officially recognized the legitimate government of Venezuela, and its new interim President, Juan Guaido.
>
> We stand with the Venezuelan people in their noble quest for freedom—and we condemn the brutality of the Maduro regime, whose socialist policies have turned that nation from being the wealthiest in South America into a state of abject poverty and despair.
>
> Here, in the United States, we are alarmed by new calls to adopt socialism in our country. America was founded on liberty and independence—not government coercion, domination, and control. We are born free, and we will stay free. Tonight, we renew our resolve that America will never be a socialist country.[43]—U.S. President Donald J. Trump (2019)

When anti-socialist Americans think of socialism, they think of Venezuela and Russia.[44] (As opposed to Scandinavian countries.)

One can make a *straw man* or a *steel man* argument. To argue against a steel man is to argue against the strongest and most accurate description of the concept in question. To argue against a straw man is to argue against a weak and simplified description.[45]

It is certainly easier to attack corrupt Venezuela as a straw man petro-state with a failed vision of socialism than Norway as a steel man social democracy whose petroleum proceeds fund its Government Pension Fund Global that "... is owned by the Norwegian people ..." and holds "... the people's money, owned by everyone, divided equally and for generations to come." Norway is not strictly socialist but its Pension Fund exhibits significant socialist flavor.[46]

---

[42]Gallup, August 13, 2018, "Democrats More Positive About Socialism Than Capitalism," available March 23, 2019 at: https://news.gallup.com/poll/240725/democrats-positive-socialism-capitalism.aspx?version=print.

[43]Remarks by President Trump in State of the Union Address, Issued on: February 6, 2019, available March 23, 2019 at: https://www.whitehouse.gov/briefings-statements/remarks-president-trump-state-union-address-2/.

[44]Pew Research Center, October 2019, "In Their Own Words: Behind Americans' Views of 'Socialism' and 'Capitalism' ", available at https://www.people-press.org/wp-content/uploads/sites/4/2019/10/PP_2019.10.07_Socialism-and-Capitalism_FINAL.pdf.

[45]Arguing against a steel man is the opposite of a arguing against a straw man. "Intentionally caricaturing a person's argument with the aim of attacking the caricature rather than the actual argument is what is meant by "putting up a straw man." Misrepresenting, misquoting, misconstruing, and oversimplifying are all means by which one commits this fallacy. A straw man argument is usually one that is more absurd than the actual argument, making it an easier target to attack and possibly luring a person towards defending the more ridiculous argument rather than the original one." [16]

[46]Because of significant state ownership of companies, Norway is perhaps the European country closest to the definition of socialist. The government owned about 60% of net national wealth in 2015, including a two-thirds stock ownership of Equinor, Norway's major oil and gas company. Norway's sovereign wealth fund, the Government Pension Fund Global that began in 1990, was created to invest revenues from oil and gas extracted from the North Sea. As stated by a video summarizing the fund: "The fund is owned by the Norwegian people ..." and "It is the

Whether a straw man or steel man, when choosing among a set of conflicting policies, we must make arguments for and against them. Increasingly, up to and ever since the *Citizens United* ruling, policies are analyzed using economic models. A policy calculated to have more benefits than costs is seen as better. Benefits can include more income, more jobs, and increased choices and degrees of freedom. Costs can include more spending, a loss of jobs, environmental impacts, and a loss of options.

Decision makers within corporations and governments run cost–benefit calculations to determine which investments and policies to pursue. This is why economic modeling is so important. For example, the U.S. EPA runs cost–benefit analyses to determine how to implement programs that affect air and water quality, and these analyses even include what is called the "value of a statistical life" that monetarily quantifies human health impacts, including premature death.[47]

Within an economic model resides some equation or algorithm that defines how to calculate benefits minus costs. Thus, running the model to maximize benefits minus costs is like maximizing the utility of the superorganism as discussed in Chap. 8. However, as that chapter's utility monster example showed, a maximized outcome for "the economy" does not guarantee a maximized outcome for each person.

What if those who benefit aren't the same as those that bear the costs? Doesn't that matter?

In its purest form *communism* is defined by citizens, without social class division, sharing property ownership for land, factories, and other natural and industrial resources. Everyone is equal, and the community collectively owns all property and distributes proceeds from everyone's work based on each person's needs. Today, people often relate communism to North Korea, Cuba, China, and the former United Soviet Socialist Republics (U.S.S.R, or Soviet Union) in which each has (or had) only one political party, the Communist Party.

It is unclear how closely an idealized communist society can be achieved and maintained in practice, particularly in a modern industrial economy. In practice the aforementioned countries do not (and did not) reach the textbook definition of communism. Practically they more closely approached *socialism* as more strictly defined as a system with state control of property. While modern-day China is governed by

---

people's money, owned by everyone, divided equally and for generations to come." As of 2019, Norway's sovereign wealth fund was valued at almost 1.1 trillion US dollars, or about $200,000 for each of Norway's 5.3 million citizens. Quotes in this passage are from the Norges Bank Investment Management, video "The fund in brief" at https://www.nbim.no/en/the-fund/about-the-fund/. Also see https://www.equinor.com/en/about-us/corporate-governance/the-norwegian-state-as-shareholder.html.

[47]Dave Merrill, "No One Values Your Life More Than the Federal Government," *Bloomberg*, October 19, 2017, available at: https://www.bloomberg.com/graphics/2017-value-of-life/. For detail see EPA documentation at https://www.epa.gov/environmental-economics/value-statistical-life-analysis-and-environmental-policy-white-paper and "Value of Statistical Life Analysis and Environmental Policy: A White Paper and Appendices A-J (2004)" at https://www.epa.gov/sites/production/files/2017-12/documents/ee-0483_all.pdf.

the one and only Communist Party, it exhibits a complex mix of socialism (state ownership), capitalism (private ownership), and market competition. Aside from post-2000 China, these countries are not known for high levels of average material standards of living and immigrants don't flock to live inside their borders. All have restrictions on freedom of expression or movement, at least more than in Western democracies. (The laws of the United States do not allow absolute and complete expression and movement for all citizens.)

Sharing one's proceeds with everyone does sap away significant incentive to innovate because you cannot accumulate material and monetary wealth based on your personal work. For example, in Cuba, ninety percent of business sales goes to the government as taxes. Only ten percent of food, products, or other sold services can be sold directly to others while keeping one-hundred percent of the proceeds. This has driven many skilled people (e.g., engineers, teachers) to take jobs within the tourism industry because the tips one receives enable a higher income. This form of taxation reduces Cuba's capability to engage in a variety of economic activities, but the government does prioritize health care and basic education. While Cubans generally have access to basic education and health care, the services are, well, basic. Many doctors have necessary skills, but they lack equipment and supplies. Children are literate, but the country doesn't produce a high proportion of engineers and scientists.

Thus, it seems communism is a unicorn—it doesn't exist in the real world. In practice, we observe some other system.

As hinted in the Powell Memo, a practical question for communism is exactly how to prevent a totalitarian ruling class from suppressing the freedom of the masses that are supposed to collectively own everything. George Orwell explained this conundrum of oppression in his 1945 book *Animal Farm*, that parallels the early history of the former Soviet Union, formed in 1917. In the book, the overly exploited farm animals (the commoners) overthrow their human owners (capitalists). Upon kicking out the humans, the animals operate the farm as a communist collective, but it doesn't last. The pigs become a ruling class and use the dogs to enforce increasingly unequal rules, just as Joseph Stalin ruled the Soviet Union as a totalitarian state. Eventually the pigs walk on two legs while coordinating with the humans because, after all, "All animals are equal, but some animals are more equal than others."

Thus, Lewis Powell, Donald Trump, and many others state, given the practical problems with achieving communist or socialist states, we should pursue capitalism and free markets. But the two concepts of free markets and capitalism have at least one thing in common with communism and socialism: they are also unicorns. They practically do not exist in their pure forms. There are no such things as free markets because by definition they are defined by legal rules and norms that restrict degrees of freedom.

This now brings us to the operative word of this subsection: *neoliberalism*.

This term captures much of the popular discontent over what's wrong with the current state of the global economy. The data presented in this book demonstrate that several trends driving this discontent (stagnant wages and a declining share of GDP going to wages) have been five decades in the making.

But what is neoliberalism? As Philip Mirowski laments, this is one of the most misused words within modern political economy:

> While it is undeniable that neoliberals routinely disparage the state … it does not follow that they are politically libertarian or … that they are implacably opposed to state interventions in the economy and society.
>
> …
>
> From the 1940s onward, the distinguishing characteristic of neoliberal doctrines and practice is that they embrace [the] prospect of repurposing the strong state to impose their vision of a society properly open to the dominance of the market as they conceive it. Neoliberals from Friedrich Hayek to James Buchanan … all explicitly proposed policies to strengthen the state. [17]—Philip Mirowski (2018)

Thus, the neoliberal ideology is *not* equivalent to neoclassical economic theory, the libertarian notion of a minimally functional government (or state), or the promotion of free markets with no barriers to entry for new companies. Libertarians realize legitimate governments must enforce rules of competition. However, as implied above by Mirowski, *neoliberals use government to legitimize their "vision of a society,"* competition be damned. This legitimacy comes not from a weak government, but a government strong enough to write and enforce laws that might be unpopular to the people. These laws tend to support markets that provide signals, such as prices, that direct our purchasing decisions. Per Lewis Powell, laws can be written to support corporations.

Chapter 8 introduced us to Friedrich Hayek and his idea for "… how to provide inducements which will make the individuals do the desirable things without anyone having to tell them what to do." [18] The book *The Road from Mont Pèlerin* discusses Hayek's ideas and explores the history of the "neoliberal thought collective" [19]. In that book, Mirowski points out that Hayek saw the economy as the ultimate information processor that was smarter than any one person or group of persons. Thus, government officials, or really any centralized authority, can't know enough to make good economic decisions. But they can set up markets to induce purchasing decisions, or might we say, reactions. To Hayek markets are distributed information systems interacting with each other and aggregating data in the most effective way to produce the "correct" signals that direct people on what to make and what to buy.

Need better decisions? Make another market! Lack information? Make another market!

To neoliberals, it's markets all the way down, and there is little to no need for people to think for themselves. The market knows better. In an extreme world with markets for everything, each of us becomes an automaton responding to price signals that might have very little to do with personal and human well-being.

But individual consumers and producers aren't the only agents that can respond to market signals. Individuals within governments can too. The last quote from Mirowski also mentions James McGill Buchanan. Buchanan was awarded the 1986

Nobel Prize for Economics for his "Contributions to the theory of political decision-making and public economics."[48]

Using what became known as *public choice theory*, Buchanan applied economic analyses to the political arena.[49] He assumed that politicians acted in their own interests. While public officials don't necessarily react to prices, they do react to bids from companies coming to their city, county, or state with jobs for their electorate. And what do politicians have to offer these bidding companies? Tax breaks.

In the market for locating corporate offices and factories, companies perform cost–benefit calculations to determine which location enables a higher profit. Politicians then compete by selling their "commodities" of lower taxes and business-friendly laws. After all, taxes are just one more cost to minimize. "The rule of law has a monetary price, and so does your corporate tax rate and regulatory environment."[50]

Ironically, it was the left-leaning economist Charles Tiebout who sparked the idea that you could determine a homeowner's "revealed preference" for tax rates and public amenities by observing where they locate. In his 1956 paper "A Pure Theory of Local Expenditures" Tiebout thought local governments could balance taxes and public services to benefit both companies and citizens [21]. It hasn't worked out this way.

Using Tiebout's approach, integrated with public choice theory, neoliberals have convinced everyone that we are all better off (i.e., higher wages, more jobs) when our governments compete in a race to the bottom of lower taxes. As Nicholas Shaxson points out in his book *The Finance Curse*:

> It's not hard to see how subversive all this [competition among governments] was. The rule of law has a monetary price, and so does your corporate tax rate and regulatory environment. Once this awesome intellectual land grab by corporate and financial interests began to enter mainstream politics in the late 1970s, it would lead inevitably to corruption, oligarchy, bank bailouts, and the growth of international organized crime.[51]—Nicholas Shaxson (2019)

One of the most recent high-profile examples of this reality was in 2017 when Amazon announced plans to build a second headquarters, HQ2, within the U.S. In response, 238 cities courted the online behemoth with incentive packages, each worth several billions of dollars. This enabled the company to gather data to inform the term sheet delivered to the mayors of the cities in which it wanted to locate all along: ultimately Long Island City in Queens, New York and Crystal City in Arlington, Virginia.[52]

---

[48]https://www.nobelprize.org/prizes/economic-sciences/1986/buchanan/facts/.

[49]https://www.neh.gov/about/awards/national-humanities-medals/james-m-buchanan.

[50]Shaxson [20, p. 44].

[51]Shaxson [20, p. 44].

[52]Amazon numbers from Nicholas Shaxson's *The Finance Curse* [20, pp. 50–51], and for further discussion of cities and countries competing to attract businesses, see in particular Chapter 2 "Neoliberalism Without Borders."

The politicians most willing to play the game can more easily gain support. After all, politicians are supposed to listen to their constituents, and the *Citizens United* ruling has now enshrined elections as a political marketplace for corporations as people. "That is, corporate dominance of elections becomes possible when political life as a whole is cast as a marketplace rather than a distinctive sphere in which humans attempt to set the values and possibilities of common life."[53]

But politicians don't only speak for the interests of cities and states within their country. Officials elected and appointed to national positions also speak for their country. Multinational companies and wealthy persons can shop around for the *country* with the lowest taxes, least transparency, and laxest regulatory environment. The Panama Papers, leaked in April 2016, showed just how extensive is this global marketplace: "The Panama Papers are an unprecedented leak of 11.5 m files from the database of the world's fourth biggest offshore law firm, Mossack Fonseca. ... The documents show the myriad ways in which the rich can exploit secretive offshore tax regimes. Twelve national leaders are among 143 politicians, their families and close associates from around the world known to have been using offshore tax havens."[54]

In the extreme case of offshore tax havens, there were no existing tax laws for local officials to change to benefit the foreign influx of money. For example, "The Panama Papers ... revealed how Mossack Fonseca ... effectively wrote the tax haven laws of Niue, a tiny Pacific island of 1500 people. Mossack Fonseca got an exclusive agreement to register offshore companies there, and this operation was soon generating 80 percent of that territory's government revenue."[55]

While this book is not focused on corruption and tax evasion, I discuss these concepts simply to recognize they exist. They are part of our reality, and we should understand their influence on society.

The foci of this book are energy and economic narratives, some of which can be expressed in mathematical models that only approximate observed economic trends. These models can be powerfully persuasive. They derive from worldviews and narratives spanning techno-optimism and techno-realism and theories ranging from neoclassical to biophysical economics. These narratives and theories assume wildly different economic rules and physical constraints for economic activity.

How can we think of the relationship between the distribution of political power and actual physical power defined as the flow of energy?

---

[53]Quote from Wendy Brown during interview with the Institute for New Economic Thinking, November 19, 2019, "How Neoliberal Thinkers Spawned Monsters They Never Imagined." Accessible at: https://www.ineteconomics.org/perspectives/blog/how-neoliberal-thinkers-spawned-monsters-they-never-imagined.

[54]Quote from *The Guardian*, April 5, 2016 "What are the Panama Papers? A guide to history's biggest data leak": https://www.theguardian.com/news/2016/apr/03/what-you-need-to-know-about-the-panama-papers. For extensive information see the Panama Papers website of the International Consortium of Investigative Journalists https://www.icij.org/investigations/panama-papers/.

[55]Shaxson [20, p. 83].

The Political Will Delusion: The Power of Physical Power

As stated earlier, political power is the ability of an individual or group to influence the behavior of others and often without the powerful group having to change its own behavior in an undesirable manner. However, in addition to controlling the flow of information via changing legal rules and social norms, one can influence the rules and norms by directly controlling the flow of energy.

Anthropologist Richard Adams separated the concept of *social power* from that of *control* :

> ...*control* over the environment is a physical matter. An actor either has it or does not. ...*Power* over an individual is a psychological facet of a social relationship ...
>
> The difference between power and control is very important ...Control is a nonreciprocal relationship in the sense that it exists between and actor and some element of the environment that cannot react rationally to shared behavioral experiences. This does not mean that the thing being controlled does not have its own peculiar behavior. A rock will act like a rock, a horse like a horse, and a stream of water like a stream of water, etc., and a corpse like a corpse. Thus control is always contingent upon understanding the nature of the object being controlled and thereby requires a set of techniques appropriate to those characteristics.
>
> Power, however, is a social relationship that rests on the basis of some pattern of controls and is reciprocal. That is, both members of the relationship act in terms of their own self-interest and, specifically, do so in terms of the controls that each has over matters of interest to the other. The behavior that results from an awareness of power is such that the actor tries to calculate what the other individual might do that could affect the actor's interests.[56]— Richard Adams (1975)

Adams goes on to say:

> It is the actor's control of the environment that constitutes the base of social power ...
>
> In speaking of "control over the environment," the word *control* refers to *making and carrying out decisions about the exercise of a technology*. The thing doing the controlling may be an individual or some social unit that has an internal power structure of its own.[57]— Richard Adams (1975)

Adams used the diagram in Fig. 9.4 to demonstrate the difference between social power and the control of physical power as the flow of energy from the environment.

Entities A and B are social entities that could be people, businesses, countries, etc. Each might have some control over the environment, X, via the use of technology in the generic sense of ideas, knowledge of the natural world, machines that both extract energy and matter from the environment and transform them for useful purposes, etc. These relationships are demonstrated via the top diagram of Fig. 9.4. If entities A and B have equal control over the environment, then neither A nor B is subordinate to the other (Fig. 9.4b). However, if A has more control over the environment than B, then B is subordinate to A (Fig. 9.4c), and A can create a situation to have more social power over B. That is to say, A is in a position to make

---

[56] Adams [22] [pp. 21–22].

[57] Adams [22, p. 13].

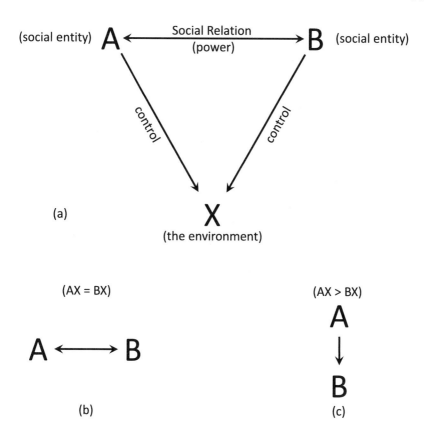

Fig. 9.4 (**a**) The basic components of power relations indicating the difference between social power (relations between social entities A and B) and control each entity has over the environment, X, via use of technology. (**b**) A and B have equal control over the environment, and neither has social power over the other. (**c**) A has more control over the environment than B, and thus has social power over B. Modified version of Figure 1 in [22]

choices that forces B to alter its behavior in an undesirable manner, but A does not have to change its own behavior in that same manner.

Thus, we cannot separate social and economic relations from their association with energy and the environment. "Power is thus clearly a relational issue between parties, but it is also a relation that exists with reference to things that can be described as external to any particular actor: the energy forms and flows, and the equivalence of values."[58]

Having control of the flow of energy resources and the technologies to extract, transport, and consume them was one way that workers were able to obtain increased political power from employers. The control of physical power by some workers

---

[58] Adams [22, p. 17].

directly led to improved working conditions and bargaining power for themselves and other workers.

Timothy Mitchell describes the historical linkage between fossil fuels and democracy in *Carbon Democracy* [2]. During the several decades before World War I, the democratic political power of workers emerged within the coal industry. In industrializing nations, such as the United States, a vast network of railroads and water ways connected coal mines, ports, cities, and power generating stations. These networks had a tremendous amount of energy flowing within them, and there were several choke points of concentration:

> Great volumes of energy [within coal] now flowed along narrow, purpose-built channels. Specialized bodies of workers were concentrated at the end-points and main junctions of these conduits, operating the cutting equipment, lifting machinery, switches, locomotives and other devices that allowed stores of energy to move along them. Their position and concentration gave them opportunities at certain moments, to forge a new kind of political power. The power derived not just from the organisations they formed, the ideas they began to share or the political alliances they built, but from the extraordinary quantities of carbon energy that could be used to assemble political agency, by employing the ability to slow, disrupt or cut off its supply.[59]—Timothy Mitchell (2013)

The battle between coal miners and owners was not civil. It was fought with blood and violence. But as coal miners persevered, and their strikes continued, eventually owners acquiesced to some of their demands. As workers felt left out of the profits from their labor, they demanded higher wages, safer conditions, and shorter hours. Their labor was needed to ensure the flow of energy that underpinned the unprecedented increase in economic growth powered by coal.

In Adams' terms, the workers had control of the environment, and, therefore, they had social power to increase their working conditions and pay. In the late 1800s U.S. coal miners went on strike two to three times more than workers in other industries.[60]

> The flow and concentration of energy made it possible to connect the demands of miners to those of others, and to give their arguments a technical force that could not easily be ignored. Strikes became effective . . . because of the flows of carbon that connected chambers beneath the ground to every factory, office, home or means of transportation that depended on steam or electric power.[61]—Timothy Mitchell (2013)

You don't have to be coal miners restricting the flow of coal to realize that control of energy provides social power. Vladimir Lenin recognized that for communism to succeed as a counter to capitalism, he must confront this link between physical and social power. He well understood the need for useful work, in the form of electricity, in the early days of the Soviet Union:

---

[59]Mitchell [2, p. 19].

[60]Mitchell [2] cites P. K. Edwards, *Strikes in the United States, 1881–1974*, New York: St Martin's Press, 1981: 106. "The strike rates per 1,000 employees for coal mining and for all industries, respectively, were 134 and 72 (1881–86); 241 and 73.3 (1887–99); 215 and 66.4 (1894–1900); and 208 and 86.9 (1901–1905)."

[61]Mitchell [2, p. 21].

Communism is Soviet power plus the electrification of the whole country. Otherwise the country will remain a small-peasant country, and we must clearly realize that. We are weaker than capitalism, not only on the world scale, but also within the country. That is common knowledge. We have realized it, and we shall see to it that the economic basis is transformed from a small-peasant basis into a large-scale industrial basis. Only when the country has been electrified, and industry, agriculture and transport have been placed on the technical basis of modern large-scale industry, only then shall we be fully victorious.[62]—Vladimir Lenin (1920)

Lenin was wondering how to compete with capitalist countries, both economically and militarily. Military organizations must spend time considering the links between force, or the exercise of control over the environment (e.g., a bomb), and the political will for peace or war. They don't have the liberty to ignore how physical resource abundance and constraints can drive people or leaders into action from hubris or fear.

U.S. Admiral Rickover, the father of the nuclear submarine, provides a poignant perspective on political will, or more specifically geopolitical power:

High-energy consumption has always been a prerequisite of political power. . . . Ultimately, the nation which control[s]—the largest energy resources will become dominant. If we give thought to the problem of energy resources, if we act wisely and in time to conserve what we have and prepare well for necessary future changes, we shall insure this dominant position for our own country.[63]—U.S. Admiral Hyman Rickover (1957)

For Rickover, just like Richard Adams, energy consumption, or control over physical power in the environment, comes first. Political, or social, power comes second. Physical power "causes," or enables, political power as the effect. We do not have to interpret Rickover's statement to mean that no other political force exists other than that from the command of physical power. But the control of physical power is a known means to social power.

Rickover is effectively applying Alfred Lotka's maximum power principle (Chap. 8) to economies. Accordingly, economic systems that promote more extractive behavior will more often survive and overcome those that do not. This concept is consistent with avoiding the internalization of long-term environmental impacts. Thus, while the maximum power principle might induce activities that are detrimental to the long-term survival of various species, ecosystems, or humanity, it posits that if you don't survive until tomorrow, you'll never get to the next year, decade, or century.

---

[62] Report on the Work of the Council of People's Commissars. December 22, 1920, Original Source: Polnoe sobranie sochinenii, 5th ed. (Moscow, 1975–79), Vol. 36, pp. 15–16. Available at website of "Seventeen Moments in Soviet History," November 30, 2019 at http://soviethistory.msu.edu/1921-2/electrification-campaign/communism-is-soviet-power-electrifi cation-of-the-whole-country/.

[63] For Delivery at a Banquet of the Annual Scientific Assembly of the Minnesota State Medical Association St. Paul, Minnesota May 14, 1957. Available at http://large.stanford.edu/courses/2011/ ph240/klein1/docs/rickover.pdf.

This is why it is so difficult to attach too much agency, or free will, to our collective decisions. Individually we make decisions that have meaning, but we're not completely in control of what we decide. Each of us is a part of our country's economic system, and at some level each country is competing with other economies. Increasingly, multinational companies shape the terms of the global economy by searching for the lowest tax rates, putting into question the independence of state governments.

Does our economy really exhibit the behavior as implied by the maximum power principle and suggested by ecologists, anthropologists, some ecological and biophysical economists, and military leaders like Admiral Rickover? We can say that *the maximum power interpretation is compatible with the data*. Each of us might not be consciously trying to increase our personal power consumption, but the data indicate this is happening at the global level.

Over time our global economy and humanity have collectively consumed energy at a higher rate to achieve higher GDP (refer back to Figs. 2.10 and 5.2). More countries are adopting capitalistic and neoclassical economic features (i.e., minimize marginal costs) into their economies because these promote higher power consumption. China is the largest and most prominent example, particularly after it joined the World Trade Organization in 2001.

The Political Will Delusion: Willpower and Temptation

The quote at the beginning of this Political Will section stated the need for more political will with regard to the Paris Agreement. It did not explicitly refer to energy consumption, but instead it referred to reduced greenhouse gas (GHG) emissions.

For the last 200 years, increasing global energy consumption has translated to increasing global GHG emissions. While this might not be the case in the future, how do we consider the conflict between our instincts to react to immediate circumstances (i.e., consume more energy now, grow the economy now) and the political will to choose a different path based upon a future goal (i.e., limit human-caused climate change)? As Daniel Dennett asks in *Freedom Evolves*:

> Where does the oomph come from to overrule our own instincts? Tradition would say it comes from some psychic force called *willpower*, but this just names the phenomenon and postpones explanation. How is "willpower" implemented in our brains?[64]—Daniel Dennett (2003)

Psychologists and economists use the term *discount rate* to describe how people make decisions, within our brains, when there are multiple options that present benefits at different points in time. Do I want one dollar now or two dollars ten years from now? Largely driven by natural selection and perhaps some idea similar to the

---

[64]Dennett [7, p. 210].

maximum power principle, humans tend to have "steep" discount rates indicating that we tend to select rewards that come sooner rather than later.

Dennett uses the story of Ulysses and the Sirens in Homer's *The Odyssey* to demonstrate the link between willpower and the idea of the discount rate. The goddess Circe warns Ulysses that during his journey home, he will sail past the Island of the Sirens. The Sirens appear to have exquisite beauty and a sweet song that lures sailors to their shores. But on approach, the sailboats crash on the rocks, and the sailors remain on the island, unwilling to leave as they listen to the song of the Sirens until they wither and die. Ulysses wants to return home to reunite with his family, but he also knows he will fall into the temptation of the Sirens as he sails by their island. He heeds Circe's advice on how to safely pass the Sirens. Upon approach of the island, Ulysses orders his men to fill their ears with beeswax, a gift from Circe, such that they will not hear the song of the Sirens. As for himself, Circe has given Ulysses a way to hear the sweet Siren song yet not be lured to his death. Ulysses orders his men to lash him to the ship's mast, but under no circumstances are they to untie him, no matter how much he commands while he listens to the Sirens' song.

The important concept related to political will is one of planning for a future time. Ulysses knows the situation that will confront him in the future, and he also knows himself. He knows his steep discount rate will not allow him to resist the Sirens when the time comes. Thus, before he is in the presence of the Sirens, he does exactly what is needed. He enacts a plan to avoid a disastrous fate. He restrains himself by ordering his men to physically tie him to the ship's mast. Thus, when his ship passes the Island of the Sirens, Ulysses' previous decision has removed his choice, or degree of freedom, to go to the island. While hearing the Sirens' song, he absolutely believes he would be happier on the island at that moment than with his family later. Only his previous decision has allowed him to experience the life he wants in the long term.

Homer likely did not know when writing *The Odyssey* that he was demonstrating the concept of the discount rate. But he knew of temptation. Temptation is the base of many stories of struggle. Perhaps no more famous quote exists than that of Saint Augustine of Hippo when praying to God: "Give me chastity and continency, only not yet."[23] As a young man, St. Augustine knew he should not lust, or at least he believed that he should not. When praying, he asked to be healed, but "not yet" because he "...feared that you [God] would hear me quickly, and that quickly you would heal me of that disease of lust, which I wished to have satisfied rather than extinguished."[65]

In the context of the energy and economic narratives, who needs more willpower? Do our political leaders need "political will" to constrain the choices for both public and private energy company investments? Do we as citizens need the "political will" to elect leaders that will constrain our energy choices to those with low-carbon impacts? Do we, as consumers, need the willpower to buy fewer products

---

[65] Book VIII, Chapter 7, *The Confessions of St. Augustine* [23].

and services? Do company executives and shareholders need the willpower to direct more investment and profits to low-carbon energy and conservation instead of to what they think is the most profitable investment option?

In considering these questions, we can consider our modern situation in light of both St. Augustine and Ulysses. Ulysses had highly reliable information from the goddess Circe, and he was certain that proper actions could avert a single bad outcome in the near term. He was not restricting the degrees of freedom for the long-term future choices that he or his crew could make once they made it home. He only needed the willpower to restrict himself during his homeward journey, not for the rest of his life. This is akin to a company or country planning for the next year or so, but as if with absolute certainty of what to expect.

On the other hand, St. Augustine struggled with the opposite time horizon as Ulysses. He sought the willpower to restrict himself from the lust he craved, not only for a while, but for the rest of his life. Immediate temptation was always too strong for him to change.

So why don't we restrict ourselves to less than our full suite of greenhouse gas-emitting energy options, much like Ulysses tied himself to his mast? The renewable energy narrative posits that all we need to do is plug our ears as we renewably sail past the Island of the Fossil Fuel Sirens. However, the problem is not that we'll quickly approach and pass the Sirens in the near future. The problem is *we're already on the island!*

Thus, St. Augustine's lust is the more apt analogy. The temptation to continuously grow the economy is ever present, and the historical data, maximum power principle, and drivers of evolution imply that our economy as a superorganism seeks more physical growth requiring higher energy consumption via the laws of physics.

But there is no God to whom to pray, or single authority that will command us, to restrict our lust for growth. To reduce greenhouse gas emissions we have to rely on ourselves, our institutions, and political officials to restrict some available energy and economic options. But countries compete for resources just as living organisms compete within ecosystems. Per evolution, a system or country that restricts its energy options more than another will in all likelihood reduce its fitness.

Physical growth on our finite planet will stop someday, but unconsciously hitting some inexact upper limit is different than consciously choosing to stay below a predetermined target.

When I think of the 2015 Paris Agreement, signed by practically all countries in 2016, in which countries agreed to non-binding "nationally determined contributions" to reduce greenhouse gas emissions, I imagine rephrasing St. Augustine's words in the context of political leaders that vote to rapidly reduce greenhouse emissions (my changes in italics):

> Give me *rapid reductions in greenhouse gas emissions*, only not yet. For I fear that *the economy* would hear me quickly, and that quickly *it* would heal me of that disease of *growth*, which I wished to have satisfied rather than extinguished.

But I don't have to rephrase St. Augustine to demonstrate the strength behind the infinite growth and substitutability economic narrative. We find this sentiment with economist William Nordhaus, who won the 2018 Nobel Prize in Economics "for integrating climate change into long-run macroeconomic analysis." On the day of announcement of his prize, he stated to his undergraduate class: "Don't let anyone distract you from the work at hand, which is economic growth."[66] The "task at hand" for Nordhaus and many other economists is growth, not some limit in atmospheric concentration of greenhouse gases.

This is not to say that Nordhaus's statement is correct or incorrect, or that I agree or disagree with his statement. This is to say that his view is that the goal is growth *now and in the future*. If achieving this goal involves reducing greenhouse gases today to mitigate climate change, then so be it. If it involves increased greenhouse gas emissions today, then so be that as well. He has an economic growth model, based on neoclassical theory discussed in Chap. 6, that runs a cost–benefit calculation, and that tells him a world with $4\,°C$ of warming optimally balances costs and benefits. Some vehemently disagree.[67]

One can argue about the structure of Nordhaus's and others' integrated assessment models (as in Chap. 6) and how they calculate benefits (i.e., economic growth) and the economic losses from climate change. Given the high-level abstraction and uncertainty of economic damages from climate change, some economists claim current modeling efforts are "close to useless as tools for policy analysis" [24].[68] One can also argue against using any such economic models at all. In all likelihood, we won't be able to tell if we ever reach some "optimal" level of warming.

To wrap up this section on political will, let's make an analogy of Nobel Laureates to the mythological stories of the Greek Gods, who regularly fought among themselves. We can contrast Nordhaus to the scientists within the Intergovernmental Panel on Climate Change (IPCC) and Al Gore who together won the 2007 Nobel Prize for Peace for emphasizing the need to limit climate change impacts by urgently reducing GHG emissions. If we are Ulysses, is the IPCC our Circe, warning us to constrain ourselves in the short term so that we will achieve the future we want? Is Nordhaus a god or a Siren tempting us with the song of immediate growth?

---

[66]Mike Cummings, *YaleNews*, "Cheers and roses from undergrads for Yale's latest Nobel laureate," October 8, 2018, https://news.yale.edu/2018/10/08/cheers-and-roses-undergrads-yales-latest-nobel-laureate.

[67]Steve Keen, " '4 °C of global warming is optimal'—even Nobel Prize winners are getting things catastrophically wrong," *The Conversation*, available November 14, 2019 at: https://theconversation.com/4-c-of-global-warming-is-optimal-even-nobel-prize-winners-are-getting-things-catastrophically-wrong-125802.

[68]The abstract from Pindyk [24] states: "A plethora of integrated assessment models (IAMs) have been constructed and used to estimate the social cost of carbon (SCC) and evaluate alternative abatement policies. These models have crucial flaws that make them close to useless as tools for policy analysis: certain inputs (e.g., the discount rate) are arbitrary, but have huge effects on the SCC estimates the models produce; the models' descriptions of the impact of climate change are completely ad hoc, with no theoretical or empirical foundation; and the models can tell us nothing about the most important driver of the SCC, the possibility of a catastrophic climate outcome. IAM-based analyses of climate policy create a perception of knowledge and precision, but that perception is illusory and misleading."

Our Nobel gods are in disagreement. The Earth systems models of the climate tell the scientists that we need to rapidly reduce GHG emissions to zero in a few decades. The economic models tell us we don't because they assume the economy grows about the same whether we transform the energy system to reduce emissions or not. Further, the Earth system models must assume the Earth is finite to model the feedbacks of GHG accumulation. The economic models generally assume the Earth is infinite.

No wonder we seem to get conflicting signals! The tragedy (pun intended) is these narratives speak past each other even when cobbled together in the integrated assessment models that force them to have dinner at the same table. The submodels speak the same language of mathematics, but their narratives are incompatible.

I write these paragraphs as a 40-something intently studying energy and economic systems and modeling, but in 2019, 16-year-old Swedish climate activist Greta Thunberg reached the same "Greek tragedy" conclusion and quickly reached the very broad audience of an assembly of the United Nations:

> People are suffering. People are dying. Entire ecosystems are collapsing. We are in the beginning of a mass extinction and all you can talk about is the money and fairytales of eternal economic growth. How dare you? ...
>
> The popular idea of cutting our emissions in half in 10 years only gives us a 50% chance of staying below 1.5 degrees [Celsius] ... How dare you pretend that this can be sold with just business as usual and some technical solutions? ... There will not be any solutions or plans presented in line with these figures here today because these numbers are too uncomfortable and you are still not mature enough to tell it like it is.[69]—Greta Thunberg (2019)

I suppose it is hard to state a lack of political will more succinctly, so let's now turn from tragedy and political will to a narrative that Thunberg railed against. This is the techno-optimistic narrative that many leaders and energy analysts promote: the idea that we can have our cake and eat it too—that we can grow the economy while decreasing energy consumption and greenhouse gas emissions.

The Decoupling and Services Delusion

Decoupling No, this isn't about breaking up with your boyfriend or girlfriend. In the context of economic growth, Chap. 4 defined a decoupled economy in two ways. *Relative* decoupling allows for increased energy and material consumption, as well as higher environmental impacts, as long as they increase more slowly than economic growth. *Absolute* decoupling restricts the energy, material, and negative environmental flows to absolutely decrease even when the economy grows.

---

[69]Greta Thunberg speech at the United Nations Climate Action Summit, September 23, 2019. Transcript available October 10, 2019 at: https://www.npr.org/2019/09/23/763452863/transcript-greta-thunbergs-speech-at-the-u-n-climate-action-summit.

A McKinsey and Company article summarizes the arguments for relative decoupling:[70]

The decoupling of the rates of economic growth (climbing steadily) and energy demand growth (ascending, but less steeply) will largely be a function of the following four forces:

1. a steep decline in energy intensity of GDP, primarily the consequence of a continuing shift from industrial to service economies in fast-growing countries such as India and China.
2. a marked increase in energy efficiency, the result of technological improvements and behavioral changes.
3. the rise of electrification, in itself a more efficient way to meet energy needs in many applications.
4. the growing use of renewables—resources that don't need to be burned to generate power—a trend with the potential not only to flatten the primary energy demand curve but also to utterly change the way we think about power.

I'll save the first item for last and start with the second of their four forces that justifies relative decoupling: energy efficiency. As discussed in Chap. 6, technological improvement might be best defined in terms of the efficiency in converting primary energy into useful work. In this sense, technology and efficiency are effectively one in the same. Data support the idea of the Jevons Paradox, in that energy efficiency is an approach to increase output in absolute, not decrease material inputs in absolute. Thus, energy efficiency and the Jevons Paradox are consistent with the concept of relative decoupling.

Discussing energy efficiency as "behavioral change" is a more daunting prospect. I can directly purchase a new and more energy efficient product, such as a light bulb, car, or air conditioner, such that I obtain the same of level of service with less energy, or more service with a similar amount of energy. If I avoid purchasing a car or as an adult live with my parents, then that might be more related to behavioral change that could lead to conservation, or absolute decoupling.

As noted in Chaps. 5 and 7, the global and U.S. economies transitioned to a different mode of operation during the 1970s, one including behavioral change from both physically and economically constrained energy supplies. One of the major changes was to focus on end-use energy efficiency itself. This change has so far promoted higher absolute growth and consumption, although many did not anticipate that effect. In addition, as summarized in Chap. 7, energy efficiency and other policy responses enacted in response to the 1970s energy constraints promoted, or at least did not prevent, increasingly unequal income distribution.

---

[70]McKinsey & Company, "The decoupling of GDP and energy growth: A CEO guide," April 25, 2019: https://www.mckinsey.com/industries/electric-power-and-natural-gas/our-insights/the-decoupling-of-gdp-and-energy-growth-a-ceo-guide?cid=other-e%E2%80%80A6.

McKinsey's third item of increased electrification is consistent with an energy and economic system attempting to reduce energy necessary for operating itself as a set of physical networks. As discussed in Chap. 5, an economy trying to grow and accumulate more "mass" of capital will seek energy efficient distribution of people, materials, and other general stuff. For example, an energy network based on electrons competes against one based on molecules. If you are distributing an energy carrier, do you want to distribute it in a stored state of chemical energy in molecules, such as gasoline in the fuel tank of your vehicle, or an unstored state of flowing electricity?

Historically, fossil fuel molecules have been cheap to extract, and the storage comes almost for free. These benefits outweighed the cost of molecule distribution. The trend toward electricity is driven by the uncertain long-term cost, environmental impact, and availability of fossil energy molecules. Because energy storage in molecules dominates transportation fuels, increased electrification largely translates to increased electric-powered travel. Electric travel might not move away from molecules completely. This depends on whether it will be cheaper to store electrons in batteries or to use electric-driven processes that make new energy molecules such as hydrogen and synthetic fuels, derived from carbon dioxide and water, as drop-in renewable replacements for fossil fuel-based diesel and gasoline.

For the fourth item on McKinsey's list, the authors refer to the company's 2019 Global Energy Perspectives Summary, which states "After more than a century of rapid growth, primary energy demand plateaus around 2030, primarily driven by the penetration of renewable energy sources into the energy mix" and "It [2030] is the first time in history that growth in energy demand and economic growth are "decoupled.""[71]

It does seem strange to refer to the future year 2030 as a time in history, but aside from that, how does their reasoning work? A "growing use of renewables," largely from wind and solar farms, affects the accounting of primary energy such that the same amount of final or end-use energy can be associated with less primary energy. Even if we demand the same quantity of electricity in a wind and solar-powered world, the primary energy we associate with that electricity can go down, just due to accounting.[72]

---

[71] McKinsey & Company, "Global Energy Perspective 2019: Reference Case, Summary," p. 9.

[72] This conclusion is due to two energy accounting artifacts. The first is related to the physical energy content method mentioned in the Appendix. The second relates to how we account for renewable energy flows that we don't convert to energy carriers. Imagine a home where say 100,000 kilowatt-hours (kWh) of solar radiation hit the property per year. With no installed solar panels, the sunlight hitting the house is converted to electricity at 0% efficiency. Now put 20% efficient solar panels on half of the house facing the sun so that 10,000 kWh/year of electricity is generated. Standard energy accounting assumes the 10,000 kWh of electricity start at 100% efficiency because it does not account all sunlight hitting the Earth as primary energy. Physically speaking, in this example, sunlight is converted to electricity at 10% efficiency after installing solar panels but 0% before.

The term "primary energy demand plateaus" is not the most accurate way to describe this effect because consumers like you and I do not directly demand primary energy resources such as oil, wind, and coal. We buy final energy carriers like electricity and gasoline. But will we eventually decrease our final demand of energy carriers, or even our consumption of total useful work? If we do, how can we know if this declining demand is most consistent with either the techno-optimistic or techno-realistic economic narrative? Should one conclusion make us feel better than the other? As the techno-optimistic narrative implies, will we happily chose to decrease energy demand because we found a better way to live? Or as the techno-realistic narrative implies, will we be forced into behavioral changes due to finite Earth constraints?

Let's return to the first item in the McKinsey decoupling list, the metric of energy intensity and its link to often misleading discussions of a service economy.

Declining primary energy intensity is nothing new, but it hasn't always been the case. Recall Fig. 5.3 comparing U.S., U.K., and global energy intensities. Figure 9.5 extends the U.K. and England data backwards for 700 years, to 1300. The industrialization of Britain shows an increasing energy intensity for about 100 years starting after the mid-1700s when heavy industrial output increased rapidly with the steam engine. The U.K.'s current decline in energy intensity started around 1890, while total energy consumption still increased through 1913, at the start of World War I.

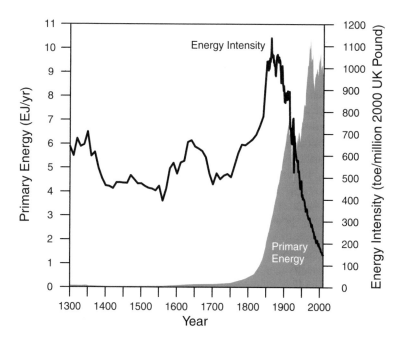

Fig. 9.5 U.K. data show that a decline in energy intensity does not necessarily translate to a decline in total energy consumption. Data from [25]

Estimated U.S. data since 1850 show an overall declining energy intensity, but with a hiatus and rise from approximately 1880 to 1920 (refer to Fig. 6.4). This period of increase also corresponds to the time of rapid expansion in construction of basic infrastructure like buildings and railroads. For example, during this time U.S. steel production increased from hundreds of thousands to tens of millions of tons per year, and railroad track miles increased from 49,000 in 1870 to 254,000 in 1916.

Global data starting in 1900 indicate energy intensity rose until around 1920, remained approximately constant through 1970, and has declined since. Due to the dominance of the U.K. and U.S. as a combined share of global GDP from the 1850s through the Great Depression (between 30% and 43%), those two countries heavily influence the global trend. Thus, both in the U.K. and the U.S., the rapid expansion of heavy industry translated to a period of constant or increasing primary energy intensity. Both the U.K. and U.S. energy intensities started declining while primary energy consumption was still increasing, and well before their primary energy consumption peaked in the first decade of the 2000s. Data to date also show declining global energy intensity with increasing primary energy consumption.

One troubling aspect of using energy intensity and the increasing dominance of service sectors as measures of progress is that the metrics depend on the definition and accounting of GDP itself. As mentioned in the description of GDP in Chap. 2, for most of history, banking services and interest on loans were seen as non-productive costs of business, not as an added contribution to GDP from the banking sector.[73] This changed in the 1990s such that the more net interest and fees a bank charges its customers, the more "productive" the banking sector appears in GDP accounting.

So, energy intensity can decline if all things stay the same, but banks charge higher interest. If the collective masses owe bankers and capitalists more and more interest on debt, then yes, people will have less of their remaining income left to buy material and energy goods. Is this the future you want, or is it the desired future of "the relatively small number of persons" that Bernays told us about, "who pull the wires which control the public mind?"

The McKinsey article also states, "Advanced economies tend to become service economies, and the energy intensity of service sectors is substantially lower than that of industrial sectors—in some cases, as low as one twentieth."

How are we to think of energy intensity as a metric for informing the structure of the economy? How do we interpret what it means to be a service economy? The McKinsey statement itself is fine, but we must include a few important caveats that are lost in many discussions.

First, energy intensity is a blunt metric, but properly interpreted, it still provides insight. By definition it is a measure of relative, not absolute, decoupling. Second, service economies don't have lower energy and materials consumption in aggregate than agricultural or heavily industrial economies. That is, evidence does not show

---

[73] See [26, pp. 106–108].

absolute decoupling of energy and material consumption for "service economies." Lastly, this is because service sectors don't exist in isolation. Services sectors are only parts of the larger coupled economy composed of many types of sectors and companies.

Let's break down these caveats.

When people refer to service economies, they refer to those with a majority of jobs or GDP in knowledge and care sectors such as business, finance, insurance, and health care instead of raw material extraction, manufacturing, and construction. By only looking at these parts, or sectors of a country's economy, we miss the holistic perspective of how they work together.

Think about your body as an example. Each organ in our body consumes energy at a certain rate, and each performs a valuable function to ensure complete health. We can think of an energy intensity, or energy consumption per mass, of different organs. Our brain consumes about 240 kilocalories per kilogram per day (kcal/kg/day), makes up 2–3% of body weight, and consumes about 20% of our resting metabolism, more than any other single organ.[74] In terms of energy per mass, some organs consume more and some less than the brain. Our heart consumes energy at higher intensity, near 440 kcal/kg/day and the skeletal muscles in our arms and legs at a lower intensity near 13 kcal/kg/day [27].

It is fantasy to think we could convert our body to a lower intensity "skeletal muscle economy" by eliminating or severely reducing the size of our heart and brain. The growth and size of mammals in relation to their metabolism, as summarized by Kleiber's Rule, occurs because lower energy intensity skeletal muscle is *predicated on the existence of the higher energy intensity heart to begin with.*

Only in *The Wizard of Oz* can Dorothy dream of a living tin man with no heart and a living scarecrow with no brain.

One can also analyze the patterns among individual countries. Using data from over 200 countries from 1991 through 2017, Blair Fix analyzed whether there was evidence for either lower fossil energy consumption or lower carbon dioxide (CO_2) emissions from countries that had a larger service sector [28]. He measured the size of the service sector in two ways: the share of total country employment in services sectors and the share of total country value added (or wages and profits) associated with services sectors. In both cases when measuring the trends between countries, the statistics show that a larger services sector is not associated with decreasing fossil fuel consumption or CO_2 emissions.

To summarize, "...the evidence indicates that a service transition does not lead to absolute carbon dematerialization."[28] The reason is that in order to provide more employment or value from services, you first need to be able to take care of more basic material needs such as housing, food, and energy provision. The catch

---

[74]Lectures of Physics in Medicine, University of Notre Dame, Chapter 2, The Energy Household of the Body. Available July 23, 2019 at https://www3.nd.edu/~nsl/Lectures/mphysics/ Medical%20Physics/Part%20I.%20Physics%20of%20the%20Body/Chapter%202.%20Energy %20Household%20of%20the%20Body/2.2%20Energy%20consumption%20of%20the%20body/ Energy%20consumption%20of%20the%20body.pdf.

is that more services-oriented economies have tended to provide those basic needs by first consuming a high quantity of energy. A service economy also first needs to build a sufficient base infrastructure during industrialization upon which the service economies later depend. Again, this is the same way our bodies grow.

In the same way that our body's organs and economy's sectors couple together, we can't look only at data within individual countries. Each country is interconnected within a global economy, just like sectors within one country and organs within one body. At the global scale the data clearly show a decline in energy intensity with an increase in total primary energy consumption for the last 50 years (see Figs. 5.2 and 5.3). That is, there is relative decoupling, not absolute decoupling of energy consumption.

The techno-optimists sometimes ignore this global trend of increasing energy consumption to focus on individual country data. Some conclude the U.S. and other rich countries have already experienced absolute decoupling.

For example, Andrew McAfee concludes:

> ...a great reversal of our Industrial Age habits is taking place. The American economy is now experiencing broad and often deep absolute dematerialization.[75]—Andrew McAfee (2019)

and

> With the help of innovation and new technologies, economic growth in America and other rich countries—growth in all of the wants and needs that we spend money on—has become decoupled from resource consumption. This is a recent development and a profound one.[76]—Andrew McAfee (2019)

McAfee justifies his conclusion by examining data collected by the United States Geological Survey (USGS) on U.S. *apparent consumption* of minerals spanning more than 100 years. He notes that apparent consumption "...takes into account not only domestic production of the resource but also imports and exports. For example, to calculate America's total apparent consumption of copper in 2015, the USGS would take the amount of copper produced in the country that year, add total imports of copper, and subtract total copper exports."[77] He presents several charts of apparent consumption of a wide variety of metals (aluminum, copper, nickel, steel, gold) and other materials (stone, cement, sand and gravel, timber, paper). These charts generally show slow growth or a constant level of apparent consumption from 1970 until the late 1990s or early 2000s before a decline in apparent consumption starting sometime in the 2000s. McAfee also recognizes the U.S.'s stagnant primary energy consumption since the mid-2000s.

---

[75]McAfee [29, p. 84].

[76]McAfee [29, p. 108].

[77]McAfee [29, pp. 78–79]. Further, the USGS "Historical Statistics for Mineral and Material Commodities in the United States": https://www.usgs.gov/centers/nmic/historical-statistics-mineral-and-material-commodities-united-states defines: Apparent Consumption = Production + Imports − Exports plus/minus (Stock Change).

Unfortunately, McAfee's conclusion of U.S. dematerialization is too hasty. Why? The answer has to do with how and where you count material consumption.

Consider buying a new car in the U.S. This car contains steel (mostly made of iron), aluminum, copper, plastics, and other materials. Further, let's assume the metal ores were mined in another country (e.g., copper in Chile). In the natural environment, the metal ores are only a fraction, often less than 5%, of the total mined material. Thus, the vast majority of total extracted material is not merely the targeted metal ore, but other rocks and soil, or mine tailings. There are two important considerations for interpreting apparent consumption. First, it ignores mine tailings. Second, mine tailings exist outside of the U.S., where you have purchased your car. Once you assume these mine tailings are associated with you, the car owner in the U.S., "the broad and often deep absolute dematerialization" disappears like a mirage.

A 2015 paper by Thomas O. Wiedmann and other authors explains the specious nature of McAfee's conclusion [30]. The apparent consumption of the USGS is similar to the concept of *domestic material consumption*, or DMC. If the metal ores are mined in the U.S. and you buy the car in the U.S., then DMC associates the mass of the mine tailings with your car purchase. However, if you mine the metal ores in another country, then DMC associates these mine tailings with the mining country that exported the metal ores, or the refined metal, to the U.S.

Wiedmann and co-authors compared DMC to another consumption-based metric, the *material footprint* (MF), that associates material usage, such as mine tailings, with the country that uses the final product, in this case the car, instead of the country that mines the raw materials of which the car is made.[78] If there were no such thing as globalized trade, we wouldn't need to distinguish material footprint from domestic material consumption.

The researchers used a worldwide data set of material trade and extraction from 1990–2008 to compare the material footprint of the U.S. and other rich countries to their GDP. Upon completing the calculations, it became apparent that *there has been no decoupling of material use*. In the authors' own words:

> The EU-27 [27 countries of the European Union], the OECD, the United States, Japan, and the United Kingdom have grown economically while keeping DMC at bay or even reducing it, leading to large apparent gains in GDP/DMC resource productivity [e.g., relative decoupling]. In all cases, however, the MF has kept pace with increases in GDP and no improvements in resource productivity at all are observed when measured as the GDP/MF. **This means that no decoupling has taken place over the past two decades for this group of developed countries** [30] (emphasis added).

---

[78]"The difference between DMC and the MF can be explained by the fact that traded goods require much more material than what is physically incorporated in them. Wealthier countries' imports of finished and semifinished products are linked to a larger amount of raw materials compared with the physical quantity traded. This also applies to metals, which are traded in the form of concentrates rather than ores. Nonexported mine tailings are included in DMC of the exporting country, where the MF allocates them to the importing (final demand) countries. DMC will therefore overestimate consumption for exporters of metals and biomass and underestimate it for importers of metals and biomass." [30].

My conclusion, which concurs with many others, including Tim Jackson in his book *Prosperity Without Growth*, is that there is practically no evidence for absolute decoupling of energy and material use.[79] Services-based economies of the rich world seem to have lower material consumption only if you neglect the material use that occurs outside of their borders.

This poses a serious question of whether absolute decoupling is possible. Can we have some kind of prosperous economy that actually consumes fewer resources?

The word *prosperity* in Jackson's book title points to a different viewpoint for services. He calls for servicization of the economy versus the usual notion of a services-based economy.[80] What he means is that to have good livelihoods while consuming fewer resources, we should focus our economies on *human services* rather than economically well-defined service sectors and jobs.

These human services could be provided by companies or governments. Instead of providing food, we should provide nutrition. Instead of providing drugs and medical tests, we should provide health. Instead of providing homes, we should provide shelter. Instead of selling vacation packages, we should provide for leisure and recreation. In addition, somehow we must value the maintenance and protection of physical and natural assets.

Yes, this is easy to write, and not so obvious how to get there. If this sounds like fantasy, Jackson specifically states he does not see this occurring within our current economic paradigm:

> So it all comes down to whether or not it's possible to implement this potential for decoupling. The most crucial question of all turns out to be about society rather than about technology. Is this massive technological transformation possible in our kind of society?
>
> To summarize massively, the answer suggested in this book … is: no. In our kind of society, in this kind of economy, it is highly unlikely that we will be able to decouple fast enough to remain within environmental limits or (ultimately) to avoid resource constraints.[81]—Tim Jackson (2017)

Why does Jackson state that we can't decouple our way to indefinite growth? As Raj Patel and Jason W. Moore point out in *A History of the World in Seven Cheap Things*, it is because of the nature of capitalism as well as the use of markets as the main basis for information and decision making. Jackson's human services are exactly those that the economic superorganism, operating within the structure of capitalism, seeks to provide as cheaply as possible:

> What every capitalist wants is to invest as little as possible, and profit as much as possible. For capitalism as a system, this means that the whole system thrives when powerful states and capitalists can reorganize global nature, can invest as little as possible, and receive as much food, work, energy, and raw materials possible.[82]—Raj Patel and Jason W. Moore (2017)

---

[79]For an extended discussion of decoupling, see Tim Jackson's *Prosperity Without Growth*, Chapter 5: The Myth of Decoupling.

[80]Jackson [31, pp. 142–143].

[81]Jackson [31, p. 164].

[82]Patel and Moore [32, p. 21].

Profits increase not only by accessing cheap nature, energy, and food, but also the human services of care (the health of people and the raising of the young), work, and the maintenance of the nation-states that keep the law and order under which capitalism can thrive. Though law and order are desirable under any socio-economic system, under capitalism companies might seek to provide as little of it as possible or charge as much as possible if their business is law and order. Privatized U.S. prison companies are examples of the latter.

Thus far, this chapter of delusions has gone from degrees of freedom and engineering consent to confidence, political will, and lastly decoupling via services. There is one services sector that deserves special attention, and for which people hold opinions spanning a wide spectrum from the most innovative to the most destructive. Some see the expansion of this sector as the reason why we can absolutely decouple material consumption from prosperity, but others see it as Exhibit A for the decline in prosperity of the middle class. Its activities and restructuring within the last several decades forced rethinking of just how the modern economy actually operates. It exposed the "flaw" in Alan Greenspan's model of the world. It is "too big to fail." I have sporadically referred to it throughout the book. It is indeed the financial services sector.

The Finance Delusion

Finance We often hear the word, but how many of us know what it really means? When I hear people speak of "financing" energy infrastructure investments, I'm reminded of the episode of the 1990s TV sitcom *Seinfeld*, "The Package." In this episode Kramer says he'll help Jerry get a refund on his broken stereo. But because the warranty on the stereo is expired, Kramer comes up with his own plan: make it appear as though the stereo was damaged during shipping. Upon opening the package Jerry receives from the post office, he and Kramer have the following exchange:[83]

> Jerry: What happened to my stereo? It's all smashed up.
> Kramer: That's right. Now it looks like it was broken during shipping and I insured it for $400.
> Jerry: But you were supposed to get me a refund.
> Kramer: You can't get a refund. Your warranty expired two years ago.
> Jerry: So we're going to make the Post Office pay for my new stereo?
> Kramer: It's just a write-off for them.
> Jerry: How is it a write-off?
> Kramer: They just write it off.
> Jerry: Write it off what?
> Kramer: Jerry, all these big companies, they write-off everything.

---

[83] October 17, 1996 episode of the TV sitcom *Seinfeld*, "The Package." https://www.seinfeldscripts.com/ThePackage.htm.

Jerry: You don't even know what a write-off is.
Kramer: Do you?
Jerry: No. I don't.
Kramer: But they do, and they are the ones writing it off.

In his scam, Kramer seems to believe that the Post Office incurs no loss of money in providing the $400 in insurance against breakage. In the context of the *Seinfeld* dialogue, a write-off is a business expense, or cost of doing business, that reduces taxable income. By reducing taxable income, there are fewer profits and thus fewer taxes paid. Thus, the write-off is not painless for the company because it is a cost that reduces profits.[84]

Many times people view some barriers to investing in energy projects just like write-offs. What might be holding back investment? Finance. What is finance? Kramer and Jerry could have told the tale in a parallel episode:

Kramer: Jerry, all these big companies, they *finance* everything.
Jerry: You don't even know what *finance* is.
Kramer: Do you?
Jerry: No. I don't.
Kramer: But they do, and they are the ones *financing* it.

This book is not about the details and various instruments of finance, but it is worth noting that finance is not one thing. It encompasses a wide array of activities. One such category is derivatives. Warren Buffett and Charlie Munger have often been quoted on what they wrote in Berkshire Hathaway's 2002 letter to shareholders:[85]

Charlie and I are of one mind in how we feel about derivatives and the trading activities that go with them: We view them as time bombs, both for the parties that deal in them and the economic system.

and

...derivatives are financial weapons of mass destruction, carrying dangers that, while now latent, are potentially lethal.

I've heard investment managers say that "finance has no moral compass." The gist of that statement is that these managers don't need to think about the consequences of their investments. They just need the rules for investment. To them it is the job of regulators and politicians to set the constraints for finance, but as this chapter's discussion of neoliberalism indicates, the revolving door for lobbyists and regulators means bankers often end up setting the rules anyway.

As Buffett and Munger further noted, the lack of constraints on derivatives is what scared them. Derivatives "... call for money to change hands at some future

---

[84]Using numbers from the *Seinfeld* episode, if the Post Office wrote-off $400 to pay for the stereo insurance claim, and their profits were taxed at 28%, then the cost of the write-off is $288, or 72% of $400, which is less than $400, but not $0.

[85]Available http://www.berkshirehathaway.com/letters/2002pdf.pdf.

date, with the amount to be determined by one or more reference items, such as interest rates, stock prices, or currency values," and to them the scary part is that "The range of derivatives contracts is limited only by the imagination of man (or sometimes, so it seems, madmen). At Enron, for example, newsprint and broadband derivatives, due to be settled many years in the future, were put on the books."[86]

Remember Enron? Enron was voted "America's Most Innovative Company" by *Fortune* magazine for six consecutive years (1995–2000) before its bankruptcy in December 2001. It was the company founded on trading energy, such as natural gas and electricity, and the company that manipulated electricity markets in California. It also manipulated its accounting so much that it fooled practically everyone and triggered the demise of its auditor, Arthur Anderson, at the time one of the top five accounting firms in the U.S. It was the company in which former CEOs Ken Lay and Jeff Skilling were convicted of conspiracy and fraud.[87]

In 2002, events like the Enron bankruptcy were on the minds of Buffett, Munger, and many other people.

On the more mundane side of finance, to which I will now turn, is loaning money to an energy company to extract energy from the environment, whether drilling for oil or building a solar farm. This is the same practice as you taking out a loan for a home or a car, but with one important difference. As a home or car owner, we pay back the loans by earning income from work and perhaps other investments. The income to pay back the loan does not directly derive from the home or car itself. When a company receives a loan for an energy project, however, it generally expects to pay back the loan from the revenue directly generated by selling the energy extracted by the project. The revenue equals the quantity of energy extracted (barrels of oil or kilowatt-hours of electricity) multiplied times the price of the energy. Profit equals revenue minus costs, and at the time of acquiring a loan, future costs and revenues are uncertain to one degree or another. This is the basic risk of debt-based finance, or taking out a loan. Just like write-offs, finance is not free, because it is not free of the risk of losing the money you have invested. Finance is a cost of investment.

The Finance Delusion: Fracking Finance

Consider the cost uncertainties in drilling for oil. First, the cost of drilling, the major cost of oil extraction, can be quite uncertain a few months out. Drilling companies hire drilling rigs by the day, and the drilling cost increases with the number of

---

[86]Berkshire Hathaway's 2002 Letter to Shareholders, available http://www.berkshirehathaway.com/letters/2002pdf.pdf.

[87]Bethany McLean and Peter Elkind, "The guiltiest guys in the room," July 5, 2006. Accessed July 24, 2019 at: https://money.cnn.com/2006/05/29/news/enron_guiltyest/. For a comprehensive story of Enron, see *The Smartest Guys in the Room: The Amazing Rise and Scandalous Fall of Enron* by Bethany McLean and Peter Elkind.

operating drilling rigs which in turn generally increases with the price of oil. Second, the price for selling oil is uncertain, and this price is influenced by an array of domestic and global economic factors. Because drilling rig costs track oil prices, extraction companies are somewhat, but not fully, shielded from oil price changes. This leads to a third uncertainty, the amount and timing of oil extraction. If you drill today, the oil comes out over several subsequent years or decades. You don't know future oil prices or output, and, thus, your ability to pay back the cost of drilling today is determined by selling oil in the future.

Data provide some confidence in one conclusion from the last decade of U.S. fracking boom: the majority of U.S. companies that primarily drill in shales (i.e., using hydraulic fracturing and horizontal drilling) don't actually make money. The consulting firm Rystad Energy tracks 40 of these companies, and over the last several years, on average only 20% of them have *positive cash flow* after paying for drilling costs.[88] That is to say, 80% of these companies spend more money than they make. In addition, the U.S. Energy Information Administration reported that out of 119 global oil and gas companies, only 15–50% of them had generated more cash from operations than their capital expenditures.[89]

(Now for a three-handed economist-type remark, usually it is just *two-handed*.) On hand one the idea of shale (or tight) oil companies spending more than they make is not all that disturbing. Companies spend money to drill today and possibly build operational capacity, and they sell oil over the next several years. Thus, for any given well, only after several years do they expect to make more money than they spend. This is the nature of waiting for a return on investment.

On hand two this *is* disturbing because this trend has been going on for several years since the drop in oil price in 2014. From any given tight oil and gas well, the rate of extraction typically falls 70–90% after the first three years [33]. Thus, the vast majority of revenue comes in the first few years, actually decreasing the uncertainty in revenue. Because of this high decline rate, it seems that after three to five years since 2014, the industry would be closer to making more money than it spends just to operate and drill new wells. As Rystad notes, "With negative cash flows, shale companies have historically relied on bond markets to finance their operations. Without additional funding and any debt refinancing, capex [capital expenditures, or spending on drilling] would have to be cut." Translation: unless these companies continue to borrow money at interest, they can't afford to pay drilling rigs to keep drilling.

This brings us to the third hand. Investors tend to want to put their money somewhere, and oil and gas is so critical that it commands investment. As noted in Chap. 4, interest rates have resided at unprecedented low levels since the Great

---

[88]Rystad Energy, "JUST 10% OF SHALE OIL COMPANIES ARE CASH FLOW POSITIVE," Press Release, May 29, 2019. Available July 24, 2019 at https://www.rystadenergy.com/newsevents/news/press-releases/Just-10-percent-of-shale-oil-companies-are-cash-flow-positive/.

[89]Energy Information Administration, "Financial Review of the Global Oil and Natural Gas Industry: First-Quarter 2019," July 2019, available at: https://www.eia.gov/finance/review/archive/pdf/financial_q12019.pdf.

Recession, so borrowing had never been cheaper. Investors couldn't make much money investing in low-interest government bonds, so they moved to corporate bonds, such as with oil and gas companies. Investors were willing to invest in growth of oil extraction, getting paid back via interest payments on debt rather than profitability of the company selling oil. As a result, U.S. tight oil extraction rose from about 0.5 million barrels per day in 2008 to close to 7 million barrels per day by the end of 2018, and total U.S. extraction climbed to more than 11.5 million barrels per day in 2018, higher than the previous peak in 1970.[90]

Low interest rates promoted debt financing that enabled investors to continue to fund drilling. If you believe that oil is still the life blood of the economy, then you have to fund oil extraction because otherwise the rest of the economy can't fully operate. Even many people holding the renewable energy narrative can agree on this point. However, when coupled with the techno-optimistic economic narrative, the renewable narrative posits we can seamlessly replace oil. When coupled with the techno-realistic narrative, it posits the transition might not be as prosperous, but that finite Earth effects force a renewable transition to occur anyway.

So, how long can shale oil companies spend more money than they make? I certainly can't tell the future, and we need much more holistic understanding of how profitable the energy sector might need to be relative to other parts of the economy. Reports show that from 2015 to the first quarter of 2019, 172 North American oil companies went bankrupt, and 100 of these occurred in the two years immediately following the oil price drop in 2014.[91] Almost all of these companies had names you have never heard of, and their assets get gobbled up by companies with names you have heard of. Keep in mind that not all oil and gas companies are solely "shale" companies. The large oil and gas companies that you have heard of (ExxonMobil, Royal Dutch Shell, Chevron, BP, etc.) don't have all their wells in the shale basket.

In the "fracking boom" we have at least the following two narratives, and they both have merit. One narrative is that hydraulic fracturing and horizontal drilling in shales spurred a new era of oil and gas production in the U.S. Absolutely. Companies drilled, and the oil came out. The other narrative is that we are scraping the bottom of the proverbial oil barrel, now going after the source rock where the oil is first formed, and in doing so many companies can't make money. Yes, we drilled and got oil, but we had to spend more money to use more materials to drill a lot more for each drop. This is also true. As with almost all boom–bust cycles from gold to oil, someone is making money, usually those who sell the tools to those prospecting for riches and those that facilitate the financial transactions.

---

[90] Energy Information Administration, "EIA adds new play production data to shale gas and tight oil reports," February 15, 2019. Accessed July 24, 2019 at: https://www.eia.gov/todayinenergy/detail.php?id=38372.

[91] Haynes and Boone, LLC., Oil Patch Bankruptcy Monitor, May 16, 2019, obtained from: https://www.haynesboone.com/publications/energy-bankruptcy-monitors-and-surveys.

Thus, the *finance delusion* associated with fracking, or hydraulic fracturing in shales, is not one that denies finance has played a critical role in the U.S. oil and gas boom, it is that it deludes us into thinking it is economically as viable as the previous one hundred years of the industry.

The Finance Delusion: You Want to Finance How Much Renewable Energy?

The finance delusion doesn't hold only for a new era of U.S. fossil extraction. Companies and governments can also finance too much renewable energy. One example is that of Georgetown, Texas, and its local electric utility, Georgetown Utility Systems (GUS). Georgetown is a relatively politically conservative city 30 miles north of Austin. In the early 2010s, Georgetown signed two major power purchase agreements (PPAs) for renewable wind and solar electricity. PPAs are contracts in which a buyer of electricity agrees to purchase all of the electricity from a developer's power plant over some period of time. Even with another existing natural gas power contract, Georgetown had purchased enough wind and solar power to declare, with some fanfare, the city was powered by 100% renewable electricity.

This declaration brought Al Gore to town as he championed Georgetown and its mayor, Dale Ross, as leading the fight against climate change. Gore even featured Georgetown and Ross in his 2017 film *An Inconvenient Sequel: Truth to Power*, a follow up to the 2006 movie *An Inconvenient Truth*.

This Gore-Georgetown connection spurred great headlines, such as "a conservative town leads the way with renewable energy." But while Gore promoted the low-carbon energy angle, Georgetown's major touted how the utility's contracts were based on business fundamentals and the low-cost electricity they locked-in for two decades. Collectively, Gore's and Ross's messages perfectly combine the narratives of renewable energy and techno-optimism.

But by 2018, all was not sitting well with GUS customers. The problem was not that Georgetown was 100% renewable. It was that Georgetown was 160% renewable![92] In 2019, the situation led GUS residential customers to pay for electricity at about 14.4 cents per kilowatt-hour (¢/kWh) instead of the 10–11 ¢/kWh that was typical across the rest of Texas. Assuming a customer consumes about

---

[92] In 2018, the Georgetown's electricity contracts amounted to annually purchasing 1070 giga-watt hours (GWh) of electricity, but the city only needed 679 GWh. Thus, Georgetown utility customers paid for 390 GWh, or 57%, more electricity than they needed. As of the middle of 2019, the city estimated excess electricity purchases would total 580 GWh, or 85% more than consumed. "FAQ Georgetown Energy Contracts" from City of Georgetown website, accessed July 3, 2019 at https://gus.georgetown.org/electric/faq-georgetown-energy-contracts/.

1000 kWh per month, we're talking about spending an extra $400–$500 per year per household.

The city was forced to respond to the controversy, as the "Rumor Control" document from the city's website stated:

> The crux of the current challenge hinges on the large amount of energy the City must clear to the market that is not currently consumed in Georgetown. Like most city-owned utilities, Georgetown contracted for more energy than it currently needs. Any energy that is not consumed by Georgetown customers must be cleared into the energy market.[93]—City of Georgetown, Texas (2019)

It's never a good thing if you need a link on your website labeled "rumor control." While it is true that other city-owned utilities sometimes buy more electricity than they need, the situation of Georgetown is a bit out of the ordinary.

So why did GUS agree to buy so much electricity? Some might say it was because city and utility officials were wined and dined by the Wall Street bankers who financed the deals. Another reason was to obtain a better PPA price due to economies of scale in building larger wind and solar farms. Further, Georgetown assumed the city would keep growing in population such that in 10 or so years their higher demand would absorb the excess electricity. Georgetown also planned to sell the excess electricity into Texas' wholesale electricity market. Georgetown had banked on forecasts that projected electricity prices would be higher than their PPA prices. Thus, they thought they had bought low and were going to sell high. As of 2019, the opposite was occurring.

The full story takes a few pages to lay out, and I've put this story in the Appendix. It explains in more detail why GUS's purchases of *low-cost* wind and solar electricity led them charging *higher electricity prices* to their customers.

Needless to say, Georgetown's electricity situation was ripe for attack for those promoting the fossil energy narrative. A lead actor was the Texas Public Policy Foundation (TPPF), a Texas-based conservative think tank that promotes markets and limited government.[94] They pounced on Georgetown's decisions as exemplary of the overreach and poor decision making by a few elected officials and bureaucrats that can occur within cities and regulated utilities like GUS.

In a summary of an August 2018 community event in Georgetown, TPPF stated "Attendees received a thorough run-down of everything energy including why renewables are unreliable, more expensive, and unsustainable. And why 100% renewable simply isn't doable."[95]

---

[93]"Rumor Control" document from City of Georgetown website, accessed July 3, 2019 at https://gus.georgetown.org/electric/rumor-control/.

[94]The Texas Public Policy Foundation website states "Through research and outreach we promote liberty, opportunity, and free enterprise in Texas and beyond.", accessed July 25, 2019 at https://www.texaspolicy.com/.

[95]"Is Georgetown Really 100% Renewable?" https://www.texaspolicy.com/is-georgetown-really-100-renewable/.

To the "more expensive" comment, see the Appendix which describes how to consider Georgetown's costs and power purchases in the context of the entire electric grid. I will not readdress the oversimplified "reliability argument" since it was discussed in Chap. 3.

The TPPF's comments on Georgetown, Texas provide an opportunity to end the chapter on a social meme. Their Vice President stated:

> ...there is a multitude of things that may distract an elected official from attending to local voters' concerns. Unelected staff may have their own agendas that they seek to impose upon the elected officials for whom they ostensibly work. Outside special interests may appeal to them. Popular issues that have little or nothing to do with their office may command their attention—for instance, city council members passing foreign policy resolutions or environmental ordinances that aren't within their scope of responsibility.
>
> In the City of Georgetown's case, the distraction from the basic services came in the form of a virtue signaling energy policy cloaked in the guise of responsible fiscal policy.[96]—Chuck DeVore (2018)

Virtue signaling is a meme describing a person expressing an opinion for the purpose of displaying his moral superiority. It has been used to mock people who post political or social stances via social media platforms, but who otherwise take no concrete action or have no direct authority to promote their view.

The term comes from the biological concept of *signaling* where a prey provides a visible signal to a predator that it is either dangerous or futile to try to eat it. For example, when a gazelle notices a stalking cheetah, it might perform a jump (called stotting or pronking) as a "...way of telling the cheetah that he sees her, has a head start, and that a chase would be futile." [34][97] Because of this communication, the cheetah smartly does not waste her energy on chasing the gazelle. We presume the gazelle and cheetah have no notion of these logical thoughts in their heads, but the data show what predators do avoid going after pronking prey.

Instead of virtue signaling, Georgetown's contracts for 160% of its electricity consumption are actually *over signaling*. Georgetown's actions are more akin to a gazelle jumping up and down so much such that it tires itself out, allowing the cheetah to more easily chase it down. In the context of the economy as an evolving superorganism, we might say Georgetown is a species that mutated so much more than the rest of the economy that it incurred higher immediate costs relative to other utilities that purchase a lower percentage of renewable energy. The utility's mutation was in the right direction of lower-cost electricity (particularly low operating costs), but perhaps a little too far for its immediate benefit in a slower changing economic environment.

To focus only on Georgetown is to be too narrow, however. For all of the higher electricity costs paid by Georgetown's utility customers as of the writing

---

[96]Chuck DeVore, "Texas Taxpayers Pay For Political Virtue Signaling With Costly Renewable Energy," *Forbes* online, December 17, 2018. Available June 20, 2019 at: https://www.forbes.com/sites/chuckdevore/2018/12/17/texas-taxpayers-pay-for-political-virtue-signaling-with-costly-renewable-energy/#5bbc9acd46a6.

[97]Lents [34, p. 279].

of this book, they have, if ever so slightly, decreased the *negative environmental externality* from carbon emissions. Since Georgetown's PPAs led to more actual physical power plants generating low-carbon electricity, they also move the needle, if ever so slightly, toward a lower-carbon grid. Because GUS bought more renewable electricity than its customers need, it also lowered wholesale electricity prices for others. In other words, electricity is more expensive for Georgetown, but for Texans overall, electricity is cheaper, thus providing a *positive economic externality* to other Texans.

What is Georgetown to do? TPPF was happy to pose three options: "As of today, residents of Georgetown who aren't pleased with paying more for their electricity for the privilege of making the dubious claim to 100% renewable power have three options: vote in a new set of elected officials who promise to focus on the basics of local governance, convince the legislature to end the electric monopoly extended to municipal government, or move out of town." Of course, these false choices aren't the only three options. By default, most residents will stay. Instead of appealing to state lawmakers, the residents could help their utility find buyers for their extra electricity. While it might seem strange that an entity promoting individual liberty suggests the invocation of action from state officials who are even further removed from individual circumstances, recall that neoliberalism promotes a strong government to enforce increased use markets, such that attacks on municipal authority are common. Remember the Denton, Texas story at the beginning of the book?

As of the writing of this book, we don't know the full value of Georgetown's renewable electricity contracts because there are still about 20 years to go to find out. Given this future uncertainty, in 2019 Georgetown's mayor Ross stuck with his earlier decisions: "I don't think I've let citizens down . . . We did what we believe was the right thing. I was very effective at spreading the message of green and our success story. I think the long term is going to prove that to be true."[98] Only time will tell.

Summary: Delusions from Free Will to Finance

This chapter discussed how an overreliance on economic narratives, or memes, such as confidence, political will, and decoupling via services and finance can distract us from understanding how the physical basis of the economy plays a role in our socio-economic outcomes and decision making. It is not that these ideas are false, but we must understand how our choices, what we see as freedom of choice among our

---

[98] Sharon Jayson, US News & World Report, "Texas City Leaders Face Wrath of Residents Over Green Energy Deal," March 28, 2019. Accessed July 25, 2019 at: https://www.usnews.com/news/cities/articles/2019-03-28/in-georgetown-texas-a-clean-energy-deal-falls-flat.

various options, or degrees of freedom, are constrained by what we understand to be the energetic and physical nature of ourselves, our world, and our economy.

While we each have a large degree of agency, or ability to choose our actions, collectively the economy follows certain patterns as outlined in Chaps. 5 and 8. The concepts of confidence, in economic growth and future social outcomes, and political will are linked to social power, and social power is linked to the technological ability to use physical power by extracting energy (at some rate) from the environment and converting it to useful work. Social power is the ability of an individual or group to influence the behavior of others, by controlling energetic processes of interest to them, without the powerful group having to change its own behavior in an undesirable manner, such as accessing fewer resources and energy.[99] Thus, a lack of political will can be seen as politicians recognizing they do not have social power over businesses' or citizens' access to resources. Social power also exists for entities such as countries, and any social power of one country over another depends on their relative abilities to extract energy from the environment. As Richard Adams stated, "It is the actor's control of the environment that constitutes the base of social power."[100]

The meme of decoupling material and energy consumption from economic growth attracts many, but at a high level the data do not support that it happens for developed economies, much less the global economy. We come to this conclusion by looking at the historical data—the data describing all economic and physical processes.

While we can try to understand the past by looking at data, we cannot do this for understanding the future. That is part of the definition of the future—there are no future data. To infer what we think is possible in the future, we must use our existing knowledge, worldviews, and models. As this book has discussed, everyone neither shares the same knowledge nor uses or believes the same worldview and models. Given that premise, the final chapter describes how we can both contemplate a wide range of future scenarios and discuss some of the important ongoing and future energy and economic questions facing society for the next several decades.

References

1. Yuval Noah Harari. *Sapiens A Brief History of Humankind*. HarperCollins, New York, NY, 2015.
2. Timothy Mitchell. *Carbon Democracy: Political Power in the Age of Oil*. Verso, London and New York, 2013.
3. Edward N. Lorenz. Predictability: Does the flap of a butterfly's wings in brazil set off a tornado in Texas?, 1972. Presented at the 139th meeting of American Association for the Advancement of Science, Washington, DC December 29.

---

[99] See [22, p. 121].

[100] Adams [22, p. 13].

4. Edward N. Lorenz. *The Essence of Chaos*. University of Washington Press, 1993.
5. Sean Carroll. *The Big Picture On the Origins of Life, Meaning, and the Universe Itself*. Dutton, Penguin Random House LLC, 2016.
6. Alfred S. Eichner. *The Macrodynamics of Advanced Market Economies*. M. E. Sharpe, Inc., 1991.
7. Daniel Dennett. *Freedom Evolves*. Viking, 2003.
8. Edward L. Bernays. The engineering of consent. *The Annals of the American Academy of Political and Social Science*, 250(1):113–120, 1947.
9. Edward L. Bernays. *Propaganda*. Horace Liveright, New York, New York, 1928.
10. Roger E. A. Farmer. *How the Economy Works: Confidence, Crashes, and Self-Fulfilling Prophecies*. Oxford University Press, Oxford, England and New York, NY, 2010.
11. Donella H. Meadows. *Thinking in Systems: A Primer*. Chelsea Green Publishing, White River Junction, Vermont, 2008.
12. Donella Meadows. Indicators and information systems for sustainable development, a report to the Balaton Group, 1998.
13. Paul Collier. *The Future of Capitalism: Facing the New Anxieties*. HarperCollins, New York, NY, 2018.
14. Donella H. Meadows, Jorgen Randers, and Dennis L. Meadows. *Limits to Growth: The 30-Year Update*. Chelsea Green Publishing, White River Junction, Vermont, 2004.
15. Jeffrey D. Clements. *Corporations are not People Why They Have More Rights Than You Do and What You Can Do About It*. Berrett-Koehler Publishers, Inc., San Francisco, CA, 2012.
16. Ali Almossawi. *An Illustrated book of Bad Arguments*. JasperCollins Publishers, New York, 2013.
17. Philip Mirowski. Neoliberalism The movement that dare not speak its name. *American Affairs*, 2(1), 2018.
18. F. A. Hayek. The use of knowledge in society. *The American Economic Review*, 35(4):519–530, 1945.
19. Philip Mirowski and Dieter Plehwe, editors. *The Road from Mont Pèlerin: The Making of the Neoliberal Thought Collective*. Harvard University Press, 2009.
20. Nicholas Shaxson. *The Finance Curse: How Global Finance Is Making Us All Poorer*. Grove Atlantic Press, 2019.
21. Charles M. Tiebout. A pure theory of local expenditures. *Journal of Political Economy*, 64(5):416–424, 1956.
22. Richard N. Adams. *Energy and structure*. University of Texas Press, Austin, Texas and London, England, 1975.
23. John K. Ryan. *The Confessions of St. Augustine*. Image Books, Doubleday Press, 1960.
24. Pindyck, Robert S. Climate Change Policy: What Do the Models Tell Us? *Journal of Economic Literature*, 51(3):860–872, SEP 2013.
25. Roger Fouquet. Divergences in long-run trends in the prices of energy and energy services. *Review of Environmental Economics and Policy*, 5(2):196–218, 2011. doi:10.1093/reep/rer008.
26. Mariana Mazzucato. *The Value of Everything Making and Taking in the Global Economy*. Public Affairs, New York, 2018.
27. Z. Wang, Z. Ying, Bosy-Westphal A., J. Zhang, B. Schautz, W. Later, S. B. Heymsfield, and M. J. Müller. Specific metabolic rates of major organs and tissues across adulthood: evaluation by mechanistic model of resting energy expenditure. *The American Journal of Clinical Nutrition*, 92(6):1369–1377, 2010.
28. Blair Fix. Dematerialization through services: Evaluating the evidence. *BioPhysical Economics and Resource Quality*, 4(2):6, Mar 2019.
29. Andrew McAfee. *More from Less The Surprising Story of How We Learned to Prosper Using Fewer Resources—and What Happens Next*. Scribner, New York, NY, 2019.
30. Thomas O. Wiedmann, Heinz Schandl, Manfred Lenzen, Daniel Moran, Sangwon Suh, James West, and Keiichiro Kanemoto. The material footprint of nations. *Proceedings of the National Academy of Sciences*, 112(20):6271–6276, 2015.

31. Tim Jackson. *Prosperity Without Growth: Foundations for the Economy of Tomorrow*. Routledge, Milton, UK and New York, NY, USA, second edition, 2017.
32. Raj Patel and Jason W. Moore. *A History of the World in Seven Cheap Things: A Guide to Capitalism, Nature, and the Future of the Planet*. University of California Press, Oakland, California, 2017.
33. David Hughes. Shale reality check Drilling into the U.S. government's rosy projections for shale gas & tight oil production through 2050. Technical report, 2018.
34. Nathan H. Lents. *Not So Different Finding Human Nature in Animals*. Columbia University Press, New York, NY, 2016.

Chapter 10
Scenarios and Trends of the Future

But anyone who believes that he can draw a blueprint for the ecological salvation of the human species does not understand the nature of evolution, or even of history—which is that of a permanent struggle in continuously novel forms, not that of a predictable, controllable physico-chemical process, such as boiling an egg or launching a rocket to the moon. [5]— Nicolas Georgescu-Roegen (1975)

The first chapter of this book noted three facts:

1. The Earth is finite.
2. The laws of nature are human constructs that describe the interactions within the natural world and are defined as being the same everywhere (per the present state of knowledge).
3. The laws of society, or legal rules and social norms, are human constructs that seek to limit human interactions to a subset of all possibilities, and they are not the same everywhere.

Even people that disagree how to interpret past energy and economic trends can still agree on these facts. This book explored these disagreements in terms of how much the first two facts influence the third. We make most of our societal laws without contemplating natural laws. Examples of such societal laws are taxes on economic activities, the legal ownership of property, the definitions of murder and self-defense, and equal human rights among genders and sexual orientations.

Nonetheless, natural laws inherently influence collective social behavior in hidden ways.

The first chapter also summarized the following sequence:

- First, physical laws and constraints describe how we can use and access energy.
- Second, energy resources physically power the economy via use in machines, buildings, and other physical capital.
- Third, our interpretations of the economy inform policy.
- Fourth, policy affects social outcomes by designing markets, regulations, and taxes that affect the distribution of money.

© Springer Nature Switzerland AG 2021
C. W. King, *The Economic Superorganism*,
https://doi.org/10.1007/978-3-030-50295-9_10

- Finally, the rules governing where, how, and when money is distributed affect energy resource extraction and consumption, leading back to the beginning.

The previous chapter noted Vladimir Lenin's quote that "Communism is Soviet power plus the electrification of the whole country." Practically all modern-day politicians and citizens recognize the value of access to electricity, and hence useful work. But, as this book argues, they don't all understand that *how* we obtain energy from the environment and *how* we convert that energy into electricity and useful work are defining features of our societies and economies, both their growth and structure.

With this thought, we can ask a fundamental question: *Are we humans freely choosing our societal and economic organization, or does our organization emerge in response to physical laws as we interact with the natural world around us?*

I've set this up as a false choice. We don't have to explain the human economy in the context of one type of laws independent of the other.

> Not only *can* we use both social rules and physical laws to assess the constraints and possibilities for future energy and economic scenarios, *we absolutely must.*

The Future Ain't What It Used to Be

The global economy has experienced many major transitions in the last 250 years. I'll summarize these into four.

The first was the transition from agrarian to fossil-fueled industrialization. It started in the United Kingdom in the late 1700s and ramped up in the United States and other Western nations in the early 1800s. This transition sparked unprecedented rates of economic growth.

The second transition spanned the two world wars, during which industrialized economies fought over control of the world's resources residing outside their individual country borders, but within the borders of their colonies. This transition marks the end of colonization. In the post-World War II era, the world knew just how destructive we could be. The atomic age had begun. While the Cold War between the U.S. and Soviet Union governed much of geopolitics, the United Nations and the Bretton Woods agreement sought stability within these newly-formed institutions for international cooperation.

The third transition occurred in the 1970s. Up to this point, rapid exponential growth seemed "normal," but it could not go on forever, and the 1970s mark the end of the most rapidly growing 30-year period in human history, the "trente glorieuses." The evidence for this transition abounds in data spanning many domains: energy, economics, and environmental. Rich and industrialized countries experienced this transition most distinctly. Industry started in earnest to take advantage of lower wages in developing countries. A new age of globalization began. The Bretton

Woods agreement ended, thus removing any direct relation between money and the stock of a physical commodity, such as gold.

This book argues that the 2000s, culminating in the 2008 financial crisis, mark a fourth major transition. This transition is marked by the end of a trend that started with industrialization: energy and food costs (as a percentage of GDP) stopped declining. This transition also marked a shift from debt-fueled growth to our current period of rich country "secular stagnation" characterized by lower GDP growth, low interest rates, and continued low wage growth. We should not necessarily interpret the current state of the economy as bad, but more as expected for a capitalist economic system reaching limits to growth. Some call this a "new normal," others call it expected.

Each of the time periods between the aforementioned transitions represents a period of some type of normality. Each transition begets a new narrative for interpreting the past and defining future possibilities. We update both our individual and collective narratives over time.

Narratives are not fixed concepts whether we summarize global economic affairs or personal experiences. Philosopher Daniel Dennett introduced the concept of the "multiple drafts" model of consciousness. The multiple drafts concept states there is not one consciousness, but many that get updated over time. Because of this, there is no one "correct" interpretation of what one experiences. It depends on *when* you ask:

> Just what we are conscious of within any particular time duration is not defined independently of the probes we use to precipitate a narrative about that period. Since these narratives are under continual revision, there is no single narrative that counts as the canonical version, the "first edition" in which are laid down, for all time, the events that happened in the stream of consciousness of the subject, all deviations from which must be corruptions of the text. But any narrative (or narrative fragment) that does get precipitated provides a "time line," a subjective sequence of events from the point view of an observer, that may then be compared with other time lines, in particular with the objective sequence of events occurring in the brain of that observer.[1]—Daniel Dennett (1991)

Just as we update our state of consciousness, over time we'll continue to observe, learn, and update our energy and economic narratives. For example, after some number of years I might update my narrative of the four major transitions in economic development since the start of the industrial era. I might think an additional transition is required, or I might decide to remove one. All of this is okay, and it helps engender humility for thinking about future options.

A Range of Futures

David Holmgren provides one useful taxonomy for considering a range of futures. Four visions, or narratives, summarize how any given person might envision the

---

[1] [2] [p. 136].

long-term future of humanity and our economy [7].[2] Holmgren is one of the founding pioneers of *permaculture*, or "permanent culture," as a systems thinking framework for designing our social systems to have similar resilience as observed in natural ecosystems. His four future scenarios, or narratives, are as follows:[3]

1. "*Techno-explosion* depends on new, large and concentrated energy sources that will allow the continual growth in material wealth and human power over environmental constraints, as well as population growth. This scenario is generally associated with space travel to colonize other planets."
2. "*Techno-stability* depends on a seamless conversion from material growth based on depleting energy, to a steady state in consumption of resources and population (if not economic activity), all based on novel use of renewable energies and technologies that can maintain if not improve the quality of services available from current systems. While this clearly involves massive change in almost all aspects of society, the implication is that once sustainable systems are set in place, a steady state sustainable society with … [little] change will prevail. Photovoltaic technology directly capturing solar energy is a suitable icon or symbol of this scenario."
3. "*Energy Descent* involves a reduction of economic activity, complexity and population in some way as fossil fuels are depleted. The increasing reliance on renewable resources of lower energy density will, over time, change the structure of society to reflect many of the basic design rules, if not details, of pre-industrial societies. This suggests a ruralization of settlement and economy, with less consumption of energy and resources and a progressive decline in human populations. Biological resources and their sustainable management will become progressively more important as fossil fuels and technological power declines. In many regions, forests will regain their traditional status as symbols of wealth. Thus the tree is a suitable icon of this scenario. Energy Descent (like Techno-explosion) is a scenario dominated by change, but that change might not be continuous or gradual. Instead it could be characterized by a series of steady states punctuated by crises (or mini collapses) that destroy some aspects of Industrial culture."
4. "*Collapse*[4] suggests a failure of the whole range of interlocked systems that maintain and support industrial society, as high quality fossil fuels are depleted

---

[2]Also see: http://www.futurescenarios.org/ and https://holmgren.com.au/future-scenarios-presentation/.

[3]http://www.futurescenarios.org/content/view/16/31/index.html.

[4]Per David Holmgren's definition, he states: "Some very influential authors such Joseph Tainter (The Collapse of Complex Societies, 1988) and Jared Diamond (Collapse: How Societies Choose to Fail or Succeed, 2005) use the term collapse to describe any ongoing reduction in complexity of the organization of civilizations. While their work is of great importance, I want to draw a distinction between what I mean by "Collapse" as the sudden failure and loss of most of the organizational complexity (such that succeeding generations retain little use or even memory of such systems) and "Descent" as a progressive if erratic process where the loss of complexity is gradual and succeeding generations have some awareness of, and knowledge from, that peak of complexity."

and/or climate change radically damages the ecological support systems. This collapse would be fast and more or less continuous without the restabilizations possible in Energy Descent. It would inevitably involve a major "die-off" of human population and a loss of the knowledge and infrastructure necessary for industrial civilization, if not more severe scenarios including human extinction along with much of the planet's biodiversity."

Holmgren's scenarios provide a broad range of futures, each associated with a key symbol: techno-explosion (vision: space colonization; movie: *Star Trek*), techno-stability (vision: renewable energy; movie: *An Inconvenient Truth, An Inconvenient Sequel: Truth to Power*), energy descent (vision: trees and transition towns; movie: I don't know of a mainstream movie representing this scenario, but there are some documentaries), and collapse (vision: chaos; movie: *Mad Max*).

These are not the only future visions, and one can easily come up with combinations. For example, the movie *Interstellar* poses that the vast majority of people live in a collapsed society (or one far down Energy Descent) while a small group of educated persons seeks to ensure the existence of *Homo sapiens* by trying to colonize other planets.

We can imagine placing any given future scenario on the two-dimensional energy and economic axes in Chap. 1. Figure 10.1 represents my placement of Holmgren's scenarios. Techno-explosion is the extreme form of the techno-optimism and infinite substitutability economic narrative. The Blue Origin and SpaceX websites succinctly state this techno-explosion vision:

> Blue's vision is a future where millions of people are living and working in space. In order to preserve Earth, our home, for our grandchildren's grandchildren, we must go to space to tap its unlimited resources and energy. If we can lower the cost of access to space with reusable

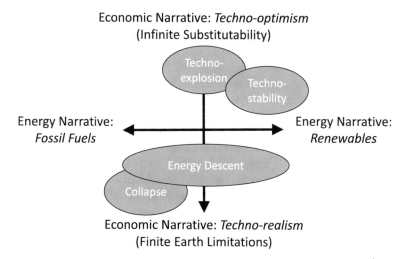

Fig. 10.1 David Holmgren's four future scenarios (narratives) placed on the two-dimensional energy-economic narratives of this book (Fig. 1.1)

launch vehicles, we can all enable this dynamic future for humanity.[5]—Blue Origin, "Our Vision" (2020)

SpaceX designs, manufactures and launches advanced rockets and spacecraft. The company was founded in 2002 to revolutionize space technology, with the ultimate goal of enabling people to live on other planets.[6]—SpaceX, "About SpaceX" (2020)

Techno-explosion need not pay heed to the energy narratives. Blue Origin recognizes the finite Earth, but instead of nurturing and living within the means of Earth, it posits that we have to leave it to save it by tapping the "unlimited resources and energy" of space. SpaceX seeks to colonize other planets to preserve our species (e.g., Julian Simon's view of "the cosmos" as the human domain of influence).

Techno-stability represents many people's vision of a future where we limit climate change impacts below a critical level by cost-effectively transitioning to a renewable energy and low-carbon economy. As Chap. 6 emphasized, because of their theoretical assumptions that neglect debt, time, and natural resources inputs, mainstream economic models present both of Holmgren's "techno" scenarios as possible, even though their assumptions and theory make them inapplicable to even ask the question. Thus, most scenarios coming from energy companies, governments, and international organizations present the case to the public that a perpetual fossil fuel or renewable/low-carbon world is possible without significant change.

The Collapse scenario is one that views fossil fuels as both limited and key to our present modern lifestyles. It also views renewables as insufficient to substitute for fossil fuels that eventually cannot continue to maintain present society. We might try to substitute renewable energy technologies, but their characteristics and costs will prove this to be a fruitless exercise at the required scale. Thus, a collapse in modern lifestyles will eventually occur.

This leaves the Energy Descent scenario. A range of activities can occur in this scenario, from maintaining the fossil fuels system as much as possible to pushing renewable energy technologies as far as they can go. However, by definition, in this scenario, all of these efforts fall short of indefinitely maintaining the current size and complexity of the economy.

The Energy Descent scenario is an appropriate description for changes that began in developed countries in the 1970s following major oil supply restrictions and price shocks. This was the beginning of what Herman Daly called "uneconomic growth."[1] There was some type of growth, but it came at the expense of too much inequality and some continued environmental impact.

The events of the 1970s triggered earnest research into today's modern wind turbines, solar photovoltaic panels, and electrochemical batteries. In some sense the world economy was on the brink of collapse in the 1970s, but it restructured itself in response. The same crisis and restructuring occurred in 2008 in the

---

[5]Blue Origin website, February 22, 2020: https://www.blueorigin.com/.

[6]SpaceX website, February 22, 2020: https://www.spacex.com/about.

Great Recession. These restructured states have generally translated to worsening livelihoods for citizens of rich countries, but they are starting from a relatively high point. For citizens of poorer and developing countries, these restructured states created opportunities for increasing livelihoods, with the economic growth in China and India being the most prominent cases.

It is important to point out the obvious: we can observe current events that exhibit tendencies from each of Holmgren's four scenarios. Techno-explosion: we do have billionaires making rockets to go into space, just like the James Bond spy movies (e.g., *Moonraker*). Techno-stability: over the last 10 years we have installed wind turbines and solar panels at increasingly rapid rates. Energy descent: over the last 50 years, rich country wage growth stagnated, income inequality increased, and both total and per capita primary energy consumption stagnated. Collapse: Venezuela since the 2008 financial crisis, and particularly since the drop in oil prices in 2015, has become a poster-child for not shifting its economy and consumer behavior from over-dependence on the sale of high-cost oil.

Each person in each generation experiences a unique set of circumstances that influence her narrative of past events and her narrative of how the future can unfold. After all, narratives are emergent beliefs that summarize a multitude of underlying processes. We use all kinds of rational arguments to support our positions among the competing narratives. As we learn more, we shift our positions.

While ultimately future details are unknowable, we can say some overall patterns, such as the scaling law patterns relating energy consumption to size of the economy (Chap. 5), are more likely than others. The remainder of this final chapter describes anticipated tendencies, trends, and battles to set the vision for humanity, our use of energy, and hence our economy.

The Battle for Control of the Superorganism

Competing Economic Memes and Models

Chapter 6 summarized arguments against neoclassical economics and its theory of growth. While many researchers, including myself, see these as reasons enough to use other existing economic approaches with more consistency in scientific and economic fundamentals, there is a larger question as to whether an alternate economic meme can supplant the neoclassical approach.

Recall that the neoclassical growth model leaves about half of economic growth unexplained by the model itself. The father of this growth model, Robert Solow, recognized this fact at its inception, and he stated as much almost 40 years after developing the initial model. He recognized ...

> ... a criticism of the neoclassical model: it is a theory of growth that leaves the main factor in economic growth unexplained. There is some truth in that observation, but also some residual misconception. First of all, to say that the rate of technological progress is exogenous is not to say that it is either constant, or utterly erratic, or always mysterious. One

could expect the rate of technological progress to increase or decrease from time to time. Such an event has no explanation within the model, and may have no apparent explanation at all. Or else it might be entirely understandable in some reasonable but after-the-fact way, only not as a systematic part of the model itself.[10]—Robert Solow (1994)

Solow posits it is possible that technological progress could be described by some "reasonable but after-the-fact way," and Chap. 6 showed that increases in energy efficiency very much seems to fit this need. In this sense, modeling technological progress as an aggregate energy conversion efficiency is a more accurate explanation than associating it with nothing specific or even nebulous ideas such as "human ingenuity."

We can systematically (somewhat tediously) measure energy conversion efficiencies that, due to the second law of thermodynamics, are limited to well below 100%. In this interpretation, "technological progress" cannot increase indefinitely. Also, since the rate of energy extraction also cannot increase indefinitely on the finite Earth, then useful work cannot increase indefinitely (useful work = energy consumption multiplied by conversion efficiency). In turn, since gross domestic product (GDP) is a proxy for useful work, GDP also cannot increase indefinitely.

I and others claim that this physically based view of economic output is more accurate. Will it win over that of neoclassical economic growth? Does a model of economic growth, including that used to inform policy, need to accurately represent economic functionality? In other words, would an economy with a more accurate economic model of itself generally prevail over an economy with a less accurate economic model of itself?

These are questions for another book, but we can form an initial hypothesis by returning to the narrative of the economy as an evolving superorganism. Assume each economy seeks to propagate its technological memes just like each biological organism seeks to propagate its genes. Thus, per the maximum power principle, the economy that accesses more energy, and transforms it more efficiently into new structures, is more fit to survive and propagate its memes.

But how do biological organisms or economies know which option enables higher power consumption? How do they know what energy input makes them more fit? We don't get any sense that they attempt to model themselves via scientific and economic calculations. Donald Hoffman, a professor of cognitive science, has studied the evolutionary impact of having "truthful," or more accurate, versus "simple," or less accurate, representations of what is really happening in the world around you.

Consider the following excerpt from Hoffman's 2010 article:

> Seeing more data takes more time. So, in the simplest version of this game, simple chooses first when competing against truth. . . .
>
>   Similarly, seeing more data takes more energy, so truth requires more energy than simple. We subtract the cost of energy from the utility that each agent gets from its territory.[9]—Justin T. Mark, Brian B. Marion, and Donald D. Hoffman (2010)

Here, lower utility is the same as lower evolutionary fitness. In addition to necessarily consuming energy in order to extract energy from the environment, an economy must also invest some amount of time to learn more about the environment. Hoffman also emphasizes that it takes energy just to gain more knowledge. His argument is that simpler rules take less time and energy to make a decision. Thus, in the context of evolution, organisms with simple rules based on relatively inaccurate descriptions of the environment can be more fit than organisms with more complicated rules based on more accurate descriptions of the environment.

What might a set of simple rules be for entities within the economy? Prices.

What does neoclassical economics, and the neoliberal paradigm, focus on? Prices.

Not only do neoclassical economics and neoliberal politics focus on markets that form prices, but in reality prices form when all agents generally lack full information about the cost and input requirements to make products. The prices might even be defined by the immediate whims of the buyer and seller. Even well-structured markets, where short-term whims play no major role, prices form using only a portion of the full costs.

Consider electricity markets, perhaps the most well-defined markets that exist. The market cost of supplying electricity from each power plant on the grid derives from the operating costs, such as paying for fuel and people to operate and maintain the power plant, and not from how much it costs to construct the power plant in the first place. Each power plant operator has a simple rule for bidding to produce electricity: bid to produce electricity at just above the marginal operating cost. If you bid below this number and are told to operate, there is a chance you will operate at a loss and be less fit. If you bid too high above this number, there is a chance that you will not be chosen to operate at all (because there are enough other power plants bidding lower costs), and again you will be less fit (you earn zero revenue but still have some costs).

To be explicit, we can express the battle of the economic memes via the following 3-part hypothesis. First, of all models of the economic system that we presently know, neoclassical economics is not the most accurate representation of economic growth and distribution. Second, neoclassical economics provides a relatively simple and teachable method to make choices that maximize immediate fitness. Third, economies organized via neoclassical economics are more fit, *ceteris paribus*, than those organized via other economic systems and rules.

For me, it is a bitter pill to swallow to even contemplate that this hypothesis might be true. I never imagined I would write that neoclassical economics might have some enhanced usefulness over more biophysically based approaches to economic modeling. At this time, neoclassical economics is clearly the most pervasive economic meme. As a cultural construct, it is winning the evolutionary game of self-replication.

Not only that, but markets based on marginal (or operating) costs, as promoted by neoclassical theory, are in one sense consistent with biological evolution. Evolution is not forward looking. In other words, evolution involves no long-term planning.

The same holds for markets as they drive economic actors to increasingly emphasize "now" over future outcomes. However, there is (at least) one important difference between biological evolution and technological change. Genetic mutations are *random*. Thus, they produce marginal, but random, changes in phenotypes. The organism tests the fitness of the mutation after the fact. In contrast, memetic mutations represented as changes in technology, from machine designs to algorithms, are not purely random. Because we have evolved to think abstractly, we have created models of the economic system. Because we have these models, we know that certain types of technological changes have higher odds of increasing the fitness the economic superorganism: those that minimize operating costs.

The Continued Trend of Lower Operating Costs

Capital owners, informed by price signals from markets, are incentivized to minimize operational costs, including costs of labor, energy, and natural resource consumption. A long as capitalism governs our socio-economic organization, we should expect operating cost minimization to continue, even in an Energy Descent scenario.

One can minimize the cost of labor by two strategies. First, automate as many tasks as possible. Second, for those tasks that prove difficult to automate, move the jobs to the countries with the lowest wages.

An economy can minimize the cost of energy consumption via a few strategies. One is to maximize energy conversion efficiency. Another is to reduce the energy required to distribute physical goods, including energy, and information. This distribution cost can be reduced by forming a business that minimizes the distribution of physical items. It takes energy to distribute things that have mass (including people), and the less mass you distribute, the less you pay for energy and materials. An economic superorganism that minimizes moving mass—people, fuels, cars, everything—serves an overall goal of accumulating more mass in total. Distribution costs are also minimized by concentrating people into cities rather than dispersing them evenly over an entire region.

How do companies minimize operating costs? By inventing and deploying physical capital, or technologies, with this purpose in mind. Energy extraction technologies that serve this purpose are those such as hydropower, wind turbines, and solar panels. Thus, many people, myself included, anticipate continued investment in these technologies in locations with good natural resource flows (rain, wind, sun). As long as the economy is growing, I also expect continued investment in these renewable technologies. Even in an Energy Descent scenario, with a shrinking economy, investment in modern renewable systems could occur, but it is not guaranteed. The same should eventually hold for energy storage technologies, such as electrochemical batteries.

Of course the cost to install these technologies matters, but once you have them, they cost very little to operate. The low operational cost of hydro, wind, and solar power is not directly related to energy conversion efficiency. It derives from minimizing the two cost categories mentioned above: labor and energy (and materials).

In this sense, the more "green jobs" associated with *installing* wind, solar, and battery systems that extract environmental energy flows, the fewer "brown jobs" associated with the continuous *extraction* of energy stocks such as fossil fuels.

Don't think of this as a statement *for* the renewable energy narrative and *against* the fossil fuel narrative. Think of it simply as an observation and description of the overall energy system as would the economy in acting as a superorganism. The superorganism is trying to maximize its net output of useful work, and it can do this by minimizing operating costs relative to output. We might think of the economic superorganism as evolving from one akin to a colony of leaf cutter ants, with relatively high labor costs to move raw materials, to one akin to a spider who first invests considerable effort in making a web infrastructure before waiting passively for its food, or energy, to be captured by its "capital."

In the extreme case of zero operational costs, the cost of the energy system would be 100% determined by capital costs. Because machines don't last forever due to the second law of thermodynamics, and they need to be maintained, we'll never fully get to this world, but someday we might get very close. A spider doesn't pay any "operational" costs to make the wind blow across its web network, but it spends energy repairing web damage caused by high winds, falling twigs, and the bugs that it catches.

In a fully capitalized electric grid, the customers' payments for electricity no longer depend on how much they actually consume. Since there are no fuel costs, customers are essentially only paying for the capital costs of the electricity infrastructure—the concrete, steel, and other materials sitting on the ground somewhere. Just like the spider, these technologies wait to perform their function. They generate electricity when the wind blows on a turbine, water flows through a dam, or sunlight shines on a solar panel. Electricity can be stored at some times and discharged at other times as needed. But fundamentally, at some instant if people want more electricity than can be generated and released from storage, then everyone can't get what they want. Because markets provide price signals based on operating costs, and grid operating costs would be close to zero, then the normal price-forming mechanism is not present to indicate a shortage and tell generators to increase power and consumers to reduce demand. The electricity provider, or grid operator, will need to follow some existing priority for determining how to throttle back supply to certain customers.

We are not used to this concept for electricity consumption, as we normally get charged money for each kilowatt-hour we consume, no matter when or how fast we consume it. However, some of us are used to "throttling" when using other network-based services: mobile data plans, for example. Most of us pay a lump sum for mobile data per month, and we're allocated a certain amount of data usage, and perhaps a maximum rate of usage (i.e., bandwidth). Some families limit wireless bandwidth in the home among multiple users so that the teenage kid doesn't use all the bandwidth playing online video games. This throttling already occurs at a neighborhood or regional level in times of stress on the electric grid, and in emergency situations it takes the form of a rolling blackout where large regions are sequentially cut off from power for a few hours at a time.

Thus, we can look for future trends in regions with high percentages of renewable electricity. In these regions electricity providers might tend to offer rate plans and provide service in a similar manner as mobile phone plans. Those who pay less will be first to get throttled, if and when the need arises. It is easy to imagine, if electricity access is seen as a right, that this might upset advocates for low-income households, but concepts already exist to assist them make home energy payments.

As this book has discussed, with the economy acting as a superorganism, it does not distribute energy equally to all of its parts. Just like biological organisms, due to internal physical constraints as well as constraints on extracting energy from the environment since 1970, the global economy has become larger by associating the average unit of GDP (and piece of capital as shown for the U.S.) with a slower rate of energy consumption (Fig. 5.2). Also, the U.S. economy shows a prominent example of wages stagnating when per capita energy consumption stagnated (see Chap. 7 and Fig. 9.3). Because most people work and earn a wage, but very few people study energy and economics, it is understandable that almost no one makes this connection between money and energy. Workers are operating costs to companies. Thus, it is much easier for political discussions to center on the debate of "people versus profits." Unfortunately, this debate usually occurs out of the context of energy constraints. While this political debate is not new, we can expect it to continue.

The Battle of Pitchforks Versus Profits

In the context of reducing operating costs, the most prevalent and ongoing battle is that between labor and capital. Karl Marx focused on the conflict between capitalists attempting to pay workers as little as possible and workers attempting to collect their "proper share" of economic proceeds. The battle between capitalists and workers lives on today, and perhaps always will. From the biophysical perspective, economic processes depend on the consumption of environmental resources, and capitalists and workers largely fight over the proceeds derived from using the natural resources that none of us created.

Just how much profit do capital owners have to allocate to labor to keep the masses from taking up their proverbial pitchforks to storm the mansions of the top 1–10%? Politicians and labor advocates often talk about the need for good paying jobs, in the energy sector or otherwise. I've heard this jobs plea provoke a skeptical statement such as: "Capital has been substituting for labor for over 500 years. Get used to it!" This statement implies that since the dawn of capitalism, the purpose of the economy has not necessarily been to create good jobs.

Ever since sugar cane production on the Island of Madeira in the 1400s, capitalists have sought higher production and lower costs by substituting capital, or property, for *paid* labor. In the context of capitalism, slaves on Madeira and U.S. plantations were treated as "capital" assets to be owned, bought, and sold, not as laborers to be paid wages. Since the outlaw of slavery, capitalism has been inventing its way back toward a system in which laborers are again paid as little as slaves.

This reduction of labor costs is why Marx thought capitalism sows the seeds of its own destruction. He thought capitalists would alienate workers to such an extent that workers eventually would revolt and destroy capitalism itself.

For at least a couple of centuries, people have predicted the eventual demise of capitalism. However, as Chap. 8 noted, over the last 50 years, more countries have tended to move toward rather than away from capitalism and markets. Of course, this does not mean that a one-world capitalist government is inevitable. I do not know whether the people will gather with their pitchforks to overtake the top 1%, whether the top 1% will suppress the masses to the point of starvation, or whether energy constraints will lead to such high debt that capitalism collapses itself. Any one or a combination of these situations could arise. They are not precluded by the laws of nature. But how can we imagine any of these situations from the viewpoint of an outside observer?

Nicholas Georgescu-Roegen offers one perspective:

> The exosomatic evolution [use of technology and energy that are separate from the human body] brought down upon the human species two fundamental and irrevocable changes. The first is the irreducible social conflict which characterizes the human species. Indeed, there are other species which also live in society, but which are free from such conflict. The reason is that their "social classes" correspond to some clear-cut biological divisions. The periodic killing of a great part of the drones by the bees is a natural, biological action, not a civil war.[5]—Nicholas Georgescu-Roegen (1975)

Think about his bee example. Honey bees don't hibernate over winter during which they feed and keep warm by consuming the honey they've made over spring and summer. Drone bees don't work. They exist to fertilize the queen bee's eggs. In winter, after the drone bees have fertilized the eggs, they are kicked into the cold to die as the hive no longer needs them. If they stayed they'd consume honey but serve no further purpose for the hive.[7] As Georgescu-Roegen implies, we outside human observers don't consider this drone neglect as genocide or mass murder. We just call it a natural response to physical constraints and a programmed result of evolution.

What would an outside observer think about how we *Homo sapiens* treat members of our species who can no longer procreate or increase economic production? This question is not in the mindset of most people, including politicians and economists, who assume energy resources and the physical environment don't affect our social decisions. One politician, however, did contemplate this question in April 2020 during the coronavirus (COVID-19) pandemic. In a television interview where he discussed the tradeoffs of stay-at-home orders between saving lives and declining economic activity, Lieutenant Governor Dan Patrick of Texas, at the age of 70, suggested that it was better for someone of his age to risk their life to keep the present and future U.S. economy strong for his children and grandchildren:

---

[7]Debbie Hadley. "Sexual Suicide by Honeybees, August 7, 2019: https://www.thoughtco.com/sexual-suicide-by-honey-bees-1968100, "How Honey Bees Keep Warm in Winter," October 7, 2019: https://www.thoughtco.com/how-honey-bees-keep-warm-winter-1968101.

But 500 people out of 29 million and we're locked down, and we're crushing the average worker. We're crushing small business. We're crushing the markets. We're crushing this country. ... there are more important things than living. And that's saving this country for my children, and my grandchildren and saving this country for all of us. And I don't want to die, nobody wants to die, but man, we got to take some risks and get back in the game, and get this country back up and running.[8]—Dan Patrick, Lt. Governor of Texas (2020)

Would an alien, observing us while orbiting Earth, conclude that any neglect of elderly and mentally or physically challenged members of our species can be explained by our genetic programming and reactions consistent with natural laws? Would an alien see these acts as social failures or productive measures to grow the economic superorganism? Of course we don't yet even know if extraterrestrial aliens exist, and thus we certainly can't answer what they would think of human society.

We can, however, imagine the minimization of labor costs at the extreme: a fully automated artificially intelligent robot society with no humans (or at least no need for humans). I will not speculate on the timing or likelihood of this ultimate outcome, as there is no consensus from experts in the field of artificial intelligence. (Refer back to Chap. 8 for the rationale for superintelligent system overtaking humanity.) But many trends point in the direction of more automation: increasingly capable artificial intelligence algorithms; increasing computational speed and ubiquity of physical devices communicating via wireless internet (e.g., 5G communications and possibly beyond); increasing automation of extraction, manufacturing, and transportation machinery (e.g., autonomous vehicles); automatic stock trading algorithms; laws that give corporations the rights of people.

Recall Richard Adams' relationship between social power and control over the environment. If human-independent self-aware artificially intelligent systems obtain more control over the physical environment than humans, then humans could become subordinate to these robot overlords.

Let's come back to the present situation, before the artificial intelligence singularity, in discussing technological "progress" via automation, employment, and pay to workers.

In his book *Prosperity Without Growth*, Tim Jackson summarizes this problem as the "productivity trap."[8]

The Productivity Trap

Jackson describes the productivity trap as follows:

> ... So there is a huge premium on any strategy that might increase the availability and the quality of employment.

---

[8]Weinberg, Tessa, 'More important things than living,' Texas' Dan Patrick says in coronavirus interview, April 21, 2020, *Fort Worth Star Telegram*, https://www.star-telegram.com/news/politics-government/article242167741.html.

At the heart of the problem [of available quality employment] lies an issue we have already identified as a key dynamic in capitalism – the pursuit of increasing labour productivity; the desire continually to increase the output delivered by each hour of working time. Though it's often viewed as the engine of progress, the relentless pursuit of increased labour productivity also presents society with a profound dilemma.

As each hour of working time becomes more productive, fewer and fewer hours of labour are needed to deliver any given level of economic output. In fact, with labour productivity continually rising, aggregate demand must rise at the same rate if the total number of employed hours is to stay the same. As soon as demand falls – or even stagnates – then unemployment rises.

With labour productivity continually rising, there is only one escape from this 'productivity trap', namely to reap the rewards in terms of reduced hours worked per employee – or in other words to share the available work amongst the workforce.[9]—Tim Jackson (2017)

As capitalism spurs innovation that saves labor costs to produce economic goods and services, the total output of the economy must increase to keep everyone fully employed. From the perspective of the finite Earth/techno-realistic and Energy Descent narratives, growth will eventually cease as technological change cannot indefinitely overcome natural resource limits. The social dilemma is that labor-saving innovation could continue (at least for a while), and we'd need fewer working hours for a stagnant or declining economic output. Thus, if employment fulfills an individual's need to feel like he or she is a worthwhile member of society, then each person would need to work fewer hours to make room for all willing employees to provide enough individual contribution.

A targeted reduction in working hours is not a crazy idea. It is part of German policy that kept unemployment from significantly rising in the Great Recession of 2008–2009. The policy is called *Kurzarbeit*, or short-time working. In this policy "If an employer wants to cut working hours to save money, the state covers up to two-thirds of the wages that staff would otherwise lose."[10] This government involvement on behalf of its citizens is compatible with supporting the country's profit-seeking companies. In contrast, facing the same recessionary pressures, companies in the U.S. tend to fire employees in the downturn before hiring back some when growth resumes. Thus, there is much less employment stability in U.S., and unemployment fluctuates much more than in Germany. Higher taxes in Germany support the *Kurzarbeit* policy that stabilizes employment. Lower taxes in the U.S. still support some social safety net, such as limited unemployment insurance that was expanded during the 2008–2009 Great Recession, but employers are forced to fire and hire more employees along with the ups and downs of the business cycle.

Even when economic times are good, economists even 100 years ago, including John Maynard Keynes, thought higher labor productivity would allow people to work less and spend time with friends and family.[11] This has happened to a degree. The average annual work hours for U.S. workers fell from 2030 in 1951 to 1770 in

---

[9][8] [p. 145].

[10]Daniel Schäfer, "Keeping the lights on," *Financial Times*, November 10, 2009 at: https://www.ft.com/content/bd1e8620-ce2e-11de-a1ea-00144feabdc0. Also see Jackson [8, p. 146].

[11]Jackson [8, p. 145].

1982. Since the 1980s, however, there has been practically no change in work hours. In 2017, the average U.S. worker worked 1760 h [3].[12]

But as this book has pointed out, since the 1970s, too many Americans have continued to work a similar or increasing number of hours with stagnant or decreasing compensation. They might justifiably want to change the economic system for the reasons Karl Marx anticipated. To the techno-optimistic narrative, this is crazy. The capitalist economic superorganism is the goose that keeps laying golden eggs, and all other options lead to worse tyrannical outcomes. The techno-realistic narrative says this goose has not discovered alchemy. It just converts energy and food into regular eggs within the bounds of natural laws. This struggle between a stagnating economy and human dignity of work poses difficult but important questions for the future: should we kill the superorganism before it kills us, and replace it with something more amenable to people? Or, is the superorganism the best way for us to survive?

Kill the Superorganism?

Evolutionary pressures drive the economy as a superorganism to consume more resources, process more information, accumulate more capital, and convert energy to useful work more efficiently. From this viewpoint, the superorganism is not about people. It's not for or against people; it's just indifferent. For most of the history of industrialization, this indifference provided generally positive unintended consequences for human prosperity. This story, supported by much data, is often promoted by the combined fossil fuel and techno-optimistic narratives. In the last half century, however, the indifference to humans has shifted toward negative unintended consequences, mostly in developed economies. This story, also supported by much data, is often promoted by the combined renewable and techno-realistic narratives.

We are caught in a conundrum. On the one hand the individualistic and profit-seeking structure of the economic superorganism is what drives innovation, creativity, and the ability to both create and solve energy and environmental problems. We seem to want this feature. On the other hand, the biophysical nature of the superorganism means that physical limits and natural laws constrain its space of solutions such that we might not be able to solve all social and environmental problems simultaneously. We might not want these physical constraints, but we have to deal with them.

---

[12]University of Groningen and University of California, Davis, Average Annual Hours Worked by Persons Engaged for United States [AVHWPEUSA065NRUG], retrieved from FRED, Federal Reserve Bank of St. Louis; https://fred.stlouisfed.org/series/AVHWPEUSA065NRUG, accessed February 29, 2020.

Historically, within the lifetimes of the current older generations, citizens, governments, and corporations banded together through events such as the Great Depression and the two World Wars. This created societal traditions and shared identities. People want to hold onto these traditions. This is a very human concept. The superorganism doesn't think much of history and tradition, it just does what it does today, using market price signals to govern individual decisions.

Jackson ends his book with a statement on the tension between human tradition and economic innovation:

> The tension between [innovation and tradition] exists for a reason. Innovation confers advantages in the evolutionary adaptation – allowing us to respond flexibly to a changing environment. This ability is more critical now than ever. But tradition and conservation also serve our long-term interests. In evolutionary terms they allowed us to build security and establish a meaningful sense of posterity.
>
> The point is not to reject novelty and embrace tradition. Rather it is to seek a proper balance between these vital dimensions of what it means to be human. A balance that has been lost in our lives, in our institutions and in our economy.[13]—Tim Jackson (2017)

The reason why a balance is needed was expressed by Carsten Herrmann-Pillath. In essence, the economic superorganism uses markets to enable more degrees of freedom. Since more degrees of freedom increase the odds of extracting more primary energy via exploration and increasing energy efficiency, then *the policy conclusion* is that:

> ...enhancing the scope of markets always and necessarily enhances and leverages the dissipation of energy. ... Second, technological knowledge is a physical phenomenon, and hence we cannot approach technological progress independent from the question how far the production and the use of knowledge itself are part and parcel of energetic dissipation in the economy. Then, we cannot view technology as a substitute for energy, as this is typically assumed in environmental and resource economics. *Thus, if neither markets nor technology are means to resolve the environmental challenges of today, those positions in ecological economics are vindicated which argue that fundamental changes of the values and institutions of capitalism are necessary to establish a sustainable global economic system* [6].—Carsten Herrmann-Pillath (2015) (emphasis added)

In stating "changes of the values and institutions of capitalism are necessary," Herrmann-Pillath expresses why Jackson and others think we increasingly face the need to balance policies that seek to increase economic growth with those that seek to increase livelihoods for people. Between World War II and the 1970s, in rich economies there were many decisions that achieved both goals. Since then, we've had a harder time finding these win-win situations. Due to the biophysical reality of how the economy operates, it is not entirely our fault. While we are partly in control of many important factors, such as how to distribute energy and money among people, we are completely in control of none of them.

Recall from Chap. 4 that both economic and physical principles inform us that we should not expect perfect income inequality. Most citizens in rich countries were satisfied with a certain level of inequality leading up to the 1980s, but they have

---

[13][8] [p. 226].

become less satisfied in the last couple of decades. We also know that tax policy affects income distribution, but that tax policy alone does not overcome physical constraints to economic growth.

If we enter an Energy Descent scenario, then a more human-centric strategy could include policies, similar to the German *Kurzarbeit*, in which the same number of workers were used to produce a declining economic output. *This policy specifically worsens labor productivity*, the exact opposite approach of the human-indifferent economic superorganism, but it has a better chance of ensuring social cohesion. To reiterate, the tradeoff is lower incentive for innovation, but in Jackson's words, this is how we might achieve prosperity without growth—human prosperity without physical and economic growth.

Climate Change

Battle over Carbon: The Price Is Not Right

Why hasn't the economic superorganism addressed the issue of climate change? After all, orthodox economic theory and the neoliberal paradigm state that when you have a problem, you make a market to address it. In the case of climate change, why not simply make a market to price greenhouse gas (GHG) emissions?

A biological organism has no choice but to be influenced by the physical "markets" governing energy exchanges with its environment. Via natural selection, evolutionary forces favor phenotypes relative to their environment. However, as far as we know, *Homo sapiens* is the only living species that contemplates its own existence 10s, 100s, or even 1000s of years into the future.

If we can control the superorganism, then we can make it contemplate our future risk to ourselves from climate change. However, if the economic superorganism acts like a biological organism, it would neither plan ahead several generations nor pursue actions that reduce immediate energy consumption.

The maximum power principle claims organisms maximize average power flow over daily to annual cycles, not centuries. Thus, while creating any price-forming market into the economic superorganism might seem natural, creating one that encourages "too much" contemplation of the future might go against the maximum power principle, and thus be rejected. This is the crux of why it is hard to establish a market price on carbon for the purpose of reducing global GHG emissions.

Nonetheless, some regional economies, such as states in the U.S. (e.g., Regional Greenhouse Gas Initiative) and countries of the European Union (EU Emissions Trading System), have established markets for GHG emissions.[14]

---

[14]The Regional Greenhouse Gas Initiative (RGGI) is the first mandatory market-based program in the United States to reduce greenhouse gas emissions. RGGI is a cooperative effort among the states of Connecticut, Delaware, Maine, Maryland, Massachusetts, New Hampshire, New Jersey,

In 1990, under the Acid Rain Program, a cap and trade system was established to reduce sulfur dioxide emissions in the U.S. Thus, in principle the same concept could exist for GHGs, but there are fundamental challenges for establishing a GHG market via a cap and trade system. The cap would be a decreasing annual limit on GHG emissions, governed by certificates that allow entities to emit GHGs.

This regulatory constraint on economic activity, or degrees of freedom, would express findings from climate science. Businesses and countries could continue economic trade, but now with the requirement to buy and sell the declining number of certificates that give the right to emit GHG emissions. But here's the rub: climate change and energy consumption involve fundamental physical processes that affect the economy. However, the neoclassical economic paradigm detaches prices from the physical nature of the economy (refer back to Chap. 6) and theorizes that market prices are based on the preferences of consumers.

Given the propensity for people to prefer consumption now versus later, and given how few people understand climate science, economic theory, and the role of physical resources in the economy, how are consumers going to perceive any proper integration of GHG emissions prices into the cost of the goods and services they buy? They might revolt because they don't understand how GHG pricing affects all of the items they buy. The 2018 *gilets jaunes*, or "yellow vests," protests in France show the difficulty in raising energy prices, at least in the short term, that would occur with pricing GHGs.[15]

The second best option to establishing a global (or regional) GHG market is a regulated carbon price. In 2019, scores of renowned economists, including 27 Nobel Laureates, signed onto the policy of a steadily increasing carbon price in the form of a revenue-neutral carbon fee whose proceeds are given back to citizens as a dividend.[16] This dividend attempts to prevent protests such as those of the yellow vest movement. Organizations have coalesced around this idea, such as Citizens Climate Lobby and the Climate Leadership Council (CLC), the latter supported by some oil and gas companies and both supported by some conservative political leaders. Why? As stated by the CLC leadership, "A well-designed carbon fee checks every box of conservative policy orthodoxy."[17]

New York, Rhode Island, and Vermont to cap and reduce CO_2 emissions from the power sector. See https://www.rggi.org/. European Union Emissions Trading System, https://ec.europa.eu/clima/policies/ets_en.

[15] Angelique Chrisafis, "Who are the gilets jaunes and what do they want?", *The Guardian*, December 7, 2018 at: https://www.theguardian.com/world/2018/dec/03/who-are-the-gilets-jaunes-and-what-do-they-want.

[16] Climate Leadership Council, Economists' Statement on Carbon Dividends, *The Wall Street Journal*, January 17, 2019 https://clcouncil.org/economists-statement/.

[17] George P. Shultz and Ted Halstead, "The winning conservative climate solution," *Washington Post*, January 16, 2020 at: https://www.washingtonpost.com/opinions/the-winning-republican-climate-solution-carbon-pricing/2020/01/16/d6921dc0-387b-11ea-bf30-ad313e4ec754_story.html.

The "revenue-neutral" idea is key for conservative support. While the government would establish a fee on GHG emissions, the fees do not increase the size of government or its revenues because the fees are to be distributed back to citizens via "equal lump-sum rebates." Thus, consumers get the price signal they need to buy low-carbon products, and while the costs of energy and other products might go up, low-income citizens are more than compensated via the dividends. Because people with more money tend to buy more things, and the production of more things triggers more emissions, I have my doubts as to whether this idea would lead to lower emissions or not. The carbon fee does provide the incentive to produce individual low-carbon products, but because the dividend shuffles money from high consuming individuals (high incomes) to lower consuming individuals (lower incomes), there is no overall incentive to consume less in total. The assumption, or hope, is that total consumption can continue to increase while total GHG emissions decrease as the dividends indirectly induce low-carbon investment faster than increases in overall economic activity. It is not obvious that this sequence will hold true, and this remains a major open research question.

At the scale of the U.S., Chinese, and global economies, so far there is neither a GHG market nor a predetermined trajectory that establishes an economy-wide carbon price. There also has been no turnaround in the trend of increasing U.S. income inequality. Thus, this is how the think tank New Consensus and a group of U.S. congresspeople arrived at the concept of the Green New Deal as a third major option to reduce GHG emissions.

In Case of Crisis: Break Glass, Enact Plan

In 2019, U.S. House Resolution 109 called for "the Federal Government to create a Green New Deal" (GND) as a vision for how to solve contemporary social and environmental problems while hedging against the worst effects of climate change.[18] It recognized the following socio-economic issues in the United States:

1. life expectancy declining while basic needs ... are inaccessible to a significant portion of the United States population
2. a 4-decade trend of wage stagnation, deindustrialization, and antilabor policies ...
3. the greatest income inequality since the 1920s,

All three issues are discussed in this book.

Inspired by the public investment of the original New Deal of the 1930s, the House Resolution "...recognizes that a new national, social, industrial, and economic mobilization on a scale not seen since World War II and the New Deal

---

[18] Accessed March 1, 2020: https://www.congress.gov/116/bills/hres109/BILLS-116hres109ih.pdf.

era is a historic opportunity ..." to create the Green New Deal to address social problems related to income inequality via large-scale investment in low-carbon and energy efficient infrastructure.

In one sense we can view the Green New Deal as a vision to inspire a plan to overcome the inability of the economic superorganism to incorporate an overarching price signal to reduce greenhouse gas emissions. If we aren't making a market, then perhaps we can make a plan, or we can just start doing stuff to reduce GHG emissions.

At the global scale, there is one example where countries coalesced around a plan to limit certain air emissions. The Montreal Protocol, adopted in 1987, phased out the use of stratospheric ozone-depleting substances, such as chlorofluorocarbons (CFCs) that were used in refrigerators and air conditioners. Since ozone in the stratosphere blocks most of the ultraviolet-B radiation from the sun, a growing hole in the ozone layer exposed more people to more radiation and risk of skin cancer. The Montreal Protocol is the only worldwide treaty signed by all member countries of the United Nations. Some think we can make a carbon reduction plan just like the Montreal Protocol formed a successful plan that reduced CFCs. Unfortunately, the scale of GHGs affecting climate is much larger than the scale of CFCs affecting the ozone layer. Climate change presents a much harder social and political problem because it presents a much harder technological problem.

Is it easier to make a carbon market or a carbon plan? For climate policy, this is a major future energy-economic battle: prices versus plans.

The more you think about it, there is little difference between low-carbon planning and setting up an information-processing market that spits out carbon prices. In some sense, the reason there is no worldwide market to price carbon is because the necessary process to define the rules of that market is itself a very grand plan. The worldwide plan is so grand it has not yet happened, despite 25 conferences to date (the Conference of Parties, 1995–2019) of the United Nations Framework Convention on Climate Change. Markets are not predetermined commandments given by the gods. They are creations of man, and historically they have promoted an increasing number of degrees of freedom to grow the economy as a superorganism.

By forcing ourselves to reduce GHG emissions, we remove some degrees of freedom.

While the UN Paris Agreement was officially signed by almost all countries in 2016, it includes no binding reductions in GHG emissions. It is a plan with no teeth. Because a binding worldwide plan is thus far unachieved, an increasing number of states, cities, businesses, and investors are committing to renewable energy and GHG reduction goals.[19] Their thought is that if country-level governments can't commit to lower carbon emissions, then maybe lower-level governments and

---

[19]America's Pledge. "Across America, states, cities, businesses, universities, and citizens are taking action to fight climate change, grow the economy, and protect public health. America's Pledge brings together private and public sector leaders to ensure the United States remains a global leader in reducing emissions and delivers the country's ambitious climate goals of the Paris Agreement." https://www.americaspledgeonclimate.com/.

companies can do it themselves. Even in the U.S., where President Trump plans to officially pull out of the Paris Agreement, in 2019 "1 in 3 Americans [lived] in a city or state that has committed to, or achieved, 100% clean electricity" by some year before 2050 [4]. This goal also holds for 12 U.S. states and six major utilities that operate across 17 states.[20]

The Green New Deal recognizes the failure to establish a carbon price via a top-down market or regulatory approach. Therefore, its proponents seek to act from the bottom-up, at the level of communities that are taking action: "...a Green New Deal must be developed through transparent and inclusive consultation, collaboration, and partnership with frontline and vulnerable communities, labor unions, worker cooperatives, civil society groups, academia, and businesses ..."[21] Community level projects are more politically popular, but it is hard to select enough effective projects to make the large impact needed to reduce the vast majority of GHG emissions by the year 2050.

An article in *The New York Times Magazine* stated this conundrum:

> The question is whether any policy is both big enough to matter and popular enough to happen.[22]—David Leonhardt (2019)

While economy-wide carbon pricing is a big enough idea to matter, it is not yet popular enough to happen.

Attacking issues of income inequality and social justice via an array of community-led low-carbon energy investments in a Green New Deal can be popular, but each project might be too slow to develop and too small to matter, even when you add them all up. Make no mistake, the accumulated concentration of GHGs in the atmospheric is indeed the sum of contributions from billions of small individual activities, and it will also take changes to billions of individual actions to reduce GHG emissions rates to below 20% of 2000 levels by 2050. In response to the vision of the Green New Deal, some organizations have started efforts to see just how to "really" act on a Green Real Deal.[23]

Absent a carbon price as simple signal for all economic actors to watch, it is unclear how any well-intentioned set of ideas, whether the Green New Deal or a carbon fee and dividend, can proactively reduce global GHG emissions. Thus, the system-wide pricing of GHG emissions is perhaps the ultimate ongoing, and future, energy and economic battle.

---

[20] See Figure 3 and Table 1 of [4].

[21] Accessed March 1, 2020: https://www.congress.gov/116/bills/hres109/BILLS-116hres109ih. pdf.

[22] David Leonhardt, "The Problem With Putting a Price on the End of the World," *The New York Times Magazine*, April 9, 2019, https://www.nytimes.com/interactive/2019/04/09/magazine/ climate-change-politics-economics.html.

[23] For example, the Energy Futures Initiative, https://energyfuturesinitiative.org/ and The Green Real Deal report: https://energyfuturesinitiative.org/grd-report and https://energyfuturesinitiative. org/s/GRD-EFI-Part-2-2.pdf.

Pricing or not, a low-carbon transition requires both building a lot of new energy infrastructure while getting rid of a lot of old infrastructure. But if governments at the state and community levels, and eventually the country level, start investing in infrastructure from which we are all supposed to benefit, who should own it and directly receive some of the proceeds? As Chap. 6 notes, the U.S. World War II manufacturing effort, often used as an analogy for a low-carbon transition, involved the U.S. government effectively paying to double the scale of U.S. manufacturing. After the war, private companies owned this capital that they didn't pay for themselves. Should the same thing happen again if we embark on a low-carbon transition? Who should own the infrastructure?

The Battle for Capital: Public Versus Private Ownership

Some ideas associated with the original New Deal and the Green New Deal strike at the heart of debate over the form of the economic system: who owns capital.

If the private ownership of capitalism is failing to address climate change and wealth inequality, then, as implied by the Herrmann-Pillath quote in this chapter, the problem might be more than the lack of a price. Perhaps the problem is the system itself.

By calling out for public ownership of energy infrastructure, the Green New Deal directly mimics the original New Deal and seeks to have all citizens benefit from a collective ownership whether that be at community, state, or national levels. But there are important differences between today and the 1930s.

First, the New Deal occurred in a United States that was relatively empty of people, relatively full of nature, and low on employment.

The abundance of untapped rivers provides one energy-related example for "full of nature." In 1930, the U.S. had a total of 7000 MW of hydropower capacity. The Hoover Dam, an iconic feat of engineering, funded and owned by the U.S. government, added only 800 MW of capacity by 1938.[24] As indicated by Fig. 4.19, most big hydropower plants were constructed in the 1950s and 1960s. In 2018 the U.S. had about 80,000 MW of installed hydro capacity, with less than 3000 MW added since 2000. Further, due to tapping out the best rivers and competing demands for water, annual hydroelectric generation has been about the same since 1974.[25]

Aside from many rivers to dam, in the 1930s the oil age was just beginning as production ramped up in East Texas. If you're fighting a major depression, it's good to have as much cheap oil and undammed river reach as you can handle.

While there is practically no scope to build new large U.S. hydroelectric stations, there is certainly sufficient scope to put a lot of people to work building wind farms,

---

[24]Using data from Energy Information Administration form 860.

[25]See Table 7.2A Electricity Net Generation: Total (All Sectors), U.S. Energy Information Administration, Monthly Energy Review https://www.eia.gov/totalenergy/data/monthly/.

solar farms, and the transmission and other electric grid infrastructure to integrate them. That said, another difference between today (early 2020) and the time of the New Deal is that U.S. is more full of people, electricity demand has plateaued (Fig. 3.5), and employment in 2019 was as high as any time in history, even though many are underemployed and ill-paid.

Converting all cars and light trucks to electric vehicles could increase electricity consumption by about 25%, but a U.S. more full of people creates higher opposition to new transmission lines, and other infrastructure, that are necessary for a 100% renewable and/or zero-carbon grid. Companies have struggled to build long-distance transmission lines across multiple political boundaries. These efforts suffer from "...the majority-minority problem which affects many ideas in a democratic society. A majority may benefit from a project such as a transmission line which helps provide renewable energy, but small minority groups may lose from such a project and will thus fight harder than the majority."[26]

Whether publicly or privately owned, it is unclear how much infrastructure local land owners and governments will tolerate when it crosses, but does not directly benefit, their territory. In the U.S., the Federal Energy Regulatory Commission (FERC) approves the siting of interstate natural gas pipelines, and thus can overrule states that oppose them. However, FERC does not have this same authority for interstate electricity transmission lines. Thus, we can expect a future battle over whether to grant FERC authority to approve transmission lines.

At perhaps an extreme form of public ownership resides the idea of the U.S. government buying private U.S. fossil fuel companies to reduce the profit-seeking incentive to extract their reserves and thus emit GHGs. In 2017 one group estimated that 1.15 trillion dollars could buy out the 25 largest U.S. oil and gas companies, plus all publicly traded coal companies.[27] Their rationale? If the U.S. Federal Reserve can spend trillions of dollars via quantitative easing, or QE, to bail out banks after the 2008 financial crisis, then why not do something similar to bail out investors in fossil fuels. To them, this is "QE for the planet."

In addition, following the 2008 financial crisis, the U.S. government did actually take ownership, partially or fully, of companies such as General Motors, insurance company A.I.G., and mortgage lenders Freddie Mac and Fannie Mae. So these precedents, and others, exist for governments of capitalist economies, even that of the U.S., to partially or fully nationalize private companies for some period of time.

At smaller scales, the idea of public and collective ownership has many appeals. For example, whether via cooperatives, where the owners are the customers, or municipal utilities, owned by local governments that are accountable to its citizens

[26]Ethan Pratt, "Clean Line Energy and America's Infrastructure Problem," July 24, 2019, https://www.energycentral.com/c/iu/clean-line-energy-and-america%E2%80%99s-infrastructure-problem.

[27]Gar Alperovitz, Joe Guinan and Thomas M. Hanna, "The Policy Weapon Climate Activists Need," *The Nation*, April 26, 2017 https://www.thenation.com/article/archive/the-policy-weapon-climate-activists-need/.

as customers, most economic benefits from energy system ownership flow to the people that use it.

However, the opposite also holds. Costly choices also affect the citizens of municipalities and owners of cooperatives. As shown by the example investments (in renewables) by the municipal utility of Georgetown, Texas (Chap. 9) and investments (in nuclear power) by regulated utilities of Georgia and South Carolina (Chap. 3), big bets can cost much more than planned. As shown by the investment and subsequent bankruptcy from a private investor buyout of a Texas power company (Chap. 3), private companies also can make big bets that go awry.

Neither public nor private investors always make good or bad decisions. Past performance is no guarantee of future results, but a common future trend is the size of investment. For energy, the packaging of smaller individual investments is becoming more favorable, and big bets are getting rarer.

The Struggle for Size: No More Megaprojects

In the case of a government buying out fossil fuel assets and companies, these would be extremely large investments in *removing* capital assets (e.g., fossil reserves) from the economy. In the opposite sense, developed economies will likely continue to face headwinds against making large single energy investments that *add* new capital, whether private or public. For energy, the era of the megaproject seems over.[28]

Certain types of energy investments have at least one characteristic in common: they can be pursued in relatively small increments less than 10s of millions of dollars, instead of a few billion dollars at a time. This holds for an individual hydraulically fractured and horizontally drilled oil or natural gas well. This holds for a solar photovoltaic panel or wind turbine. This holds for storage systems from batteries that store electricity to tanks that store propane. This holds for smart grid devices and algorithms that turn electrical devices off at times of peak electricity demand and turn them on at times of low demand. This even holds for natural gas power plants that can be installed in increments of 10s of MW.

In economies with no more growth in energy consumption, it is too risky to plan for one large and expensive energy generation or extraction project. Even if investing the same amount of money in aggregate, you can minimize financial risk by investing in multiple small investments distributed among several projects. Thus, there is less chance that any given investment puts an investor into bankruptcy or insolvency.

In the last several years, the U.S. has seen dozens of coal power plant retirements, and these occur in relatively large chunks of 100s or 1000s of MW at a time (the entire U.S. has about 1,100,000 MW of power plant capacity). Nuclear power plants

---

[28] Jeffrey Tomich, "Is the era of the utility megaproject over?", *EE News*, August 3, 2017, https://www.eenews.net/energywire/2017/08/03/stories/1060058301.

will be up for retirement in the next few decades, and a few have already been decommissioned. These also come in chunks of 1000s of MW. In all likelihood these will not be replaced with new coal or current-generation nuclear power plants. A series of the smaller investments, of the types mentioned in this section, will fill in the gaps.

In Case of No Price or Plan to Stop Bleeding, Apply Pressure

Since there is not yet a large enough market to set a carbon *price*, and there is not yet a grand enough binding carbon-reducing *plan*, another p-word describes a third approach: *pressure*.

Consumers are increasingly using whatever social influence they have to pressure private companies and investors to disclose their exposure to climate change and make choices consistent with lowering GHG emissions. In response, companies are increasingly investing in low-carbon energy supplies to power their operations.

These types of activities fall into the "environment" aspect of the so-called environmental, social, and governance, or ESG, investing. Even BlackRock, in 2019 the world's largest investment manager, jumped on the ESG train. Depending on your viewpoint, investment managers are either late to the station or they're added very much needed inertia to low-carbon efforts from the investing community.

In a letter to shareholders, BlackRock CEO Laurence Fink stated his firm was increasingly including ESG criteria into their investment products.[29] Some environmental advocates were not impressed, as they stated that "BlackRock continues to be the largest global investor in coal, oil, and natural gas extraction …"[30] When you are the world's largest investor, you have a good chance to also be the world's largest investor in fossil fuels. Seemingly in response to pressure from "climate activists, investors, legislators, and other thought leaders," BlackRock's 2020 client letter announced the beginning of a major divestment from "thermal coal."

Environmental groups like the Sierra Club are still skeptical and want investors like BlackRock to develop more definitive low-carbon thresholds for investment, to act on faster time lines, and to vote for pro-climate shareholder resolutions (rather than merely abstain).[31]

---

[29]Larry Fink's Chairman's Letter to Shareholders from BlackRock's 2018 Annual Report. Accessed March 7, 2020 at: https://www.blackrock.com/corporate/investor-relations/larry-fink-chairmans-letter.

[30]Sierra Club press release, "BlackRock CEO Larry Fink Faces Protest at Annual Shareholder Meeting for Lack of Action on Climate Change," Thursday, May 23, 2019, accessed March 2, 2020 at: https://www.commondreams.org/newswire/2019/05/23/blackrock-ceo-larry-fink-faces-protest-annual-shareholder-meeting-lack-action.

[31]Sierra Club, "BlackRock Responds to Demands for Stronger Climate Action with Bold New Commitments," January 14, 2020, accessed March 7, 2020 at https://www.sierraclub.org/

How fast can even the largest investment firms be "pressured" to accelerate a low-carbon energy transition? This is a great question and a fundamental energy-economic trend to watch going forward. Whether investment management firms are tentative or realistic, BlackRock doesn't overplay its "constructive role" when it states government action is still "required:"

> A successful low-carbon transition will require a coordinated, international response from governments aligned with the goals of the Paris Agreement, including the adoption of carbon pricing globally, which we continue to endorse. Companies and investors have a meaningful role to play in accelerating the low-carbon transition. BlackRock does not see itself as a passive observer in the low-carbon transition. We believe we have a significant responsibility – as a provider of index funds, as a fiduciary, and as a member of society – to play a constructive role in the transition.[32]—BlackRock (2020)

Summary

So here we have it. Practically all countries of the world signed the Paris Agreement in 2016 to limit GHG emissions enough to have a good chance to limit global warming to $1.5\,°C$, but they won't make any binding commitments. The biggest investors, biggest energy companies, and most famous economists claim we should set up some sort of carbon price, but it hasn't happened even though these are among the firms and individuals that many people believe have legislators under their thumbs. Somehow collectively we don't create the low-carbon system that practically all individual companies and countries claim to desire.

There seems to be a paradox. The low-carbon energy solutions appear at hand, yet the economy does not take the steps to actually lower greenhouse emissions. The paradox exists only if we force a false choice between the endpoints of the each of the energy and economic narratives of this book. The paradox vanishes if we think of the global economy as a superorganism.

The most certain way to reduce greenhouse gas emissions is to reduce consumption of physical resources, but we seem unwilling (so far) to self-impose this constraint. One of the main reasons is because the techno-optimistic and infinite substitutability economic narrative, which dominates economic thinking, and thus, also policy design, says we don't have to. It does not contemplate physical constraints on long-term growth.

---

press-releases/2020/01/blackrock-responds-demands-for-stronger-climate-action-bold-new-commitments. BlackRock 2020 Client Letter, "Sustainability as BlackRock's New Standard for Investing," accessed March 2, 2020 at: https://www.blackrock.com/corporate/investor-relations/blackrock-client-letter.

[32]BlackRock 2020 Client Letter, "Sustainability as BlackRock's New Standard for Investing", accessed March 2, 2020 at: https://www.blackrock.com/corporate/investor-relations/blackrock-client-letter.

In contrast, the techno-realistic narrative assumes the finite Earth can and will eventually constrain increases in consumption and economic growth. In the short term we try to grow the economy by substitution and increasing our options, but in the long-run physical constraints restrict both growth and our options for growth.

From an evolutionary perspective, each entity (person, company, country) within the economic superorganism competes against the others within a physical world, and in doing so seeks to remove constraints on itself. This is how the superorganism considers the techno-optimism narrative, and why it ignores the energy narratives of fossil fuels versus renewable energy. It minimizes constraints by using some combination of all types of energy technologies. At the same time the superorganism realizes its physical nature, and will not be surprised at an end to growth. It expects it. The difficult questions relate to whether or not we should plan for the end of growth and if so, what such a plan even looks like. Plans can look like additional constraints on options, but one can also enact plans to remove as many constraints as possible.

There are tradeoffs between short-term versus long-term thinking, between markets versus plans, between applying versus removing economic constraints, and between worldviews that consider the economy as a physical system versus those that don't. Too often people use the energy and economic narratives to speak past each other rather than engage in thoughtful conversations on these tradeoffs.

I hope this book better enables these conversations.

References

1. Daly, H.E.: Uneconomic growth in theory and fact. http://www.feasta.org/documents/feastareview/daly.htm (1999). Online; accessed 1-February-2020
2. Dennett, D.C.: Consciousness Explained. Back Bay Books (1991)
3. Feenstra, R.C., Inklaar, R., Timmer, M.P.: The next generation of the Penn world table. American Economic Review **105**(10), 3150–3182 (2015). Available for download at www.ggdc.net/pwt
4. Friedman, M.: Progress toward 100% clean energy in cities and states across the U.S. (2019). https://innovation.luskin.ucla.edu/wp-content/uploads/2019/11/100-Clean-Energy-Progress-Report-UCLA-2.pdf
5. Georgescu-Roegen, N.: Energy and economic myths. Southern Economic Journal **41**(3), 347–381 (1975)
6. Herrmann-Pillath, C.: Energy, growth, and evolution: Towards a naturalistic ontology of economics. Ecological Economics **119**, 432–442 (2015). https://doi.org/10.1016/j.ecolecon.2014.11.014. http://www.sciencedirect.com/science/article/pii/S0921800914003589
7. Holmgren, D.: Future Scenarios: How Communities Can Adapt to Peak Oil and Climate Change. Chelsea Green Publishing, White River Junction, VT (2009)
8. Jackson, T.: Prosperity Without Growth: Foundations for the Economy of Tomorrow, second edition edn. Routledge, Milton, UK and New York, NY, USA (2017)
9. Mark, J.T., Marion, B.B., Hoffman, D.D.: Natural selection and veridical perceptions. Journal of Theoretical Biology **266**(4), 504–515 (2010). https://doi.org/10.1016/j.jtbi.2010.07.020. http://www.sciencedirect.com/science/article/pii/S0022519310003772
10. Solow, R.M.: Perspectives on growth theory. The Journal of Economic Perspectives **8**(1), 45–54 (1994). http://www.jstor.org/stable/2138150

Appendix A

How Do We Count Energy and Power?

The reason we can count energy is that we have postulated it as an idea. This idea has now translated to physical laws that enable us to quantify energy in different forms. Thus, there are different mathematical formulas to calculate quantities of these different forms of energy. Our goal is often to perform useful work as a final *energy service*, and thus the better we understand how to quantify different forms of energy, the better we can design machines that transform these forms of energy into useful work instead of dissipated heat.

Example forms of energy are chemical, gravitational, kinetic, magnetic, and electrical (a simple and informative reference is Table 1.1. in [16]). Chemical energy is associated with breaking apart the chemical bonds in materials (e.g., via combustion, or burning, hydrocarbons such as oil, natural gas, and coal as well as burning carbohydrates such as wood) and separating dissimilar materials (e.g., separating iron and oxygen when starting with iron ore). Electrical energy describes potential energy that is created by separating and storing charged particles (e.g., "opposite" charges attract) such as in a capacitor and an electrochemical battery. Electrochemical batteries, such as the lead-acid and lithium-ion batteries in our cars and mobile devices, combine both chemical and electrical concepts. Gravitational energy (on Earth) is related to moving objects, including fluids like water, closer or further from the center of the Earth. In addition, the moon creates tides on Earth as the moon's gravity attracts the ocean's water. Electromagnetic energy describes the work required to pull two magnets apart once they are stuck together as well as the sunlight that creates life on Earth via plant photosynthesis and drives the photoelectric effect in a photovoltaic solar cell. Kinetic energy describes the energy within flowing fluids, such as the wind and water in a river, and an object in motion such as a car moving along the road.

© Springer Nature Switzerland AG 2021
C. W. King, *The Economic Superorganism*,
https://doi.org/10.1007/978-3-030-50295-9

Common units of power are the *watt* (W) and the *horsepower* (hp).[1] One horsepower was originally used to describe what it sounds like. It approximates the physical power that a draft horse could provide, say when pulling a plow or a cart. In the United States we still use this to quantify engines in cars and planes. A Honda Civic engine is rated at about 200 hp and Tesla Model S at about 520 hp. Each engine on an Airbus 380 is rated at 1800 hp. A healthy man can only sustain about 100 W, or 0.13 hp, for more than a few hours [16].

Units of energy are obtained by multiplying a unit of power by a unit of time. Perhaps the most common unit is kilowatt-hour (kWh), or 1000 W times 1 h of time. When we pay for electricity used in our homes, we pay some number of cents for each kWh. In the United States, depending on how you operate your house and where you live, a household might consume 6000–14,000 kWh per year. Other common energy units are the British Thermal Unit (Btu) and Joule (J).[2]

Because we have both formulas and units to calculate energy stored in different forms, we can quantify how much energy we "consume" of each form. If you are paying attention, you would note that because of the first law of thermodynamics, the word "consume" is actually inappropriate for discussing energy. If we track the change in each form of energy involved in a process, we end up with the same total quantity of energy both before and after. Knowing this, scientists derived the concept of *exergy*. Exergy is defined as a quantity that incorporates the concepts of both the first and second laws of thermodynamics such that it is not mathematically conserved like energy. "Exergy is the correct thermodynamic term for 'available energy' or 'useful energy,' or energy capable of performing mechanical, chemical or thermal work" [2].[3] Thus, it is more accurate to say we consume exergy, but for the purposes of this non-technical book, I use the much more common term energy.

Energy Data and Accounting

Three common data sets that estimate how much primary energy is consumed by human activity are those of the U.S. Department of Energy's Energy Information Administration (EIA), the International Energy Agency (IEA) of the Organization for Economic Co-operation and Development (OECD), and the BP Statistical

---

[1] 1 horsepower ~750 W.

[2] Technically one British Thermal Unit is the heat that will raise the temperature of one pound of water by one degree Fahrenheit. The amount of heat from burning a typical kitchen match is about 1 Btu and 1000 J. 1 watt (W) = 1 Joule/second = 1 J/s. 1 Btu = 1055 J.

[3] Exergy is defined as the potential of a resource or system to perform physical work as compared to a reference condition of its surroundings. The reference condition has a defined temperature, pressure, and other characteristics that enable one to compare the energy content of a resource to that of the reference condition with the assumption that any system at reference conditions by definition no longer has potential energy to perform useful work and is at equilibrium with its surroundings.

Review of World Energy (BP). One interesting point about comparing the data in these data sets is that *the methods for counting primary energy are not the same*. That is to say, if you were discussing with your neighbor (like we all do!) how much energy was consumed in the world last year, and one of you used data from the EIA and the other used data from IEA, each of you would be justified in saying that you were correct even though the numbers are different.

Unfortunately, there is no single number that is fundamentally more correct. The narratives of energy strike again! This time purely from accounting.

The explanation begins by recalling energy is an abstract mathematical accounting principle. At some point in counting primary energy, one must assume a way to estimate the "primary energy equivalent" of a quantity of electricity generated from non-combustible resources. This point is key to having an informed discussion comparing fossil and renewable energy narratives. If more energy is better, and not everyone agrees that it is, then how you count energy becomes quite important.

There are normally one of the three methods used to estimate "primary electricity" as primary energy: the partial substitution method, the direct equivalent method, and the physical energy content method which is a hybrid of the first two [6, 8]. The partial substitution method gives non-thermal "... electricity production an energy value equal to the hypothetical amount of the fuel required to generate an identical amount of electricity in a thermal power station using combustible fuels" [8].[4] The EIA and BP use this partial substitution method. Essentially, this method assumes that the amount of primary energy associated with solar power is the same amount of heat energy required as input into an average thermal power plant.

The direct equivalent method assumes that all technologies that generate electricity without combustion (i.e., nuclear fission, solar photovoltaics, concentrating solar power, wind, geothermal, hydropower) have the "engineering equivalent" translation of one kilowatt-hour (kWh) to 3.6 megajoules (MJ) of energy.[5] Thus, even though the generation of heat is a precondition for electricity generation from nuclear, concentrating solar, and geothermal power plants, this quantity of heat is ignored in translating electricity to a unit of heat (e.g., joules or British thermal units).

The physical energy content method is a hybrid method that the IEA uses to overcome the physical inconsistency of the direct equivalent method regarding nuclear, concentrating solar (sunlight to heat), and geothermal-based electricity. Thus, the IEA quantifies the amount of heat that was needed before converting some of that heat into electricity. The IEA assumes a thermal conversion efficiency to translate 1 kWh of generation into a quantity of joules. For example, nuclear power is assumed to have a thermal efficiency of 33%, so 1 kWh of generation is equal to $(1 \text{ kWh})(3.6 \text{ MJ/kWh})(1/33\%) = 10.9 \text{ MJ}$. For non-thermal technologies (e.g., wind, photovoltaics, and hydropower) the IEA's physical content then uses the

---

[4][p. 137].

[5]1 megajoule = 1 million joules. 1 kWh = 1 kilowatt-hour = 1000 watt-hours = (1000 watts)(1 h) = (1000 Joules/s)(1 h)(3600 s/h) = 3,600,000 J = 3.6 MJ.

direct equivalent method to quantify primary energy content in that "... the normal physical energy value of the primary energy form is used for the production figure. For primary electricity, this is simply the gross generation figure for the source" [8].[6] Thus, 1 kWh = 3.6 MJ.

To see how this primary equivalence issue plays out, let's compare data sources. A specific example is to ask: How many megajoules (MJ) of primary energy should we associate with one kilowatt-hour (kWh) generated from a hydropower dam? Consider world hydropower generation in 2012. EIA, BP, and IEA report 3646, 3685, and 3760 terawatt-hours (TWh).[7] These numbers are all very similar (within 3% difference) as expressed in the native units (watt-hours) of measured electricity from the hydropower plants, and we can consider them equivalent for our purposes here. Each database also lists the hydropower generation in thermal primary-equivalent units, and these translate to 36.6, 35.0, and 13.3 exajoules (EJ)[8] for the EIA, BP, and IEA, respectively.[9] The last of these three quantities is almost one third the value of the other two.

Thus, the same (for practical purposes) amount of listed electricity in TWh from each data set converts to approximately 2.6 and 2.7 times more primary energy in the BP and EIA databases, respectively, as it does in that of the IEA. There are also other differences in estimating primary energy from nuclear and geothermal electricity, but to date, these differences have not caused too much rancor within energy studies, largely because primary energy has been dominated by burning fossil fuels and biomass-derived fuels (>85% of primary energy, however you look at it) since industrialization.

If the world moves more toward non-combustible renewable electricity sources, the difference in the statistical methods could become more of a concern. Consider the world moving toward 100% of its electricity production coming from non-thermal technologies such as wind, solar, and hydropower. Ignore for now the practicality and other characteristics that this world might have. (That discussion occurs throughout the book!) This future world has no combustible fuel-based electricity generation. What is the amount of primary energy (in EJ) associated with electricity? The IEA, using the physical energy content method, would maintain consistency in staying with its accounting method. BP, which converts non-thermal electricity from units of TWh to primary energy in units of Mtoe[10] "... on the basis of thermal equivalence assuming 38%" conversion efficiency of heat to electricity would thus estimate primary energy from electricity that is 2.6 times higher (1/0.38 = 2.6) than the IEA. The EIA has traditionally estimated the heat content for non-

---

[6][p. 137].

[7]1 terawatt-hour is 1 trillion watt-hours or 1×10^{12} W-h.

[8]1 exajoule is 1×10^{18} joules.

[9]EIA: 3646 TWh is converted using 9516 Btu/kWh and 1055 J/Btu to equate to 36.6 EJ. BP: 3685 TWh is equated to 834 million tonnes of oil equivalent (Mtoe), and using 41.9 gigajoules (GJ) per Mtoe one obtains 35.0 EJ (NOTE: 1 GJ is equal to 1×10^{9}). IEA: 3756 TWh is also listed as 315.8 Mtoe which translates to 13.3 EJ.

[10]1 Mtoe = 1 million tonnes of oil equivalent. One toe is equal to approximately 41.9 gigajoules (GJ).

combustible renewable electricity by adjusting their conversion factor each year. The adjustment each year is to account for the changing efficiency of thermal power plants in converting combustible fuels to electricity: "The fossil-fuels heat rate is used as the thermal conversion factor for electricity net generation from noncombustible renewable energy (hydro, geothermal, solar thermal, photovoltaic, and wind) to approximate the quantity of fossil fuels replaced by these sources" [5, Table A6]. For example, the efficiency of thermal power plants is reported as 24% for 1950 such that non-combustible primary renewable electricity was assumed to be equivalent to burning 14.8 MJ of fuel for every kWh generated, or 1/0.24 = 4.1 times the value as IEA would report. In 2014, the EIA reports this *heat rate* as 10.0 MJ/kWh such that thermal power plants are assumed to be 36% efficient, and thus each kWh generated from non-combustible wind, solar photovoltaic, and hydropower are assumed to be equal to 10 MJ of primary energy, or 2.8 times higher than the IEA would estimate.[11]

Primary Energy Stocks versus Primary Energy Flows

Given that there are different forms of energy, the different accounting methods for primary energy production amount to trying to fit a round peg into a square hole, in two ways. The first set of round pegs are the different forms of "primary energy." Electricity from hydropower dams, wind turbines, and photovoltaic panels is driven by primary power *flows* of water, air, and sunlight, respectively. The water, wind, and sun each have their own characteristics and distribution across the planet. These flows occur continuously, and to take advantage of them, you need a device that extracts them as you are exposed them. Unlike an advertisement for General Electric wind turbines, in which a boy goes to great length to capture wind in a jar that his grandfather in turn uses to blow out the candles on his birthday cake, we can't store wind in a jar.[12]

Distinctly different from primary flows are *stocks* of primary energy resources such as fossil fuels, biomass, and radioactive materials (e.g., uranium). Stocks are accumulated flows. For example, fossil fuels are buried accumulations of biomass that absorbed sunlight millions of years ago. While we don't have stored wind, it seems Mother Nature did store sunlight for us.

So, flows are round, and stocks are square. Yet we want to add them together in our statistical databases. In order to add a quantity of flow to a quantity of stock you

---

[11] Source: EIA website for Monthly Energy Review, Table A6, accessed March 13, 2017. The definition of heat rate is the quantity of fuel energy input divided by the (usually net) electricity generation output from a power plant expressed in units such as Btu/kWh and MJ/kWh. For 2014, the Table A6 of the EIA Monthly Energy Review lists a heat rate for U.S. thermal electricity, and thus non-combustible electricity, as 9510 Btu/kWh which is equal to 10.0 MJ/kWh.

[12] GE "Jar" advertisement featuring Catch The Wind (Donovan): https://www.youtube.com/watch?v=sj2YR922xBk. Viewed May 28, 2018.

need to assume a way to convert from one form to another. Thus, the second set of round pegs is that composed of different ways of converting flows (that we don't burn) to stocks (that we do burn or use to create heat).

One implication of this primary energy accounting conundrum is that if you want to make it appear that your country, or the world overall, is continuing to produce and consume a higher quantity of energy, then you can use the method that converts non-thermal electricity into a larger amount of primary energy. If you want to make it look like the world is consuming less energy as more wind, solar, and hydropower are generated in substitution for fossil-fueled electricity, then use the "physical energy content" method of the IEA that converts non-thermal electricity into its minimum engineering equivalent amount of primary energy. If we conceptually replace 9 MJ of coal with 3.6 MJ of wind power, both of which translate to 1 kWh of electricity at my house, then one can say we consumed less primary energy while delivering the same amount of secondary energy. Thus, some researchers use the physical energy content method to justify statements that the world can convert to 100% renewable energy while consuming less total primary energy per year [3, 4, 9, 11, 12]. On the other hand, if we socially and politically see more power consumption as "good," then there might be pressure to conform to an accounting method that maximizes the calculated number.

Which primary energy accounting method is correct? This is a question with no consensus answer. I have a very hard time associating a characteristic of one form of energy used in a certain technology (e.g., chemical energy of natural gas released as heat when burned in power plants) to quantify another form of energy that drives a completely different technology (e.g., gravitational potential energy released from water falling across turbines in hydropower dam). If we did reach a world only powered by hydropower, wind turbines, and solar panels, why would I estimate the energy production of the world based on the concept of burning fuels? Thus, the physical energy content method is most conceptually consistent to me. However, there are good arguments for the partial substitution method in order to perform consistent country-scale accounting [7].

The unfortunate conclusion is that you can probably use whichever method you like to answer a particular question, just as long as you consistently interpret your accounting. This is "unfortunate" because an outside person has to understand the accounting framework in order to interpret the results, and this will become more important if renewable electricity becomes more dominant in our energy mix. For example, many researchers want to relate changes in total primary energy consumption to changes in gross domestic product (GDP), and the answer might change depending upon what energy data you use! See Chaps. 5 and 6 for discussion of how energy consumption and production relates to the growth and structure of the economy. It turns out that explaining GDP using the useful work consumption has more explanatory power than explaining GDP using primary energy consumption, and there are both physical and economic reasons why we should expect this to be the case [1, 2, 13]. This finding means that many energy-economic analyses are talking past each other by not analyzing the most appropriate type of data (see Chap. 6).

With Friends Like This, Who Needs Enemies? Renewable Energy Narrative (Extended Explanation from Chap. 3)

To give you a feel for how researchers try to translate assumptions in models into insightful outcomes, I highlight one of the modeling points of contention within the Jacobson et al. [11] study. This centers on the assumptions for how hydropower generation would operate within the presented 100% wind, water, and sunlight (WWS) scenario. It is necessary to go through the numbers in the paper to explain the disagreement. It is important to note that Clack et al. describe the hydropower modeling as an *error* and Jacobson describes it as an *assumption*. This leaves the lay reader to become confused on how bad an assumption can be before it becomes an error.

Table 2 of Jacobson et al. [11] shows the amount of hydropower electricity generated over 6 years (2050–2055) as 2413 TWh. The hydropower footnote of Table 2 states "The capacity factor for hydropower from the simulation is 52.5%, which also equals that from ref. 22." where reference 22 is another paper by Jacobson [10]. Knowing total energy generation and capacity factor enables the calculation of installed hydro capacity as 87.5 GW, and this is the same stated "installed capacity" for 2050 in Table S2 of [11]. Clack et al. [4] call this a modeling error because Figure 4B of Jacobson's paper [11] indicates the need (on the simulated days indicated) for approximately 1300 GW of instantaneous power output from hydropower. To produce 1300 GW at any instant, the installed capacity must exist and be accounted for. This is a very large difference between 1300 GW and 87.5 GW of hydropower capacity as the 1300 GW of hydropower capacity is "... approximately 9 times the theoretical maximum instantaneous output of all [presently U.S.] installed conventional hydropower and pumped storage combined" [4]. Quantities adding up to 1300 GW of hydropower capacity do not appear in Jacobson et al. [11], even when including 57.7 GW of pumped hydropower storage capacity listed in Table S1. Thus, Clack et al. [4] call this a modeling error because "The hydroelectric production profiles depicted throughout the dispatch figures reported in both the paper [11] and its supplemental information routinely show hydroelectric output far exceeding the maximum installed capacity ...". In short, the model results contradict the stated quantities and thus the capabilities used within the model.

Jacobson et al. [11] do not agree that the 1300 GW of hydropower capacity represents a modeling error. They state "... The value of 1300 GW is correct, because turbines were assumed added to existing reservoirs to increase their peak instantaneous discharge rate without increasing their annual energy consumption ..." as stated in footnote 4 of the original paper [12].[13] However, footnote 4 of Table

---

[13]"Clack et al. (1) [4] then claim incorrectly that the 1300 GW drawn in figure 4B of Jacobson et al. (2) [11] is wrong because it exceeds 87.48 GW, not recognizing that 1300 GW is instantaneous and 87.48 GW, a maximum possible annual average [table S2, footnote 4 in Jacobson et al. (2) [11] and the available LOADMATCH code]. The value of 1300 GW is correct, because turbines were

S2 of [11] only references that hydropower is limited by "...its annual power supply ..." and is used for "...peaking ..."[14] Jacobson's papers make no mention of a quantity relating to 1300 GW of installed capacity.[15] In addition, since Jacobson et al.'s rebuttal agrees that 1300 GW is a correct number for installed hydropower, then they indeed have made a couple of errors. First, the capacity factor of hydropower capacity is really 3.5%, and not the stated 52.5%.[16] To state an annual generation of 2413 TWh/year you can use 87.5 GW at 52.5% capacity factor or 1300 GW at 3.5% capacity factor, but not 1300 GW at 52.5% capacity factor as done in the paper. Second, since Jacobson et al. do not ever explicitly state a quantity of hydro capacity near 1300 GW, they seemingly neglect the additional cost of approximately 3.4 trillion dollars.[17] Jacobson et al. [12] reply that the cost to modify the existing hydropower facilities is only ~3% of the entire wind, water, and solar power system that they modeled, and thus is too small to negatively impact their conclusions.[18]

In a Clack et al. rebuttal to Jacobson et al.'s rebuttal, they state that Jacobson et al. are "Purposefully refusing to acknowledge clear mistakes. This is most clearly seen in this exchange from the discussion of installed capacity of hydropower in the Jacobson et al. models" [14]. In addition they conclude that Jacobson et al.'s response "...confirms that the [hydropower capacity modeling] error is actually more severe than this." This last statement sums up the argument on the hydropower

---

assumed added to existing reservoirs to increase their peak instantaneous discharge rate without increasing their annual energy consumption, a solution not previously considered. Increasing peak instantaneous discharge rate was not a "modeling mistake" but an assumption consistent with Jacobson et al.'s (2) table S2, footnote 4, and LOADMATCH, and written to Clack on February 29, 2016" [12].

[14]Footnote 4 of Table S2 of [11]: "Hydropower use varies during the year but is limited by its annual power supply. When hydropower storage increases beyond a limit due to non-use, hydropower is then used for peaking before other storage is used."

[15]Jacobson's 50 state roadmap for 100 % WWS states 91.7 GW of total hydropower capacity being needed in Table 2 [10].

[16]Capacity factor of 1300 GW of hydropower capacity = (2413 TWh × 1000 GW/TW)/(1300 GW × 8760 hour/year × 6 years) = 3.5%.

[17]Clack et al. [4] calculate this $3.4 trillion using Jacobson et al.'s [11] stated cost to add hydro capacity at $2.82 million/MW: (1300 GW total future capacity—87.4 GW of existing capacity) × ($2820 million/GW) = $ 3,400,000 million = $3.4 trillion.

[18]From [12]: "Jacobson et al. (2) only neglect the cost of additional turbines, generators, and transformers needed to increase the maximum discharge rate. Such estimated cost for a 1000-MW plant (23) plus wider penstocks is ~385 (325–450)/kW, or ~14% of hydropower capital cost. When multiplied by the additional turbines and hydropower's fraction of total energy, the additional infrastructure costs ~3% of the entire wind, water, and solar power system and thus doesn't impact Jacobson et al.'s (2) conclusions." Source (23) noted by [12] is International Renewable Energy Agency (2012) Renewable energy technologies: Cost analysis series, Vol. 1. Available at www.irena.org/documentdownloads/publications/re_technologies_cost_analysis-hydropower.pdf. Accessed February 10, 2014. Note: Jacboson et al.'s rebuttal implies the additional hydro capacity is ($385/kW)(1300–87.5 GW) = 0.5 trillion $, which is, as they stated in [12] approximately 3% of the their calculated 14.6 trillion $ total capital cost in their 2015 article [11].

modeling question. Jacobson et al.'s explanations of their model's hydropower capacity were sufficiently answered in their journal papers or the responses to Clack et al.'s critiques.

In a July 12, 2017 Greentech Media interview, Jacobson further addresses the hydro modeling critique, and he says that increasing the discharge rate (of water from existing dams) is one option, but not the only option.[19] He does recognize it would be a policy decision to discharge at these high rates, but he does not agree that there was an error in modeling. Jacobson states it is a legitimate question as to what is the upper limit for adding hydro capacity to existing dams, and then states he has many scenarios available on his website that require no additional hydro capacity.

Georgetown, Texas 100% Renewable Electricity (Extended Explanation from Chap. 9)

To knock down the narrative against Georgetown's renewable energy purchases, we now have to explain the difference between the wholesale electricity market price and the "price" in Georgetown's Power Purchase Agreements (PPAs). I put the last price in quotes, because the PPAs really represent the *cost* of electricity.

Here is how to understand why Georgetown Utility Services could simultaneously have *low-cost* PPAs *and* have to pass *higher electricity prices* to their customers.

The prices in PPAs really are meant to cover the *entire cost* of building and operating a power plant. This is called the levelized cost of electricity, or LCOE. The LCOE includes two major categories of costs. First is the amortized capital cost. This is like a home mortgage when you borrow the money up front to buy the house and pay back the bank using constant monthly payments for 15–30 years. The power plant developer borrows the money to build the power plant, and Georgetown owes them a constant amount of money each month.

The second cost is the operational cost one must spend for each generated kilowatt-hour. These operational costs include paying for fuel and employees to operate the plant. Since wind and solar farms have no fuel costs, their operational costs are a small fraction of the total costs. Since natural gas and coal power plants need fuel, their operational costs are a larger fraction of their total costs. With this notion of operational and capital costs, we can relate Georgetown's woes to the electricity market.

In general, market prices are determined *only by operational costs*, not by capital costs. In the case of electricity markets, the wholesale market price at any given time of the day is determined by the operational cost of the last, or marginal, power plant that needs to turn on to provide the next incremental demand of electricity. For

---

[19] July 12, 2017: https://www.greentechmedia.com/articles/read/an-interview-with-mark-jacobson-about-100-percent-renewable-energy.

example, assume we have two power plants each rated at 100 megawatts (MW), and each with an amortized capital cost of 3 ¢/kWh. Power plant A has an operational cost of 2 ¢/kWh, and power plant B has an operational cost of 3 ¢/kWh. Thus the levelized cost of electricity is 5 ¢/kWh for power plant A and 6 ¢/kWh for power plant B. If the power demand is 90 MW, then only power plant A needs to turn on, and the market price is 2 ¢/kWh. However, if the power demand increases to 110 MW, now power plant B also has to turn on, and the market price is determined by the operational cost of power plant B at 3 ¢/kWh

Now imagine you are Georgetown and you have wind and solar contracts that represent an LCOE of 4 ¢/kWh. This full cost is *lower than the LCOE of both power plants A and B*. But, when you sell your electricity into the market, you are not selling at LCOEs of 5 or 6 ¢/kWh for power plants A and B, you are selling at their operational costs of 2 or 3 ¢/kWh. *So while your **cost** of electricity is lower, you have to sell your electricity at a loss due to the lower market **price**.* While this explains the crux of Georgetown's problem, they still projected that the market prices were going to be higher than their PPA costs. The reason why this turned out not to be the case was because of, perhaps ironically, there was so much investment in oil extraction as well as solar and wind power!

The power plant dictating marginal prices in Texas is usually a natural gas power plant, and increasingly the wind and solar plants affect marginal prices. Because there is so much drilling for oil, and much of the U.S.'s natural gas is "associated" with oil extraction (meaning operators get natural gas when they only really want to extract oil), the high supply of natural gas has kept its price low. Low natural gas prices translate to low electricity market prices. This is exactly the conclusion that the Department of Energy came to in its 2017 "Staff Report to the Secretary on Electricity Markets and Reliability".[20] The Trump administration asked for this report, many thought to justify a claim that renewable electricity hampered grid reliability by, among other things, depressing wholesale electricity prices that in turn forced the closure of many coal power plants while also putting economic pressure on nuclear plants. It turns out high natural gas production and low natural gas prices were more responsible for coal plant closures than wind and solar power.

Georgetown's electricity situation was ripe for attack for those promoting the fossil energy narrative. A lead actor was the Texas Public Policy Foundation (TPPF), a Texas-based conservative think tank. Their website states "Through research and outreach we promote liberty, opportunity, and free enterprise in Texas and beyond".[21] TPPF promotes markets and limited government, thus they pounced on Georgetown's decisions as exemplary of the overreach and poor decision making by

---

[20]DOE Staff Report to the Secretary on Electricity Markets and Reliability, available July 24, 2019 here https://www.energy.gov/downloads/download-staff-report-secretary-electricity-markets-and-reliability and directly here https://www.energy.gov/sites/prod/files/2017/08/f36/Staff%20Report%20on%20Electricity%20Markets%20and%20Reliability_0.pdf.

[21]Texas Public Policy Foundation, accessed July 25, 2019 at https://www.texaspolicy.com/.

a few elected officials and bureaucrats that can occur within regulated utilities like GUS.

In a summary of an August 2018 community event in Georgetown, TPPF stated "Attendees received a thorough run-down of everything energy including why renewables are unreliable, more expensive, and unsustainable. And why 100% renewable simply isn't doable".[22] There is no need for me to readdress this oversimplification of the reliability argument (discussed in Chap. 3, Section "The Reliability Discussion: It's About Time"). I will restate that electric grid operators have to date effectively integrated the operational characteristics of all power plants to reliably operate the grid on the time frames of seconds to days. I have confidence they will continue to do so, balancing tradeoffs along the way as needed based on engineering capabilities.

To the "more expensive" comment, on an annualized energy basis such as levelized cost of electricity (LCOE), wind and solar power can *now* be cheaper than coal and nuclear. Depending on the situation and location, the LCOE of a natural gas power plant might be more or less expensive. Via the Full Cost of Electricity project, coordinated by my office of the Energy Institute at the University of Texas at Austin, we made an LCOE calculator so that people could have discussions about the cost of electricity by inputting their own numbers into a map-based online tool (Fig. A.1).[23] The calculator can be accessed at https://energy.utexas.edu/calculators. It allows the user to change assumptions for fuel costs, installation costs, and the cost of carbon dioxide emissions from the life cycle of the power plant. As a reminder, in this paragraph I'm discussing the cost of electricity over the lifetime of the power plant, not the instantaneous price of electricity in the market.

A presentation posted online from TPPF's 2018 forum is oversimplified and avoids the necessary context of the electric grid as a unified system.[24] The goal of electric grid operators is to maintain reliable operation of the grid, not each individual power plant. One of the TPPF's arguments against renewables was that ERCOT peak demand in 2018 was 72–74 gigawatts (GW), and that there was a "gap" between that demand and the generation at peak (about 3 or 4 GW) from installed wind and solar farms. The presentation further stated that "gas, coal, and nuclear" power plants met the challenge of serving the rest of the load. This simplified categorization distracts and misses the point, but in a consistent manner for an organization that focuses on individual liberty. Why not also convince people to focus on individual power plant technologies instead of the "community" of power plants that work together on the grid? TPPF's argument is like stating the following:

---

[22]"Is Georgetown Really 100% Renewable?" https://www.texaspolicy.com/is-georgetown-really-100-renewable/.

[23]For the full suite of white papers produced within the Full Cost of Electricity project, see the Energy Institute website, https://energy.utexas.edu/, and the project website, https://energy.utexas.edu/policy/fce.

[24]Presentation used in Texas Public Policy Foundation Georgetown, Texas forum, August 15, 2018: https://files.texaspolicy.com/uploads/2018/10/03132637/Georgetown-Presentation.pdf.

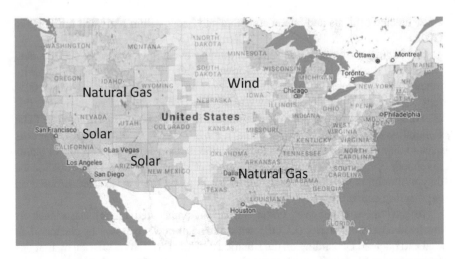

Fig. A.1 A map showing the power plant type with the lowest levelized cost of electricity, by county, using data representative of costs in 2018. Purple = solar photovoltaics; green = wind; orange = natural gas combined cycle. Map generated on October 1, 2019 using default values from "Levelized Cost of Electricity (LCOE) in the United States by County" at https://energy.utexas. edu/calculators. Methodology described in Rhodes et al. [15]

(1) There are five types of power plants, A, B, C, D, and E.
(2) If I add up the contributions from only D and E, they aren't enough.
(3) In order to serve all the load, we had to use types A, B, and C also.

We can try to install enough wind (D) and solar (E) to meet 100% of the demand 100% of the time, but we're pretty sure that will be more expensive than if we use a combination of all the power plants, including various storage and demand response technologies. We're also very sure that installing only coal (A), nuclear (B), or natural gas (C) power plants will be more expensive than the mix of all options. I can make the argument above with any two types of power plants for Statement 2, and the other three types of power plants for Statement 3. The grouping can be anything you want depending on the narrative you want to tell. Any objective analysis makes it clear that if we just want the cheapest electricity, the real argument is about the mix of power plant technologies. Anyone worth their salt in studying electric grid operations knows that the cheapest mix includes significant fractions of electricity from wind and solar farms. Recall that even those that promote the renewable energy narrative don't agree on the likelihood of creating an energy system based on 100% renewable electricity (Chap. 3 Section "With Friends like This, Who Needs Enemies?—Renewable Energy Narrative").

References

1. Ayres, R.U.: Sustainability economics: Where do we stand? Ecological Economics **67**(2), 281–310 (2008). Times Cited: 2

2. Ayres, R.U., Warr, B.: Accounting for growth: the role of physical work. Structural Change and Economic Dynamics **16**, 181–209 (2005)
3. Brown, T., Bischof-Niemz, T., Blok, K., Breyer, C., Lund, H., Mathiesen, B.: Response to burden of proof: a comprehensive review of the feasibility of 100% renewable-electricity systems. Renewable and Sustainable Energy Reviews **92**, 834–847 (2018). https://doi.org/10.1016/j.rser.2018.04.113. https://www.sciencedirect.com/science/article/pii/S1364032118303307
4. Clack, C.T.M., Qvist, S.A., Apt, J., Bazilian, M., Brandt, A.R., Caldeira, K., Davis, S.J., Diakov, V., Handschy, M.A., Hines, P.D.H., Jaramillo, P., Kammen, D.M., Long, J.C.S., Morgan, M.G., Reed, A., Sivaram, V., Sweeney, J., Tynan, G.R., Victor, D.G., Weyant, J.P., Whitacre, J.F.: Evaluation of a proposal for reliable low-cost grid power with 100% wind, water, and solar. Proceedings of the National Academy of Sciences **114**(26), 6722–6727 (2017). https://doi.org/10.1073/pnas.1610381114. http://www.pnas.org/content/114/26/6722.abstract
5. EIA: Annual energy review 2011, doe/eia-0384(2011) (2012)
6. GEA: Global Energy Assessment—Toward a Sustainable Future. Cambridge University Press (2012)
7. Giampietro, M., Mayumi, K., Şorman, A.H.: Energy Analysis for a Sustainable Future: Multiscale integrated analysis of societal and ecosystem metabolism. Earthscan from Routledge (2013)
8. IEA: Energy statistics manual. Tech. rep., International Energy Agency of the Organisation for Economic Co-operation and Development, (2005). http://www.iea.org/publications/freepublications/publication/energy-statistics-manual.html
9. Jacobson, M.Z., Delucchi, M.A.: Providing all global energy with wind, water, and solar power, Part I: Technologies, energy resources, quantities and areas of infrastructure, and materials. Energy Policy **39**(3), 1154–1169 (2011). https://doi.org/10.1016/j.enpol.2010.11.040
10. Jacobson, M.Z., Delucchi, M.A., Bazouin, G., Bauer, Z.A.F., Heavey, C.C., Fisher, E., Morris, S.B., Piekutowski, D.J.Y., Vencill, T.A., Yeskoo, T.W.: 100% clean and renewable wind, water, and sunlight (wws) all-sector energy roadmaps for the 50 united states. Energy Environ. Sci. **8**, 2093–2117 (2015). https://doi.org/10.1039/C5EE01283J
11. Jacobson, M.Z., Delucchi, M.A., Cameron, M.A., Frew, B.A.: Low-cost solution to the grid reliability problem with 100% penetration of intermittent wind, water, and solar for all purposes. Proceedings of the National Academy of Sciences **112**(49), 15,060–15,065 (2015). https://doi.org/10.1073/pnas.1510028112. http://www.pnas.org/content/112/49/15060.abstract
12. Jacobson, M.Z., Delucchi, M.A., Cameron, M.A., Frew, B.A.: The united states can keep the grid stable at low cost with 100% clean, renewable energy in all sectors despite inaccurate claims. Proceedings of the National Academy of Sciences **114**(26), E5021–E5023 (2017). https://doi.org/10.1073/pnas.1708069114. http://www.pnas.org/content/114/26/E5021.short
13. Kümmel, R.: The Second Law of Economics: Energy, Entropy, and the Origins of Wealth. Springer (2011)
14. Qvist, S.A.: Response to Jacobson et al. (june 2017). Tech. rep. (2017). http://www.vibrantcleanenergy.com/wp-content/uploads/2017/06/ReplyResponse.pdf
15. Rhodes, J.D., King, C., Gulen, G., Olmstead, S.M., Dyer, J.S., Hebner, R.E., Beach, F.C., Edgar, T.F., Webber, M.E.: A geographically resolved method to estimate levelized power plant costs with environmental externalities. Energy Policy **102**, 491–499 (2017). https://doi.org/10.1016/j.enpol.2016.12.025. http://www.sciencedirect.com/science/article/pii/S0301421516306875
16. Smil, V.: Energy in Nature and Society: General Energetics of Complex Systems. The MIT Press, Cambridge, Mass. (2008)

Index

© Springer Nature Switzerland AG 2021
C. W. King, *The Economic Superorganism*,
https://doi.org/10.1007/978-3-030-50295-9

Printed in the United States
By Bookmasters